# DESIGN OF ELECTRICAL MACHINES

**K G Upadhyay**

Assistant Professor
Electrical Engineering Department
Madan Mohan Malviya Engineering College
Gorakhpur (UP)

**NEW AGE INTERNATIONAL (P) LIMITED, PUBLISHERS**

**LONDON • NEW DELHI • NAIROBI**

Bangalore • Chennai • Cochin • Guwahati • Hyderabad • Kolkata • Lucknow • Mumbai

Visit us at **www.newagepublishers.com**

**Copyright © 2008, New Age International (P) Ltd., Publishers**
Published by New Age International (P) Ltd., Publishers
First Edition: 2008
**Reprint: 2018**

**All rights reserved.**
No part of this book may be reproduced in any form, by photostat, microfilm, xerography, or any other means, or incorporated into any information retrieval system, electronic or mechanical, without the written permission of the publisher.

## GLOBAL OFFICES

- **New Delhi** **NEW AGE INTERNATIONAL (P) LIMITED, PUBLISHERS**
7/30 A, Daryaganj, New Delhi-110002, (INDIA)
**Tel.:** (011) 23253771, 23253472, **Telefax:** 23267437, 43551305
**E-mail: contactus@newagepublishers.com • Visit us at www.newagepublishers.com**
- **London** **NEW AGE INTERNATIONAL (UK) LTD.**
27 Old Gloucester Street, London, WC1N 3AX, UK
**E-mail: info@newacademicscience.co.uk • Visit us at www.newacademicscience.co.uk**
- **Nairobi** **NEW AGE GOLDEN (EAST AFRICA) LTD.**
Ground Floor, Westlands Arcade, Chiromo Road (Next to Naivas Supermarket)
Westlands, Nairobi, KENYA, **Tel.:** 00-254-713848772, 00-254-725700286
**E-mail: kenya@newagepublishers.com**

## BRANCHES

- **Bangalore** 37/10, 8th Cross (Near Hanuman Temple), Azad Nagar, Chamarajpet, Bangalore- 560 018
**Tel.:** (080) 26756823, **Telefax:** 26756820, **E-mail: bangalore@newagepublishers.com**
- **Chennai** 26, Damodaran Street, T. Nagar, Chennai-600 017, **Tel.:** (044) 24353401, **Telefax:** 24351463
**E-mail: chennai@newagepublishers.com**
- **Cochin** CC-39/1016, Carrier Station Road, Ernakulam South, Cochin-682 016, **Tel.:** (0484) 2377303, **Telefax:** 4051304
**E-mail: cochin@newagepublishers.com**
- **Guwahati** Hemsen Complex, Mohd. Shah Road, Paltan Bazar, Near Starline Hotel, Guwahati-781 008,
**Tel.:** (0361) 2513881, **Telefax:** 2543669, **E-mail: guwahati@newagepublishers.com**
- **Hyderabad** 105, 1st Floor, Madhiray Kaveri Tower, 3-2-19, Azam Jahi Road, Near Kumar Theater, Nimboliadda Kachiguda, Hyderabad-500 027, **Tel.:** (040) 24652456, **Telefax:** 24652457
**E-mail: hyderabad@newagepublishers.com**
- **Kolkata** RDB Chambers (Formerly Lotus Cinema) 106A, 1st Floor, S N Banerjee Road, Kolkata-700 014
**Tel.:** (033) 22273773, **Telefax:** 22275247, **E-mail: kolkata@newagepublishers.com**
- **Lucknow** 16-A, Jopling Road, Lucknow-226 001, **Tel.:** (0522) 2209578, 4045297, **Telefax:** 2204098
**E-mail: lucknow@newagepublishers.com**
- **Mumbai** 142C, Victor House, Ground Floor, N.M. Joshi Marg, Lower Parel, Mumbai-400 013
**Tel.:** (022) 24927869, **Telefax:** 24915415, **E-mail: mumbai@newagepublishers.com**
- **New Delhi** 22, Golden House, Daryaganj, New Delhi-110 002, **Tel.:** (011) 23262368, 23262370, **Telefax:** 43551305
**E-mail: sales@newagepublishers.com**

**ISBN: 978-81-224-2282-5**
**C-17-07-10615**

Printed in India at Paramanand Offset, Delhi.
Typeset at Pagitek Graphics, Delhi.

**NEW AGE INTERNATIONAL (P) LIMITED, PUBLISHERS**
7/30 A, Daryaganj, New Delhi-110002
Visit us at **www.newagepublishers.com**
**(CIN: U74899DL1966PTC004618)**

# DESIGN OF ELECTRICAL MACHINES

# Preface

The development in technology has contributed in advancement of electrical machines design, over the past few decades. The development of digital computer has also provided the additional advantages for achieving economical and optimum design of different machines.

The main purpose of any textbook is to provide single source coverage of the full spectrum of conventional and computer aided design of electrical machines. This book is an outcome of the author's experience of teaching electrical machines and electrical machine design courses at the College of Technology, G.B. Pant University, Pantnagar, and at MMM Engineering College, Gorakhpur, U.P., India. Experiences shared at Nanyang Technical University, Singapore, a multinational company, Bangkok, Asian Institute of Technology, Bangkok and Alemaya University, Ethiopia, Africa has also been helpful in bringing out this book. It is designed for undergraduate students of electrical engineering in accordance with the syllabi of Indian universities/institutions. It will be specially helpful for U.P. Technical University students. It is also helpful for AMIE students.

Beginning with the principles of the designs, chapter 1 highlights the design principle, limitations and recent trends in electrical machine designs. Chapter 2 is devoted to discuss important features related to electrical machine design such as effect of fringing of flux, types of damper windings, different type of stator leakage flux, slot leakage reactance, calculation of pole leakage flux and different types of stator and rotor slots. Chapter 3 provides the detail of magnetic, conducting and insulating materials used in the electrical machines. Heating and cooling of electrical machines are explained in chapter 4. In this chapter, the heat dissipation by radiation, conduction and convection, volume of air and liquid required, types of enclosures, methods of ventilation, heating and cooling curves, ratings of electrical machines and methods of measurement of temperature rise are discussed.

In chapter 5, construction principle, estimation of main dimensions, damper winding design, regulation, temperature rise and other design aspects of 3-phase synchronous machines are discussed. Chapter 6 provides, construction and design steps of induction motors. Chapter 7 highlights construction and design of d.c. machines. It also highlights construction, important features and design of d.c. machines.

Chapter 8 deals with the constructional features and design procedure of transformer. Calculation of e.m.f. per turn, different dimensions of transformers, no load current, leakage reactance, regulation and tank size are given. Computer aided design are discussed in brief in chapter 9 and 10. General principles of computer aided design are discussed in chapter 9, whereas chapter 10 is devoted to computer aided design of transformer, alternator, induction machine and d.c. machines.

Solved numerical examples and questions with short answers, at the end of chapters, are included in the book. Unsolved problems are also given at the end of the chapters to enhance the students understanding of the topics presented.

I offer my most humble and profound indebtness to Prof. D.S. Chauhan, Ex Vice-chancellor, U.P. Tech. University, Lucknow for his deep concern and inspiring discussions both for my academics and for my personal welfare. I feel proud to be enriched by stimulating discussions and constant encouragement from Dr. S.N. Singh, Associate Professor, Electrical Engineering Department, IIT Kanpur.

I gratefully acknowledge the inspiring discussions and suggestions extended to me by Prof. S.C. Srivastava, Elec. Engg. Deptt., IIT Kanpur, Dr. G.K. Singh, EED, IIT Roorkee, Dr. K.N. Srivastava, A B B Sweden and Mr. K.S. Verma, KNIT Sultanpur.

I am extremely grateful to Shri M.M. Tripathi, senior design engineer, DOEACC, Gorakhpur for providing me continuous support.

I gratefully acknowledge the support provided by several of my colleagues and friends who have contributed to the development of the text. My thanks are also due to New Age International Pvt. Limited, especially its editorial and production team, for their utmost cooperation in bringing out the book on time.

I gratefully acknowledge the affection and cooperation received from my elder brother Shri K.L. Upadhyay, younger brother Mr. Hare Krishan Upadhyay, and my brother-in-laws Shri Rajendra Pandey, CEO and Chairman, American Lumber Co. Ltd., Bangkok, Thailand and Shri Ram Ajor Pandey.

Finally no words are adequate to express my indebtness to my family members for their pains and sufferings. I acknowledge with utmost warmth the unending support, love and affection received from my wife **Meena** and daughters **Kritika** and **Astha.**

**K.G. UPADHYAY**

# *Contents*

*Preface* v

## SECTION I: GENERAL

**1. Introduction** **3**

1.1 Principles of Design 3
1.2 Types of Electrical Machines 5
1.3 Limitations in Design 6
1.4 Recent Trends in Design 7
1.5 Outline of the Text 7

**2. Important Features Related to Electrical Machine Design** **9**

2.1 Introduction 9
2.2 Effect of Fringing of Flux 11
2.3 Types of Damper Windings 13
2.4 Different Types of Stator/Armature Leakage Flux 15
2.4.1 Slot Leakage Flux 15
2.4.2 Tooth Top Leakage Flux 16
2.4.3 Zig-zag Leakage 16
2.4.4 End Winding or Overhang Leakage Flux 16
2.4.5 Harmonic or Belt or Differential Leakage Flux 17
2.4.6 Skew Leakage Flux 17
2.4.7 Peripheral Leakage Flux 17
2.5 Calculation of Slot Leakage Reactance for Different Type of Slot 17

2.5.1 Slot Leakage Reactance for Parallel Sided Slots with Single Layer Winding 17
2.5.2 Slot Leakage Reactance for the Parallel Sided Slots with Double Layer Winding 21
2.5.3 Slot Leakage Reactance for Tapered Slot 21
2.5.4 Slot Leakage Reactance for Round Slot 22
2.5.5 Estimation of Tooth Top, Overhang, Zig-Zag and Harmonic Leakage Reactances 22
2.6 Calculation of Pole Leakage Flux ($\phi_{el}$) 22
2.7 Different Types of Stator and Rotor Slots 26
*Unsolved Questions 28*

**SECTION II: CONVENTIONAL DESIGN OF MACHINES**

**3. Magnetic, Conducting and Insulating Materials in Electrical Engineering** **31**

3.1 Introduction 31
3.2 Magnetic Materials 32
3.2.1 Diamagnetic Materials 32
3.2.2 Paramagnetic Materials 33
3.2.3 Ferromagnetic Materials 33
3.2.4 Materials for Solid Core 33
3.2.5 Sheet Steels 34
3.3 Classification of Conducting Materials 36
3.3.1 Good Conducting Materials 36
3.3.2 High Resistivity Materials 39
3.3.3 Materials Used for Solders 39
3.3.4 Materials for Contacts 40
3.3.5 Carbon Materials 40
3.3.6 Superconducting Materials 41
3.4 Insulating Materials 42
3.4.1 Gases as Insulating Materials 43
3.4.2 Vacuum as Insulation 44
3.4.3 Liquid Insulating Materials 44
3.4.4 Solid Insulating Materials 45
3.4.5 Different Insulating Materials Available 47
3.4.6 Insulating Materials Used in Electrical Machines 48
3.4.7 Specific Application of Class F and H Insulation 49
3.4.8 Effect of Moisture on Insulation 49
3.4.9 Protection of Insulation from Moisture 49
*Questions with Short Answers 50*
*Unsolved Questions 53*

**4. Heating and Cooling of Electrical Machines 55**

4.1 Introduction 55
4.2 Different Cooling Systems 56
4.2.1 Air Cooling 56
4.2.2 Open Circuit Cooling 56
4.2.3 Closed Circuit Cooling 56
4.2.4 Hydrogen Cooling 56
4.2.5 Liquid Cooling 56
4.2.6 Direct Cooling 56
4.3 Heat Dissipation by Radiation, Conduction and Convection 56
4.3.1 Heat Dissipation by Radiation 56
4.3.2 Heat Dissipation by Conduction 57
4.3.3 Heat Dissipation by Convection 58
4.4 Volume of Air and Liquid 60
4.4.1 Volume of Air 60
4.4.2 Liquid Required 61
4.5 Temperature Rise-time Curve 61
4.5.1 Machine Under Heating 61
4.5.2 Machine Under Cooling 64
4.6 Types of Enclosures for Rotating Electrical Machines 65
4.6.1 Open Type Machine 65
4.6.2 Protected Type Machine (P) 65
4.6.3 Drip, Splash and Hose Proof Type Machines 65
4.6.4 Pipe or Duct Ventilated Type Machines 66
4.6.5 Totally Enclosed, Special Enclosure Type Machines 66
4.7 Methods of Ventilation 67
4.7.1 Induced and Force Ventilation 67
4.7.2 Radial Ventilating System 68
4.7.3 Axial Ventilating System 69
4.7.4 Radial and Axial Ventilation in One Machine 69
4.7.5 Multiple-inlet Ventilation Method 70
4.7.6 Cooling of Totally Enclosed Machines 70
4.7.7 Cooling of Turbo-alternators 72
4.7.8 Direct Cooling (of Turbo-alternator) 74
4.7.9 Cooling of Hydrogenerators 76
4.8 Heating and Cooling Curve and Ratings of Electrical Machines 76
4.8.1 Heating and Cooling Curves for Different Operating Conditions 76
4.8.2 Standard Ratings of Electrical Machines 76
4.8.3 Different Rating (Duty) of Machine 77
4.8.4 R.M.S. Horse Power Rating 78
4.8.5 Selection of Motor Capacity for Different Ratings 78
4.9 Methods of Measurement of Temperature Rise 80
4.9.1 Thermometer Method 80
4.9.2 Resistance Method 80

4.9.3 Embedded Temperature Detector Method (ETD) 80
*Solved Problems 81*
*Questions with Short Answers 84*
*Unsolved Problems 90*

**5. Construction, Principle and Design Aspects of 3-phase Alternator 92**

5.1 Introduction 92
5.2 Construction of Alternator 92
5.2.1 Stator Construction 94
5.2.2 Rotor Construction 95
5.3 Magnetic Portions of Three-phase Alternator 99
5.4 E.M.F. Equation of an Alternator 99
5.5 Distribution Factor and Pitch Factor 100
5.5.1 Distribution Factor 100
5.5.2 Pitch Factor 101
5.6 Flux Density Distribution Curve: Salient Pole Machine 102
5.7 Specific Electric and Specific Magnetic Loadings 104
5.7.1 Specific Electric Loading 104
5.7.2 Magnetic Loading: Specific Magnetic Loading $\overline{B}$ or $B_{\delta_1}$ 105
5.8 Output Equation and Its Derivation 106
5.8.1 Output Equation 106
5.8.2 Derivation of Output Equation 107
5.9 Estimation of Main Dimension (D and L) 108
5.9.1 Salient Pole Alternator (Water Wheel Generator) 108
5.9.2 Steps for Estimation of D and L of Salient Pole Alternator 110
5.9.3 Main Dimension of Turbo-Alternator 112
5.10 Runaway Speed 112
5.11 Effective Length of Alternator ($L_e$) 113
5.12 Length of Air Gap (δ) 114
5.13 Selection of Stator Slots 114
5.14 Estimation of Turns Per Phase, Total Conductors, Conductors Per Slot and Their Corrected Values 115
5.15 Design of Teeth and Slots 118
5.15.1 Estimation of Size of Teeth and Slots 119
5.15.2 Width of Slot 120
5.15.3 Height of Teeth and Slot 120
5.16 Phase-Current 120
5.17 Sectional Area of Stator Conductor and Slot Insulation 120
5.18 Stator Yoke Design 122
5.19 Stator Copper and Iron Losses, Weight of Iron and Copper 124
5.20 Design of Damper Windings 127
5.21 Rotor Design of Salient Pole Synchronous Alternator 131
5.21.1 Pole Profile Drawing of Salient Pole Alternator 132
5.21.2 Height of Pole Yoke 134
5.21.3 Height of the Pole Core of Salient Pole Machine 135

5.22 Determination of Open Circuit Characteristics (OCC) by Design Data 136
5.22.1 MMF for Air Gap 136
5.22.2 Pole Leakage Flux 140
5.23 Stator Leakage Reactance 142
5.23.1 Slot Leakage Reactance 142
5.23.2 Tooth Top Leakage Reactance 143
5.23.3 End Winding Leakage Reactance 144
5.24 Determination of Full Load Field mmf 144
5.25 Short-Circuit Characteristics 146
5.26 Design of Field Winding of Salient Pole Alternator 146
5.27 Field Losses, Total Loss, Efficiency and Regulation 148
5.28 Temperature Rise in Alternator 149
5.28.1 Temperature Rise of Field Coils 149
5.28.2 Temperature Rise of Stator 150
5.29 Rotor Design for Turbo-alternator 152
*Solved Problems on Alternator* 155
*Problems with Short Answers* 187
*Unsolved Problems* 189

**6. Construction and Design Steps of Induction Motors** **192**

6.1 Introduction 192
6.2 Construction 193
6.2.1 Stator 193
6.2.2 Rotor 193
6.2.3 Stator Winding 195
6.2.4 Rotor Winding 195
6.2.5 Sliprings 195
6.2.6 Shafts and Bearings 195
6.2.7 Frame 196
6.3 Output Equation 196
6.4 Selection of Flux Density $\left(\overline{B}\right)$ and Specific Electric Loading 197
6.5 Estimation of Main Dimensions 197
6.6 Effective Length of Machine ($L_e$) 198
6.7 Estimation of Turns Per Phase ($N_{ph}$), Total Conductors ($Z_T$) and Conductors Per Slot ($N_c$) etc. and Their Corrected Values 199
6.8 Air Gap Length of Motor 201
6.9 Cross-Sectional Area of Conductor 201
6.10 Stator Slot Design 202
6.11 Tooth Design 203
6.12 Height of Stator Yoke 204
6.13 Outer Diameter 204
6.14 Stator Winding Resistance, Copper Loss, Iron Loss, Stator Weight etc. 205
Length of Mean Turn 205
Stator Iron Loss 206

6.15 Rotor Design 206
6.15.1 Rotor Diameter 206
6.15.2 Number of Rotor Slots 206
6.15.3 Total Number of Rotor Conductors 207
6.15.4 Rotor Current 207
6.15.5 Sectional Area of Rotor Conductor 208
6.15.6 Rotor Slot Design 209
6.15.7 Rotor Teeth Design 210
6.15.8 Rotor Yoke Design 210
6.15.9 Rotor Copper Loss 210
6.16 Efficiency 211
6.17 Concept of Flattened Flux Density $\left(B_{30^\circ} \text{ or } B_{60^\circ}\right)$ 212
6.18 No Load Current 213
6.18.1 Magnetizing Current $\left(I_\mu\right)$ 213
6.18.2 Estimation of Loss Component of the Current ($I_c$) 215
6.19 Estimation of Performance of Induction Motor 217
6.19.1 Estimation of Leakage Reactance 217
6.19.2 Calculation of Resistance 220
6.19.3 Short-Circuit Current and Power Factor 221
6.19.4 Dispersion Coefficient 221
6.20 Construction of Circle Diagram from Design Data 221
6.21 Stator Temperature Rise 223
*Solved Problems on Three-phase Induction Motors and Overall Design of Induction Motor* 224
*Questions with Short Answers* 244
*Unsolved Questions* 246

**7. Construction, Important Features and Design of D.C. Machines** **248**

7.1 Introduction 248
7.2 Types of D.C. Machines 249
7.2.1 Homopolar Machine 249
7.2.2 Heteropolar Machine 249
7.3 Construction of D.C. Machines 249
7.3.1 Field System 249
7.3.2 Armature 253
7.3.3 Commutator 255
7.3.4 Brushes 255
7.3.5 Bearings 256
7.4 Important Features of D.C. Machines 256
7.4.1 Factors Affecting the Choice of Specific Magnetic Loading ($\overline{B}$) 256
7.4.2 Factors Affecting the Choice of Specific Electric Loading $\left(\overline{ac}\right)$ 256
7.4.3 Selection of Air Gap Length ($\delta$) 257
7.4.4 Effect of Armature Reaction 257

7.4.5 Armature Winding Terms 258
7.4.6 Single Layer and Double Layer Winding 259
7.4.7 Types of Armature Winding 259
7.4.8 Winding Pitches 261
7.5 Output Equation 261
7.6 Determination of $P_a$ for Different Machines 263
7.7 Selection of $\overline{B}$ and $\overline{ac}$ 264
7.7.1 Selection of Specific Magnetic Loading ($\overline{B}$) 264
7.7.2 Selection of Specific Electric Loading $(\overline{ac})$ 264
7.8 Factors Effecting the Selection of Number of Poles 265
7.8.1 Frequency 265
7.8.2 Weight of Iron Parts 265
7.8.3 Weight of Copper 267
7.8.4 Commutator Length 267
7.8.5 Labour Charges 268
7.8.6 Flash Over 268
7.8.7 Important Features for Selection of Number of Poles 269
7.9 Selection of Core Length and Diameter 269
7.9.1 Factors for Selection of Core Length 269
7.9.2 Selection of Armature Diameter 270
7.10 Estimation of Main Dimension (D & L) – (Separation of D and L) 270
7.11 Ventilating Ducts 272
7.12 Calculation of Air Gap Length 272
7.13 Details of Armature Design 273
7.13.1 Armature Winding 273
7.13.2 Number of Armature Conductors 274
7.13.3 Number of Commutator Segments 274
7.13.4 Factors Effecting the Choice of Number of Slots 275
7.13.5 Estimation of Corrected Values 276
7.13.6 Cross-Sectional Area of Armature Winding Conductor 276
7.13.7 Selection of Current Density 276
7.13.8 Limitation on Conductor Size 277
7.13.9 Insulation of Armature Winding 277
7.13.10 Dimensions of the Slot 277
7.13.11 Resistance of Armature and Armature Voltage Drop 278
7.13.12 Copper Weight and Copper Loss in Armature Winding 279
7.13.13 Height of Armature Core 279
7.14 Armature Temperature Rise 281
7.15 Design of Field System 281
7.15.1 Sectional Area of Pole 281
7.15.2 Design of Height of Pole 282
7.15.3 Yoke Design 285
7.16 Design of Series Field Winding 286
7.17 MMF of the Magnetic Circuit and OCC 286

7.18 Design of Shunt Field Winding 288
7.19 Design of Commutator 290
7.20 Length and Design of Interpoles 292
7.20.1 Length of Interpoles 292
7.20.2 Design of Interpoles 292
7.21 Compensating Winding 295
7.22 Losses and Efficiency 295
*Solved Problems 297*
*Problems with Short Answers 332*
*Unsolved Questions 334*

**8. Constructional Features and Design Procedure of Transformers 335**

8.1 Introduction 335
8.2 Types of Transformers 335
8.2.1 Core Type Transformers 336
8.2.2 Shell Type Transformers 337
8.3 Distribution and Power Transformers 337
8.3.1 Distribution Transformers 337
8.3.2 Power Transformers 338
8.4 Core and Yoke 338
8.5 Stepped Core and Yoke 339
8.6 Comparative Characteristics of Core and Shell Type Transformers 340
8.7 Tank and Accessories Like Conservator, Breather, Bushings 340
8.8 Output Equation 341
8.8.1 Single-phase (1-$\phi$) Core Type 343
8.8.2 Output Equation of Single-phase (1-$\phi$) Shell Type 344
8.8.3 Output Equation for a 3-phase (3-$\phi$) Core Type Transformer 345
8.8.4 Output Equation of a 3-phase (3-$\phi$) Shell Type Transformer 346
8.9 Window Space Factor, Current Density and Flux Density 347
8.10 EMF per turn ($E_t$) 348
8.11 Different Dimensions of Transformer 350
8.12 Steps to Design a Transformer 353
8.13 Estimation of NO Load Current ($I_0$) 354
8.14 Determination of Number of Turns and Length of Mean Turns of Windings 356
8.15 Resistance of H.V. Winding and L.V. Winding 358
8.16 Leakage Reactance 359
8.17 Regulation 365
8.18 Estimation of Losses and Efficiency of a Transformer 366
8.19 Design of Tank 367
8.19.1 Main Dimensions of Tank 369
*Solved Problems on Transformers 371*
*Questions with Short Answers 383*
*Unsolved Problems 386*

# SECTION III: COMPUTER AIDED DESIGN OF ELECTRICAL MACHINES

**9. Principles of Computer Aided Design** 391

9.1 Introduction 391
9.2 Advantages and Limitations of Computer Aided Design 392
9.2.1 Advantages of Computer Aided Design 392
9.2.2 Limitations of CAD 393
9.3 Different Approaches for Computer Aided Design 393
9.3.1 Analysis Method of Design 393
9.3.2 Synthesis Method of Design 393
9.3.3 Hybrid Method of Design 395
9.4 Optimization 396
9.5 Standardisation of Design 396

**10. Computer Aided Design of Electrical Machines** 397

10.1 Introduction 397
10.2 Computer Aided Design of 3-phase Alternator 397
10.2.1 Programme and Flow Chart 405
10.2.2 Flow Chart for Overall Design of 3-phase Alternator 408
10.3 Computer Aided Design of 3-phase Induction Motor 409
10.3.1 Flow Chart for Overall Design of 3-phase Induction Motor 415
10.4 Computer Aided Design of D.C. Machines 416
10.4.1 Flow Chart for Overall Design of D.C. Machines 421
10.5 Computer Aided Design of Transformer 422
10.5.1 Flow Chart for Overall Design of Transformer 422
MATLAB Program of Design Problem, Answer and Discussion 423

**Appendices** 432

*Appendix A: Tables for Standard Sizes and Insulation Covering 432*
*Appendix B: Standard Specifications 440*

**Bibliography** 441

# *Section I: General*

# Introduction

This chapter provides principles of design, types of electrical machines, limitations in design and recent trends in electrical machine design. A chapter-wise detail outline of the text has also been presented.

## 1.1 PRINCIPLES OF DESIGN

There are different specific tasks to be performed by the electrical machines. The design of electrical machines/equipments for the specific applications is based upon the application of theoretical scientific concepts, technology and related inventions. The suitable design depends upon the proper adjustment of iron portion, copper, air gap, insulation, ventilation and cooling of machine. The selection of conducting, magnetic and insulating materials has important role in machine design. We have to explore the availability of the each type of material, its characteristics, which should be according to the specifications, the performance limitations and its cost.

When a machine is to be designed for a specific application then it is preferred that the designed machine should fulfill the required specifications. But it is very difficult to design a machine which meets all the specifications and requirements. The machine should be cheaper, durable, reliable and should perform according to requirements. The manufacturing costs, operating and running costs of the machine are considered. Design of a machine means that the cost should be minimum. If the cost is minimized, it may not be a durable machine. So the important criteria for good design is to get the minimum losses for given cost. If only cost is considered and we try to minimize it, this may result in a machine which have higher operating and maintenance cost. The losses will be more. If we try to minimize the losses in the machine, it may become more costly. So there should be a balance between cost and the losses to have an optimum design of the machine for a particular application.

An electric machine is an electro-mechanical energy conversion device, which converts mechanical energy into electrical energy and vice versa. When the machine converts mechanical energy into electrical energy it is called as generator. When the machine converts electrical energy into mechanical energy it is called as motor. A part of energy is converted to heat. This energy is lost and cannot be recovered. An electrical machine can be designed to operate either as a generator or as a motor. Faraday's law of electromagnetic induction states that e.m.f. induced in a closed electric circuit is equal to the rate of change flux linkages

Flux linkages, $\Psi = N\,\phi$

where, $N$ = number of turns of coil

$\phi$ = flux linking with all the turns

∴ The e.m.f. induced in the coil

$$e = -\frac{d\psi}{dt}$$

or

$$e = -N\frac{d\phi}{dt}$$

The negative sign in above expression indicates that the current produced due to induced e.m.f. always opposes the change in flux linkages. The change in the flux linkages can be achieved by

(*i*) a stationary coil with respect to flux but magnitude of flux varying with respect to time.

(*ii*) the coils moving through a flux which is constant with respect to time.

(*iii*) the coils moving through a flux with varying magnitude with respect to time.

When a conductor moving at right angles to a uniform, stationary and constant magnetic field an e.m.f. is induced and is expressed by

$$e = B\,l\,v \text{ volts}$$

where, $B$ = flux density of magnetic fields (Wb/m$^2$)

$l$ = length of conductor perpendicular to the magnetic field (m)

$v$ = velocity of conductor (m/sec)

The force produced by the interaction of magnetic field and current carrying conductor is given by the Biot-Savart's law and can be expressed by

$$F = B\,l\,I$$

where, $B$ = flux density of magnetic fields (Wb/m$^2$)

$l$ = length of conductor perpendicular to magnetic field (m)

$I$ = currents (Amp)

When a conductor is placed on a rotor with radius $r$ and the torque produced by the current in the conductor is given by

$$\text{Torque} = F \times r$$

$$\text{Torque, } T = B\,l\,I \cdot r \text{ Newton-metre}$$

This is an electro-magnetic torque.

$$\text{Power} = \text{Torque} \times \text{Angular velocity}$$

$$= T \cdot \omega$$

It is, therefore, concluded that the design and construction of generators and motors are based upon the facts given below:

(*i*) An e.m.f. is induced in a conductor or a set of conductors subjected to a magnetic field in such a manner that either conductor moves and cuts the stationary lines of flux or it (stationary conductors) is cut by the varying lines of flux. In both the cases there will be a relative motion between the conductor and the magnetic field. This is the characteristics of the generator action.

(*ii*) When a conductor or a set of conductors are placed in a magnetic field and an electric current is supplied through the conductor then conductor observes a mechanical force. This is the characteristics of the motor action.

(*iii*) The armature of an electrical machine carries the conductors in which the e.m.f. is induced. The field portion of the machine produces the magnetic field. The armature may be rotating with stationary field or the armature may be stationary with rotating field. Usually in large machines the armature is stationary (called stator) and the field is rotating (rotor). Relative advantages and disadvantages are discussed in detail in 5.2 Chapter 5.

## 1.2 TYPES OF ELECTRICAL MACHINES

There are two different types of machines based on the nature of e.m.f. and current produced:

(*a*) a.c. machines.

(*b*) d.c. machines.

The a.c. machines can be further classified as

(*i*) Synchronous machines.

(*ii*) Induction machines.

A synchronous machine has normally a field winding on the rotor and three-phase winding on the stator. The field winding is supplied by a separate d.c. source. The field winding produces a field and due to rotation of rotor, the field moves in space at the speed of the rotor. This rotating field links with the three-phase stator conductors and hence voltage is induced in them.

The magnitude of the voltage will depend upon the strength of field (magnetic field), number of turns and frequency (corresponding to poles and r.p.m.) for an alternator. For a synchronous motor three-phase supply is given to the stator. Since the three-phase winding is distributed in space in the stator at 120° apart and the currents in three-phases are varying with time so a rotating field (at a synchronous speed, $N_s = \frac{120f}{P}$) is produced in the air gap of the machine. In the rotating field fictitious North and South poles are produced rotating at synchronous speed ($N_s$) and behave as North and South poles were being rotated at synchronous speed. The field produced by the rotor conductor current reacts with the stator field. Initially the stator poles interact with rotor poles and since rotor is at standstill and stator poles are rotating at synchronous speed. So if North pole of stator field is in contact with South pole of rotor at any instant then attractive force will be observed by rotor and it will try to move in the direction of stator rotating field. But soon after that South pole of stator field will come in contact of South pole of rotor and repulsive force will act upon rotor. Hence a pulsating force acts upon a stationary rotor and so it does not move. That is why a three-phase synchronous motor is not self starting.

We can rotate the rotor of the motor by some other means at or reacts the synch. speed so that North and South poles of stator and rotor fields (opposite poles) are coupled (magnetically locked) with each other. The motor keeps on rotating at synchronous speed ($N_s$). The other way to make it self starting may be to provide the damper winding on the rotor of the motor. Damper winding on the rotor behaves like the cage rotor of three-phase induction motor. So three-phase synchronous motor with damper winding is started as induction motor. It picks up the speed near to synchronous speed and when opposite poles of stator fields and rotor fields come in contact (polar region), the speed of rotor suddenly jumps from induction motor speed to synchronous speed and poles are coupled (magnetically locked) with each other and rotor rotates at synchronous speed. The three-phase synchronous motor runs only at synchronous speed or not at all.

An induction machine has polyphase winding on the stator as well as on the rotor. The three-phase supply to the stator winding produces a rotating magnetic field (r.m.f.) in the air gap. This field links with the rotor conductors and due to change in time an e.m.f. is induced in the rotor winding and hence a current is produced in the rotor winding. The field produced by the rotor current reacts with the stator field and tries to oppose the cause by which it is being produced. The cause due to which e.m.f. is induced in rotor conductor is relative speed between rotor and stator field. So induction motor starts running in the direction of r.m.f. to minimize the cause (relative speed).

A d.c. machine has stationary poles on the stationary part. The field winding is provided for poles which produces a stationary field in the air gap of the machine. The armature winding is placed on the rotor of the machine.

## 1.3 LIMITATIONS IN DESIGN

When a machine has to be designed and constructed, the choice of suitable materials and manufacturing technology becomes important. The following considerations are required which impose the limitations on the machine design:

(*i*) ***Saturation of the Magnetic Circuit:*** The saturation of the magnetic circuit disturbs the straight line characteristics of magnetisation (*B-H*) curve resulting in increased excitation required and hence higher cost for the field system.

(*ii*) ***The Temperature Rise Over the Ambient Temperature:*** The excessive temperature rise may cause insulation failure. The life of the machine depends upon the life of the insulation. If the machine is continuously operated above the specified temperature limit, the life of the insulation and hence the life of the machine will be reduced. By providing proper ventilation and cooling, the temperature rise can be kept within the safe limit.

(*iii*) ***Insulation:*** The insulating properties and the strength of the insulating materials are considered on account of breakdown due to excessive voltage gradients set up in the machine.

(*iv*) ***Mechanical Strength:*** The machine should have the ability to withstand centrifugal forces and other stresses.

(*v*) ***Efficiency:*** The efficiency of the machine should be high for low running cost. The specific magnetic and electric loading should be low to achieve high efficiency. With low value of magnetic and electric loadings, the size of machine will be larger and hence more capital cost (initial investment).

(*vi*) Type of commutation desired whether ideal, under or over commutation.

(*vii*) ***Power Factor:*** Power factor required in a.c. machines whether it is low or high.

The best design would be one which attains the maximum advantage of better performance and high efficiency in the least possible cost. Sometimes the life, heating, rating and even efficiency of the machine have to be changed to bring down the cost of design and manufacturing.

## 1.4 RECENT TRENDS IN DESIGN

The electrical machine design depends upon several factors like mechanical stresses, magnetic forces, temperature rise and cooling medium adopted for the machine.

Taking into consideration the different parameters and operating conditions, the designer has to design a machine which should be best suitable. It should not merely fulfil the specifications. The machine designed should be rugged, simple, efficient, economic and safe. Designers should have sufficient technical knowledge and should be able to visualize the simultaneous inter-relationships of all the parameters and their effect. The visualization can be based on wide practical experiences. Sometime machine may have to operate in isolation but sometime it is used in a system with so many machines. The machine which has to operate in such system cannot be designed in isolation but the design of machine depends upon the optimization of the system performance.

For a particular application, several set of design of machine may need to be done to find the optimum designed machine. In finding the optimum design of the machine many iterations may be required to incorporate the changes in parameters till the satisfactory performance in the machine is obtained. These calculations with indefinite iterations are manually not possible. These calculations can be easily done by means of a digital computers. By using computer the data can be easily varied several times to get the optimum design. So the computer has become a powerful and important tool in designing the electrical machines.

## 1.5 OUTLINE OF THE TEXT

Chapter 1 highlights the principle of design, limitations in design and recent trends in design. Important features like effect of fringing of flux, types of damper windings, different type of stator leakage flux, calculation of slot leakage reactance for different type of slots, calculation of pole leakage flux, different types of stator and rotor slots are described in chapter 2. Chapter 3 starts with the description of magnetic conducting and insulating materials. These materials are separately discussed with their specific applications. In chapter 4 different cooling systems, heat dissipation by radiation, conduction and convection, volume of air and liquid required, temperature rise time curve, types of enclosures for rotating electrical machines, different methods of ventilation, rating of electrical machines, methods of measurement of temperature rise etc. are explained.

Chapter 5 is devoted to the description of construction, principle and design aspects of three-phase synchronous machine. Derivation of output equation, estimation of main dimensions, effective length, design of stator teeth and slots, winding design, rotor design with damper winding design, determination of open circuit characteristics (OCC), temperature rise in alternator and rotor design of turbo-alternator are discussed in this chapter.

Chapter 6 provides the details of construction of induction machines. This chapter also provides the output equation, estimation of main dimensions, effective length of machine, stator teeth and slot design, winding design, outer diameter, rotor design, efficiency, flattened flux density, no load current, estimation of performance of induction motor, construction of circle diagram from design data, stator temperature rise etc.

Chapter 7 describes the construction and some features of d.c. machines. Output equation and its derivation, design of field system, design of commutator, design of interpoles, compensating winding and losses and efficiency are also given in this chapter.

Chapter 8 elaborates different types of transformers, stepped core and yoke, output equation, window space factor, e.m.f. per turn and different dimensions of transformers. The steps to design a transformer, estimation of no loud current, leakage reactance and regulation, design of tank etc. are also provided in this chapter.

In chapter 9, general principles of computer aided design are discussed. The different approaches for computer aided design, optimization and standardization of design are also discussed in this chapter.

Chapter 10 provides the computer aided design of transformer, three-phase alternator, three-phase induction machines and d.c. machines. Different design problems for each machines are solved with the help of computer programme in C. The related flow chart for these problems are given. The flow chart for overall design of the machines are also provided in this chapter.

Some tables which are required in the design of different machines are given in appendices.

2

# Important Features Related to Electrical Machine Design

## 2.1 INTRODUCTION

The different parts of a machine in the magnetic circuit are air gap, stator tooth, stator yoke/core, rotor tooth/pole core and shoes and rotor yoke. These parts will be separately mentioned for the different machines to estimate the ampere-turns required to produce the flux. The magnetic circuit for any machine consists of series and parallel circuits. The flux in the air gap is determined as per the requirements of machine design like voltage, armature ampere conductor etc. The flux distribution in different parts of the machine is found and ampere-turns required to produce that flux in each part is calculated and then total ampere-turns required for all parts of the machine to have flux in the air gap at the time of open circuited condition can be calculated. Since the size of teeth is not uniform so the flux density varies across its height and hence the ampere-turn per meter also varies across its height. The ampere-turns per meter are calculated at the top, middle and bottom of the tooth and then by Simpson's rule, the average value of ampere-turn per meter for tooth are calculated.

With the help of flux in each part of the magnetic circuit of the machine the flux density in each part can be calculated and checked whether it is within the permissible limit or not. If flux density will be more, losses will be more, temperature rise will be more and efficiency will decrease. So if the flux density in any portion is more, the size of that portion can be modified to get the optimum value of flux density. The flux density should not be very low since the machine size will not be fully utilized and hence uneconomic operation is obtained.

Copper losses in the windings is calculated for different machines by knowing current in conductor (phase current), length of mean turn, number of turns and the number phaser etc. The size of iron parts of the machine are designed and the flux density in the respective parts are also known. With the help of sizes available, the weight of different portions *e.g.*, teeth, stator core, pole core, pole yoke etc., can be calculated. The curves are available which gives specific iron loss in watts per kg in any portion of machine with respect to the value of flux density in that portion. If the flux density changes, the specific iron loss *i.e.*, the iron loss per kg weight of iron will also be changed. Hence the iron losses for different portions of the machine are separately calculated since each portion of the machine has

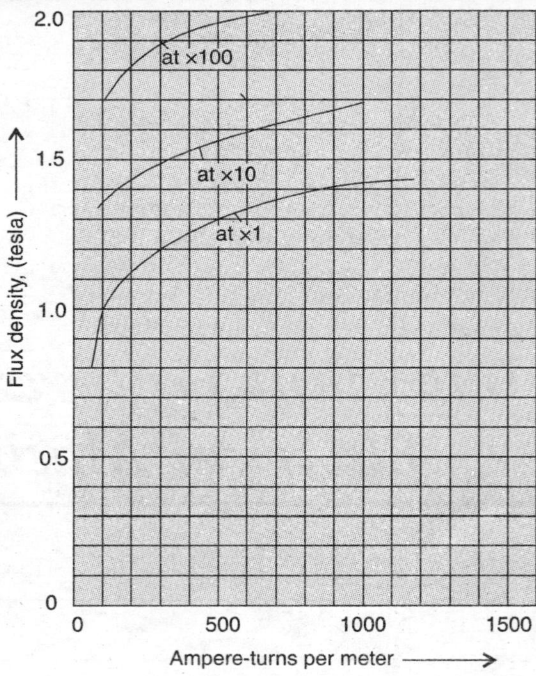


**Fig. 2.1** Flux density vs ampere-turns per meter curve for non-oriented steel.

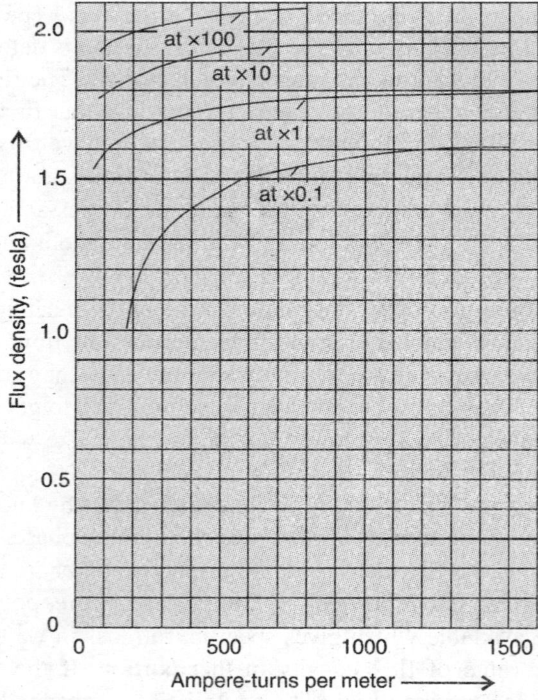


**Fig. 2.2** Flux density vs ampere-turns per meter curve for CRGO steel grade 123.

different value of flux density. By knowing the iron loss in watts per kg weight of iron for any portion and also the weight of that portion, the total iron loss in that portion can be calculated. Finally, the total core loss or iron loss in all the parts of the machine can be found out by calculating iron losses in each parts.

The calculation of total ampere-turns required, copper losses, core losses, pole leakage flux, and other parameters required in designs of transformer, 3-phase alternator, 3-phase induction motors and d.c. machines are discussed in respective chapters. In this chapter some important features which are relevant to the design of different machines are discussed.



## 2.2 EFFECT OF FRINGING OF FLUX

The design of 3-phase alternator, 3-phase induction motor, d.c. machines are discussed in subsequent chapters. While designing a machine the effective axial length and effective length of air gap are calculated. The iron surface along the air gap is not uniform and there are radial ventilating ducts provided along its length. The fringing of flux lines occurs across the teeth and ducts of the machine and effects the axial length (the length through which the magnetic field is being associated) and air gap length of the machine. How the fringing of flux affects these values are discussed below.

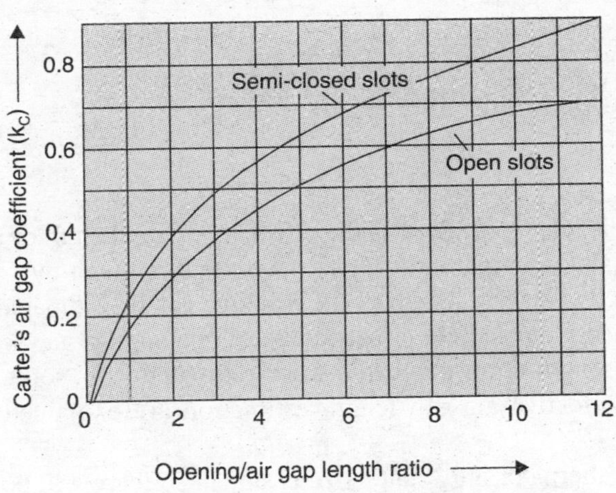


**Fig. 2.3** Carter's air gap coefficient.

There may be two types of following construction in 3-phase alternators:

(*a*) Salient pole alternator.
(*b*) Cylindrical rotor alternator.

In case of 3-phase salient pole alternator the poles have uniform surface of iron and stator are slotted. If the flux lines from poles are linking to the teeth through the width of teeth at the top only, the flux distribution can be shown by Fig. 2.4. But there is fringing of flux around the tooth top in a slotted stator/armature. Flux links not only to the top of the teeth but also along its sides of the height as shown in Fig. 2.5.

The reluctance of air gap when there is fringing of flux is more than that in the core of a smooth armature but lesser than that when the flux is to be confined only through the tooth top. So the air gap length observed from magnetic field point of view will be different than the actual length of air gap or

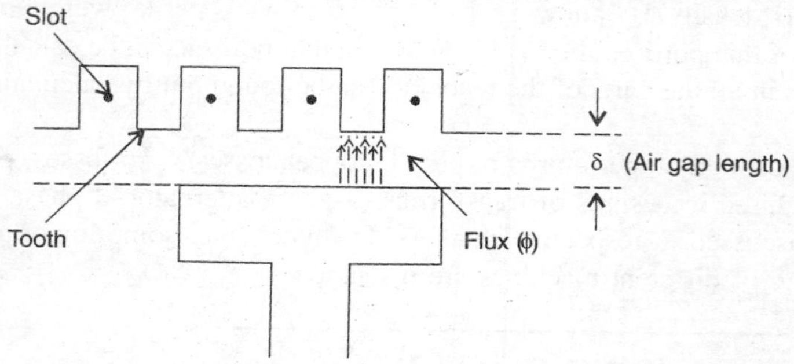


**Fig. 2.4** When no fringing of flux.

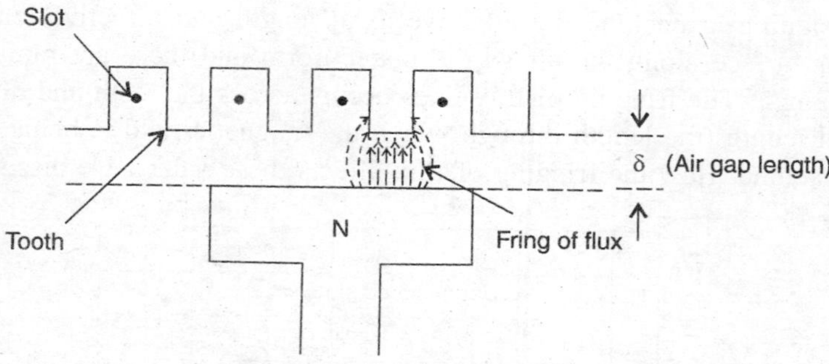


**Fig. 2.5** Fringing of flux.

physical air gap length. This length is called as the effective air gap length of the machine. The fringing of flux depends upon the slot opening and air gap length and is given by Carter's air gap coefficient given in Fig. 2.3. The effective air gap length of the machine can be estimated with the help of Carter's air gap coefficient (obtained from the curve between slot opening/air gap length ratio vs Carter's air gap coefficient). Effective length of air gap can also be estimated from some expressions available for respective machine. The effective air gap length estimation are explained while the machines are designed.

In case of salient pole alternator only stator size is slotted so there will be only one Carter's air gap coefficient for stator side. Effective air gap length, $\delta' = k_c \cdot \delta$ where $k_c$ = Carter's air gap coefficient. Whereas when cylindrical rotor alternator have to be designed, stator and rotor both are slotted so the fringing of flux is effected from both sides and hence Carter's air gap coefficient for stator as well as for rotor should be found to calculate air gap coefficient $k_c = k_{c_1} \cdot k_{c_2}$

where, $k_c$ = Carter's air gap coefficient

$k_{c_1}$ = Carter's air gap coefficient for stator

$k_{c_2}$ = Carter's air gap coefficient for rotor

So effective length of air gap, $\delta' = k_c \cdot \delta$

when the radial ventilating ducts are provided across the length of machine, it is shown in Fig. 2.6, $l_1, l_2, l_3, ..., l_n$ are the core stacks of machine lengths,

$b_v$ = length of ventilating duct

$b'_v$ = effective length of ventilating ducts

$n_v$ = the number of ventilating ducts

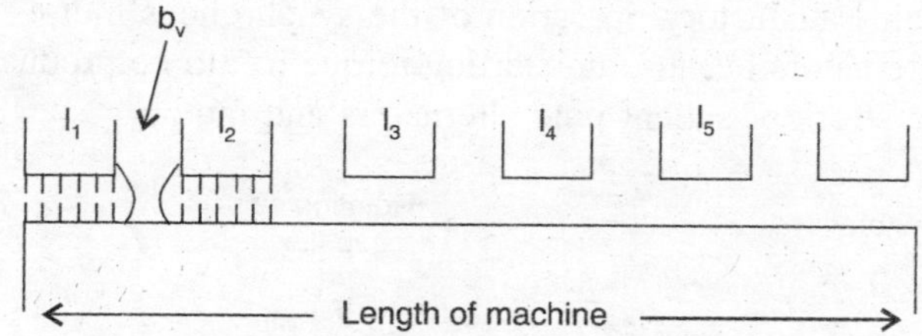


**Fig. 2.6** Effect on width of ventilating ducts.

So when the ventilating ducts are provided the effective length can be estimated by estimating effective width of ventilating ducts, $b'_v$ with the help of Carter's air gap coefficient. In case of ducts Carter's air gap coefficient depends upon width of duct to air gap length ratio and can be obtained from Fig. 2.3. The effective length of ventilating ducts can also be calculated from expression available for the machine.

Effective length of ventilating ducts = $b'_v$

Effective length of machine, $L_e = L - n_v\, b'_v$

where, $L$ = overall length of the machine. The effective lengths of respective machines are explained and calculated while design of the machine is done.

## 2.3 TYPES OF DAMPER WINDINGS

Damper windings are used in synchronous generator to

(*i*) suppress the inverse rotating fields.
(*ii*) damp out the oscillations in the machine.

and damper windings are used in synchronous motors to obtain mainly the following characteristics:

(*i*) to provide the starting torque.
(*ii*) to prevent hunting in the machines.

If the speed of the rotor is equal to the speed of rotating magnetic field (r.m.f.) developed by the stator *i.e.*, the relative speed between the rotor and r.m.f. is zero, the synchronous machine operates satisfactorily. When there is any disturbance and deviation in speed, the machine tries to maintain its stability which depends upon synchronising power. Unloaded synchronous motor may be assumed to have zero load angle at no load with armature resistance neglected and can be shown as in Fig. 2.7.

When the load on the motor is increased gradually the load angle also increases accordingly. But if the load varies in large amount (large step) suddenly, inspite of varying gradually, the motor must slow down momentarily *i.e.*, the motor speed must become less than synchronous speed to meet the load change. The load angle starts increasing. The rotor oscillates or swinges due to sudden load applied on motor and search for or hunt for its new equillibrium space position.

Damper bars form the starting winding for synchronous motors. For good starting torque, damper bars should have high resistance and for large damping effect near synchronous speed, the damper bars should have low resistance. Therefore, a compromise between good starting torque and good damping effect can be made to obtain satisfactory operation of the synchronous motor. Low resistance damper winding can be used in alternators because no starting torque has to be produced. Damper winding is being provided in the pole stress of salient pole alternators and motors.

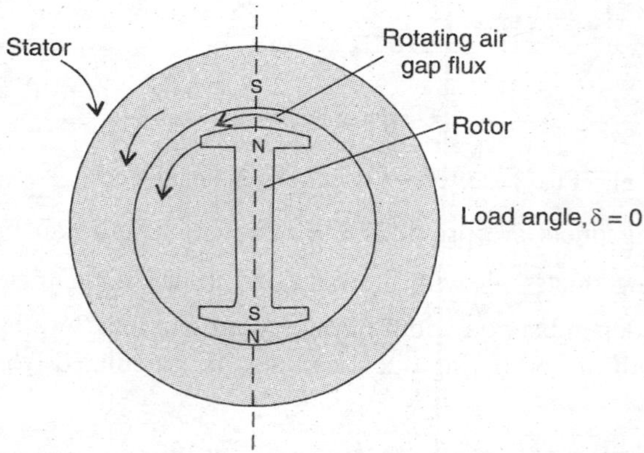


**Fig. 2.7** Load angle at unloaded synchronous motor.

Damper windings are not used on turbo-generators. The solid-steel rotor cores of such machines, provide path for eddy currents, especially in the quadrature axis, where the iron may form an equivalent rotor circuit, thus producing the same effects as those of damper bars.

The damper winding can be made by low resistance copper, brass or aluminium bars embedded in slots in the pole shoes of salient pole machines. The ends (projected out from machine length) are tapered and are connected by a short circuiting ring on both ends of the machine. This is same like squirrel cage winding in squirrel cage induction motor. The two types of damper windings can be done.

(*i*) Open type or non-connected type.
(*ii*) Connected type.

These two types of windings are shown in Fig. 2.8 (*a*) and (*b*).

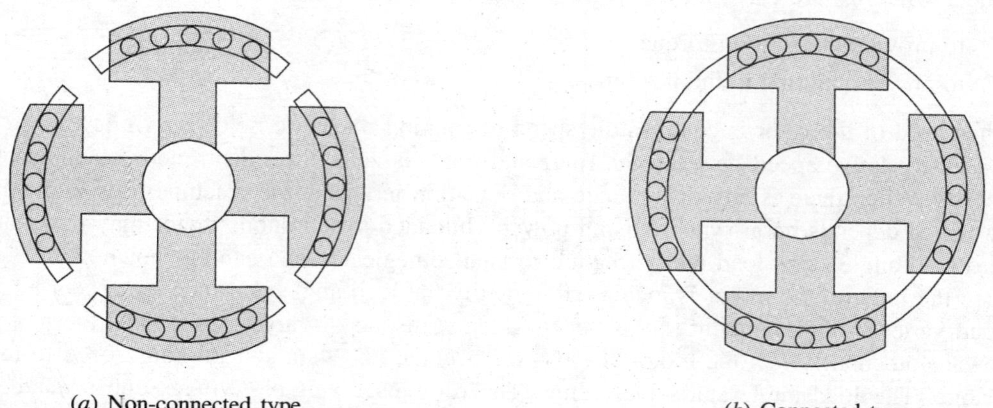

(*a*) Non-connected type (*b*) Connected type

**Fig. 2.8** Different type of damper winding.

When the rotor is running at synchronous speed, the relative speed between damper bars and rotating air gap flux is zero. Due to zero relative speed, there will be no flux-cutting across the rotor damper winding and e.m.f. generated in damper bars is zero consequently no damping torque is developed. The damper winding is effective only during rotor hunting, when rotor speed is changing from synchronous speed.

When the rotor speed becomes more than or less than the stator field (r.m.f.) then there will be a relative velocity between rotor (damper winding conductors) and stator field. An e.m.f. will be induced in the rotor damper winding and current flows in damper bars. This current sets up its own field which opposes the cause (relative speed) by which it is produced and hence it tries to minimize the relative speed between rotor and stator field. So if the rotor speed is less than the stator field, it will try to accelerate the rotor speed and if the rotor speed is more the stator field, it will try to retard the rotor speed. With this phenomenon damper winding helps in reducing the hunting or oscillations in the machine.



## 2.4 DIFFERENT TYPES OF STATOR/ARMATURE LEAKAGE FLUX

There are several ways in which leakage of flux occurs in the stator/armature having distributed a.c. or d.c. windings. The different types of leakage flux can be classified as below:

(*i*) Slot leakage flux.
(*ii*) Tooth top leakage flux.
(*iii*) Zig-zag leakage flux.
(*iv*) End winding or Overhang leakage flux.
(*v*) Harmonic or belt or Differential leakage flux.
(*vi*) Skew leakage flux.
(*vii*) Peripheral leakage flux.

### 2.4.1 Slot Leakage Flux

When the flux crosses the slot from one tooth to the next tooth and returns through the iron portion at the bottom of the slot and linking the portion of the winding conductor below it is called as slot leakage flux. Slot leakage flux in the stator/armature is shown in Fig. 2.9.

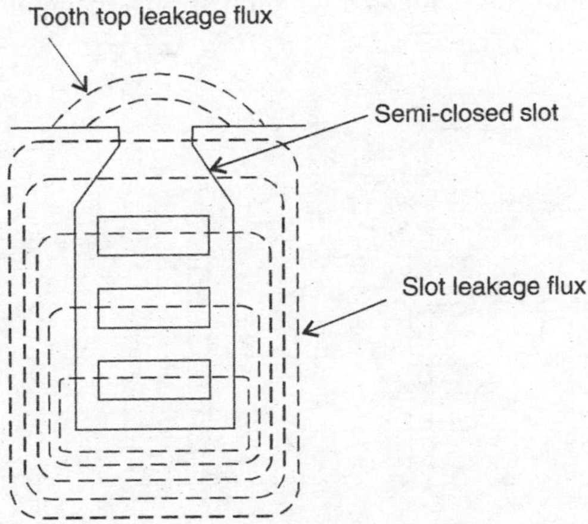


**Fig. 2.9** Slot leakage and tooth top leakage in stator/armature.

### 2.4.2 Tooth Top Leakage Flux

When the flux passing from top of one tooth links with the top of another tooth then this flux is called as tooth top leakage flux. When the air gap length of the machine will be higher, the tooth top leakage flux will be more. This is the reason that tooth top leakage flux is very important in the machines with large air gap length. The tooth top leakage flux is considerably enough in case of synchronous machines and d.c. machines. But tooth top leakage flux is normally negligible for induction machines since they have small air gap length and hence the tendency of tooth top leakage is very low. Figure 2.9 shows the tooth top leakage flux.

### 2.4.3 Zig-zag Leakage

The flux which passes from one tooth to another in a random manner is called as zig-zag leakage flux.

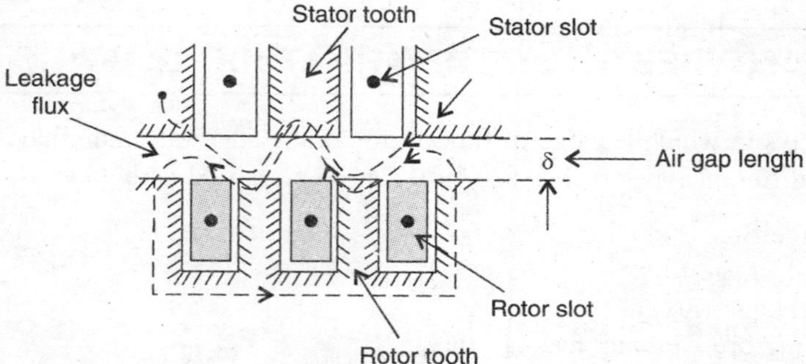


**Fig. 2.10** Zig-zag leakage flux.

The magnitude of zig-zag leakage flux depends upon the length of air gap of the machine and also on the relative positions of the tips of teeth. The zig-zag leakage flux is shown in Fig. 2.10.

### 2.4.4 End Winding or Overhang Leakage Flux

The end portions of the winding (projected out beyond axial length of machine) produces a separate leakage of flux and hence it is called as end winding or overhang leakage flux. The magnitude of the end winding leakage flux depends upon the end winding arrangement, end covers with conducting and

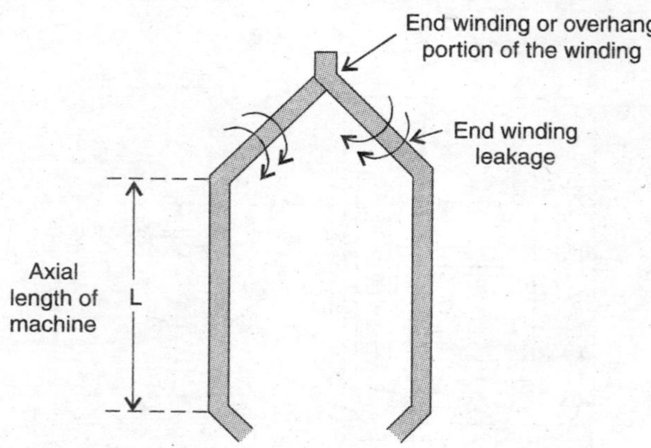


**Fig. 2.11** End winding leakage flux.

magnetic characteristics and other metal masses in its vicinity. End winding leakage flux is shown in Fig. 2.11.

### 2.4.5 Harmonic or Belt or Differential Leakage Flux

When primary and secondary winding m.m.f. magnitude and distributions are not similar then unbalanced components causes harmonic fluxes. These harmonic fluxes rotates at their own synchronous speed and hence causing a fundamental frequency reactive voltage drop in the primary. Due to difference of magnitude and distributions of m.m.fs., there exists a net m.m.f. which causes the leakage of flux and is called as harmonic leakage flux. When the m.m.fs. of the windings are equal in magnitude and distributions, there is no harmonic leakage flux. The harmonic leakage flux or differential leakage flux can be neglected for normal windings of the machine.

### 2.4.6 Skew Leakage Flux

This type of leakage flux occurs when the slots in the machine are skewed. The skewing of slots are usually done in squirrel cage induction motors to eliminate harmonic torques and noise. With skewed slots the voltage induced in rotor conductors are reduced and so decrease in mutual flux causing a difference between total flux and mutual flux.

### 2.4.7 Peripheral Leakage Flux

The flux passing through peripheral length of air gap without linking any windings or iron portions is called as peripheral leakage flux. This flux may be assumed to be negligible for different machines.

Slot leakage, tooth top leakage, end winding leakage and zig-zag leakage are usually calculated to calculate the leakage reactance in the machines because other leakage fluxes are small in magnitude and can be neglected.

Chapter 2

## 2.5 CALCULATION OF SLOT LEAKAGE REACTANCE FOR DIFFERENT TYPE OF SLOT

### 2.5.1 Slot Leakage Reactance for Parallel Sided Slots with Single Layer Winding

Calculation of slot leakage reactance $(X_{sl})$ for parallel sided slots with single layer winding are considered as given in Fig. 2.12.

Following assumptions are made while calculating the slot leakage inductance:

(1) The leakage path is straight across the slot and around the iron path at the bottom of the slot.
(2) The permeance of air path only is considered.
(3) The reluctance offered by iron part is zero and

$$\mu_{Fe} = \infty$$

(4) The current in the conductors in the slot is uniformly distributed.

Let, $h_1$ = height of conductor in the slot

$h_2$ = free space for insulation

$h_3$ = tapered portion of the slot for winding support wedges

$h_4$ = tooth width

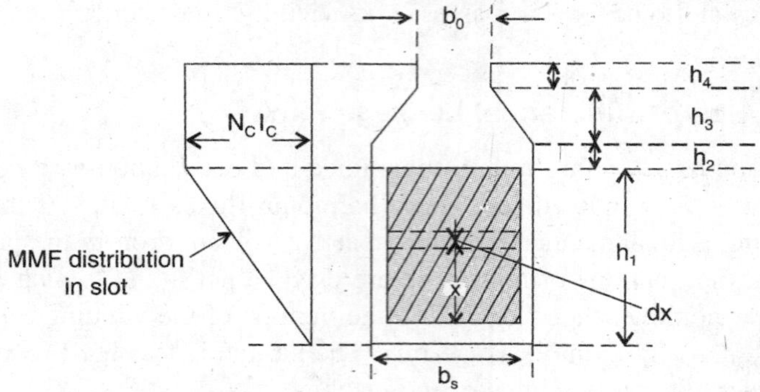


**Fig. 2.12** Parallel sided slot with single layer winding.

$b_s$ = width of slot

$b_0$ = opening width of slot

$N_c$ = number of conductors per slot

$I_c$ = conductor current

$L$ = length of conductor portion in the slot of machine

The permeance of an elementary portion ($dx$) considered at a height $x$ as shown in Fig. 2.12 from the bottom of conductor

$$= \frac{\mu_0 L dx}{b_s} \qquad \text{[From the expression of permeance as } \frac{\mu_0 A}{l}\text{]}$$

$MMF$ at height $x = \dfrac{x}{h_1} \cdot N_c I_c$

Flux in the elementary portion $dx$

$$d\phi_x = MMF \times \text{Permeance}$$

$$= \frac{x}{h_1} \cdot N_c I_c \times \frac{\mu_0 L dx}{b_s}$$

$$= \frac{N_c I_c \mu_0 L}{h_1\, b_s} \cdot x dx$$

Number of the conductors linking with flux in the elementary portion $dx$

$$= \frac{x}{h_1} \cdot N_c \text{ conductors}$$

So flux linkage due to $\dfrac{x}{h_1} \cdot N_c$ conductors

$$\Delta\psi_x = \frac{x}{h_1} \cdot N_c \cdot \frac{N_c I_c \mu_0 L}{h_1 b_s} \cdot x dx$$

or $$\Delta\psi_x = \frac{N_c^2 I_c \mu_0 L}{h_1^2 b_s} \cdot x^2\, dx$$

Since the flux linkages are defined as the actual flux multiplied by the number of conductors associated with it.

or $$\Delta\psi_x = \frac{N_c^2 I_c \mu_0 L}{b_s} \cdot \left(\frac{x}{h_1}\right)^2 dx$$

∴ Total flux linkages for the height of conductor ($h_1$) in the slot can be obtained by

$$\psi_1 = \int_0^{h_1} \Delta\psi_x$$

$$= \int_0^{h_1} \frac{N_c^2 I_c \mu_0 L}{b_s} \cdot \left(\frac{x}{h_1}\right)^2 dx$$

$$= \frac{\mu_0 N_c^2 I_c L}{b_s} \int_0^{h_1} \left(\frac{x}{h_1}\right)^2 \cdot dx$$

$$= \mu_0 N_c^2 I_c L \cdot \frac{1}{b_s} \cdot \frac{h_1}{3} = \mu_0 N_c^2 L I_c \cdot \frac{h_1}{3b_s}$$

$$= \mu_0 N_c^2 I_c L \cdot \frac{h_1}{3b_s} \qquad ...(2.1)$$

Flux linkage for the free space portion ($h_2$):

Permeance of slot of height ($h_2$) $= \mu_0 \times \frac{\text{area}}{\text{length}} = \mu_0 \cdot \frac{Lh_2}{b_s}$

*MMF* at height above $h_1 = N_c I_c$

Flux linking through height $h_2 = N_c I_c \mu_0 \cdot \frac{L h_2}{b_s}$

Flux linkage for height $h_2$ can be obtained by flux linking through height $h_2$ multiplied by the number of conductors ($N_c$) linking with it.

∴ Total Flux linkage for height $h_2 = N_c I_c \mu_0 \cdot \frac{L h_2}{b_s} \cdot N_c$

$$= \mu_0 N_c^2 I_c L \cdot \frac{h_2}{b_s} \qquad ...(2.2)$$

Flux linkages for portion $h_3$:

The mean length can be calculated as

$$= \frac{b_0 + b_s}{2}$$

Flux linking through height $h_3 = N_c I_c \mu_0 \cdot \dfrac{L h_3}{\dfrac{b_0 + b_s}{2}}$

Flux linkage for this portion ($h_3$) can be obtained by multiplying flux linking through height $h_3$ with number of conductor ($N_c$) linking with it.

∴ Total flux linkage in portion ($h_3$)

$$= \mu_0 N_c^2 I_c L \cdot \frac{2h_3}{b_0 + b_s} \qquad \text{...(2.3)}$$

Flux linkages for portion $h_4$:

Length in this case will be $b_0$. Flux linkage for this portion can also be estimated as in portion $h_2$ and $h_3$ so flux linkage in portion $h_4$ is

$$= \mu_0 N_c^2 I_c \, L \cdot \frac{h_4}{b_0} \qquad \text{...(2.4)}$$

Total flux linkages can be estimated by equations (2.1), (2.2), (2.3) and (2.4) and is given by

$$\psi = \mu_0 N_c^2 I_c \, L \left[ \frac{h_1}{3b_s} + \frac{h_2}{b_s} + \frac{2h_3}{b_0 + b_s} + \frac{h_4}{b_0} \right]$$

Total flux linkage of all the slots of one phase

$$\psi_{sl} = Pq \mu_0 N_c^2 I_c L \left[ \frac{h_1}{3b_s} + \frac{h_2}{b_s} + \frac{2h_3}{b_0 + b_s} + \frac{h_4}{b_0} \right]$$

where, $P$ = number of poles and $q$ = number of slots/pole/phase

Number of turns per phase, $N_{ph} = \dfrac{N_c P q}{2}$

or $$N_c = \frac{2N_{ph}}{Pq}$$

∴ $$\psi_{sl} = 4\mu_0 \frac{N_{ph}^2}{Pq} \cdot L I_c \left[ \frac{h_1}{3b_s} + \frac{h_2}{b_s} + \frac{2h_3}{b_0 + b_s} + \frac{h_4}{b_0} \right]$$

So slot leakage inductance

$$L_{sl} = \frac{\psi_{sl}}{I_c} = 4\mu_0 \frac{N_{ph}^2}{Pq} \cdot L \left[ \frac{h_1}{3b_s} + \frac{h_2}{b_s} + \frac{2h_3}{b_0 + b_s} + \frac{h_4}{b_0} \right] \text{ henry}$$

∴ Slot leakage reactance

$$X_{sl} = 2\pi f L_{sl} \qquad ...(2.5)$$

Slot leakage reactance can be calculated by equation (2.5).

### 2.5.2 Slot Leakage Reactance for the Parallel Sided Slots with Double Layer Winding

Calculation of slot leakage reactance ($X_{sl}$) for parallel sided slots with double layer winding are considered as given in Fig. 2.13.

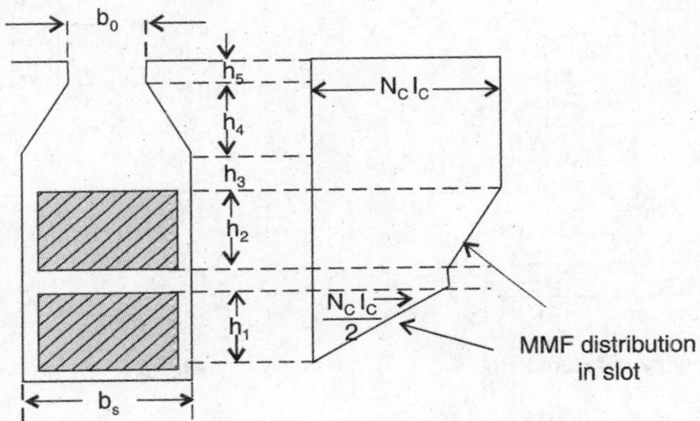


**Fig. 2.13** Parallel side slots with double layer winding.

All the dimensions of the slots can be explained as in 2.5.1. The flux linkages in different portion of the slot can be calculated in a similar way as it is done for slot with single layer winding in 2.5.1 *i.e.*, by knowing flux linking a particular portion and the number of conductor in that portion. Then total flux linkages of all the slots of one phase, $\Psi_{sl}$ can be calculated. It can be given by the expression.

$$\Psi_{sl} = 4\mu_0 \frac{N_{ph}^2}{Pq} \cdot LI_c \left[ \frac{2h_1}{3b_s} + \frac{h_2}{4b_s} + \frac{h_3}{b_s} + \frac{2h_4}{b_0 + b_s} + \frac{h_5}{b_0} \right]$$

∴ Slot leakage inductance can be given by, $L_{sl} = \dfrac{\Psi_{sl}}{I_c}$

or
$$L_{sl} = 4\mu_0 \frac{N_{ph}^2}{Pq} \cdot L \left[ \frac{2h_1}{3b_s} + \frac{h_2}{4b_s} + \frac{h_3}{b_s} + \frac{2h_4}{b_0 + b_s} + \frac{h_5}{b_0} \right]$$

So slot leakage reactance

$$X_{sl} = 2\pi f L_{sl} \qquad ...(2.6)$$

### 2.5.3 Slot Leakage Reactance for Tapered Slot

The details of the slot is given in Fig. 2.14.

Slot leakage inductance for tapered slot can be given by

$$L_{sl} = 4\mu_0 \frac{N_{ph}^2}{Pq} \cdot L \left[ \frac{2h_1}{3(b_{s_2} + b_s)} + \frac{2h_2}{(b_{s_1} + b_{s_2})} + \frac{2h_3}{b_{s_1} + b_0} + \frac{h_4}{b_0} \right]$$

and slot leakage reactance is

$$X_{sl} = 2\pi f L_{sl} \quad ...(2.7)$$

### 2.5.4 Slot Leakage Reactance for Round Slot

The details of the round slot is given in Fig. 2.15. It can be concluded that the slot leakage inductance of round slot is independent of the diameter.

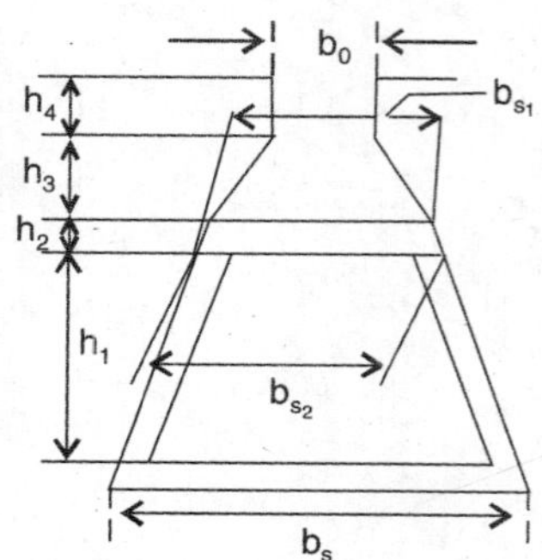


**Fig. 2.14** Details of tapered slot.

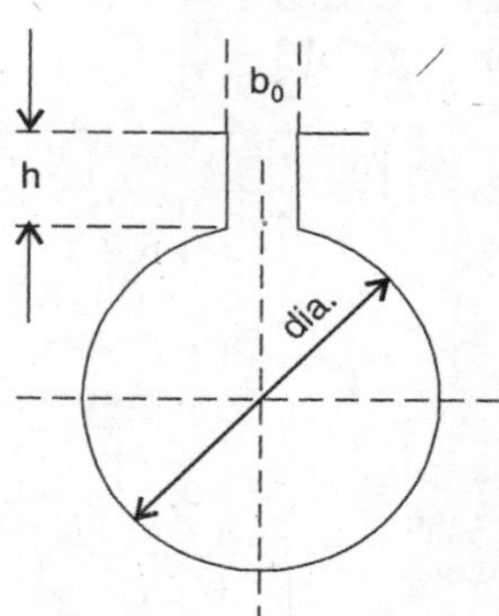


**Fig. 2.15** Details of round slot.

Slot leakage inductance is calculated by

$$L_{sl} = 4\mu_0 \frac{N_{ph}^2}{Pq} \cdot L\left[0.66 + \frac{h}{b_0}\right] \quad ...(2.8)$$

### 2.5.5 Estimation of Tooth Top, Overhang, Zig-Zag and Harmonic Leakage Reactances

The leakage reactances due to tooth top leakage, zig-zag leakage, end winding leakage and harmonic leakage fluxes are calculated when the design of respective machines are done. When overhang leakage reactance is to be calculated then factor, $k$ is needed. Factor $k$ depends upon coil span to pole pitch ratio and can be obtained from Fig. 2.16. Similarly, when harmonic leakage reactance has to be calculated the factor $k_{h_1}$ for stator and $k_{h_2}$ for rotor should be known. These factors can be obtained from Fig. 2.17. These factors are used to calculate the leakage reactances.

## 2.6 CALCULATION OF POLE LEAKAGE FLUX ($\phi_{el}$)

The interpolar space in the field system of salient pole synchronous machine offers a parallel path to the lines of flux leaving the pole and entering the stator in the same way as a slot offers a parallel path for the lines of flux passing through the tooth. A part of the flux do not enter the stator but take a path from pole shoe to adjacent pole shoe, pole core or body to pole core and also a very small portion from pole core to pole yoke as given in Fig. 2.18. The reluctance of the interpolar space is larger still some flux take path through this space. It varies from 10% to 20% of useful flux ($\phi$). Since the useful flux only enters the stator side and induces e.m.f. in the stator winding. The flux which passes through interpolar space is a waste and is called as pole leakage flux ($\phi_{el}$). So the pole core, pole shoe and pole yoke have more flux than the useful flux.

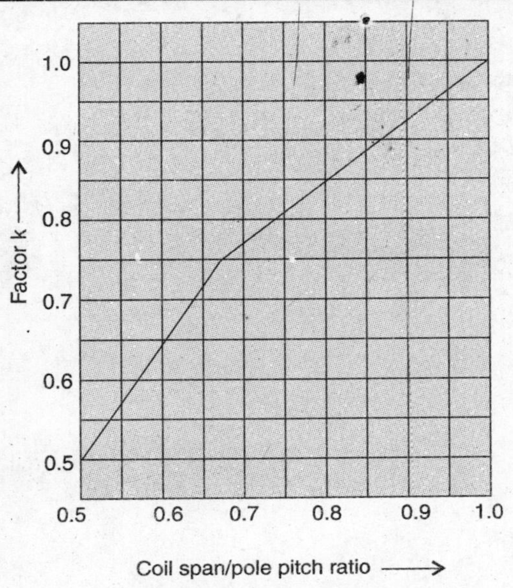


Fig. 2.16 Curve for factor *k*.

Fig. 2.17 Curve for $k_{h_1}$ and $k_{h_2}$.

Chapter 2

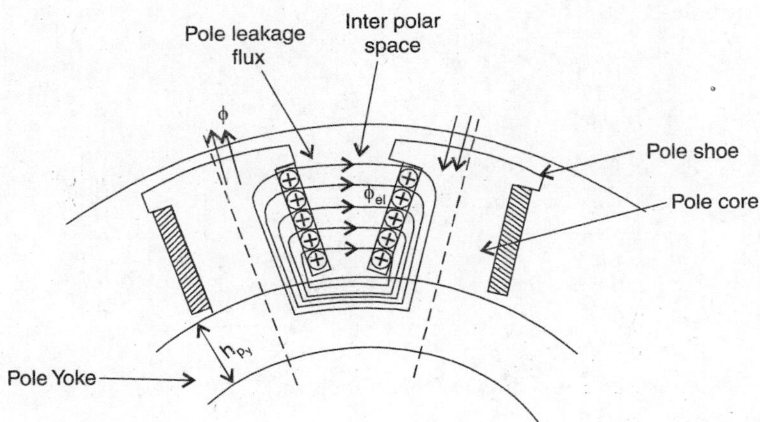


Fig. 2.18 Pole leakage flux ($\phi_{el}$).

The different dimensions of the pole and the interpole space of a salient pole machine is given in the Fig. 2.19. The m.m.f. which acts upon the air gap, stator teeth and stator yoke/core is responsible in determining the magnitude of the pole leakage flux.

The steps to calculate the pole leakage flux are explained as below.

**Calculation of pole leakage flux ($\phi_{el}$):**

Pole leakage flux consists of the following leakage flux between the pole pair:

$\phi_i$ = 2 × Leakage flux from pole shoe to pole shoe end or between inner surfaces of pole shoes.

$\phi_{ii}$ = 2 × Leakage flux from pole side to pole side end or between end surfaces of pole shoes.

$\phi_{iii}$ = 2 × Leakage flux from pole core to pole core end or between inside surfaces of pole cores.

$\phi_{iv}$ = 2 × Leakage flux from pole core side to pole core side or between end surfaces of pole cores.

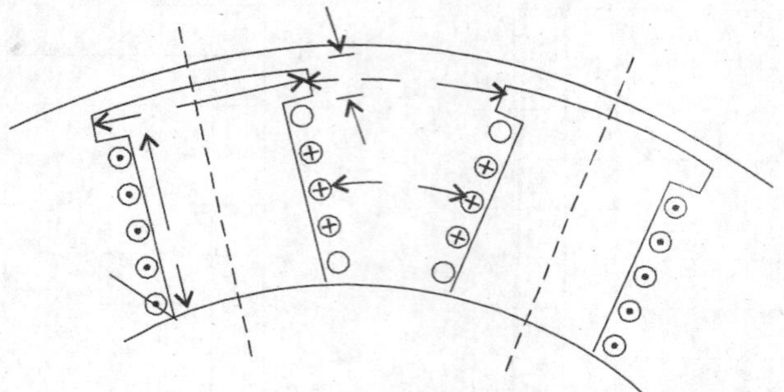

**Fig. 2.19** Different dimensions of pole.

So pole flux leakage $\phi_{el}$ will be

$$\phi_{el} = \phi_i + \phi_{ii} + \phi_{iii} + \phi_{iv}$$

Total flux per pole = $\phi + \phi_{el}$.

**Estimation of $\phi_i$:**

$a_{ps}$ = distance between pole shoes

$a_{pcm}$ = distance between pole cores at middle of core

$h_{pc}$ = height of pole core

$b_{pc}$ = width of pole core

$$b = \frac{h_1 + h_2}{2}$$

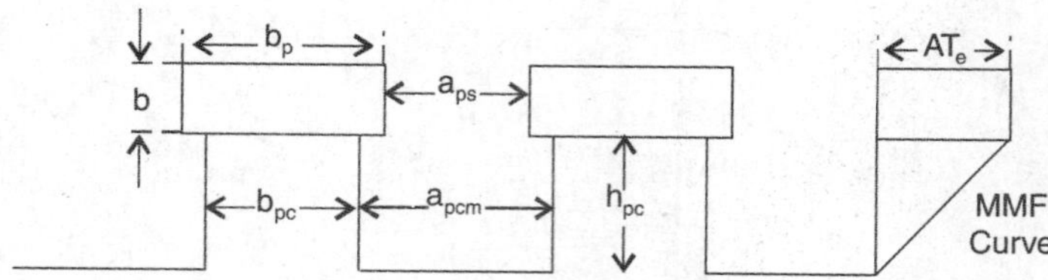


**Fig. 2.20** Details of distances.

The details of distances are shown in Fig. 2.20.

$\phi_i = 2 \times MMF \times$ Permeance

$$\text{Permeance } = \frac{\mu_0 A}{l} = \frac{\mu_0 L_p b}{a_{ps}}$$

$$\text{So, } \phi_i = 2 \times AT_e \times \frac{\mu_0 L_p b}{a_{ps}}$$

**Estimation of $\phi_{ii}$ :**

It is assumed that the flux path between surfaces of the pole shoes go through a part of the way in arc of circle of radius $y$ and part of the way in straight line along the distance $a_{ps}$. Take both sides (ends) of the machine into consideration. The path of the flux in arc of circle of radius $y$ and in distance $a_{ps}$ is shown in Fig. 2.21.

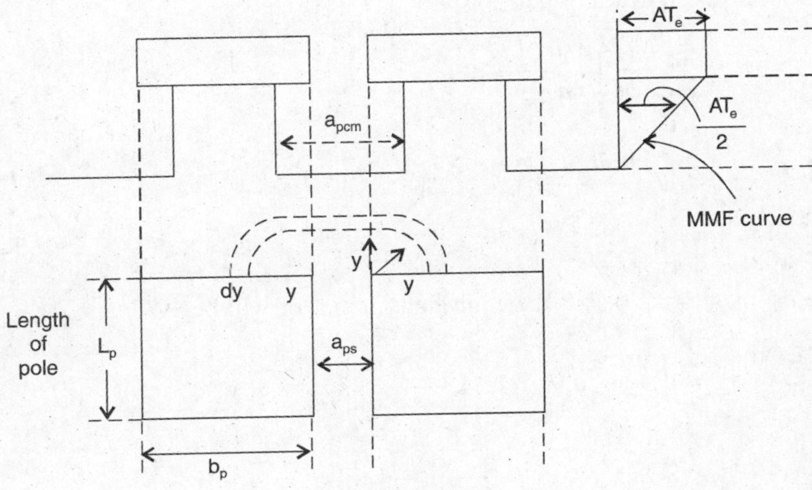


**Fig. 2.21** Path of flux $\phi_{ii}$.

Chapter 2

Let us consider the elementary flux, $dy$, at distance $y$

Elementary area $= bdy$

Flux through elementary area

$$d\phi_{ii} = 2AT_e \frac{\mu_0 bdy}{a_{ps} + \pi y}$$

Total flux for both ends of machine

$$\phi_{ii} = 2\int_0^{b_p/2} 2AT_e \frac{\mu_0 bdy}{a_{ps} + \pi y}$$

$$= 4AT_e \mu_0 b \left[ \ln \frac{(a_{ps} + \pi y)}{\pi} \right]_0^{\frac{b_p}{2}}$$

$$= 4AT_e \frac{\mu_0 b}{\pi} \left[ \ln \frac{a_{ps} + \frac{b_p}{2}\pi}{a_{ps}} \right]$$

$$= 4AT_e \frac{\mu_0 b}{\pi} \left[ \ln \left( 1 + \frac{b_p \pi}{2a_{ps}} \right) \right]$$

**Estimation of $\phi_{iii}$ :**

It can be estimated similar as $\phi_i$, but the equivalent $a_{pcm}$ (ampere-turn at the middle of the pole core) will be half of $AT_e$ as shown in Fig. 2.21.

$$\phi_{iii} = 2\left(\frac{AT_e}{2}\right)\frac{\mu_0 h_{pc} L_p}{a_{pcm}}$$

$$= AT_e \frac{\mu_0 h_{pc} L_p}{a_{pcm}}$$

**Estimation $\phi_{iv}$ :**

It can also be estimated like $\phi_{ii}$ but the equivalent m.m.f. of will be $AT_e/2$.

$$\phi_{iv} = 4\left(\frac{AT_e}{2}\right)\mu_0 h_{pc} \cdot \ln\left(1+\frac{b_{pc}\pi}{2a_{pcm}}\right)$$

$$= 2\,AT_e\,\mu_0 h_{pc} \cdot \ln\left(1+\frac{b_{pc}\pi}{2a_{pcm}}\right)$$

Total leakage flux,

$$\phi_{el} = \phi_i + \phi_{ii} + \phi_{iii} + \phi_{iv}$$

Total flux per pole, $(\phi_p) = \phi + \phi_{el}$.

## 2.7 DIFFERENT TYPES OF STATOR AND ROTOR SLOTS

When a slot area has been decided for the stator (specially in induction motor) the slot for that area may have the different alternatives like more height with less width or less height with more width. The slot leakage flux is directly proportional to slot height (depth). So the deeper slots, in induction motors, have more leakage reactance and hence less torque. Similar effect is obtained when induction motors have wider slots. Further, the slots may be open, semi-closed or closed type as shown in Fig. 2.22 (*a*), (*b*) and (*c*). The reluctance offered by closed slots is lowest and reluctance offered by semi-closed slots are less than the reluctance offered by open slots. So the open slots have leakage reactance less than

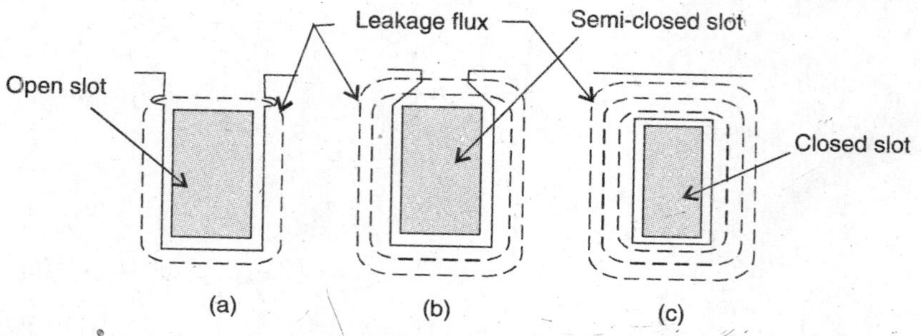


**Fig. 2.22** Different types of slots.

that of semi-closed slots. The closed slots have leakage reactance more than that of semi-closed slots. The flux distribution in open slots is not uniform whereas in closed slots the flux distribution is uniform. But leakage reactance in closed slots are quite high. Usually semi-closed slots are being used to optimise the leakage reactance.

In squirrel cage induction motors, deep bar rotor slots are used. When deep bar rotor slots are used then the starting torque of induction motors are increased. The starting torque is increased by increasing the effective resistance of the stator at starting. In squirrel cage induction motors deep bar rotor slots are shown in Fig. 2.23.



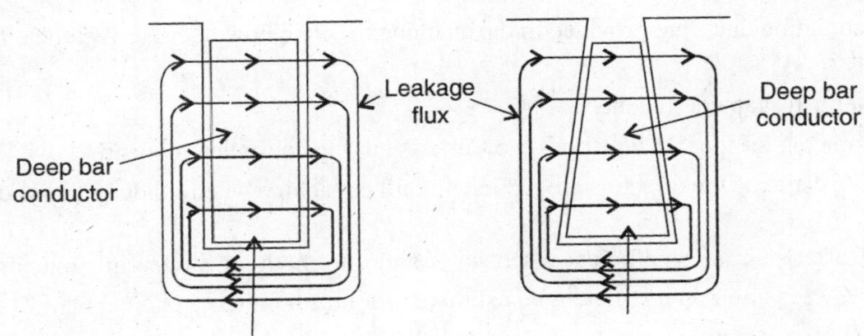


**Fig. 2.23** Deep bar rotor slots.

Due to very high (infinite) permeability of iron all the leakage flux lines passes just below the bottom of slot. The effect of deep bar rotor conductor is explained by considering a number of layers of conductors like top and bottom layers in Fig. 2.24. The leakage inductance of bottom layer is fairly larger than that of the top layer because the bottom layer conductors are associating large leakage flux lines. So the leakage reactance of the bottom layer of winding conductor is more than that of top layer of winding conductor. At the time of starting of the machine the frequency of rotor e.m.f. is equal to the supply frequency. So the distribution of rotor current between parallel connected top and bottom layers depends upon the leakage reactance. The current in top layer of conductor is larger than that in bottom layer at the time of starting. Therefore more current in top layer and this results in increase in effective resistance of rotor and thereby increasing the starting torque. Since the change in rotor current distribution depends upon the inductive effect, the effective resistance becomes a function of the frequency and also the height of the rotor bar. So a deep bar rotor can have an effective resistance at standstill (supply frequency) much larger than its d.c. resistance. Now when the motor starts running, the frequency of rotor current becomes lesser and lesser and so the effect of leakage reactance in distributing the current in the rotor bar (between top and bottom layer) gradually decreases. Finally, when motor picks up the rated speed the current distribution is mainly determined by the resistance of

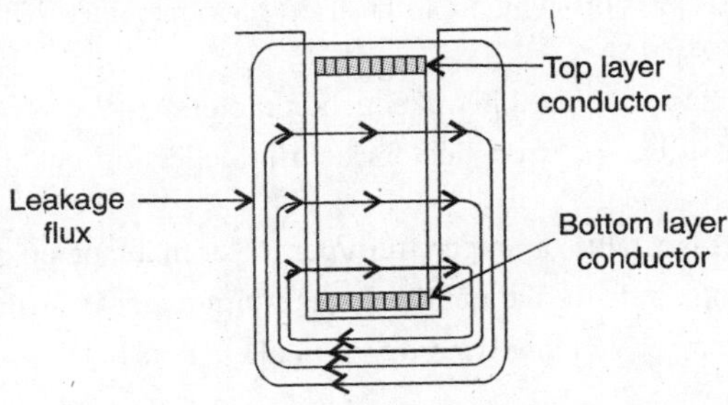


**Fig. 2.24** Rotor slot with different layer conductor.

top and bottom layer of the windings. Since under running condition, the current distribution depends upon the d.c. resistance of the two layers, so the current distribution is uniform between the two layers and so good running condition of the machine is also achieved.

## UNSOLVED QUESTIONS

1. Explain fringing of flux.
2. Discuss how does the fringing of flux effect the length of air gap of the machine?
3. The radial ventilating ducts are provided in the machine. Explain how does it effect the effective axial length of the machine?
4. What is Carter's air gap coefficient?
5. On what parameters of the machine does the Carter's air gap coefficient depend?
6. Discuss how to estimate the effective air gap length of the machine when salient pole synchronous machines are used?
7. Explain the Carter's air gap coefficient when stator and rotor both are slotted in a machine.
8. How can the Carter's air gap coefficient be estimated for a turbo-alternator?
9. Explain what is the effective length of the machine?
10. What are different objective to provide a damper windings in a synchronous generator and a synchronous motor?
11. Explain the characteristics of a damper winding to be selected for a salient pole alternator and discuss the reason.
12. Discuss why the damper winding are not used in turbo-alternators?
13. How the damping effect is obtained in a turbo-alternator? Explain.
14. Explain the different types of damper windings used in synchronous machines.
15. Explain that how a damper winding helps in reducing the oscillations/hunting in the machine?
16. Write the different types of leakage of fluxes.
17. Explain zig-zag leakage reactance. Discuss ends winding leakage and harmonic leakage reactance.
18. What is leakage reactance of a machine?
19. Explain slot leakage, tooth top leakage.
20. Discuss skew leakage flux and peripheral leakage flux.
21. Explain pole leakage flux.
22. What are the different pole leakage fluxes? Mention their name according to the portions of the pole they links.
23. Discuss the different types of slots which can be used in a machine. Why semi-closed slots are usually preferred over open and closed slots?
24. Why deep bar rotor conductors are used in an induction motors?
25. Explain how does the deep bar rotor conductors help in achieving better starting torque and also good running conditions?
26. Discuss how does the average value of ampere-turn/per meter in a tapered teeth be estimated?
27. Explain the use of Simpson's rule to find out the ampere-turn/meter in teeth of the machine.
28. Why the low resistance damper winding is preferred in alternators?

# *Section II: Conventional Design of Machines*

3

# Magnetic, Conducting and Insulating Materials in Electrical Engineering

## 3.1 INTRODUCTION

There are different parts in any machine and they should possess different characteristics to perform the required job for a particular part or portion. When we have to construct a 3-phase alternator, 3-phase induction motor, transformer or d.c. machines then laminated portion of machine, winding conductors, aluminium bars, insulating material of different size and quality etc. are required. It is preferred to design a machine, which fulfills the parameters like reduced size, cost, weight, increased efficiency, reliability etc., up to maximum extent. The characteristics of the materials used to manufacture the different part of a machine should be optimally good to achieve the better performance. If the materials are of high grade, the machine of reduced weight, cost, high efficiency can be obtained. The materials used to construct the above mentioned machines are generally classified into conducting materials, magnetic materials and insulating materials.

The materials, which allow the continuous passage of an electric current when subjected to electric potential difference, are called as conductor. The conductor is said to be better if the current density for a given potential difference is higher. High conductivity in metals depends upon the presence of free or conduction electrons. These electrons are able to move freely and do not belong to any particular atom. The good conductors of electricity are silver, copper, aluminium etc. The heat conduction in conductors is also through free electrons similarly as electrical conductivity. A relationship between the electrical conductivity and thermal conductivity exists.

Metals and alloys with low resistivity are used as conductors as well as electrical contacts. Alloys with high resistivity are used for resistors and as heating elements. The materials called as super conductor shows almost zero resistivity when they are operated below transition temperature. These materials have special properties and are used as supermagnets etc.

Materials, which can be converted into a magnet or can be attracted towards magnets, are called as magnetic materials. Magnetic materials can set-up magnetic field and generate electric power. All materials are magnetic since they show some response to magnetic field. The ferromagnetic materials have tendency to align them along with the magnetic field. Iron, cobalt, nickel etc. are known as magnetic materials. The alloys can be obtained by a combination of these metals with other materials with the help of certain heat treatment process. Generally the magnetic materials can be divided into two classes, permanent or hard magnetic materials and soft-magnetic materials. The hard magnetic materials are those, which have strong tendency to remain magnetized in the direction of applied field on removal of external magnetic field. In the second type of materials they become demagnetized on removal of an applied field. The direction of magnetization can also be easily changed in soft magnetic materials. Hard materials are used for making permanent magnets while soft materials are used for electromagnets.

The materials, which provide electrical insulation between two parts or the portion of the machine, are called as insulating materials. The electrons are bound so tightly to atomic nuclei in insulating materials such that the electrical conduction by electrons cannot occur. Gaseous, liquid and solid insulating materials are available with different properties. Material for an application can be selected after checking all the properties and making a cost comparison of different alternatives available. The solid insulating materials can be selected on the basis of their thermal capacity.

The materials used in different machines should possess the following properties:

Electrical properties — Conductivity, resistivity, dielectric permeability, dielectric strength, etc.
Magnetic properties — Permeability, coercive force, etc.
Mechanical properties — Elasticity, plasticity, ductility, toughness, hardness, brittleness, malleability, etc.
Thermal properties — Specific heat, thermal expansion, thermal conductivity, etc.
Physical properties — Density, porosity, structure, etc.
Chemical properties — Corrosion resistance, composition, etc.
Optical properties — Colour, light transmission, light reflection, etc.
Acoustical properties — Sound transmission, sound reflection, etc.

Suitable combination of the properties mentioned above is required in the selection of a material for an application. The materials are affected by mechanical loading, temperature increase, electric and magnetic fields, exposure to radiation etc. A large variety of materials are available and more than one material may be suitable for an application. The different requirements may be classified into the requirements like

(*i*) Properties of the material.
(*ii*) Fabrication requirements.
(*iii*) Economic considerations.

Classification of materials, their properties, materials available and applications of conducting, magnetic and insulating materials are discussed in this book.

## 3.2 MAGNETIC MATERIALS

The materials, which can be magnetized, are called as the magnetic materials. All the magnetic materials have some magnetic properties. The degree to which they have the magnetic properties depends upon their relative permeability. The magnetic materials can be divided into following category.

### 3.2.1 Diamagnetic Materials

The materials which have their relative permeabilities slightly less than unity are categorised as diamagnetic materials. Bismuth, silver, lead, copper and water are some of the examples of the diamagnetic materials.

### 3.2.2 Paramagnetic Materials

The materials which have their relative permeabilities slightly greater than unity like air, aluminium and palladium are called as the paramagnetic materials.

### 3.2.3 Ferromagnetic Materials

The materials, which have their relative permeabilities much greater than unity are called as ferromagnetic materials. Nickel, cobalt, iron, steel, permalloy and super-permalloy are the examples of the ferromagnetic materials. The diamagnetic materials and paramagnetic materials have negligible application in the field of electrical engineering. The ferromagnetic materials are being frequently used in electrical engineering. The ferromagnetic materials can be classified as hard magnetic materials and soft magnetic materials.

***(1) Hard Magnetic Materials:*** Magnetic materials, which have large hysteresis loop area and hence large energy loss per cycle of magnetization are classified as hard magnetic materials. The magnetization curve for hard magnetic materials is shown in Fig. 3.1 (*a*). Carbon steel, tungsten steel, cobalt steel and hard ferrites are categorized as the hard magnetic materials. These materials are suitable for making the instruments and devices, which require permanent magnets.

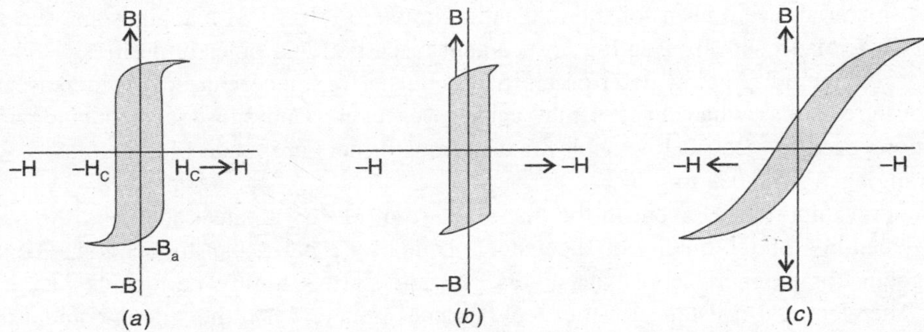


**Fig. 3.1** Hysteresis loops for hard and soft magnetic materials

***(2) Soft Magnetic Materials:*** Some magnetic materials have steep magnetization curve as given in Fig. 3.1(*b*). These materials have relatively small and narrow hysteresis loop and hence small energy loss per cycle of magentization. These materials are called soft magnetic materials. Silicon steel, nickel iron alloys and soft ferrites are the soft magnetic materials. These materials can be used for the construction of cores of electrical machines, transformers, electromagnets, reactors, relays etc.

The soft magnetic materials are useful in electrical engineering for making different electrical machines. The use of soft magnetic materials depends upon whether they carry a unidirectional flux or reversal of flux direction. The soft magnetic materials can be categorized for the use on the basis of the type of flux.

### 3.2.4 Materials for Solid Core

These materials are cast iron, cast steel and ferro-cobalt. These materials are used for making yokes of d.c. machines, cores of d.c. electromagnets and relays etc. They should have high permeability and the

*B-H* curve provides a criterion for the choice of material. The high value of density at low value of ampere-turns will reduce the cross-sectional area and as a result the weight of the materials will also be reduced. Mechanical considerations necessitate alloying the material resulting into the increase in the mechanical strength. But alloying reduces the permeability of the materials. The properties of cast-iron, cast steel and forged steel are discussed below.

***(1) Cast iron:*** The magnetic properties of cast iron are poor. It has a low relative permeability and is mainly used in making yokes of d.c. machines. The use of cast iron reduces the cost of the machine but increases the weight of the machine. This is not preferred to be used in the machines in which the mechanical stresses may be more.

***(2) Cast steel:*** It has better magnetic properties than cast iron. Low carbon steels are used for stationary parts, which is not subjected to large mechanical stresses. Since the increase in carbon content increases the mechanical strength but reduces the magnetic permeability. Hence the carbon content can be increased for those parts of magnetic circuit, which need better mechanical qualities at the cost of the permeability.

***(3) Forged steel:*** High mechanical strength is required for the rotor of the turbo-alternators. Forged steels are normally used for the rotors of turbo-alternators. The rotors are being forged and machined from a cast ingot.

### 3.2.5 Sheet Steels

Earlier sheet materials were used for the magnetic circuits of electrical machines and the cores of the transformers was made with iron with a low content of carbon and other impurities. This suffers with disadvantage of ageing. Ageing deteriorates magnetic performance caused by increase in coercive force and hysteresis loss which in turn may cause overheating and subsequent breakdown. Presently the laminations used in electrical machines and in transformers are made of silicon steel with content of silicon ranging from 0.3% to 4.5%.

Silicon is an important element in the manufacture of electrical sheet steel and the percentage of this varies depending on the quality of sheet steel needed for a particular application. The silicon steel is advantageous for magnetic cores due to its properties. It is highly permeable *i.e.*, it is easy to magnetize. The electrical resistance of the steel becomes greater when the silicon content is increased. This reduces eddy current loss. Sheet steel with silicon content also reduces hysteresis loss in the cores of the machine.

The addition of silicon reduces the effect of ageing. Addition of silicon also provides reduced hysteresis loss and increased resistivity and hence reduced eddy current loss. But the addition of silicon has some drawbacks like increase in the percentage of silicon reduces the permeability at higher flux density and also reduces the ductility. When the silicon content is above 5%, the alloy thus obtained is very brittle and cannot be punched or sheared. When the silicon content is low the alloy thus obtained is ductile but have high losses. The variety of electrical sheet steels are available in which silicon content varies from low value (0.3 per cent) to high value (4.5 per cent). Sheet steel with low silicon content possess good permeability at high flux densities, high ductility but high losses. Sheet steels with high silicon content have lower permeability, lower ductility and less losses.

It is required to achieve higher output to weight ratio in electrical machines. So the iron parts of the electrical machines should be able to work satisfactorily at higher values of flux density hence magnetic material should have a high saturation flux density. The presence of silicon becomes disadvantageous. In rotating electrical machines sheet steels with low silicon content should be used and such steels are called as dynamo grade steel.

Sheet steels with about 0.5% silicon are employed in small machines where iron losses can be tolerated. But in turbo-generators the highest possible efficiency is desired and hence in order to keep down iron losses, sheet steel laminations with high silicon content (about 4.5%) are used. Sheet steels possessing higher silicon content (4.5%) are called as 'transformer grade steels'. These are mainly used in transformers as the magnetizing current is not so important. This steel is also called high resistance steel due to its high resistivity and low eddy current loss.

**Cold Rolled Grain Oriented (C.R.G.O.) Sheet Steel:** If iron portion of the machine is working in saturation region the m.m.f. required for iron parts of magnetic circuit will be very large. In order to keep down the size of the machine it is necessary to use a high flux density. So it is needed to develop a material, which can work at high flux densities without requiring a large m.m.f. and also without causing excessive iron loss.

The hot rolled sheets have the constituent crystals in a random manner. The crystallographic axes of the individual crystals do not align themselves in the direction of rolling, with sheet surface or with each other and hence hot rolled sheets, have a low value of permeability. The rolling direction of the material is magnetically superior as compared with other directions in the sheet. Cold rolled grain oriented (C.R.G.O.) sheet steel was developed due to this property. Grain oriented electrical sheet steel is produced by cold reduction methods and is available in coils or sheets. This cold reduced material has strong directional magnetic properties, the rolling direction being the direction of highest permeability and also the direction of lowest hysteresis loss.

This material is suitable for use in transformers and also for large turbo-alternators because the axis of the core can be made to correspond with the rolling direction of the sheet and hence full use is made of high permeability low loss direction of the sheet. Generally 0.35 mm thickness of laminations

**Table 3.1** Hot rolled sheet steel (dynamo grades)

| *Grade* | *Alloys* | *Special alloys* | *Medium resistance* |
|---|---|---|---|
| Density (gm/cm$^3$) | 7.82 | 7.78 | 7.72 |
| Silicon percentage | 0.3 | 0.85 | 1.75 |
| Resistivity, μΩ/cm$^3$ | 15 | 21 | 33 |
| Temp. coeff. of resistance/ °C | 0.00338 | 0.00250 | 0.00165 |
| Stacking factor (0.50 mm sheets) | 0.95 | 0.95 | 0.95 |

**Table 3.2** Hot rolled sheet steel (transformer grade)

| *Different characteristics* | *Grade 92* | *Grade 86* | *Grade 80* |
|---|---|---|---|
| Density (gm/cm$^3$) | 7.55 | 7.55 | 7.55 |
| Silicon percentage | 4.0 | 4.0 | 4.0 |
| Resistivity, μΩ/cm$^3$ | 60 | 60 | 60 |
| Temp. coeff. of resistance/ °C | 0.00075 | 0.00075 | 0.00075 |
| Stacking factor (0.35 mm sheets) | 0.92 | 0.92 | 0.92 |

is used for transformer, while for rotating machine cores the thickness of laminations used is about 0.50 mm. Cold rolled steel sheets are used in transformer construction where losses have to be kept low. The silicon present in hot rolled steel for transformers is usually 4 to 4.5%, while in cold rolled steel laminations used for transformers it is usually 3 to 3.5%. For transformers, the stampings are cut to the size suitable for shearing. For rotating machines cores, the stampings are punched.

The properties of some of different grades of magnetic materials, which are used for dynamos and transformers, are given in Table 3.1 and 3.2. The cold rolled grain oriented transformer grade magnetic material properties are shown in Table 3.3.

**Table 3.3** Cold rolled grain oriented sheet steel (transformer grades)

| *Different properties* | *Grade 51* | *Grade 46* | *Grade 41* |
|---|---|---|---|
| Thickness (mm) | 0.35 | 0.30 | 0.28 |
| Density (gm/cm$^3$) | 7.65 | 7.65 | 7.65 |
| Silicon percentage | 3.1 | 3.1 | 3.1 |
| Resistivity, $\mu\Omega$/cm$^3$ | 48 | 48 | 48 |
| Stacking factor | 0.97 | 0.97 | 0.97 |

## 3.3 CLASSIFICATION OF CONDUCTING MATERIALS

The conducting materials can be classified into following category:

(a) **Good Conducting Materials:** These materials have very low resistivity. They are used as conductors in electrical machines and other apparatus. They are also used for transmission and distribution of electrical energy. Copper and aluminium are widely used in different field of application.

(b) **High Resistivity Materials:** These materials are used to make resistances and heating devices, thermocouples etc. These materials must withstand high temperatures for long time without melting. They are generally alloys of metals like constantan, nichrcome, etc. and are being used as resistance.

(c) **Materials Used for Solders:** Tin-lead solders are usually being used in electrical engineering to join copper, bronze, brass, etc.

(d) **Carbon Materials** are also available and are being used for different applications in the machines and equipments.

(e) **Superconducting Materials.**

### 3.3.1 Good Conducting Materials

Good conducting materials should fulfill the following properties:

(*i*) Good Conductivity.

(*ii*) Low temperature coefficient.

(*iii*) Mechanical strength sufficient to withstand with different forces.

(*iv*) Easy manufacturing in wire shape *i.e.*, good drawability.

(*v*) Corrosion resistant and easily solderable.

For different applications, different combination of properties is needed. The most important conductors are copper and aluminium but some time, brass, bronze and steel is also used in electrical machines and equipments. Some of the following conducting materials are discussed below:

***(a) Copper:*** Copper wires of different gauges as required in the machine are used for windings of the machine. Aluminium is also being used as conductor in a limited way for smaller and medium size transformer windings because of the cost and availability consideration of copper. Aluminium is cheaper and available in India. Copper has however, better characteristics than aluminium as far as its use as conductor is considered and is more commonly used in electrical machines.

Copper windings have different types of insulation coverings depending on the permissible temperature limits, allowed at particular applications of machines such as cotton covering, enamel covering, paper covering, varnish bonded glass fibre covering etc. Round or rectangular conductors are usually being used. Standard specifications for conductors are given so that they will be available in required size and characteristics.



In the process of manufacturing of copper, the ores are crushed and then calcined in a furnace. The silica and a small quantity of coke are mixed with the calcined ores. The mixture is melted in a blast furnace. The melted metal is oxidized in Bassemer converter. After oxidization process blister copper is obtained. Impurities in blister copper are removed by melting it in the presence of air in a reverberatory furnace. By this process 99.7% pure copper is obtained. Hundred per cent pure copper can be obtained by the process of electrolysis.

Copper conductors used in the electrical engineering have high strength and hardness, excellent corrosion resistance, high electrical and thermal conductivity, non magnetic properties, excellent machinability, excellent resistance to fatigue, easy joining by soldering etc. Copper is a strong metal, hard, durable and highly ductile and has strong resistance against corrosion, oxidation and pitting. When a cold bar of copper is drawn the copper obtained is hard. When copper is first heated to about 300–350 degree centigrade and then cooled, hard copper becomes soft and has 2–3% more conductivity than hard copper. Hard copper is springy whereas, soft copper is not springy and can be easily drawn in any shape.

Cold drawn copper is used for overhead transmission lines, bus-bars etc. where high mechanical strength is required. Soft copper due to its flexibility is used as conductor in cables, windings and coils. Rectangular copper bars are also used in electrical machines. Hard copper is used for commutator segments in electrical machines, which can withstand wear due to the brushes as compared with soft copper.

When the copper is mixed with cadmium, chromium, silver and tellurium in small percentage the increase in the mechanical strength of copper is obtained but the conductivity is reduced. It may be used for rotor bars, transmission line conductors and collector wires. Copper with 0.7 to 1.0 per cent cadmium has greater strength under both static and alternating stresses and have better resistance to wear, making it suitable for contact.

Copper with silver is a special type of high conductivity copper, which is used for rotor conductors of large turbogenerators and commutators. The different properties of copper is shown in Table 3.4.

***(b) Aluminium:*** These days aluminium is replacing copper due to a number of economic and engineering reasons. It is being used in the shape of wires for motors and also used for making busbars. Alloys of aluminium with magnesium, silicon, iron, etc. have greater mechanical strength, so it can be used as overhead transmission lines. These alloys have comparatively high conductivity. Steel cored aluminium cables with aluminium wires around a core of steel wire are also being used. The steel reinforced aluminium conductor (A.C.S.R) is used for long span transmission lines due to its greater mechanical strength as well good electrical characteristics.

**Table 3.4** Different properties of copper, aluminium and steel

| *Properties* | *Density (20°C-gm/cm²)* | *Electrical resistivity (Ω -mm²/m)* | *Thermal conductivity (Cal/cm-sec °C)* | *Temperature coefficient of resistivity (°C⁻¹)* |
|---|---|---|---|---|
| Copper | 8.94 | 0.01682 | 0.923 | 0.00411 |
| Aluminium | 2.703 | 0.02828 | 0.503 | 0.00403 |
| Steel | 7.8 | 0.13 | – | 0.005 |

Aluminium wires are resistant to corrosion. In open air the aluminium is coated with thin film of aluminium oxide which protects the aluminium from further oxidation.

In the process of the extraction of aluminium from bauxite, the ore is purified and then dissolved in fused cryolite of double fluoride of aluminium and sodium, $AlF_3 3NaF$. The solution is then sent to an electric furnace and then aluminium is separated by electrolysis process.

Aluminium is 3.5 times lighter than copper. Pure aluminium is softer than copper and hence it can be easily rolled into thin sheets. It has low mechanical strength, hence it is difficult to make a thin wire. The resistivity of aluminium is about 1.65 times greater than that of resistivity of copper. In view of the resistivity, mechanical strength and density, an aluminium conductor should have a cross-sectional area 1.6 times the copper conductor and the weight of the aluminium should be equal to 0.48 times the weight of copper for the same resistance and length of a wire to be made.The various properties of aluminium are given in Table 3.4.

***(c) Brass:*** Brass is an alloy of copper and zinc. It has greater mechanical strength and wear resistance than that of copper but much lower conductivity. Brass with 30% zinc has 25 per cent conductivity as compared to copper. Brass has good weldability and solderability. It can be easily made to a desired shape and has good resistance to corrosion. Usually it is used in the manufacture of electrical apparatus and instruments as a current carrying and structural material.

***(d) Bronze:*** Bronze are the copper alloys containing tin. Cadmium, beryllium and other metals. It has fairly high conductivity. Cadmium copper (0.9% Cd) has about 85–95% conductivity as compared to copper. Bronze containing 0.8% Cd has 50–80% conductivity as compared to copper. All types of bronze have high mechanical strength. Copper with cadmium is used for making contact wires and commutator segments. Beryllium bronze is generally used for making current carrying springs, brush holders, sliding contacts and knife-switch blades to be used for different applications.

Bronze with 1.25 to 10 per cent tin and a small quantity of phosphorus has very good winding characteristics, hardness, low coefficient of friction and good corrosion resistance and is called as phosphorous bronze. Springs, diaphragms and bearing plates are made by phosphorous bronze. Bronze with less than 8 per cent aluminium have good hot and cold rolling properties, high strength, good wear resistance and good corrosion resistance. Bronze with 25–30 per cent Nickel have excellent corrosion resistance, hard, tough and ductile. Bronze with 1.5 to 3 per cent silica and bronze with aluminium and silicon have properties of good alloys, like high strength and good working characteristics in cold and hot conditions.

***(e) Steel:*** Steel is not usually used as a conductor due to its low electrical conductivity. Although it has good mechanical properties and low cost. Steel with 0.1 to 0.15 per cent carbon is used for wire. Its resistivity and losses are much higher for alternating current than that of d.c. current because of the magnetic properties of steel.

Steel has easy corrosion characteristics by moisture and heat. Overhead steel conductors are galvanized to prevent corrosion. In this process the steel is cleaned and is dipped in molten zinc. The layer of zinc on the steel protects steel from rusting. Overhead conductors made of steel are used to transfer small amounts of power due to less electrical conductivity. Aluminium conductors are used for high voltage systems, long spans reinforced with steel gives high tensile strength to overhead lines. Some properties of steel are shown in Table 3.4.

### 3.3.2 High Resistivity Materials

Alloys of different metals give materials of high resistivity. According to the applications, these materials can be classified into three groups.

(*a*) for precision electrical measuring instruments and for standard resistances and resistance boxes.
(*b*) for resistance elements for all kinds of rheostats and other control devices.
(*c*) for furnaces heating device and loading rheostats.

The materials used for the above applications should posses high resistivity and good mechanical strength and should have high resitivity and good mechanical strength.

Materials used for precision electrical instruments and standard resistances should have stable value of resistance with changing value of time and temperature. The value of temperature coefficient of resistivity should be low. The materials used for rheostats may possess large temperature coefficient and thermoelectromotive force but should have low cost and permissible working temperature. These materials are used in bulk quantity. Constantan (60% Cu, 40% Ni) is most commonly used.

The material used for heat elements in heaters, furnaces etc. should have stable heat characteristics *i.e.,* material should have high melting point. The element should be able to withstand corrosion in the atmospheric or other environment in which it is being operated.

**Table 3.5** Characteristics of different high resistivity materials

| *High resistivity materials* | *Specific weight (g/cm²)* | *Resistivity (ohm-mm²/m)* | *Temperature resistance coefficient ($10^{-3}$ per °C)* | *Working temperature (°C)* | *Uses* |
|---|---|---|---|---|---|
| Manganin (86% Cu, 12% Mn, 2% Ni) | 8.4 | 0.42–0.48 | 5–30 | 60–70 | Precision instruments |
| Constantan | 8.9 | 0.48–0.52 | 5–25 | 450–500 | Rheostats, other control devices |
| Nichrome (76.5% Ni, 20% Cr, 1.5% Mn, and 2% Fe) | 8.4–8.5 | 1.0–1.1 | 10–20 | 1100 | Heating elements |

Platinum is being used in laboratory electrical furnaces at a temperature of 1300°C. It is material, which have high melting point of 1770°C and have negligible corrosion effect. The most commonly used material for high temperature is nichrome. It can be used up to a temperature of 1100°C.

Various characteristics of different materials with high resistivity is given in Table 3.5.

### 3.3.3 Materials Used for Solders

An alloy used to join two or more pieces of metals is called as solders. The melting point of a solder should be lower than the melting point of materials to be joined. The molten solder joins the two pieces of metals together and this process of joining the two metals are known as soldering.

Solders can be classified as soft solders, which have melting point lower than 400°C and hard solders, which have melting point higher than 400°C.

An alloy of tin and lead can be used as soft solder. The most popular combination of tin and lead is 50% tin, 50% lead. The tin-lead solder and used to join copper, bronze, brass, lead-tinned iron, zinc, etc.

An alloy of copper and zinc are used as hard solder. It melts at a very high temperature and is being used for joining brass, copper, iron, and steel. There are two varieties of hard solder like brazed solder and silver solder. Brazing is soldering at high temperatures.

### 3.3.4 Materials for Contacts

The electrical contact means a practicable junction between two conductors used in electrical engineering, which are carrying currents. Materials used for making contacts operate under the different conditions *i.e.*, the contacts may have to make and break the electrical circuits very frequently.

The suitable operation of electrical contact depends upon some factors like contact resistance, contact force, voltage and current. All contacts should offer the minimum resistance as possible to the flow of current. The performance and life of contact materials basically depends upon the magnitude of voltage and current, which the contacts have to make and break.

Contacts being used to make or break the circuits should be able to withstand arcing or spark-over, whenever, they are separated or brought together. The quality of contacts deteriorate with time due to

(*a*) mechanical wear in the machine.
(*b*) corrosion due to oxidation and other chemical reactions with the surroundings.
(*c*) erosion due to fusing evaporation, and wear of the working surfaces during service. Due to corrosion, contact surfaces acquire a film of oxide, which has low conductivity and hence reduces the effectiveness of electrical contacts.

Different materials for electrical contacts may include copper, molybdenum, nickel palladium, platinum, silver, tungsten etc.

**Table 3.6** Properties of different materials for electrical contacts

| *Properties* | *Nickel* | *Platinum* | *Palladium* | *Molybdenum* | *Silver* | *Tungsten* |
|---|---|---|---|---|---|---|
| Melting point (°C) | 1456 | 1773 | 1554 | 2625 | 960 | 3410 |
| Density (gm/cm$^3$) | 8.90 | 21.45 | 12.0 | 10.2 | 10.49 | 19.3 |
| Electrical resistivity (micro-ohm-cm) | 6.84 | 9.83 | 10.8 | 5.17 | 1.59 | 5.5 |

Various properties of different materials used to make electrical contacts are given in Table 3.6.

### 3.3.5 Carbon Materials

Carbon is available in different forms *i.e.*, the diamond with extreme hardness and graphite with grey and soft. Carbon materials used in electrical engineering are manufactured from graphite and other forms of carbon. Graphite is available in nature as a mineral with high content of carbon (up to 95%).

It is crystalline in structure and has a very high melting point (approx. 3900°C). Pure carbon is a semiconductor and have negative temperature coefficient of resistivity. The conductivity of carbon is slightly less than that of metals and their alloys. Carbon is extensively used as brushes for electrical machines, carbon electrodes for electric furnaces, electrolytic baths, arc welding, non wire resistors, arc light, battery cell elements microphone powders etc.

The brushes made with carbon are almost universally used for current collection from the rotating parts of electrical machines. Carbon and graphite are cotinuously used for sliding electrical contacts because the use of it as a contact material have several advantage like

(*a*) Maintaining its properties under high temperature. High instantaneous temperatures exist at the point of rubbing contacts. Carbon retains its properties under such conditions and remains solid even above the temperature of 3000°C.

(*b*) Carbon has low density and hence it is lighter than several metals, only magnesium is comparable with carbon.

(*c*) Carbon cannot be welded to metals under the conditions in which metals weld to one another, like in the heat of an electric arc.

The properties of different materials, which may be used as brushes are shown in Table 3.7.

**Table 3.7** Properties of different materials used for brushes

| *Brush materials* | *Electrical resistivity ($10^{-3}$ ohm-cm)* | *Contact drop* | *Usually used in* |
|---|---|---|---|
| High resistivity, Resin bonded carbon | 5–30 | High | Fractional H.P. motors |
| Hard, low resistivity baked carbon | 4 | Low | Cranes |
| Electrographite | 4 | Medium | d.c. machines |
| High speed electrographite | 6 | Medium | Large power machine |
| Copper and bronze graphite | 0.5–0.003 | Low | a.c. and low voltage d.c. machines |

### 3.3.6 Superconducting Materials

The state of material in which it has zero resistivity is called superconductivity. Superconducting phenomenon was discovered by a Dutch Scientist, Heike Kamerlingh Onnes at the university of Laiden. He observed that the resistivity of mercury vanished completely below 4.2 K, the transition from normal conductivity occurring over a very narrow range of temperature of order of 0.05 K. The temperature at which there is a transition from normal state to superconducting state is called transition temperature. Transition temperature for some materials is shown in Table 3.8.

From the time the discovery of superconductivity in mercury, several thousand superconductivity materials have been found. These include pure metals as well as their alloys. Many of them become superconducting under special conditions like high pressure, thin films, amorphous state etc. Superconductivity has also discovered in some organic systems.

By the use of superconducting materials much high current and current densities are possible with no power loss but machines have to work under special conditions. By the use of superconductors the machine size will be reduced considerably, with practically negligible loss and the machine efficiency will reach to an exceptionally high level. But the use of superconductors in making the electrical machines, transformers and rotating machines depends on the comparative gain in reduction of power loss against the cost for provision of special conditions.

**Table 3.8** Materials with their transition temperature

| *Materials* | *Transition temperatures* |
|---|---|
| Mercury | 4.16 |
| Niobium | 8.00 |
| Tin | 3.73 |
| Uranium | 0.80 |
| Zinc compound | 0.82 |
| $Nb_2Zr$ | 10.8 |
| $Nb_3Al$ | 18.0 |
| $Nb_3Sn$ | 18.1 |
| $Pb_2Au$ | 7.0 |

## 3.4 INSULATING MATERIALS

These materials offer high resistance to flow of current. The use of insulating materials in electrical machines confine the flow of current into the specified paths. The insulation part in any part of the machine or electrical system may fail during operation. The electrical and mechanical stresses and the temperature of the machine have effects on the life and performance of insulating materials. So the operating temperature of any portion of the machine should not exceed the permissible limit continuously for long time. For a particular engineering application the choice of several insulating materials may be available. The insulating material should be selected by considering the following factors so that it can operate satisfactorily.

**Electrical Characteristics:** The insulating materials should possess the characteristics to give suitable performance of an electric circuit. There is an analogy between electrical properties and the mechanical properties.

**Mechanical Strength:** The insulating materials should have suitable tensile strength, elongation, tensile modulus, compressive modulus, impact strength etc. The material chosen should have optimum characteristics for all the requirements.

**Thermal Properties:** It depends upon the temperature and time of application. It can operate in the device for its required lifetime. Properties of materials change at lower temperatures such as tensile strength gets greater, elongation smaller and also materials become brittle. Hence material should have optimum response for these characteristics.

**Environmental Effects:** It depends on the environment in which the machine has to work. Environment includes the effect of air, light, ultraviolet rays, acid and alkali fumes, humidity etc., that affects the machine in different ways.

The dielectric strength of an insulating material is the most important properties from the electrical application point of view. It is that value of voltage at which the electrical rapture (failure) of an insulating material occurs in practical use. It can be defined under specific condition and is measured in volts per unit thickness of the material. For insulating materials the loss tangent and resistivity are also important phenomenon. Insulating materials are good if they have a low dissipation factor (loss tangent), high insulating resistance, good dielectric strength and high mechanical strength. They should possess high thermal conductivity and high thermal capability. A good insulating material should have low permittivity providing a high discharge stress, free from gaseous insulation to avoid discharge, homogenous to avoid local stress concentration. They should be free from moisture and ionic contamination to limit loss tangent and electrochemical deterioration. Insulating materials should also have resistance to erosion by discharge and resistance to thermal and chemical deterioration.

The different insulating materials used in electrical engineering may be gases including vacuum, liquids and solids.



### 3.4.1 Gases as Insulating Materials

Gases and air are being used as an insulator. The brief description of different such insulating materials are presented here.

***(a) Air as Insulating Material:*** Air is freely available and hence it is the most important insulating material as compared with all other dielectric gases. The air provides insulation between the overhead transmission lines. Air is generally used as insulating material when voltages are not very high. Leakage of current through air is much less than that through solid or liquids insulating materials under the same operating conditions.

The electric strength of air is small when the distance between the electrodes is small. This is about 3 to 5 kV/mm. The discharge voltage of air between two electrodes increases very slowly with the increase in the distance between the electrodes. The electric strength of air is considerably affected by pressure. The electric strength of gases increases considerably with increase in the pressure and the electrical strength of air and other gases become low at lower pressure.

***(b) Nitrogen as Insulating Material:*** Nitrogen is used as an insulating material in electrical equipments. It fulfills the purposes of chemical and electrical properties. Nitrogen is being used in transformers to replace air, since air causes oxidation and oxidation reduces the life of equipments. In oil filled transformers with free breather, the oxidation of oil leads to sludge formation and hence degradation of the electrical and mechanical properties of the equipments. Chemically inert nitrogen is being used in transformers to replace oxidizing atmosphere. Nitrogen in transformers is sealed. Nitrogen under pressure along with oil treated papers is used in gas pressure cables. The different properties of Nitrogen as compared to air is given in Table 3.9. The different properties of air are taken as unity.

***(c) Hydrogen as Insulating Material:*** Hydrogen is very light gas and possess the properties and due to these properties it is suitable as a coolant and insulation in electrical machines in place of air. The important characteristics of hydrogen as compared to air at equivalent temperature and pressure with all the parameters for air taken equal to unity is given in Table 3.9.

Using hydrogen as a coolant can increase the rating and efficiency of electrical machines. Organic insulation is less susceptible to ageing in hydrogen because there is no oxidizing effect. These days usually hydrogen cooling are applied to large turbogenerators and synchronous condensers. Inert gases like argon, neon, mercury and sodium vapours are used in various electronic tube and equipments.

***(d) Carbon Dioxide as Insulating Material:*** Oxide gases have little use in electrical devices. Carbon dioxide is being used as an insulator in few fixed capacitors and is also used in oil filled high voltage equipments like cables and transformers. The carbon dioxide has dielectric constant equal to 1.000985 at 0°C. The dielectric strength of carbon dioxide is almost comparable to that of air.

The different properties of carbon dioxide with respect to air is given in Table 3.9. All the properties of air are taken as reference and are kept at unity.

**Table 3.9** Different characteristics of nitrogen, hydrogen and carbon dioxide

| *Properties* | *Nitrogen* | *Hydrogen* | *Carbon dioxide* |
|---|---|---|---|
| Density | 0.97 | 0.07 | 1.52 |
| Thermal conductivity | 1.08 | 6.69 | 0.64 |
| Thermal capacity | 1.05 | 14.35 | 0.85 |
| Electric strength | 1.00 | 0.60 | 0.90 |

(*e*) ***Electronegative Gases as Insulating Material:*** The electronegative gases have quite better insulating characteristics as compared to other gases like nitrogen and air. If the dielectric strength of air is taken as unity, the dielectric strengths of $SF_6$ is 2.35 times and $CCl_4$ is 6.63 times. The electronegative gases are non-inflammable and non-explosive. These materials include halogenated gasses such as halides of carbon and sulphur. Such gases are sulphur hexafluoride and various freon gases. Sulphur hexafluoride is being used as a coolant. It is also used in high voltage circuit breakers, cables, measuring instruments etc.

Sulphur hexafluride ($SF_6$) has odourless, colourless, non-toxic and non-inflammable characteristics. It is five times heavier than air. This has low solubility in water and can be liquefied by compression. It has cooling properties better than that of air or nitrogen. During normal operating conditions it is chemically inert and completely stable. It does not react with quartz up to 500°C. It does not react with metals, ceramics, rubber, resins etc below 150°C.

The presence of sulphur in the molecule may cause corrosivity. Use of $SF_6$ increases the interrupting capacity of the circuit breakers and hence they can be used where the frequency of circuit interruption is high. $SF_6$ has electronegative properties and its relatively large molecules have much affinity for free electrons with which they combine and make the gas filled break in the circuit much more resistant to dielectric breakdown up to a temperature of 2000 K.

### 3.4.2 Vacuum as Insulation

Vacuum *i.e.*, low pressure as practically possible have merit for several applications in electrical engineering. Vacuum is suitable voltage insulator up to stress level of $10^7$ V/cm near to the electrode surface. Its insulation characteristic decreases less than $10^5$ V/cm when the gap in the circuit increases more than few centimeters.

### 3.4.3 Liquid Insulating Materials

Liquid insulating materials are normally used as an insulating and heat transfer medium. Usually, liquid insulating materials are used with solid insulating materials. The liquid insulating materials should have following characteristics:

(*i*) Arc quenching characteristics should be good.

(*ii*) It should be non-flammable.

(*iii*) It should have high electric strength, impulse strength and volume resistivity.

(*iv*) This must have low dielectric dissipation factor.

(*v*) It should possess high specific heat and thermal conductivity.

(*vi*) It should have good chemical stability and gas absorbing properties.

(*vii*) It must have low viscosity, low density, low volatility low solvent power and high flash point.

Different type of liquid insulating materials with their temperature characteristics and applications are given in Table 3.10.

**Table 3.10** Temperature limits of different liquid insulating materials

| *Different liquids* | *Temperature limits* | *Applications* |
|---|---|---|
| Petroleum (mineral) oils | – 50 to 110°C | All equipments |
| Askarels | – 50 to 110°C | Transformers, switchgear |
| Silicon liquids | – 90 to 200°C | Transformers |
| Halogenated hydrocarbons (except askarels) | – 50 to 200°C | Electric equipment |
| Synthetic hydrocarbons | – 50 to 110°C | Cables |
| Organic easters | – 50 to 110°C | Electronic |
| Vegetable oils | – 20 to 100°C | Few applications |



Electrical properties of mineral oils synthetic hydrocarbons and askarels are shown in Table 3.11.

Mineral oils are normally used as insulation and heat transfer medium. Mineral insulating oil is widely used for insulations in electrical engineering applications.

**Table 3.11** Different properties of liquid insulating materials

| *Different properties* | *Mineral oils* | *Synthetic hydrocarbon* | *Askarels* |
|---|---|---|---|
| Breakdown voltage, kV, 2.5 mm | 30–50 | 40–60 | 20–45 |
| Dielectric dissipation factor, 90°C | 0.001–0.005 | less than 0.0005 | 0.02–0.05 |
| Resistivity GΩm, 90°C | 20–2000 | 1.0–600 | greater than 10 |
| Permittivity | 2.1–2.5 | 2.1 – 2.2 | 4.8–5.3 |

Mineral oils are used in H.T. oil filled cables, capacitors and transformers with a.c. supply, where the oil has to act as a heat transfer medium with low viscosity. Medium viscosity oils are used in switchgear and circuit breakers. High viscosity oils are used in gas pressure and solid cables and cable sealing ends. The mineral insulating oils, which have been properly dried and purified, have a permittivity of 2.1 to 2.2, which is independent of the basic hydrocarbon constitution.

### 3.4.4 Solid Insulating Materials

***Classification of Solid Insulating Materials:*** The insulating materials are classified into the following classes as given in Table 3.12. This classification of materials mainly based on the temperature in which that type of insulating material can suitably work. This is given as per I.S. 1271-1958 and I.S. 1281. This is also called thermal classification of the insulating materials. Table 3.13 shows the materials in each class. The classification is given below:

**Table 3.12** Classification of insulating materials on the basis of temperature

| *Class of insulation* | *Temperature* |
|---|---|
| Class Y or 0 | 90°C |
| Class A | 105°C |
| Class E | 120°C |
| Class B | 130°C |
| Class F | 155°C |
| Class H | 180°C |
| Class C | above 180°C |

**Table 3.13** Different materials available in different class of insulation

| *Class of insulation* | *Different insulating material available* |
|---|---|
| Y | Cotton, natural silk, regenerated cellulose fibre, cellulose potato fibre, polyamide, fibre, paper and paper products, press board, vulcanized fibro wood, Aniline, formaldehyde resins, urea-formaldehyde resins. |
| A | Cotton, natural silk, regenerated cellulose fibre, cellulose acetate fibre, paper and paper products, press board, vulcanized fibre root. |
| E | Wire resin on polyvinyl formal, polyurethane or epoxy resins, mouldings with cellulose fillers, cotton fabric laminates, paper laminates, cross-linked polyester resins, cellulose triacetate film, polyethylene terephthalato film, polyethylene terephthalato fibre, varnished polyethylene, terephthalate textile. |
| B | Glass fibre, Asbestos, varnished glass fibre, textile, varnished asbestos, built-up mica glass fibre laminates, asbestos laminates, mouldings with mineral fillers, polymono-chloro-trifluoroethylene. |
| F | Glass fibre, asbestos, varnished glass fibre, textile, varnished asbestos, built-up mica. |
| H | Glass fibre, asbestos, varnished glass fibre textile, varnished glass fibre laminates textile, varnished asbestos built-up mica, silicon elastomer. |
| C | Mica, porcelain and other ceramics, glass quartz. |
| | *Note:* Maximum operating temperature may be limited by physical, chemical or electrical proprieties as operating temperature. |
| | Treated glass fibre textile, treated asbestos, built up mica. Polytetraflouroethylene. |

The insulating material must possess the following characteristics:

(*i*) The insulating materials must be able to withstand with high temperature fluctuations without influence from humidity, dirt or corrosive chemical surrounding.

(*ii*) The insulating materials should withstand mechanical stress.

(*iii*) The space required by the insulating materials should be as less as possible so that active core and copper materials in the machine could be used in the best possible manner.

(*iv*) The insulation should be able to dissipate heat generated in the machines for proper cooling.

(*v*) The insulation should be able to withstand the transient high voltage surges in the supply system.

### 3.4.5 Different Insulating Materials Available

Some of the insulating materials, which are available and are being used in electrical engineering, are given below:

***(1) PVC Tape:*** PVC tape is a flexible sheet insulator made from PVC on which an adhesive has been added. PVC tape is pressure sensitive and has neat appearance and is available in different colours. The PVC tapes are used for repairing and insulating the cables wires.

***(2) Black Tape or Friction Tape:*** Cotton tape with an adhesive insulating bituminous compound on one or both sides is called black tape. It is used as replacement of the protective braid of cables. It is also used as the finishing tape over the rubber and other insulating tapes.

***(3) Rubber Tape:*** It is used to insulate joints in cables. Rubber tape is made of unvulcanized rubber and have cloth backing which readily vulcanize with rubber insulation with slight pressure to form a solid sheath. It is water proof. It should be covered with black tape which protects against abrasion and wear.

***(4) Cotton Tape:*** Cotton tape is fine plain white cotton used in electrical machinery for wrapping of armature and field coils. It is fibrous and absorbs moisture. So it should be suitably impregnated. The thickness of this tape is normally 0.15mm.

***(5) Silk Tape:*** Silk has similar characteristics as cotton but due to fine texture and higher tensile strength, it has superior insulating characteristics as compared to cotton. It is also impregnated with varnish or insulating oil.

***(6) Press Boards:*** The paper and press boards are used for the insulation of winding and cable conductor. This is also used as primary dielectrics in capacitors, backing for mica insulation. Press boards are also used in slot insulation of electrical machines and transformer insulation.

The suitable characteristics of paper insulation are thickness, apparent density, finish, porosity, tensile strength and tearing resistance. Dielectric strength, dielectric loss and permeability, depends upon its density.

***(7) Mica:*** This has high dielectric strength and the resistance to heat, but due to its brittle nature, it is very difficult to use it in natural state. Large size sheets are made in the form of small splitting and bonded with suitable varnish. The insulating and thermal properties of the mica sheets are dependent upon the nature and the extent of the bonding substance used. The bonding materials which are used are, shellac, alkyd resin, epoxy resin and silicon. Resin bonded mica is normally used for class *B* insulation and silicon bonded mica for class *H* insulation.

***(8) Micanite Sheets:*** Micanite sheets are made by mica splittings and bonded with suitable bonding materials. This is supported on one or both sides by paper, cloth or glass. This increases the flexibility of mica for use in insulation of armature coils and slots. The characteristics of the sheets depend upon the nature and amount of bonding and supporting materials used.

***(9) Asbestos:*** Asbestos is normally used in electrical engineering in the form of paper, tape, cloth and board. It is used as filler material for natural and synthetic insulating resins. It is impregnated with a liquid or solid in all dielectric applications.

***(10) Composite Materials:*** Few of the insulating materials have good electrical characteristics but lack some physical properties and few materials may possess good physical and thermal properties but poor electrical characteristics. This difficulties can be solved by mixing/overlapping one or more materials together to form composite insulating materials. There are several composite materials, which have different dielectric properties. They are asbestos polyethylene fibre, asbestos paper, polyesterin resin etc. Class *F* insulations are obtained from micaceous products.

### 3.4.6 Insulating Materials Used in Electrical Machines

***(1) Power and Distribution Transformers:*** The different materials used in power and distribution transformers for different parts are given below in Table 3.14.

**Table 3.14** Insulating materials for different parts of power and distribution transformers

| *Different parts* | *Insulating materials* |
|---|---|
| Low voltage coil to ground | thick radial spacers or tubes etc. |
| High voltage coil to low voltage coil | pressboard, paper or porcelain etc. |
| Turn to turn | paper, glass tapes etc. |
| Layer to layer and coil to coil | craft paper, glass fabric, press board etc. |
| Liquid as insulating and cooling medium | mineral oil, air, nitrogen etc. |
| Bushings | porcelain, phenolic bonded tubes etc. |

***(2) Rotating Machines:*** The insulating materials used for rotating machines are:

Conductor — enamel cotton covering (class *A*); fibre glass coated with an organic varnish (class *B*); fibre glass coated with silicon varnish (class *H*).

Slot liner — mica, fibre glass, cotton tapes, synthetic resin, fabric tapes.

***(3) High Voltage Insulation:*** Initially bitumen was used with mica as a thermoplastic material of class *B*. These days epoxy novolac resins are being used. In class *F* insulation, mica folium and glass fabric enriched by epoxy resin are being used. Presently high voltage electrical machines use epoxy mica glass tapes. These tapes are made of different type of resin *i.e.,* resin rich or the resin poor. Class *H* insulation characteristics can be achieved when composite materials with polyamide filaments and mica are used.

***(4) Sleeves:*** Varnished fibre glass sleeves are used for leads taken out from machines. The varnish used in a particular application of machine decides the thermal classification.

***(5) Slot Wedge:*** Backelised fabric wedge for class *B* insulation, epoxy glass cloth laminate for class *F* insulation and silicon resin bonded glass fabric laminate for *H* class insulation are the different materials used.

***(6) Inter-Turn Insulation:*** The different materials used are polyester enamel or varnished double glass for class *B* insulation, esterimide enamel or polyurethane varnished double glass for class *F* insulation.

***(7) Coil to Coil and Phase to Phase Insulation:*** Fibre glass baked mica is used as a coil separator for class *B* insulation and polyamide paper with polyester film baking is used as the coil separator for class *F* insulation in the slots of the machine.

***(8) Winding Overhangs:*** Isophthalate varnished glass and wet varnish treated glass tape is used for class *B* and *F* insulation. Glass fabric with silicon elastomer coating is used for class *H* insulation.

### 3.4.7 Specific Application of Class F and H Insulation

Class *F* and *H* motors are available only for specific use like, the places where ambient temperature is high, cooling facilities not available at the location of mounting of motors, limitations of space hence smaller size of motors, duty cycle for motor such as plugging, reversing, frequent starting etc. Class *F* and class *H* insulations affect the design of machine, which are discussed as below.

(*i*) When a motor is required to operate at higher temperature then the frame size of the motor should be increased. Due to this the materials such as copper and iron in the motor will be increased. In spite of doing this, the class of insulation such as class *F* or *H* insulation can be used without increasing the size of the machine.

(*ii*) When the machine is designed with Class *F* and *H* insulation to reduce the size of the machine then this reduce the use of active materials in the motor and hence reduce the cost. The disadvantage of this is that the performance of the motor deteriorates, *e.g.*, the power factor, efficiency, starting torque are lower than that of normal size and class of insulation. The running costs of motor increases and also temperature in the motor increases.



### 3.4.8 Effect of Moisture on Insulation

The insulating material placed in a humid atmosphere absorbs moisture. First water vapour is absorbed on the surface, then water vapour diffuse causing the reduction of moisture concentration gradient then water vapour is absorbed in the region of low vapour concentration. The higher humidity in air creates troubles in electrical insulation. The failure of insulation may take place. The effect of moisture causes some changes in the electrical, physical, mechanical and chemical characteristics.

The electrical characteristics of the insulating materials depends on a number of factors like, chemical, molecular and sub-molecular structure, moisture etc. The moisture may cause sudden changes in the material structure and its chemical composition by dissolving of impurities contained in dielectrics. The electrical strength of materials decreases when moisture is absorbed. The decrease in strength may not be proportional to the amount of the moisture absorbed. Usually it depends on how the moisture is distributed along the surface of the insulating material.

### 3.4.9 Protection of Insulation from Moisture

Following methods can do the protection of insulation against moisture:

***Impregnation:*** The windings of low voltage machines are impregnated with backing varnishes and with compounds. The impregnating solidifies the windings, increases thermal conductivity, improves electrical and mechanical strength and heat resistance of windings.

***Hydrophobic:*** The insulating materials are water proofed (or rendered hydrophobic) to protect the insulating materials against moisture. This treatment is usually more effective for polymers containing hydroxyls. Some inorganic materials are also rendered hydrophobic to decrease the effect of moisture in these materials.

***Sealing:*** Hermetic sealing is normally used to protect insulation against moisture. It helps in maintaining adequate insulation properties and protects them against mechanical damage. Hermetic sealing is usually affected by potting in compound or by impregnation and coating. A number of sealing methods are adopted such as moudling, injection, encapsulation, dipping etc. An insulating material should be completely dried before it is sealed.

The compounds, which are normally used in sealing the insulation of low voltage machine are polyster-styrene, styrene, polyurethane, silicon base compounds etc.

## QUESTIONS WITH SHORT ANSWERS

**Q. 1. Give the difference between dynamo grade and transformer grade steel.**

**Ans.** The basic difference between the dynamo grade and transformer grade steel is based on the silicon content in steel used as magnetic material.

In rotating electrical machines steel with low silicon content is used; the silicon content may be of the order of 0.5% only. Such steels are termed as dynamo grade steels. But in turbo-alternator the silicon content is high to keep the losses less.

When the silicon content is increased in the steel, the resistivity of steel increases. Steels with silicon content of 4 to 5% are called as transformer grade steels and are used in transformers. On account of high resistivity due to more silicon, eddy current losses are low.

**Q. 2. Distinguish between soft and hard magnetic materials with their applications.**

**Ans.** Magnetic materials with narrow hysteresis loop are called as soft magnetic materials because hysteresis losses are small. These are used in the manufacture of electrical machines, transformers and many kinds of electrical apparatus, instruments and devices.

Magnetic materials with large hysteresis loops are called hard magnetic materials. These materials are used in electrical machines of low power rating, and in all kinds of instruments and devices which need permanent magnets to set up magnetic fields of their own.

**Q. 3. Explain that CRGO sheets have superior magnetic properties in comparison to those of the hot rolled steel.**

**Ans.** CRGO sheets have superior magnetic properties in comparison to those of hot rolled steel. CRGO steel is manufactured by a series of cold reductions and intermediate annealings. The cold reduced material has strong directional magnetic properties, the rolling direction being the direction of highest permeability. This direction is also the direction of lowest hysteresis loss. So the CRGO sheets have less hysteresis loss than hot rolled steel.

**Q. 4. Write the name of conductors having high and low resistivity with their applications.**

**Ans.** Conductors having low resistivity are copper and aluminium. These are used for making all types of windings required in electrical machines, apparatus and device as well as for conductor of transmission and distribution systems.

Conductors having high resistivity are michrome, manganin, carbon etc. These are used where it is needed to dissipate electrical energy as heat *i.e.*, in starting and regulating devices for motors, rheostats, heating elements and other similar applications in electrical engineering.

**Q. 5. Explain superconductivity and its scope of application in the design of electrical machines.**

**Ans.** Materials exhibiting zero value of resistivity at some conditions are known as superconductors. The conductivity of such materials is infinite and hence called as superconductivity. A large number of metals become superconducting below a particular temperature. This temperature is known as transition temperature.

With the introduction of superconducting materials, much higher current densities will be possible and machines can work at low temperatures. The machine size will be considerably reduced and hence there may be much scope of superconducting materials in the design of electrical machines.

**Q. 6. How the joints in a transformer effects the losses?**

**Ans.** When C.R.G.O. sheets are used to make the transformer cores the flux path is in oriented direction. But this is not at the corners of the cores and hence it increases the loss and no load current. This losses could be reduced by adopting the metred joints for the transformer as given in Fig. A.

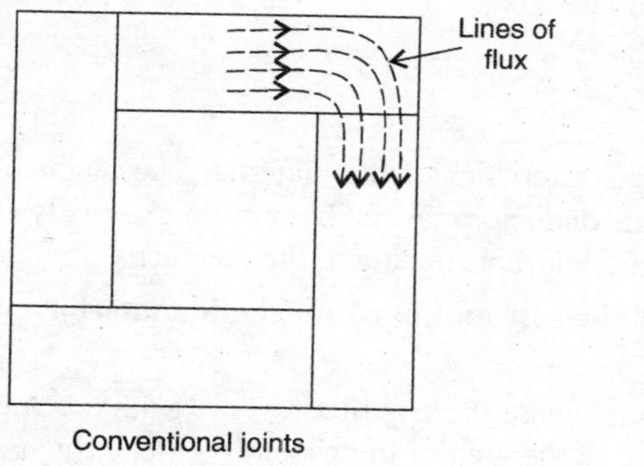


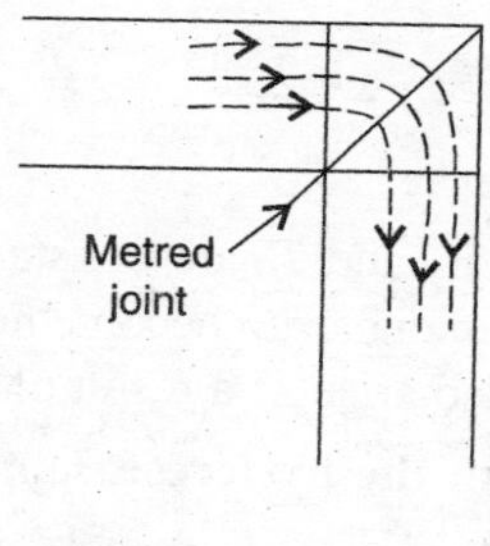


**Fig. A**

**Q. 7. Write the classification of insulating materials with their thermal rating and also give some examples in each class of insulating materials.**

**Ans.** Classification of insulating materials on the basis of thermal rating are given as below.

| *Class of materials* | *Temperature range* | *Materials* |
|---|---|---|
| Y (formerly 0) | 90°C | Cotton, silk, paper, press board, wood, PVC vulcanized natural rubber etc. |
| A | 105°C | Cotton, silk and paper impregnated in a liquid dielectric. |
| E | 120°C | Moulding powder, plastics etc. |
| B | 130°C | Mica, glass fibre, asbestos, etc. |
| F | 155°C | Glass fibre, asbestos etc. |
| H | 180°C | Silicon elastomers and combinations of materials. |
| C | above 180°C | Mica, porcelain, glass and quartz with or without an inorganic binder. |

**Q. 8. Explain that the transition from the superconducting state to normal conducting state is reversible for a superconducting material.**

**Ans.** Suppose a superconductor has a temperature $T < T_c$. If a magnetic field $H$ is applied, the material remains superconducting until a critical field $H_c$ is reached such that $H > H_c$ then, the material is in normal state. The transition from the superconducting to the normal state under influence of a magnetic field is reversible as is clear from the Fig. The function $H_c$ $(T)$ follows with good accuracy a formula of the form

$$H_c(T) = H_c\left(1 - \frac{T^2}{T_c^2}\right)$$

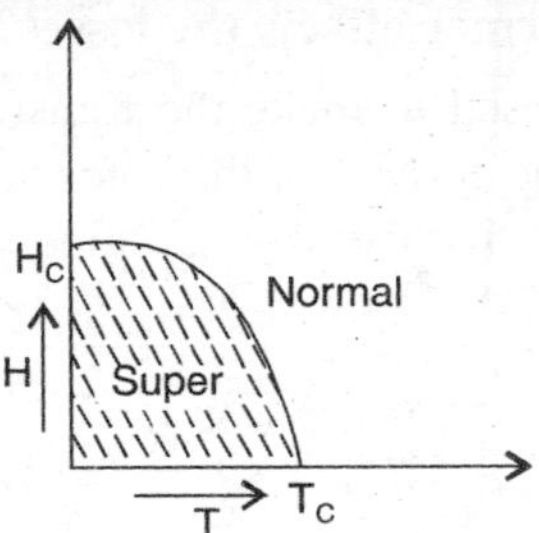


where $H_c$ and $T_c$ are constants characteristics of the material. The magnetic field causing a superconductor to become normal conductor may not be only an externally applied field but it may also arise as a result of electric current flowing in the conductor.

**Q. 9 Explain the hysteresis loops of the soft and hard magnetic materials, with their use in design.**

**Ans.** Magnetization of a material occurs where there is flux reversal with a certain loss of energy. The loss per cycle is proportional to the area of the hysteresis loop which depends upon the quality of material. The loss due to hysteresis for a material operating in an alternating field are usually expressed in watt/kg or watt/m$^3$.

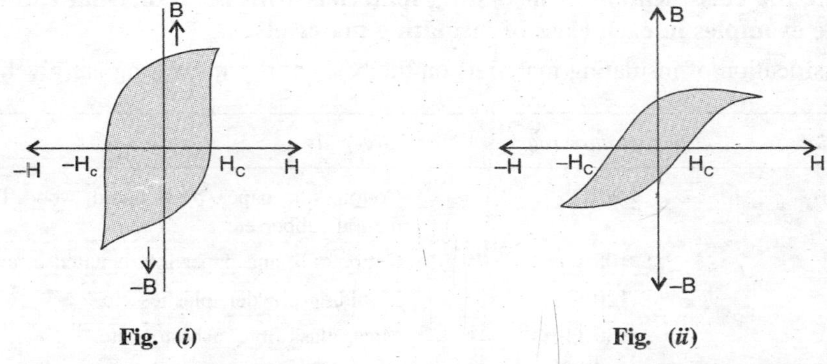


**Fig.** (*i*) **Fig.** (*ii*)

Hysteresis loss per m$^3$, $W_h$ = (area of loop) × $f$. A hard material will have larger area of hysteresis loop hence more losses in material whereas a soft material such as cast steel will have lesser area of hysteresis loop and hence less losses in the material. Hysteresis loop for cast iron and cast steel respectively are shown in Fig. (*i*) and (*ii*).

**Q. 10 Explain that how can a coil made of superconductor produce a very high flux density in a machine without iron core?**

**Ans.** Very large value of flux density can exist in air or space but in order to produce a strong field in air, the conductors have to carry a large value of current. The high value of current can be obtained by high value of current density in small conductor area so that volume of machine does not increase. The small conductor area offers high value of conductor resistance and hence high ohmic loss resulting in large temperature rise. Thus very high temperature rise can not be permitted and hence large current densities in small conductor area should not be allowed. But if the resistivity of the material selected is very low or approximately equal to zero, like in superconducting materials, the ohmic loss produced would be zero and hence no temperature rise due to ohmic loss.

So the superconducting materials having zero resistivity and hence zero resistance can permit very high value of current densities and hence high value of flux densities can be produced in air giving very small machine volume because of absence of iron parts in the machine. Supercooled coils can produce flux densities of about 10 Wb/m$^2$ or even higher value in the air.

**Q. 11 Explain the directional variations of iron loss and magnetizing current of a specimen of cold rolled gain oriented steel.**

**Ans.** The directional variations of iron loss and magnetizing current of a typical specimen of cold rolled grain oriented steel is given in Fig. (*i*).

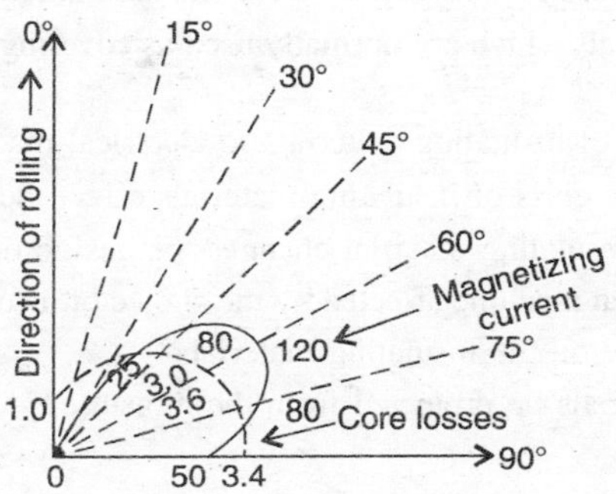


**Fig. (*i*)**

The iron loss and magnetizing current in the directions of rolling for a given peak flux density are taken as unity. As shown in given figure the relative iron loss and magnetizing currents are respectively 3.6 and 120 for magnetization on an axis at 60° with respect to the direction of rolling. The use of C.R.G.O. is highly advantageous if its directional properties are suitably utilized. C.R.G.O. steel can be used to higher flux densities than hot rolled sheet steel. C.R.G.O. steel give much reduced iron loss and much better magnetization curve than hot rolled steels. C.R.G.O. steel is costlier than hot rolled steel but it is preferred to use in electrical machines like transformers.

## UNSOLVED QUESTIONS

1. Write the name of conducting materials that can be used as conductors for the windings of electrical machines.
2. What are the different properties of the conducting materials which decides whether the material is suitable or not suitable for a particular application?
3. Discuss that why copper is the most suitable as a conductor for the windings of an electrical machine?
4. What is the role of magnetic materials in an electrical machine?
5. Mention the different properties that a good magnetic material should possess when used in electrical machines.
6. How does the quality of magnetic material effect the design of electrical machines?
7. How does the specific magnetic loading effect the design of electrical machines?
8. Why the iron parts of a machine are laminated?

Chapter 3

**9.** Mention various losses occurring in magnetic materials used in electrical machines.

**10.** Explain how does the eddy current loss depend upon the thickness of lamination used in the electrical machines?

**11.** How is the hysteresis loss optimized in the electrical machines? Discuss.

**12.** Explain how is the reduction of eddy current loss achieved in magnetic materials in the electrical machines?

**13.** Mention that on what basis the insulating materials used in electrical machines are classified?

**14.** Discuss the use of Class *'F'* and Class *'H'* insulating materials.

**15.** Write the name of insulating materials that can work above 180°C temperature.

**16.** What is effect of temperature on the life of the insulating materials used in electrical machines?

**17.** Write different insulating materials normally used for the laminations of the core of electrical machines.

**18.** Mention the insulating materials which are normally used as covering of the conductors of the windings of electrical machines.

**19.** Discuss the causes of failure of insulating materials in electrical machines.

**20.** How does the selection of the class of insulating materials effect the design of the machine?

**21.** How a change in a class of insulating material changes the design of electrical machine? Explain.

**22.** How is the size of the electrical machine affected by the choice of insulating material used for the machine?

**23.** What is the effect of moisture on the insulating materials?

**24.** How can the insulating materials be protected from the moisture?

# 4

# *Heating and Cooling of Electrical Machines*

## 4.1 INTRODUCTION

When any machine whether generator, motor or transformer are operating then there are losses occurring in the machine in various forms. These losses are in the form of eddy current loss, hysteresis loss, ohmic loss, friction and windage loss etc. These losses occurs in different parts of the machine. Due to these energy losses in any portion of the machine, the temperature rises in that portion and machine part gets heated. The temperature rise in any portion of the machine should be within the specified limit for that portion so that it can work satisfactorily. As the size or rating or in some cases quality deteriorates of the machine then the rate of temperature rise also changes. The temperature rise depends on the relation between heat produced by losses in any portion of the machine and the heat dissipated by that portion. When the losses are more, heat produced will be more and temperature rise may be more and it may effect the proper operation of the machine and if the temperature rise is too high it may damage the machine. Nowadays different cooling facilities are being provided in the machines for dissipation of heat produced in different portions of the machines so that it can work satisfactorily within the required temperature limit. The final temperature rise is reached when the rate of heat produced due to losses are equal to heat dissipated. Heat produced in the machine are to be dissipated in surroundings. Heat dissipation to the surrounding air is by radiation, conduction and convection. When the machines are smaller, natural cooling by means of convection and radiation may be sufficient. But for larger machines, fans are fixed on the rotor of the machines and definite air flow through ducts are produced in the machine. For every large machines, other external means of cooling are adopted like mostly transformers are oil immersed and in that case heat from the winding coils and the transformer cores is conducted to the oil, then it passes to the tank walls by conduction and convection and then by the tank walls the heat is dissipated to surrounding air by means of conduction and radiation.

## 4.2 DIFFERENT COOLING SYSTEMS

### 4.2.1 Air Cooling

In this case, no additional material arrangement is provided and only air is in contact with the different parts of the machine *i.e.*, the windings, cores and other parts, for cooling it.

### 4.2.2 Open Circuit Cooling

This is the method of cooling in which the coolant is drawn from the medium, surrounding the machine. The coolant passes through the machine to cool it and then again returns to the surrounding medium.

### 4.2.3 Closed Circuit Cooling

This is the method of cooling in which a primary coolant is circulated in a closed circuit through the machine and, if necessary, through a heat exchanger. Heat is transferred from the primary coolant to the secondary coolant either through the structural parts or through the heat exchanger.

### 4.2.4 Hydrogen Cooling

In this cooling system, the hydrogen is being used for cooling the windings, cores and other parts of the machine.

### 4.2.5 Liquid Cooling

In this cooling system, liquids are being used for cooling the different parts of the machine. The different parts may be cooled either by immersing in the liquid or by circulation of liquid across it.

### 4.2.6 Direct Cooling

In this cooling method, the coolant comes in direct contact with the different portion of the machine to be cooled.

## 4.3 HEAT DISSIPATION BY RADIATION, CONDUCTION AND CONVECTION

### 4.3.1 Heat Dissipation by Radiation

The heat which is being dissipated by radiation from a machine surface depends upon the temperature of that surface and also on other characteristics such as colour, surface roughness etc. For a small spherical radiating body inside a large and on black spherical cavity, loss due to radiation per unit surface may be obtained by Stefan-Boltzmann's law:

Power loss due to radiation

$$P = 5.7 \times e\left(T_1^4 - T_2^4\right) \times 10^{-8} \text{ watts/m}^2 \quad \text{...(4.1)}$$

where, $e$ is the emissivity of the surface.

The emissivity of a perfect black surface is unity and it is always less than unity for others.

$T_1 = (\theta_1 + 273)$

$= $ Absolute temperature of the heat emitting surface

$\theta_1 = $ Temperature of the heat emitting surface (°C)

$T_2 = (\theta_2 + 273)$

$= $ Absolute temperature of the ambient medium absorbing the heat

$\theta_2 = $ Temperature of the ambient medium (°C)

For example, a transformer tank may be considered as a small spherical body, radiating heat and the heat radiated from the transformer tank can be calculated.

Equation (4.1) can be rewritten as

$$P = 5.7 \times e(T_1 - T_2)\left[T_1^3 + T_1^2\,T_2 + T_1\,T_2^2 + T_2^3\right] \times 10^{-8}$$

the term $\left(T_1^3 + T_1^2\,T_2 + T_1\,T_2^2 + T_2^3\right)$ can be represented by a factor $k$ as its variation is very less within the conventional temperature limits. So, it can be rewritten as

$$P = 5.7 \times 10^{-8} \times ek\,(T_1 - T_2)$$

or $$P = \lambda_r \theta \text{ watts/m}^2 \qquad \text{...(4.2)}$$

where, $\lambda_r$ is the specific heat dissipation or emissivity of the surface in watts per unit area of the surface at 1°C temperature rise. $\lambda_r$ depends upon the amount of temperature difference so it is not constant. The specific heat dissipation by radiation can be represented by Table (4.1) when the ambient temperature is assumed at 20°C. When the temperature of the ambient medium increases, the specific dissipation increases slightly for same rise of temperature.

**Table 4.1** Specific heat dissipation ($\lambda_r$) at ambient temperature of 20°C

| *Temperature rise (θ°C)* | *$\lambda_r$ in watts/m²-°C* |
|---|---|
| 5 | 5.03 |
| 20 | 5.42 |
| 40 | 6.05 |
| 80 | 7.30 |

If the total surface area by which the heat is being dissipated is known then the total heat dissipated by that surface can be calculated as

$$P_{tr} = P \times \text{Total surface area}$$

$$P_{tr} = PS = \lambda_r \theta S$$

or $$P_{tr} = \lambda_r\,\theta S \text{ watts} \qquad \text{...(4.3)}$$

Some apparatus like transformer, used at outdoor, may absorb heat radiation from the sun by insolation.

### 4.3.2 Heat Dissipation by Conduction

The heat dissipated by conduction per unit area of the surface is

$$P = \frac{\theta_1 - \theta_2}{\rho t} \text{ watts/m}^2 \quad \text{...(4.4)}$$

where, $\theta_1$ = the higher temperature at one side of surface

$\theta_2$ = temperature at next side of surface

$\rho$ = thermal resistivity of the material, ($\Omega$ – m),

thermal conductivity (W/°C – m) will be $\sigma = 1/\rho$

$t$ = length of medium, (m)

If the area of surface through which heat is to be dissipated is $S$, then the total heat dissipated is

$$P_t = \frac{S(\theta_1 - \theta_2)}{\rho t} \text{ watts} \quad \text{...(4.5)}$$

The temperature difference ($\theta_1 - \theta_2$), across the conducting medium can be calculated as

$$\theta = \theta_1 - \theta_2 = P_t \frac{\rho t}{S} \quad \text{...(4.6)}$$

**Table 4.2** Thermal resistivity of different materials

| *Different materials* | *ρ (Ω – m)* | *Different materials* | *ρ (Ω – m)* |
|---|---|---|---|
| Air (still) | 20 | Mica | 3 |
| Cotton cloth | 14 | Sheet steel (Parallel to lamination) | 0.02 |
| Micanite | 8 | | |
| Compressed paper | 8 | Sheet steel (across lamination) | 0.05 to 0.1 |
| Paper | 7.5 | | |
| Transformer coil | 6.25 | Brass | 0.01 |
| Pressboard | 6 | Aluminium | 0.005 |
| Varnished cloth | 5 | | |
| Mica tape | 0.6 to 6.6 | Copper | 0.0026 |

The different values of thermal resistivities of different materials is being shown in Table (4.2). The thermal resistivity of air is far more than the thermal resistivity of insulating materials used in the machines. If there is air pocket present in an insulation of the machine, it may dangerously affect the heat dissipation. The heat dissipation will be less and this may lead to the large temperature rise of that portion of machine.

### 4.3.3 Heat Dissipation by Convection

When any portion of machine gets heated due to various losses then gas or liquid particles in contact with the heated body become lighter. The particles becoming lighter rises giving place to other comparatively cooled particles which also get heated and rise. This process goes on and heat is being dissipated from the heated surface by particles and is called as convection. The convection are of two types viz. natural convection and artificial convection.

***(i) Natural Convection:*** When gas or liquid particles are in contact with the heated body in natural way, the particles are heated and rise and due to this process the heat is dissipated. This process is called as natural convection. Heat dissipated by natural convection per unit area of surface $S$ is

$$P = \lambda_c \theta \quad ...(4.7)$$

where, $\lambda_c$ is the specific heat dissipation by natural convection.

Total heat dissipated by natural convection is given by

$$P_{tc} = \lambda_c \theta S \quad ...(4.8)$$

where, $S$ is the surface area of heated body.

***(ii) Artificial Convection:*** In present trend, the heat developed in the different portions of the machines are dissipated by artificial circulation of cooling medium. The turbo-alternators are being cooled by circulating hydrogen and transformer tank are cooled by forced air circulation across the tank. This type of cooling process is called as cooling by artificial convection. In this process, the heat dissipation increases due to increase in convection. Heat dissipated by artificial convection per unit area of surface $S$ is

$$P = \lambda'_c \theta \quad ...(4.9)$$

where, $\lambda'_c$ is the specific heat dissipation of surface applied to forced/blast circulation and is given as

$$\lambda'_c = \lambda_c \left(1 + k\sqrt{V_r}\right) \text{ watts/m}^2\text{-°C}$$

where, $\lambda_c$ = specific heat dissipation by natural convection

$V_r$ = relative velocity between surface on which air is blasted and air (m/sec)

$k$ = a constant which depends upon the type of surface

For uniform surface $k = 1.3$

Non-uniform surface $k = 0.5$

Total heat dissipated by surface

$$P_{tc} = \lambda'_c \theta S \quad ...(4.10)$$

Total heat dissipated by radiation and convection both is given by

$$P = P_{tr} + P_{tc} \quad ...(4.11)$$

From equations (4.3), (4.8) and (4.10), the total heat dissipation by radiation and convection both can be estimated. From equation (4.11), the total heat dissipated,

$$P = P_{tr} + P_{tc} = (\lambda_r + \lambda_c) \times \theta \times S \text{ watts}$$

$$P = \lambda \, \theta \, S \quad ...(4.12)$$

where, $\lambda$ is the specific heat dissipation due to radiation and convection both or emissivity due to both.

Equation (4.11) is explained by Newton's law of cooling. Newton's law of cooling is applicable suitably for cases where the body is applied by a uniform air.

Total specific loss dissipation due to radiation and convection both at 40°C can be obtained by Table (4.3) for different type of surface.

**Table 4.3** Specific loss dissipation due to radiation and convection both

| *Type of surface* | *λ watts/m²-°C* |
|---|---|
| Varnished tape | 15.0 |
| Cotton tape | 12.4 |
| Oil paint | 13.0 |
| Aluminium paint | 10.8 |
| Tarnished metal | 9.1 |
| Polished metal | 8.2 |

## 4.4 VOLUME OF AIR AND LIQUID

### 4.4.1 Volume of Air

If the total loss to be dissipated is known and the temperature rise of the cooling machines is also known, it is possible to find out the amount of cooling medium required. If the air is the cooling medium, the quantity of air can be calculated for heat to be dissipated and accordingly the power required for the fan to provide this volume of air can also be calculated. The volume of air required to dissipate this heat loss to be dissipated can be calculated by

$$Q = \frac{P}{C_p \theta} \cdot V \cdot \frac{(\theta_1 + 273)}{273} \times \frac{760}{H} \times 10^3 \text{ m}^3/\text{sec.}$$

For air, $C_p = 995$ J/kg – °C, $V = 0.775$ m³,

So,

$$Q = 0.78 \frac{P}{\theta} \times \frac{(\theta_1 + 273)}{273} \times \frac{760}{H} \text{ m}^3/\text{sec.} \quad \text{...(4.13)}$$

where, $Q$ = volume of air in m³/sec

$P$ = heat loss to be dissipated in kW

$\theta$ = temperature rise in 0°C of cooling air

$\theta_1$ = inlet temperature of air in °C

$H$ = barometric height in mm of Hg

$C_p$ = specific heat of air at contact pressure in, J/kg-°C

$V$ = volume of 1 kg of air in m³

If the volume of air required has to be supplied by a fan having efficiency of $\eta$ and at a pressure $p$, then the power $P_f$ required for the fan to provide this quantity of air is given by

$$P_f = \frac{pQ}{\eta \times 10^3} \text{ kilowatt} \quad \text{...(4.14)}$$

where, $\eta$ = efficiency of fan, usually between (0.2 and 0.4)

$p$ = pressure in newtons/m²

The equation (4.14) can be used to calculate the amount of gas or hydrogen required for cooling when the cooling medium is gas or hydrogen. While using equation (4.14), then thermal conductivity, density and specific heat should also be taken into consideration. Specific heat of hydrogen at constant pressure is 3.4 cal/gm-°C and 1 kg of dry gas occupies a volume of 11.3 m$^3$ at 273 K and 760 mm of mercury. Volume of hydrogen required to dissipate power (P) in kilowatt is given by

$$Q = 0.8 \times \frac{P}{\theta} \times \frac{(\theta_1 + 273)}{273} \times \frac{760}{H} \text{ meter cube/sec}$$

### 4.4.2 Liquid Required

If water is being used as the medium to cool air in closed circuit machines or to cool oil in a transformer and the specific heat of water is one then the quantity of water required is given as

$$Q = 0.240 \times \frac{P}{Q} \text{ litres/sec} \quad \text{...(4.15)}$$

The difference in inlet and outlet temperature of water medium can be taken as 8 to 10°C. When the oil is being used as the cooling medium the quantity of oil can be calculated as

$$Q = 0.240 \times \frac{P}{C_p \theta} \text{ litres/sec} \quad \text{...(4.16)}$$

where, $C_p$ is the specific heat of oil usually lying between 0.4 to 0.5.

Chapter 4

## 4.5 TEMPERATURE RISE-TIME CURVE

The electric machines like alternators, induction motors, d.c. machines and transformers can be considered as a homogeneous body in which heat is being developed at a uniform rate. The heat dissipation is at a rate proportional to the temperature rise.

### 4.5.1 Machine Under Heating

In a short time $dt$, the heat developed in the machine is $bdt$. The heat stored by the machine is $m\,C_p\,d\theta$. The temperature rise in time $dt$ is $d\theta$.

Heat dissipated = $S\lambda\, dt\, \theta$

So, Heat developed = Heat stored + Heat dissipated

or

$$bdt = m\,C_p\,d\theta + S\lambda\,dt\,\theta \quad \text{...(4.17)}$$

or

$$\frac{b}{mC_p} = \frac{d\theta}{dt} + \frac{S\lambda}{mC_p} \cdot \theta$$

where, $b$ = heat developed in machine in watts or joules/sec

$C_p$ = specific heat in J/kg-°C

$m$ = weight of active part of machine in kg

$\theta$ = temperature rise at any time $t$ in °C

$\theta_m$ = maximum temperature rise under heating condition in °C

$t$ = time in seconds

$S$ = cooling surface in m$^2$

$c$ = cooling coefficient = $\frac{1}{\lambda}$

$\lambda$ = specific heat dissipation or emissivity in watts/m$^2$-°C

$\theta_c$ = final steady temperature rise in °C when cooling

$\tau_h$ = heating time constant in second

$\theta_l$ = lowest temperature under cooling condition in °C

$\tau_c$ = cooling time constant

$Q_i$ = initial temperature rise over ambient in °C

***Derivation of Equation for Heating:*** From equation (4.17) we get,

$$dt = \frac{d\theta}{\dfrac{b}{mC_p} - \dfrac{S\lambda\theta}{mC_p}} \quad \text{...(4.17}a\text{)}$$

Let,

$$\frac{b}{mC_p} - \frac{S\lambda}{mC_p} \cdot \theta = u$$

So

$$\frac{du}{d\theta} = -\frac{S\lambda}{mC_p} \cdot 1$$

$$\therefore \quad d\theta = -\frac{mC_p}{S\lambda} \cdot du$$

Now, putting this value of $d\theta$ in above equation (4.17$a$), we get

$$dt = \frac{-\dfrac{mC_p}{S\lambda} \cdot du}{u}$$

$$\int dt = -\frac{mC_p}{S\lambda} \int \frac{du}{u}$$

or

$$t = -\frac{mC_p}{S\lambda} \cdot \log_e u + k$$

On putting the value of $u$, we get

$$t = \frac{-mC_p}{S\lambda} \log_e \left( \frac{b}{mC_p} - \frac{S\lambda}{mC_p} \cdot \theta \right) + k \quad \text{...(4.17}b\text{)}$$

where, $k$ is constant of integration and can be estimated by putting initial conditions *i.e.*, when $t = 0$ then $\theta = \theta_i$, putting this in above equation in (4.17*b*)

we get, $$0 = -\frac{mC_p}{S\lambda}\log_e\left(\frac{b}{mC_p} - \frac{S\lambda}{mC_p}\cdot\theta_i\right) + k$$

or $$k = \frac{mC_p}{S\lambda}\log_e\left(\frac{b}{mC_p} - \frac{S\lambda}{mC_p}\cdot\theta_i\right)$$

Substituting the value of $k$ in (4.17*b*), we get the expression for $t$ as,

$$t = -\frac{mC_p}{S\lambda}\log_e\left(\frac{b}{mC_p} - \frac{S\lambda}{mC_p}\cdot\theta\right) + \frac{mC_p}{S\lambda}\log_e\left(\frac{b}{mC_p} - \frac{S\lambda}{mC_p}\cdot\theta_i\right)$$

$$= -\frac{mC_p}{S\lambda}\log_e\frac{\dfrac{b}{S\lambda} - \theta}{\dfrac{b}{S\lambda} - \theta_i} \qquad \text{...(4.17}c\text{)}$$

when machine temperature rises to its maximum steady temperature, there is no further increase in its temperature and so the heat developed is equal to heat dissipated. So at $t = \infty$, $d\theta$ (temperature rise) = 0

$\theta = \theta_m$, and so from equation (4.17)

$$\theta_m = \frac{b}{S\lambda}$$

The above equation (4.17*c*) can now be written as

$$t = -\frac{mC_p}{S\lambda}\log_e\frac{\theta_m - \theta}{\theta_m - \theta_i},$$

where, $\dfrac{mC_p}{S\lambda}$ is called the heating time constant $\tau_h$.

So, $$t = -\tau_h \log_e\frac{\theta_m - \theta}{\theta_m - \theta_i}$$

or $$\frac{\theta_m - \theta}{\theta_m - \theta_i} = e^{-t/\tau_h}$$

or $$\theta = \theta_m\left(1 - e^{-t/\tau_h}\right) + \theta_i e^{-t/\tau_h} \qquad \text{...(4.17}d\text{)}$$

when, $t = 0$, $\theta = 0$ and when $t = \infty$, $\theta = \theta_m = \dfrac{b}{S\lambda}$ from equation (4.17) at $t = \infty$, $d\theta = 0$ *i.e.*, when machine is started from cold conditions, $\theta_i = 0$.

Now putting the value of $\theta_i = 0$ in equation (4.17*d*), we get an expression as

$$\theta = \theta_m \left(1 - e^{-t/\tau_h}\right) \quad \text{...(4.18)}$$

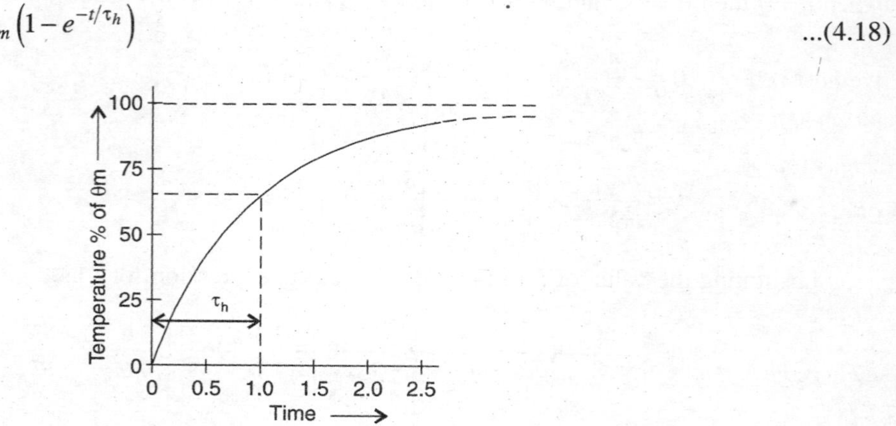


**Fig. 4.1** Temperature rise vs time curve.

During heating time constant (defined as the time taken by the machine to reach 0.632 of its final steady temperature rise) $\tau_h$ *i.e.*, when $t = \tau_h$, the machine will reach 0.632 × Maximum temperature rise $(\theta = \theta_m (1 - e^{-1}) = 0.632\ \theta_m$. The value of $\tau_h$ can be obtained as

$$\tau_h = \frac{mC_p}{S\lambda} \quad \text{...(4.19)}$$

The nature of curve obtained between temperature vs time is shown in Fig. (4.1).

At $t = \infty$, $\theta = \theta_m$, *i.e.*, the maximum steady temperature rise. There is no further increase in temperature and hence the rate of heat generation is equal to heat dissipation. Hence, heat stored, $mC_p d\theta = 0$.

### 4.5.2 Machine Under Cooling

When there is reduction of losses or the machine is stopped and hence losses are not occurring, the temperature of the machine will be reduced and the curve obtained between temperature vs time will be an exponential curve and can be shown by Fig. (4.2).

$$\theta = \theta_i\, e^{-\left(\frac{t}{\tau_c}\right)} \quad \text{...(4.20)}$$

The above equation can be obtained from the equation of temperature rise as in case of heating of machines,

*i.e.*, $$\theta = \theta_c\left(1 - e^{-t/\tau_c}\right) + \theta_i e^{-t/\tau_c}$$

when the machine is stopped, there is no losses and hence no heat produced so there is no temperature rise while cooling.

So, $\theta_c = 0$ and hence $\theta = \theta_i\, e^{-t/\tau_c}$ ...(4.20*a*)

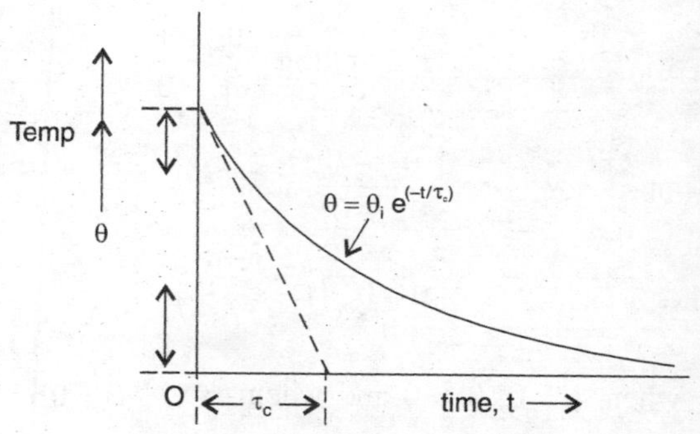


**Fig. 4.2** Cooling-time curve.

where, $\theta_i$ = initial temperature rise over ambient in °C

$\tau_c$ = cooling time constant

If $t = \tau_c$ (Cooling time constant, defined as the time taken by the machine to fall in temperature to 0.368 of the initial value of temperature $\theta_i$, of machine)

So when $t = \tau_c$,

Then, from equation (4.20*a*), we get

$$\theta = \theta_i e^{-1}$$

$$= 0.368\ \theta_i$$

This satisfy the above definition of cooling time constant.

It is assumed that the cooling is proportional to the temperature drop. Maximum temperature reached by a continuous rated machine,

$$\theta_m = \frac{b}{S\lambda} = \frac{Cb}{S}$$

where, $C$ = cooling coefficient = $\frac{1}{\lambda}$.



## 4.6 TYPES OF ENCLOSURES FOR ROTATING ELECTRICAL MACHINES

The different types of enclosure can be explained in accordance with IS : 4722–1968.

### 4.6.1 Open Type Machine

In this type of machines, there is no restriction to ventilation except that are needed to provide good mechanical strength in the machine. There are several type of open machines like Open Pedestal (OP) machine and Open End Bracket (OEB) machines.

(*a*) *Open Pedestal (OP) Machine:* This type of open machine has pedestal bearing mounted on the bed plate.

(*b*) *Open End Bracket (OEB) Machine:* In this type of machine, the bearings form an integral part of the end shields. The air is in contact with stator and rotor by providing a wide opening in the end shields.

### 4.6.2 Protected Type Machine (P)

In such machines, the rotating and other sensitive portions of the machine are protected from any contact which may lead to the accident. The mechanical protection is being provided like screen of wire mesh, expanded metals or by any other suitable cover.

### 4.6.3 Drip, Splash and Hose Proof Type Machines

(*a*) *Drip Proof (DP) Machine:* In this type of machines the end shields are so designed to exclude vertically falling water or dirt.

(*b*) *Splash Proof (SPLP) Machine:* The end shields are so designed to exclude falling water or dirt at any angle between vertical and 100°C.

(*c*) *Hose Proof (HSP) Machine:* The end shields are so designed to enclose the machine so that it exclude the water during its operation as well as during rest and also during the washing with help of nozzle at pressure of 3.5 kg/cm$^2$.

### 4.6.4 Pipe or Duct Ventilated Type Machines

The end shields in these machines are so designed so that the ventilating air may be conveyed into the machine or from the machine through pipes or ducts. Apertures are provided for the connection to pipes through which air can come in and go out. In such machines there may be provision for inlet duct only, provision for inlet and outlet ducts both or there may be provision to outlet ducts only. There are several type of arrangements which can be made to cool the machine. The different arrangements are Forced-drought (PVFD) and Induced-drought (PVID) etc., with air by external means.

### 4.6.5 Totally Enclosed, Special Enclosure Type Machines

The totally enclosed machines are protected by means of enclosures without openings but are not necessarily air tight. In machines with special enclosures the machines can work suitably in a desired circumstances without any further protection.

**Totally enclosed type machines are:**

(*a*) *Totally Enclosed (TE) Machine*: Air enclosed in such machines has no connection with external air but machine is not necessarily air tight.

(*b*) *Total Enclosed Fan Cooled (TEFC) Machine:* In these machines, cooling is provided by a fan which is driven by motor itself. The external air is blown through the cooling surface.

(*c*) *Totally Enclosed Separately Air Cooled (TESAC) Machine:* In such machines, the fan blowing the external air to the cooling surface are driven separately.

(*d*) *Totally Enclosed Water Cooled (TEWC) Machine:* In this type machines the cooling is augmented by water cooled or by other liquid cooled surfaces embodied in the machine itself.

(*e*) *Totally Enclosed Closed Air or Gas Circuit Machines:* Totally enclosed machines with special provision for cooling the air or gas enclosed in machine to cool it by passing it through the external coolers. The different coolers are used given as

(1) Air (CACA)

(2) Water (CACW)

(3) CGGW. In this the machine is cooled by gas and the cooling gas is circulated through water-cooled gas coolers.

**Special enclosure machines are:**

(*a*) *Weather Proof (WP) Machine:* This type of machine should work satisfactorily in a weather conditions specified by the purchaser.

(*b*) *Water-Tight (WT) Machine:* The machines should work satisfactorily when immersed in water without any damage or leakage. The machines should be immersed to a depth of not less than 1 m and/or with an external water pressure of 0.1 kg/cm$^2$ for one hour.

(*c*) *Submersible Machine:* This type of machines are capable of working for a quite long period when submerged under certain water head.

(*d*) *Flame-Proof Machine (FLP):* Such machines can be used in hazardous atmosphere in accordance with IS: 2148–1962. These machines may be used in mines, petroleum plants etc.

## 4.7 METHODS OF VENTILATION

### 4.7.1 Induced and Force Ventilation

Self ventilation and separate ventilation are classified in two category:

(*i*) Induced ventilation, (*ii*) Forced ventilation

***(i) Induced Ventilation:*** When the fan develops a pressure inside the machine lower than the atmospheric pressure then the ventilation of machine is induced due to air sucked into the machine due to higher external pressure. The air is pushed out to the atmosphere by the fan. Figure 4.3 shows the induced ventilation with internal fans and Fig. 4.4 shows the induced ventilation with external fans.

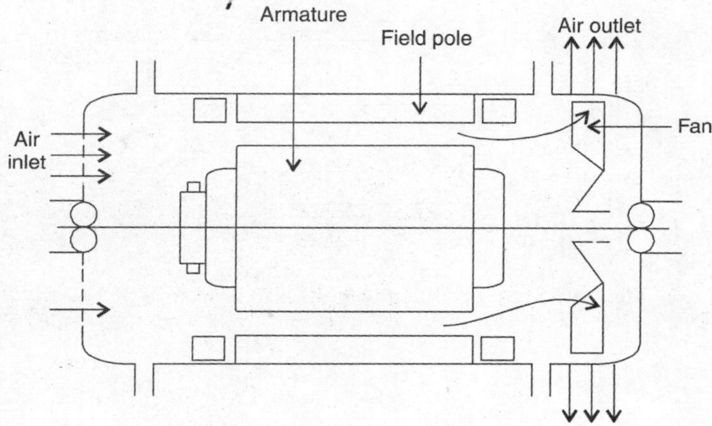


**Fig. 4.3** Induced ventilation with the help of internal fans.

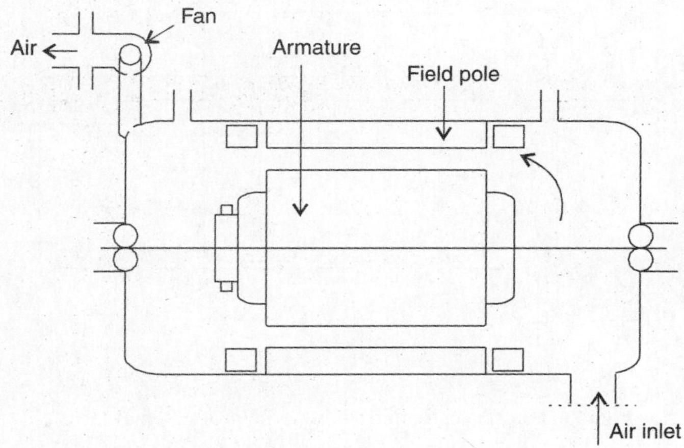


**Fig. 4.4** Induced ventilation with the help of external fans.

***(ii) Forced Ventilation:*** When the fan sucks the air from the atmosphere and blows into the machine then this process is the forced ventilation. The air pushed into the machine is pushed out to the atmosphere. Figure 4.5 shows the forced ventilation with internal fans and Fig. 4.6 shows the forced ventilation with external fans.

In forced ventilation, the temperature of cooling air rises due to losses in the fan. In induced ventilation cold air comes into the machine. So the amount of air required in forced ventilation is more as compared with induced ventilation. The noise in the machine is optimized by using smaller specific loadings with reduced losses and smaller fan diameter. In induction motors with radial ducts in stator and rotor, forced self ventilation is adopted.

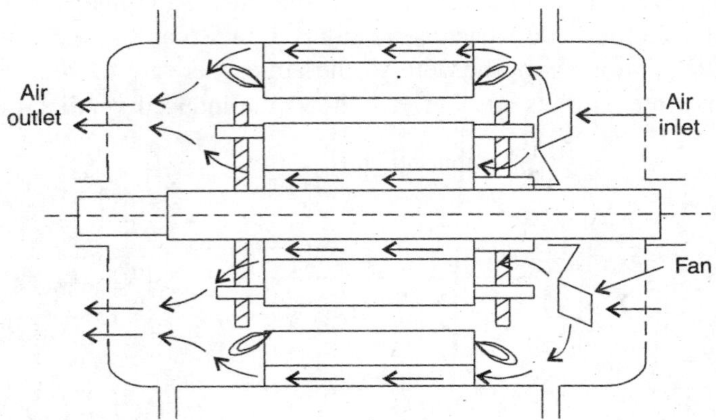


**Fig. 4.5** Forced ventilation with the help of internal fans.

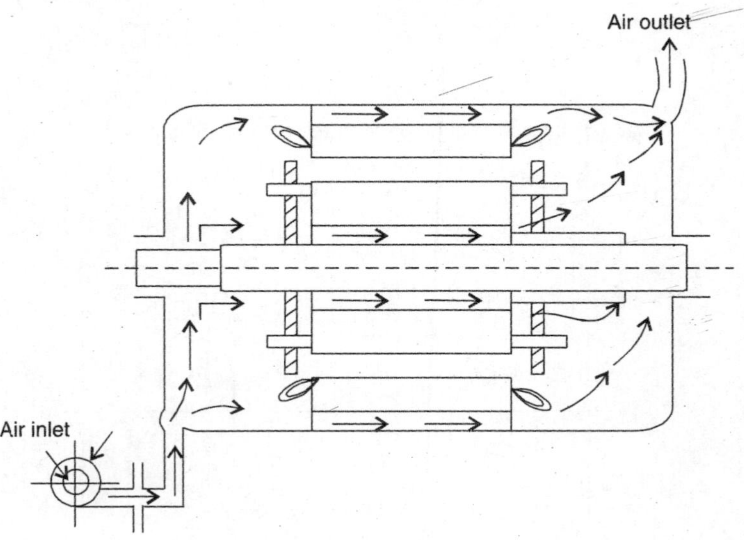


**Fig. 4.6** Forced ventilation with the help of external fans.

### 4.7.2 Radial Ventilating System

This is mostly employed because the rotation of rotor induces a natural centrifugal motion of the air. The movement of the air can be increased by providing rotor fans. Fig. 4.7 shows the radial system

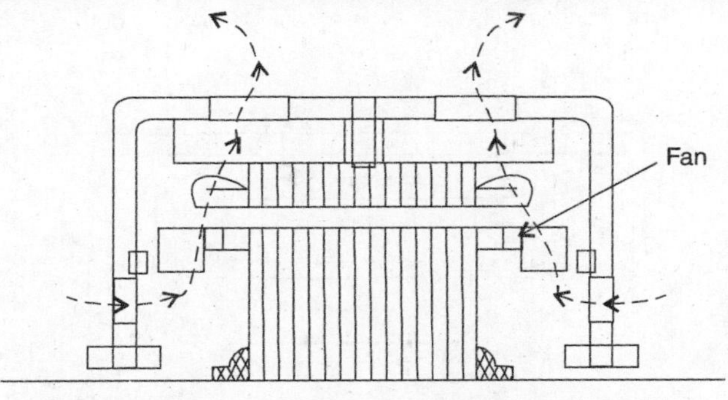


**Fig. 4.7** Radial ventilation for small machines with small core length.

for small machines with small core lengths and it is suitable for machines up to 20 kW. The end shields are so arranged to guide air over the overhang and the back of the core. For larger rating machines with longer core length, the axial length of core is subdivided by providing the ventilating ducts along the axial length of the machine as shown in Fig. 4.8 and the air through the radial ducts are parallel to the overhang. High rate of heat dissipation is possible in the air gap of the machine.



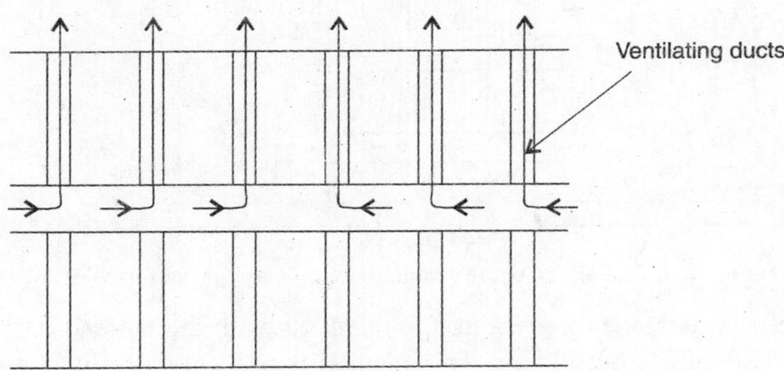


**Fig. 4.8** Radial ventilating ducts for long core machines.

### 4.7.3 Axial Ventilating System

Axial ventilation is suitable for high speed machines. In high speed machines solid rotor construction with spider is used. This is done to avoid the centrifugal stresses. Due to the restriction of centrifugal stresses, radial ventilating ducts are not provided. To provide proper ventilation to the machine axial ventilation is used. Machines with larger heat dissipation have holes punched in core to form through ducts. This increases the cooling surface and better cooling effect is achieved. But at the same time it requires large core diameter to incorporate the holes. Figure 4.9 shows the method of axial ventilation of a small machine with plain core.

### 4.7.4 Radial and Axial Ventilation in One Machine

Radial and axial ventilation both are usually used in large motors and small turbo alternators. Figure 4.10 shows the arrangement of combined radial and axial ventilation. The air is drawn into the

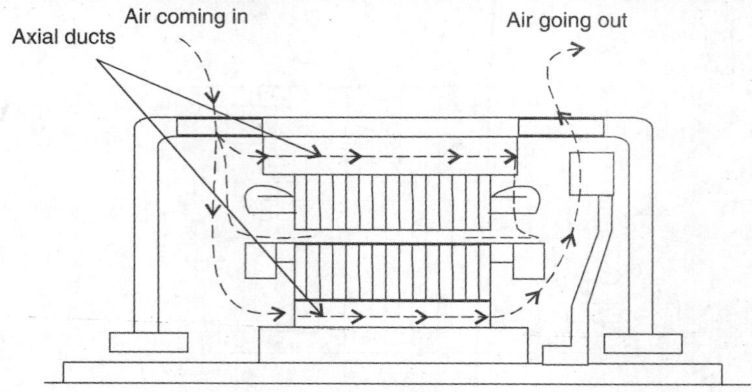


**Fig. 4.9** Axial ventilation method.

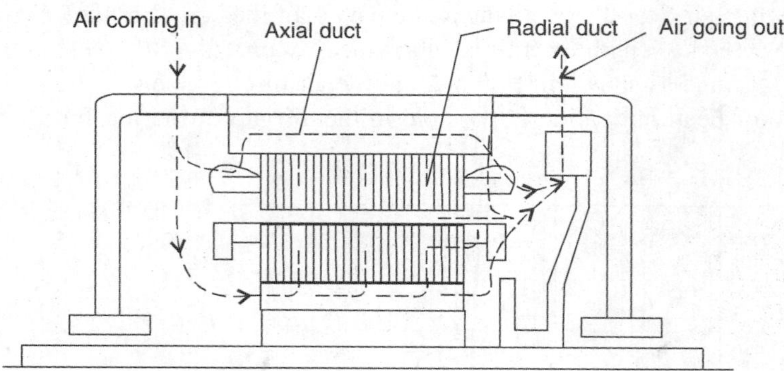


**Fig. 4.10** Radial and axial ventilation employed in one machine.

machine from one end and is made to pass through the ducts with the fan end of the rotor spider. The fan mounted on the shaft forces out the air. This method is used because by use of axial ventilation only, the area of axial ducts required to carry sufficient quantity of cooling air becomes more which leads to more iron loss and hence axial and radial ventilation both are employed.

### 4.7.5 Multiple-inlet Ventilation Method

Axial ventilation cannot be used for turbogenerator with long lengths since the cooling air becomes hot till it reaches up to central part of machine. So multiple inlet method is preferred in large length machines. In this method, the stator frame and stator core are segregated into a number of compartments. These compartments are used as inlet and outlet chambers alternately. The air is forced radially inwards at the inlet and it is forced radially outwards at the outlet. Each part of the machine should get the cool air. The multiple inlet method of ventilation is shown in Fig. 4.11.

### 4.7.6 Cooling of Totally Enclosed Machines

Totally enclosed machine has to work in an atmosphere where the air inside the machine has no connection with outside. If the machine is left without any ventilation, the ventilation in such cases will be very poor and hence the machine rating will have to keep low. This type of machine *i.e.*, totally

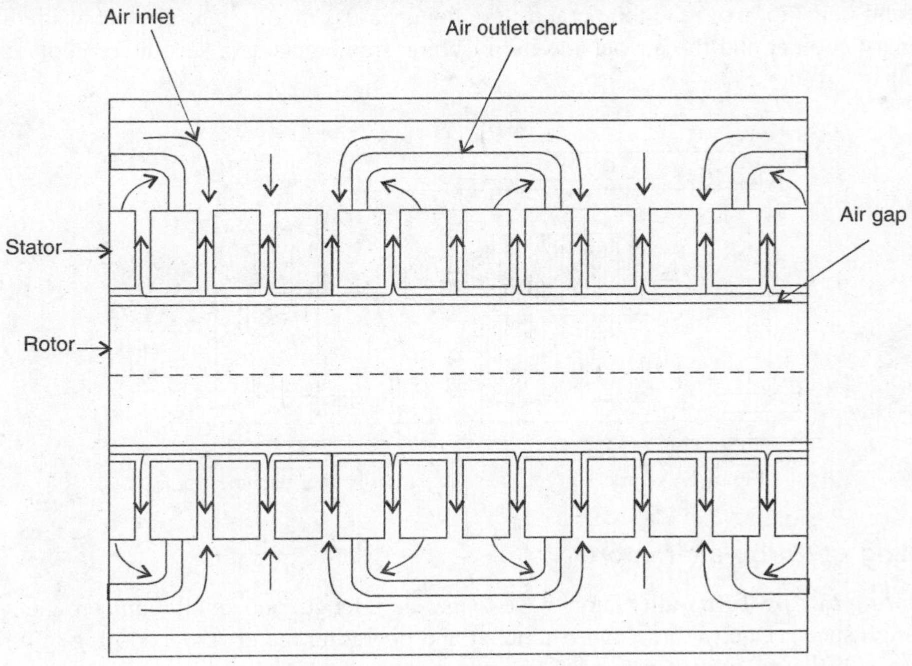


**Fig. 4.11** Multiple inlet method of ventilation.

enclosed machine with any ventilation will be uneconomical than the open and protected machines. So in case of totally enclosed machines, the heat produced inside the machine is dissipated through the external surface of the machine. The heated outer surface of the machine is cooled by forced air on its surface by a fan mounted on the shaft outside the machine frame. Figure 4.12 shows a type of totally

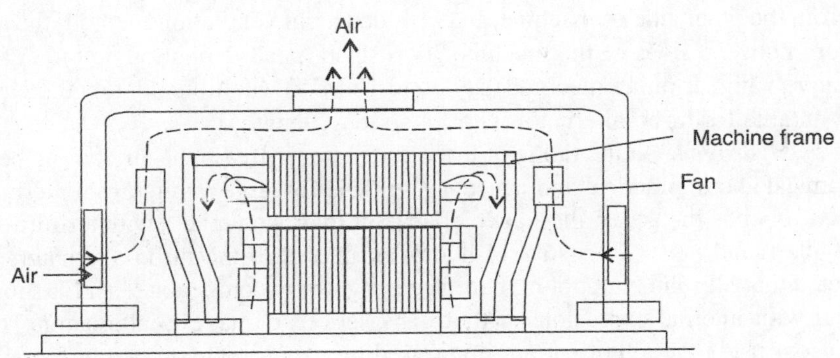


**Fig. 4.12** Ventilation of totally enclosed machine.

enclosed machine in which a fan is mounted on the shaft outside the machine frame. This fan blows air on the caracass through a space between the main housing and a thin cover plate. This ventilation is suitable up to 25 kW machines. Another type of ventilation arrangement is provided as shown in Fig. 4.13. It is suitable for larger machines. In this arrangement a fan is provided inside the machine which circulate air inside the machine frame. Circulation of air inside the frame improves the heat dissipation and helps to avoid the temperature gradients across the air gap. The air inside the machine acts as primary coolant and the air outside the machine frame acts as a secondary coolant.

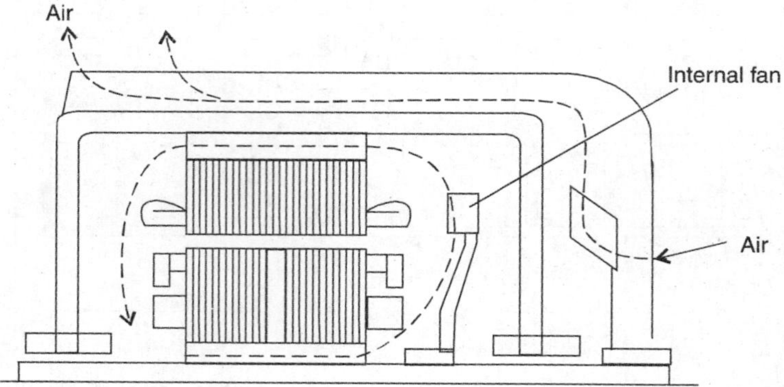


**Fig. 4.13** Ventilation with air circulation inside machine frame.

### 4.7.7 Cooling of Turbo-alternators

**(a)** ***Cooling by Air:*** Turbo-alternators have longer core length and small diameter and are operated at high speed. Due to these constructional and operating features, the cooling of turbo-alternator is a difficult task rather than cooling a water wheel generator. Air cooling method is used for small turbo-alternators. The various methods of air cooling are:

(*i*) One sided axial ventilation
(*ii*) Two sided axial ventilation
(*iii*) Multiple inlet method

In one sided axial ventilation, the air is forced to enter in the machine from one side only and it leaves from the other side of machine. In two sided axial ventilation method, the air is forced to enter from both the sides of the machine. Two sided axial ventilation method can be used for comparatively higher rating machine (say up to 10 MW). Multiple inlet method as explained in 4.7.5 is suitable for turbo-alternator with longer core lengths.

**(b)** ***Cooling by Hydrogen:*** As the rating of turbo-alternator increase the air cooling becomes highly uneconomical and impractical and hence not suitable for large rating turbo-alternator (above 50 MW). Hydrogen is the better alternative of air. But hydrogen mixed with air forms an explosive over a wide range *i.e.*, 4% to 76% of hydrogen in air. So the turbo-alternators should be so constructed that all joints in cooling path is gas tight, to avoid leakage and it should be able to withstand with internal explosions without any serious damage to alternator. The probability of explosion inside the alternator are minimised by using the hydrogen at a pressure of about 200–300 kN/m$^2$ since due to any leakage the hydrogen will be evacuated to atmosphere. The method of hydrogen cooling is shown in Fig. 4.14.

Proper methods should be followed for filling and emptying the hydrogen, checking the purity of hydrogen by thermal conductivity measurements etc. There are several advantages associated with hydrogen cooling method. They are,

(*i*) Increased rating

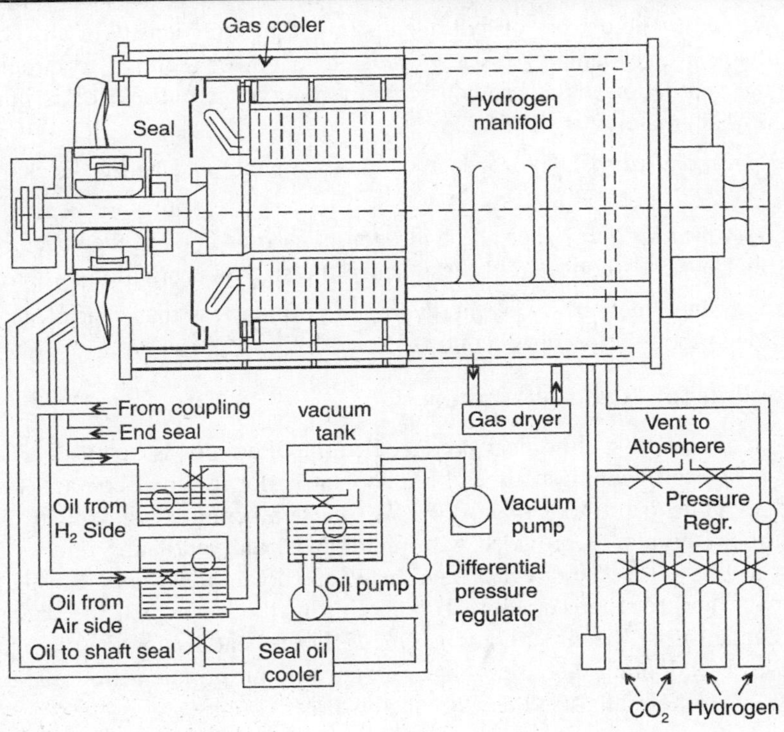


**Fig. 4.14** Method of hydrogen cooling.

Chapter 4

(*ii*) Increased life
(*iii*) Improved efficiency
(*iv*) Elimination of fire hazards
(*v*) Reduced size for ventilation
(*vi*) Reduced noise

***(i) Increased Rating:*** When hydrogen is used as coolant then the rate of heat transfer from machine to outside becomes more. So the heat produced in the turbo-alternator can be very quickly and effectively removed and hence the machine can be loaded more than that it can be loaded when air cooling method is adopted. This characteristic is achieved since heat transfer coefficient of hydrogen is 1.5 times more and thermal conductivity is seven times more than that of air. An increased output can be achieved for a given machine. The increase in rating is about 30% at 2 atm and 40% at 3 atm.

***(ii) Increased Life:*** When air is used as coolant, the high voltage winding insulation breakdown may occur because corona discharge takes place. Due to corona discharge, ozone, nitric acid and other chemical compounds are formed which deteriorates the insulation and hence the insulation life is decreased. The thermal conductivity of air is quite less than the hydrogen and insulation. The thermal conductivity of hydrogen is almost equal to that of insulation. It has already been explained that there should not be any air pocket in insulation. But if there is any air pocket in insulation, by using hydrogen as coolant, the local temperature rise risk can be avoided since the heat conduction through hydrogen will be as good as through the insulation.

***(iii) Improved Efficiency:*** The density of hydrogen is 0.07 times the density of air. The power required to circulate hydrogen is considerably less as compared to power required to circulate air. Since by using hydrogen, the ventilation losses are considerably reduced and efficiency is improved (by about 0.8% for a 100 MW alternator).

***(iv) Elimination of Fire Hazards:*** Hydrogen is not inflammable and hence checks the expansion of fire in the machine.

***(v) Reduced Size for Ventilation:*** When the hydrogen is used as coolant the space required for same cooling effect as that by air is considerably less hence the size of ventilation system is reduced.

***(vi) Reduced Noise:*** Since the density of the hydrogen is quite less than that of air and hence when rotor runs in hydrogen then the noise is reduced.

### 4.7.8 Direct Cooling (of Turbo-alternator)

In conventional cooling methods, the heat produced in the alternator is being dissipated to a coolant which is totally outside the winding coils and the coolant is not in direct contact with conductor. For machines of higher rating (say more than 100 MW) the temperature gradient in the coil insulation becomes high. Turbo-alternator is provided with radial and axial ventilating ducts with multiple inlet cooling as shown in Fig. 4.15. The direct contact between coolant and the coil insulation/conductor is required to dissipate the heat more quickly. The cooling method in which the conductor is in direct contact with the coolant is called direct cooling. Due to faster heat dissipation, the machine rating can be further increased by adopting direct cooling. The temperature gradient in coils, teeth are reduced. Hydrogen and water are normally used as coolant for direct cooling.

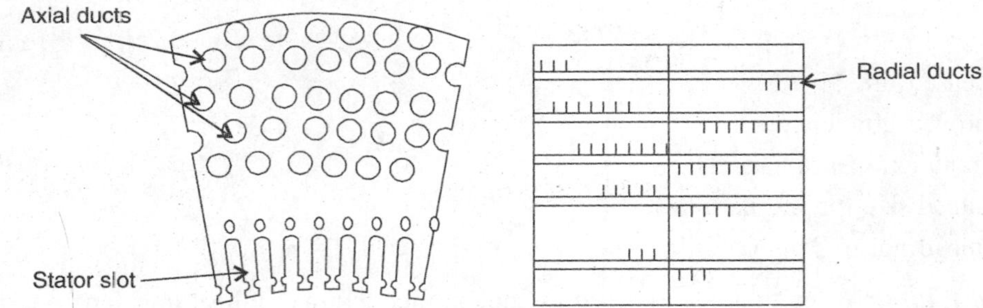


**Fig. 4.15** Radial and axial ducts.

***(a) Direct Hydrogen Cooling:*** Stator and rotor conductors are made hollow to provide direct hydrogen cooling. Figure 4.16 shows the hollow rotor conductors. The rotor conductors have rectangular ducts or tubes. Hydrogen is pumped from one end of the machine to the other side. The hydrogen gas is passed through flexible insulating connections in the tubes. The tubes in the slot are electrically connected at the overhang by copper bars to make inlet and outlet. The hollow conductors are made of hard drawn silver bearing copper with synthetic resin bonded glass-cloth laminated insulation. By applying direct hydrogen cooling of stator and rotor windings with hydrogen cooled stator cores higher electric loading up to 500 MW can be achieved. For the alternators of even higher ratings, the hydrogen cooling causes much loss and hence direct water cooling method is adopted in such machines.

***(b) Direct Water Cooling:*** The stator cores of highest possible turbo-alternators are hydrogen cooled and stator and rotor conductors are direct water cooled. The viscosity of pure water is very small and

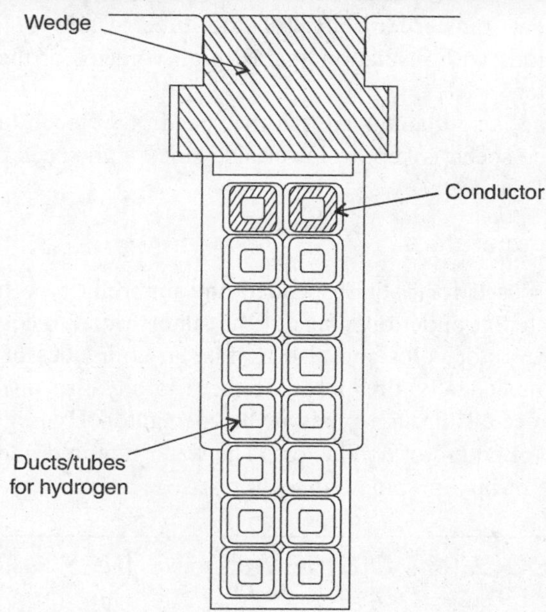


**Fig. 4.16** Hollow rotor conductors with hydrogen cooling.



it can be circulated in small tubes by providing a pressure head of optimum height. The rotor slot with conductor and ducts for direct water cooling is shown in Fig. 4.17.

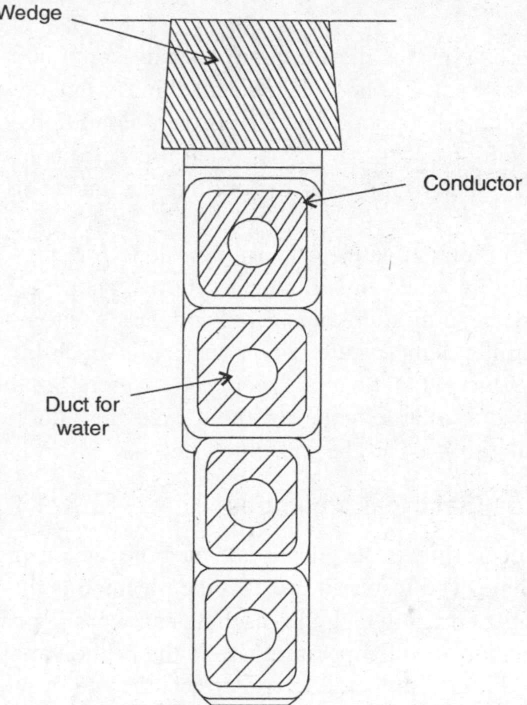


**Fig. 4.17** Direct water cooling for rotor.

Water has very high heat transfer capacity. But the direct water cooling has problems to make flexible water-tube connections with insulation against high voltages of the windings and to maintain low conductivity of the water.

The height of water head is so maintained that the speed of water in tubes is maintained between 1.5 m/sec and 2.5 m/sec. The speed is kept to lower side since high speed of water may cause erosion and cavitation.

### 4.7.9 Cooling of Hydrogenerators

Closed circuit ventilation is used for both suspended and umbrella type low speed hydrogenerators. The water coolers are mounted at outlet openings of the stator frame to cool the air. Air circulation is caused by the rotation of the rotor poles and also by fans at both sides of the rotor. Alternators with closed circuit ventilations are usually fire proof. If there is any fire inside the generator, water or gaseous carbonic acid is injected through pipes in the generator. Hydrogenerators with large rating have direct water cooled stator and rotor conductors. The water required in direct water cooling method is only 1/4th of the amount of air for same temperature rise.

## 4.8 HEATING AND COOLING CURVE AND RATINGS OF ELECTRICAL MACHINES

### 4.8.1 Heating and Cooling Curves for Different Operating Conditions

The heating and cooling curves for an electrical machine, when it is operating in different conditions is shown in Fig. 4.18.

The heating and cooling curves for the machine working continuously on full load is shown by curve 1 and 1′. The cooling curve corresponds when the machine is shut down after working continuously on full load or when the machine is operated under different conditions of loading.

Heating and cooling conditions of the machine, when it is operated on full load for a short period followed by a rest period sufficiently long so as to cool the machine to its initial temperature is shown in curve 2.

Heating and cooling conditions in case of intermittent load for short period on full load followed by no load and the cycle being repeated indefinitely is shown by curve 3. The temperature rise of the machine becomes practically uniform after some time, and varies within a range. Heating and cooling curves indicate that the maximum temperature rise of curve 3 is much lesser than the final temperature rise of curve 1. Similarly the curve 2 and curve 1 can be compared. So, the machine used for intermittent loading can be overloaded to a larger extent without exceeding the limit of final temperature rise, caused by the insulating material used in the machine.

### 4.8.2 Standard Ratings of Electrical Machines

There are different losses occurring in an electrical machine which produces heat and causes the temperature rise in the machine. The insulation part in the machine is the most vulnerable to temperature rise. The maximum temperature at which the machine can work depends upon the type and quality of insulation used in the machine. The temperature rise is the prime factor in determining the machine rating and specific operating limits. Temperature rise effects the life of winding insulation and also the insulation in other portion of the machine. Higher temperature oxidise and carbonise the insulating materials gradually and hence it effects the life of the machine as well as restricts the output of the machine.

### 4.8.3 Different Rating (Duty) of Machine

The rating of an electrical machine is power output which is based upon certain definite conditions assigned to it. The rating is mainly decided by the temperature rise in the machine. Machine is rated on the basis of its thermal characteristics due to temperature rise. A machine can always be overloaded to some extent provided the temperature rise in the machine does not exceed the permissible value. Different ratings are defined for the machine depending upon the type of duty (Load) and its duration. The majority of motor are designed for continuous load since when they are used in industrial application as drives and they have constant load continuously. Under these conditions of loading, motor attains its final steady temperature rise *i.e.*, all portion of the machine attains thermal equilibrium and there is no further rise of temperature in the machine. Whereas in some applications, the motor is stopped before it reached to its final steady temperature rise *i.e.*, motor is working intermediate periodic duty.

The different type of ratings which can classified on the basis of load and its duration are:

(*i*) Continuous rating (Continuous duty).

(*ii*) Short time rating (Short time duty).

(*iii*) Intermittent rating (Intermittent duty).



***(i) Continuous Rating:*** In continuous duty, the duration of load on machine is fairly long and so the continuous rating of a machine can be defined as the output which a machine should give continuously without its temperature rise above permissible limits. All portion of the machine attains its maximum final steady temperature rise after certain time of its start. When the machine attains its maximum final steady temperature or thermal equilibrium, then there will be no further increase in the temperature of the machine operating under some loading conditions. The heating and cooling curves for continuous duty machine is shown in Fig. 4.18 by curve 1 and 1′. Continuous rating motors are being used in fans, pumps, industrial drives etc. Continuous rating machines can also be over loaded by 15–20% for short duration (say 1 hr. to 1.5 hrs.). But if the machine rating is continuous maximum, no over-load should be applied.

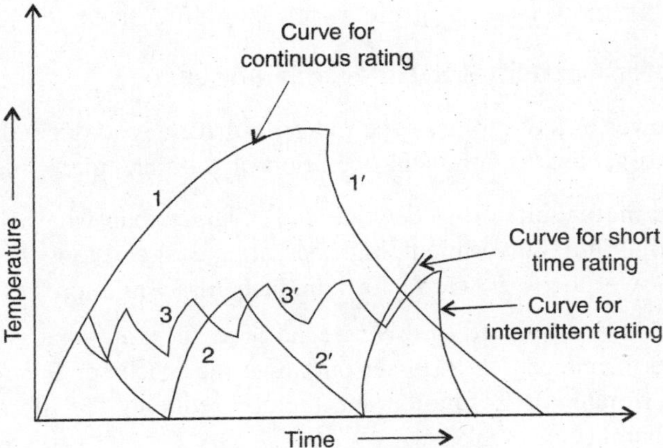


**Fig. 4.18** Heating and cooling curve for different conditions.

***(ii) Short Time Rating:*** The short time rating of a machine can be defined as its output at which the machine can work for a specified period without exceeding the maximum specified temperature rise in the machine. In short time duty machines, the working period of the machine is very short so that the temperature rise in the machine is below its maximum permissible limit of temperature rise and the no load period or shut down period or rest period is so long so that the machine reaches to its cold condition. Curve 2 and 2′ in Fig. 4.18 shows the heating and cooling curves respectively for short time rating machine. Heating curve 2 indicates that the temperature rise in this case is quite low as compared to the temperature rise in case of continuous loading. So the higher losses in the machines can be safely tolerated in short time loading of the machine for the same maximum specified temperature rise. It means that for the machines with same dimensions, the output of the machine with short time rating will be more than that of the machine with continuous rating. Short time duty machines are used in navigation lock gates, railway turn table etc.

***(iii) Intermittent Rating:*** The intermittent rating of machine means operating condition during which the machine works for short time load periods alternate with rest periods *i.e.*, machine works for a short period followed by a period of rest or no load. Curve 3 and 3′ in Fig. 4.18 shows the curve of heating and cooling for intermittent loading conditions. Curve 3 shows that the temperature rise in the machine with this rating becomes uniform after some time and varies within a range. It can also be seen that the maximum temperature rise, attained on full load is much lesser than the final steady temperature rise attained for the continuous operation at full load. Thus, for the same dimensions, and same cooling conditions, a machine with intermittent loading have larger output than a machine with continuous loading. The applications of this type of loading are in lifts, cranes, cutting tools, tool drives etc.

### 4.8.4 R.M.S. Horse Power Rating

In different type of machine loading discussed, the load during a particular period remains constant whereas there are some applications in which the machine is required to drive a load which is constant for some period and after that it may have to drive another value of constant load for another fixed time and then followed by no load or rest period. So the load on the motor changes in cyclic order.

Such loading are in rolling mills, cranes, hoists etc. The heating of machine in such cases is proportional to the torque requirement of the load or the horse power delivered to the load. For this type of load the r.m.s. horse power rating of the motor should be given.

### 4.8.5 Selection of Motor Capacity for Different Ratings

The selection of horse power or kW rating of the motor to drive a load depends on initial investments and operating cost etc., to achieve economical operation of the machine.

(*a*) If the size of the motor selected to drive a load is larger than what is required, it will lead to operation with high initial investment, high operating cost and low efficiency. If induction motors are used at lower loads, power factor is reduced drawing more reactive power.

(*b*) If the size of the motor selected to drive a load is lesser than what is required, it will lead to unsatisfactory performance due to lesser output of the machine. The temperature rise may be more which deteriorates the insulation and thereby reducing the life of machine and also increasing the probability of breakdown.

(*c*) Motors designed for short time ratings and intermittent ratings should have higher current density than that in motors designed for continuous rings because the motors may be allowed to reach the maximum temperature rise permissible quickly than the motor for continuous duty.

So to select proper or optimum horse power or kW rating of motors, the different loading conditions should be known. This means that the continuous loading or other loadings should be known. The load diagram of motor may be used to select it. The selection of motor capacity for variable loads can be done with the help of following methods.

***(i) Average Loss Estimation Method:*** The losses $(P_1, P_2, \ldots\ldots, P_n)$ of motor are estimated for different duration $(t_1, t_2, \ldots\ldots, t_n)$ of the load. The average losses are

$$P_{ave} = \frac{P_1 t_1 + P_2 t_2 + P_3 t_3 + \ldots\ldots\ldots + P_n t_n}{t_1 + t_2 + t_3 + \ldots\ldots\ldots + t_n} \quad \ldots(4.21)$$

If the average losses estimated by equation (4.21) are approximately equal to the losses of the selected motor at rated condition, the motor may be suitable for that particular drive and hence it can be selected. If the difference between two losses are larger, it should not be selected. The other motor should be taken and average losses should be estimated and then again compare it with loss at rated condition and this process is continued to get a condition when difference of the losses are minimum. When the losses are equal or nearly equal the motors can be selected.

***(ii) Equivalent Current Estimation Method:*** Equivalent current of the motor can be estimated with currents $(I_1, I_2, \ldots\ldots, I_n)$ in different known time duration $(t_1, t_2, \ldots\ldots, t_n)$ by

$$I_{eq} = \sqrt{\frac{I_1^2 t_1 + I_2^2 t_2 + I_3^2 t_3 + \ldots\ldots\ldots + I_n^2 t_n}{t_1 + t_2 + t_3 + \ldots\ldots\ldots + t_n}} \quad \ldots(4.22)$$

The equivalent current estimated by equation (4.22) should be compared by the rated current of the motor. The equivalent current should be preferably less than the rated current or at most it should be equal to the rated current.

***(iii) Equivalent Torque Estimation Method:*** The equivalent torque for each duration of loadings should be estimated with torque $(T_1, T_2, \ldots\ldots, T_n)$ for different duration $(t_1, t_2, \ldots\ldots, t_n)$ by

$$T_{eq} = \sqrt{\frac{T_1^2 t_1 + T_2^2 t_2 + T_3^2 t_3 + \ldots\ldots\ldots + T_n^2 t_n}{t_1 + t_2 + t_3 + \ldots\ldots\ldots + t_n}}$$

A motor having torque nearly equal to the equivalent torque may be selected.

***(iv) Equivalent Power Estimation Method:*** The equivalent power of selected motor is estimated from expression

$$k_{eq} = \sqrt{\frac{k_1^2 t_1 + k_2^2 t_2 + k_3^2 t_3 + \ldots + k_n^2 t_n}{t_1 + t_2 + t_3 + \ldots + t_n}}$$

where, $k_{eq}$ is the equivalent power and $k_1$, $k_2$, $k_3$, ..., $k_n$ are powers for different intervals $(t_1, t_2, \ldots\ldots, t_n)$. The motor can be finally selected if the equivalent power is nearly equal to motor power.

## 4.9 METHODS OF MEASUREMENT OF TEMPERATURE RISE

There are three methods as per IS: 422–1968 of determining the temperature rise of windings and other parts. They are classified as below:

(*a*) Thermometer method
(*b*) Resistance method
(*c*) Embedded temperature detector method

### 4.9.1 Thermometer Method

In this method, a thermometer is used for the accessible surfaces of the completed machine. This method is applied only to a point of the surface of the machine. The temperature at one point only can be measured *i.e.*, at the point of surface where thermometer is placed. Alcohol thermometers should be used in place of mercury thermometers where there is any varying or moving magnetic field since mercury thermometers are unreliable in these conditions.

### 4.9.2 Resistance Method

This method is used to determine the temperature of the winding. The temperature of the winding is estimated by increase in winding resistance. The resistance is measured at both cold and hot conditions and average temperature rise is measured by use of the resistance temperature coefficient. This is applicable to winding only. The temperature rise $(\theta_2 - \theta_1)$ of the winding can be calculated by (used for copper conductor only).

$$\frac{\theta_2 + 235}{\theta_1 + 235} = \frac{R_2}{R_1}$$

where, $\theta_2$ = temperature of the winding at the end of the test in °C
$\theta_1$ = temperature of the winding at the start of measurement in °C
$R_2$ = Resistance of the winding at the end of the test
$R_1$ = Resistance of the winding at cold

The following alternative formula is used for practical applications:

$$\theta_2 - \theta_a = \frac{R_2 - R_1}{R_1}(235 + \theta_1) + \theta_1 - \theta_a$$

where, $\theta_a$ is the temperature of cooling medium at the end of the test in °C.

### 4.9.3 Embedded Temperature Detector Method (ETD)

Embedded temperature detectors are resistance thermometers or thermocouples built into the machine during the construction at the points which are not accessible when the machine is completed.

At least six detectors are suitably distributed at suitable points of stator length where the highest temperature are likely to occur. The embedded detectors are protected from contact with cooling medium.

The detector is located between the insulated coil sides within the slot when the machine has two coil sides per slot. The detectors are located between insulated coil sides at the point where the highest temperature is likely to occur in the machines having more than two coil sides per slot.

The embedded detectors indicate the temperature of one internal point. These detectors do not necessarily give the hot spot temperatures *i.e.*, the highest temperature. But if the detectors are placed at locations of highest temperature then they indicate the hot spot temperatures.

## SOLVED PROBLEMS

**Q. 1** Calculate the volume of cooling air in $m^3$/sec required to carry off the losses in a 12 MW generator having an efficiency of 98.0%. The inlet and outlet temperature of air may be taken at 15°C and 35°C. Barometer reads 750 mm. Find power required to drive the fan to provide this circulation of air at pressure 1000 newton/$m^2$; efficiency of fan = 35%.

**Solution:**

The efficiency of 12 MW generator is 98.0%.

$$\text{Input power} = \frac{12000}{0.98} = 12244.8 \text{ kW}$$

Losses = 12244.8 – 12000 = 244.8 kW

Volume of air required per second

$$= 0.78 \times \frac{P}{\theta} \times \frac{(\theta_1 + 273)}{273} \times \frac{760}{H}$$

$$= 0.78 \times \frac{244.8}{(35-15)} \times \frac{(15+273)}{273} \times \frac{760}{750}$$

= 10.20 $m^3$/sec.

Power required for the fan

$$P_f = \frac{P \times Q}{\eta \times 10^3} = \frac{1000 \times 10.20}{0.35 \times 10^3}$$

= 29.14 kW.

**Q. 2.** Initially the temperature of transformer is 26°C. After two hours operation on full load it is 56°C and after four hours run it is 71°C.

(*a*) Calculate the maximum final temperature rise with full load on the transformer.

(*b*) Estimate heating time constant.

(*c*) How much time will it take after start for the transformer to reach 5/6th of its final steady state temperature rise?

**Solution:**

After two hours the temperature rise = 56°C – 26°C = 30°C

After four hours, the temperature rise is = 71°C – 26°C = 45°C

The equations can be given as

$$30 = \theta_m\left(1 - e^{-2/\tau_h}\right) \quad ...(i)$$

and

$$45 = \theta_m\left(1 - e^{-4/\tau_h}\right) \quad ...(ii)$$

where, $\theta_m$ = maximum temperature rise in °C

$\tau_h$ = heating time constant

From equations (*i*) and (*ii*), we get

$$1 + e^{-2/\tau_h} = 1.5$$

or $\tau_h$ = 2.9 hours

(*a*) $\theta_m = \dfrac{30}{1 - 0.5} = 60°\text{C}$

(*b*) Heating time constant $\tau_h$ = 2.9 hours

(*c*) 5/6th of maximum temperature rise = 60 × 5/6 = 50°C

∴ we can write

$$50 = 60\left(1 - e^{-t/2.9}\right)$$

or $t$ = 5.18 hours

∴ Transformer will take 5.18 hours to reach up to a temperature of 50°C.

**Q. 3** A radiating body of spherical surface with coefficient of emissivity 0.82 is put in a hall. The temperature of the body is 65°C and that of hall is 20°C. Calculate the heat radiated from the body in watt/m$^2$.

**Solution:**

Emissivity, $e$ = 0.82

$$T_2 = 273 + 20 = 293 \text{ K}$$

$$T_1 = 273 + 65 = 338 \text{ K}$$

Using equation (3.1), we get the heat radiated

$$P = 5.7 \times 10^{-8} \times \left[(338)^4 - (293)^4\right] \times 0.82$$

= 263.1 watt/m$^2$.

**Q. 4** When a motor runs at its continuous rating, its final temperature rise is 75°C. It has heat time constant of 0.75 hours.

(*i*) Calculate the temperature rise after 1 hour of the start of the motor and running continuously on the load.

(*ii*) Calculate the maximum steady temperature if the temperature rise in one hour rating is 75°C.

(*iii*) How much time the motor will take to a temperature rise from 50°C to 75°C if it is working at its one hour rating?

**Solution:**

(*i*) $\theta = 75\left(1 - e^{-1/0.75}\right)$

$= 75(1 - 0.285)$

$= 53.62°C$

(*ii*) $75 = \theta_m \left(1 - e^{-1/0.75}\right)$

$$\theta_m = \frac{75}{0.715} = 104°\text{C}$$

(*iii*) Time taken up to 75°C is one hour. Time taken to rise to 50°C.

$$50 = 104\left(1 - e^{-1/0.75}\right)$$

or $t$ = 29.5 minutes.

So time taken to reach from 50°C to 75°C

= 60 – 29.5 = 30.5 minutes.

**Q. 5** A 45 MVA turbo-alternator running at its rated load at 0.82 power factor. Volume of cooling air required is 35 $m^3$/sec. The temperature of air at inlet is 20°C and at outlet it is 50°C. Mercury reading is 760 mm. Assume specific heat of air 995 J/kg-°C and volume of air as 0.775 $m^3$/kg. If the air is to be cooled by water and in this process the temperature rise of water is 10°C, calculate

(*i*) Efficiency of the machine.

(*ii*) Amount of water required to cool the air in litre/sec.

**Solution:**

Given, $\theta_1 = 20°C$

$\theta = 50°C - 20°C = 30°C$, $H = 760$ mm

$Q = 35$ $m^3$/sec., $C_p = 995$ J/kg-°C

Volume of air, $V = 0.775$ $m^3$

The volume of air required to dissipate the heat loss is given by

$$Q = \frac{P}{C_p\theta} \times V \times \frac{(\theta_1 + 273)}{273} \times \frac{760}{H} \times 10^3 \text{ m}^3\text{/sec} \quad ...(i)$$

Now, putting all the given values in equation (*i*), we get

$$35 = \frac{P}{995 \times 30} \times 0.775 \times \frac{(20 + 273)}{273} \times \frac{760}{760} \times 10^3$$

or Power loss $P = \dfrac{35 \times 0.78 \times 30 \times 273}{293} = 763$ kW

Alternator output is 45 MVA

$$\text{Output} = 45 \times 10^3 \times 0.82 \text{ kW}$$

$$= 36.9 \times 10^3 \text{ kW}$$

$$\text{Efficiency} = \frac{\text{Output}}{\text{Output + Losses}} = \frac{36.9 \times 10^3}{36.9 \times 10^3 + 763}$$

$$= \frac{36900}{37663} = 0.979$$

or $\eta = 97.9\%$

(*ii*) Amount of water required to cool the air is given by

$$Q = 0.240 \times \frac{P}{\theta} \text{ litre/sec.}$$

$$= 0.240 \times \frac{763}{10} = 18.3 \text{ litre/sec.}$$

**Q. 6** An induction motor has to drive the loads at a constant speed but different loading pattern of 140 kW for 25 minutes then rest for 10 minutes, 100 kW for 20 minutes and then rest for 10 minutes. This operating sequence has to be continued indefinitely. Calculate the suitable capacity of a continuously rated motor for the above load cycle. Motors of 105, 110, 120 kW are available. The ratio of maximum power to nominal power should not exceed by 1.6.

### Solution:

Equivalent power from equation

$$k_{eq} = \sqrt{\frac{k_1^2 t_1 + k_2^2 t_2 + k_3^2 t_3 + \ldots + k_n^2 t_n}{t_1 + t_2 + t_3 + \ldots + t_n}}$$

$$= \sqrt{\frac{(140)^2 \times 25 + 0 \times 10 + (100)^2 \times 20 + 0 \times 10}{25 + 10 + 20 + 10}} = 103.4 \text{ kW}$$

The motor of 105 kW can be selected. The ratio of maximum power to nominal power $= \frac{140}{105} = 1.33$ which is suitable.

## QUESTIONS WITH SHORT ANSWERS

**Q. 1 Write the equation representing the Newton's law of cooling and state the operating conditions under which the law is strictly applicable.**

**Ans.** When machine starts working, losses are produced in various parts of the machine and hence temperature rises. After sometime machine attains a steady temperature rise *i.e.*, thermal equilibrium. At this stage there is no temperature rise and heat produced is equal to heat dissipated by radiation and convection. Thus by Newton's law the total heat dissipated by radiation and convection both is

$$P = (\lambda_r + \lambda_c)\,\theta S$$
$$= \lambda\theta S$$

where, $\lambda$ = specific heat dissipated due to radiation and convection both or emissivity due to both .

$\lambda_c$ = specific heat dissipation by natural convection

$\lambda_r$ = specific heat dissipation by radiation

The Newton's law of cooling is strictly applicable for cases where the body is acted upon by a uniform current of air. It is therefore, applicable to natural cooling only for a restricted range of temperature.

**Q. 2 Explain the difference between radiation and insolation. In which machine do these two phenomena affect the temperaturerise?**

**Ans.** By radiation process heat is dissipated from the surface by means of heat waves and depends upon the temperature of the surface and other characteristics like colour, roughness etc.

The machines used for outdoor duty absorbs radiation from the sun by insolation. The outer atmosphere of the earth receives about 1.3 kW/m$^2$ and under suitable weather conditions approximately 1.0 kW/m$^2$ can reach directly on the earth surface. In case the absorption factor of the machine surface is high, it may absorb large energy by insolation and hence temperature of the machine may rise by few degree. In transformers working at outdoor, these two phenomena of radiation and insolation effect the temperature rise.

**Q. 3 Explain cooling of turbo-alternators using hydrogen as cooling medium.**

**Ans.** The cooling of large alternators is one of the complex problems in electrical engineering. Being high speed machines the diameter as compared with those of hydro-electric generators are much smaller. The problems of cooling arise because large size turbo-alternators are characterized by long core lengths and small diameters. The method of cooling normally used for large turbo-alternator is hydrogen cooling. The use of hydrogen for the cooling has the following advantages:

(*i*) Quiter operations resulting in reduction in windage noise, since the density of hydrogen is quite less.

(*ii*) Reduced maintenance because the casing is sealed and dirt cannot enter.

(*iii*) The elimination of fire hazards.

(*iv*) Longer service life of the insulation.

(*v*) Smaller size of coolers.

(*vi*) Increased life and also improved efficiency.

**Q. 4 Explain how hydrogen cooling of an alternator stator effects the following characteristics of the machine:**

**(*a*) Increase in rating** **(*b*) Increase in efficiency**

**(*c*) Increase of working life** **(*d*) Lesser noise**

**Ans.** (*a*) Hydrogen has a heat transfer coefficient 1.5 times and thermal conductivity 7 times than that

Chapter 4

of air. When hydrogen in used as a coolant, the heat is more rapidly taken up from the machine parts and given out. Therefore heat generated is more effectively dissipated and the active material can be loaded more than that is possible with air cooling and hence rating of the machine is increased by hydrogen cooling.

(*b*) The density of hydrogen is 0.07 times the density of air and therefore the power required for circulation of hydrogen is quite less than that of the power required for an equivalent quantity of air. Thus windage losses are reduced which are a major portion of total losses in a high speed machine and hence efficiency of the machine is increased.

(*c*) The thermal conductivity of hydrogen is usually 7 times that of air and is of the same order as the winding insulation. When insulation pockets are filled with hydrogen, heat conduction through them will be as good as through winding insulation and consequently high local temperature rises are not there which are the main sources of insulation breakdown and hence the working life of machine is increased with use of hydrogen cooling.

(*d*) The noise produced in hydrogen cooled machine is less as the rotor moves in a medium of smaller density. When rotor moves in hydrogen having very low density as compared to air, lesser noise is produced.

**Q. 5 Write the use of each of the following types of insulating material in electrical machinery:**

**(*a*) Silicon** **(*b*) Synthetic resin enamel**
**(*c*) Glass fibre** **(*d*) Micafolium**
**(*e*) Epoxy resin** **(*f*) Insuline**

**Ans. (*a*)** ***Silicon:*** These are used in dry transformers, traction motors, mill motors and miniature aircraft machines operating over a temperature range of 200°C to – 40°C.

**(*b*)** ***Synthetic resin enamel:*** Bonded paper, cotton and glass fibre with synthetic resin laminates have good electrical properties and used for insulation purpose in machines.

**(*c*)** ***Glass fibre:*** Fibrous glass coverings are normally used for windings and tapes which are required to operate in the class *B* temperature range.

**(*d*)** ***Micafolium:*** It is used for wrapping of conductors and consists of mica splittings and air dried. It may be moulded directly as conductors then rolled and compressed between heated plates to solidity the material and to remove air from it.

**(*e*)** ***Epoxy resin:*** These materials are of importance in casting potting, laminating adhesive and varnishing applications, and in the encapsulation of small transformers.

**(*f*)** ***Insulin:*** This is a Kaolin mixture which is being used for spraying on one or both sides of the laminations. The total thickness of coating per lamination is 0.01 to 0.025 mm.

**Q. 6 Differentiate between open-circuit and closed-circuit ventilation.**

**Ans.** ***Open-circuit and Closed-circuit Ventilation:*** In open circuit ventilation heat is given up directly to the cooling air through the machine which is being continuously replaced whereas in closed circuit ventilation heat is transferred to the cooling medium through an intermediate coolant circulating in a closed circuit through the machine and a cooler.

**Q. 7 Explain radial and axial ducts.**

**Ans.** ***Radial and Axial Ducts:*** The ducts provided in axial direction are called as axial ducts. This system of ventilation is suitable for machines of medium output and high speed. The heat transfer with axial ventilation is non-uniform and the iron losses are more.

The system in which ducts are provided in radial direction are called radial ventilating ducts. The system in most commonly used because the movement of rotor induces natural centrifugal movement of air in radial directions. For machines with large lengths radial ventilating ducts are provided by dividing the core into stacks of 8 to 10 cm thick. The advantages of radial system are minimum energy losses for ventilation and uniform temperature rise of the machines in axial direction. The length of machine is increased by providing radial ducts.

**Q. 8 What is difference between direct and indirect cooling?**

**Ans.** ***Direct and Indirect Cooling:*** When the cooling medium used comes in direct contact of the heated parts and dissipate the heat the method is called as direct cooling. When the coolant does not come in direct contact with the heated parts of the machine but takes away the heat of the primary coolant in a heat exchanger is called as indirect method of cooling.

**Q. 9 Explain splashproof and drip-proof machines.**

**Ans.** ***Splashproof and Drip-proof:*** A machine in which the opening for ventilation are so protected as to exclude vertically falling water or dirt is called as drip proof machine and a machine in which the ventilating openings are so designed that drops of liquid or solid particles falling on or reaching any part of them any angle between the vertically downward direction and 100° from that direction cannot enter in the machine are called splash proof machines.

**Q. 10 Write the different method of calculation of motor rating for variable loads.**

**Ans.** The different methods normally used for calculation of motor rating for variable loads are given below:

(*i*) Method of average losses.
(*ii*) Equivalent current method.
(*iii*) Equivalent torque method.
(*iv*) Equivalent power method.

By any one method given above the suitable rating of motor, to be selected, can be calculated.

**Q. 11 Write different types of enclosures used in electrical machines.**

**Ans.** The different types of enclosures used are:

**(*a*)** ***Open Pedestal Type:*** Rotor and stator ends are freely in contact with atmosphere air. The rotor are carried on pedestal bearings mounted on the bed plate.

**(*b*)** ***Protected or End Cover Type:*** The screen protection are provided or the protection is done by providing fine mesh covers.

**(*c*)** ***Drip Proof:*** Constructed to exclude falling water or dirt.

**(*d*)** ***Pipe Ventilated:*** End covers are closed except for flanges aperture in pipes along which the cooling air is drawn.

(*e*) Totally enclosed.

**Q. 12 Discuss the rating of an electric machine. How the temperature rise affects the selection of machine rating?**

**Ans.** The rating of electrical machines is mainly decided by the considerations of temperature rise. A machine can always be overloaded to a certain extent provided the temperature rise does not exceed the specific limit. Three ratings classified on the basis of load and its duration are

(*a*) Continuous rating
(*b*) Short time rating
(*c*) Intermittent rating

These ratings depend upon the load and its duration such that when machine works for particular rating the maximum temperature does not exceed the specified limit. To ensure reliable operation, the electrical machine and the equipment are designed for a suitable temperature of insulating material used in machines. The design of an electrical machine is limited by the restrictions imposed by the insulating materials. For the same size of machine if insulating material with large permissible temperature rise is used, the rating of the machine can be increased or more output can be obtained from the machine. Thus by additional cost of insulating material of better quality, more output can be taken from the machine. The economical operation of the machine can be obtained.

**Q. 13 Mention the desired properties of insulating materials to be used in electrical machines.**

**Ans.** The insulating material whether liquid or solid should have the following properties:

(*i*) High dielectric strength.
(*ii*) High insulating strength.
(*iii*) Good heat conductivity.
(*iv*) Less dielectric loss.
(*v*) Effect of moisture should be less.
(*vi*) Should not deteriorate with time even for repeated heat cycle.
(*vii*) It should have sufficient mechanical strength to withstand with vibration and other forces developed in the machine during operation.
(*viii*) Liquid insulating materials should not evaporate or volatilize.
(*ix*) Solid insulating materials should have high melting point.

**Q. 14 Define thermal resistivity used for calculating heat dissipation by conduction.**

**Ans.** Thermal resistivity (ρ) is defined as the thermal resistance of material having surface area of 1 $m^2$ and 1 m length of medium. It is denoted by ohm-meter (Ω-*m*). A material having a longer value of thermal resistivity, dissipates less amount of heat.

**Q. 15 Write different cooling systems for rotating electrical machines.**

**Ans.** According to IS: 4722–1968, the different cooling systems can be classified into:

**(*i*)** ***Natural Cooling:*** The machine is cooled by natural air circulation caused by either rotating parts of the machine or the temperature difference. The machine is cooled without the use of fan.

**(*ii*)** ***Self Cooling:*** The machine is cooled by air circulation by providing a fan on the rotor and hence driven by the machine itself.

**(*iii*)** ***Separate Cooling:*** The machine is cooled by circulation of air by a fan fixed externally and not driven by the machine. It may also be cooled by a cooling medium other than air and the motion in coolant are provided with external arrangements.

**Q. 16 Write the precautions to be taken for hydrogen cooled machines.**

**Ans.** Following precautions should be taken for hydrogen cooled machines:

(*a*) The hydrogen when mixed with air forms an explosive mixture. So the frame of hydrogen cooled machine should be made strong enough to withstand possible internal explosion without suffering serious damage.

(*b*) All joints in cooling circuit of the machine should be gas tight.
(*c*) Oil film shaft seals should be used to prevent leakage of hydrogen.
(*d*) To reduce the risk of explosion in the machine casing the hydrogen is maintained above atmospheric pressure in the machine so that any leakage is always from machine to atmosphere. The hydrogen is used at a pressure of 200–300 kN/m$^2$.
(*e*) When filling and emptying the casing of machine, an explosive hydrogen air mixture must be avoided.

**Q. 17 Define and discuss the heating time constant.**

**Ans.** The heating time constant can be defined as the time taken by the machine to attain 0.632 of its final steady temperature rise.

The heating time constant of the machine is

(*a*) directly proportional to the total heat capacity of the machine.
(*b*) inversely proportional to the total heat dissipation from its cooling surface.
(*c*) inversely proportional to the total heat dissipation. The value of specific heat dissipation is large for well ventilated machines and so their heating time constant is small. The heating time constant is large for machines with poor ventilation. Specific heat dissipation increases with forced cooling, as a result heating time constant decreases with this type of cooling. Heating time constant increases in direct proportion to the linear dimension of the machine. So large size machines have a larger heating time constant.



**Q. 18 Write the various methods adopted to cool a transformer.**

**Ans.** The various methods of cooling generally adopted in transformers are given as below:

(*a*) Air natural cool (AN type).
(*b*) Air blast cooled (AB).
(*c*) Oil immersed, self cooled (ON type).
(*d*) Oil immersed, water cooled (OW).
(*e*) Oil immersed, air blast (OB).
(*f*) Oil immersed, forced oil cooled with air coolers (OFN).

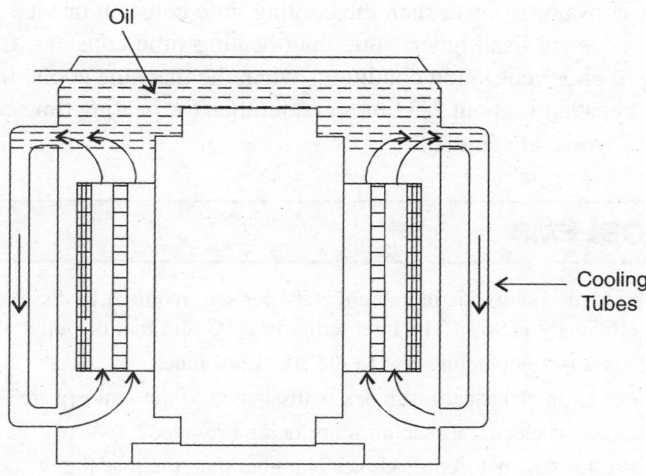


**Fig. A**

(*g*) Oil immersed, forced oil cooled with air blast (OFB).

(*h*) Oil immersed forced oil cooled with forced water cooler (OFW type).

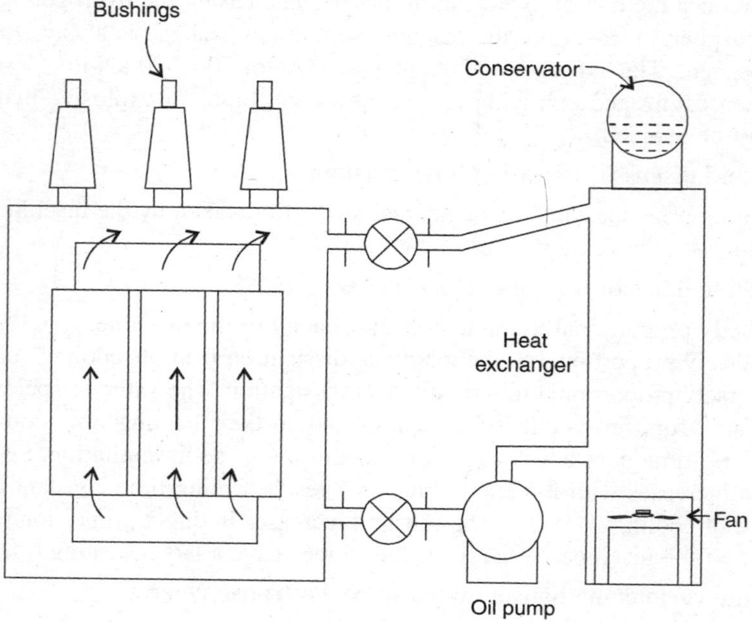


**Fig. B**

Figures *A* and *B* show the method used for cooling arrangements for oil immersed, self cooled (ON) type and oil immersed, forced oil cooled with air blast (OFB) type of transformers respectively. ON type of cooling arrangements are provided for small rated transformers whereas OFB are used for large rated transformers.

**Q. 19. Explain why heating time constant is kept lower than cooling time constant?**

**Ans.** Heating time constant is lower than the cooling time constant or vice versa it can be said that cooling time constant has higher value than heating time constant. Cooling time constant is larger owing to poor ventilation conditions when the machine cools. In self cooled motors the cooling time constant is about 2-3 times greater than the heating time constant because cooling conditions are worse at standstill.

## UNSOLVED PROBLEMS

1. Calculate the volume of cooling air in cubic meters per sec. required to dissipate the losses of a 10 MW generator whose efficiency is 96%. The inlet temp. is 15°C and that of outlet air is 35°C. **(P.U. 1979)**
2. Write the main causes of temperature rise in electrical machines.
3. How the heat produced in electrical machines is dissipated to the atmosphere? Explain.
4. Why enclosure in case of electrical machines are being provided?
5. The temperature rise for the enclosed machines is higher than the machine without enclosure for the same output. Discuss the above statement.

6. Calculate the volume of cooling air in cubic meter per sec. required to dissipate the losses of a 12MW generator whose efficiency is 96%. The inlet temperature is 14°C and that of outlet air is 34°C.
7. Suggest the types of enclosure when an electrical machine is to be installed in damp situations.
8. Write the type of enclosure which is most suitable in dusty atmosphere.
9. For an electrical machine to be used in coal mines, suggest a suitable enclosure for this machine.
10. Mention various methods that are commonly used for the cooling purposes of rotating electrical machines.
11. A 15 MVA transformer has an iron loss of 75 kW and a copper loss of 100 kW at full load. The tank dimensions are 3.5 × 3.0 × 1.5 metres. The transformer oil is cooled by 2.4 litres of water per second through a cooling coil. Estimate the average temperature rise of the tank, if the difference of temperature of water at the inlet and the outlet is 16°C. The specific loss dissipation from the tank walls is 12 W/m$^2$-°C. **(AMIE S 1991)**
12. Explain the nature of the temperatures rise-time curve for an electrical machine.
13. The inner dimensions of the former of field coil of a d.c. generator are 150 × 250 mm$^2$. The former is 2.5 mm thick. Calculate the heat conducted across the tonner from winding to core if there is an air-space of width 1.0 mm between the former and the pole core. The thermal conductivity at former and air is 0.166 and 0.05 W/m-°C respectively. The height of the winding is 200 mm and the temperature rise is 40 °C. **(AMIE S 1991)**
14. A 40 MVA turbo-alternator delivers continuous rated load at 0.6 power factor. The following cooling air measurements are taken:

    In take air temperature ($\theta_1$) = 20°C,

    Outlet air temperature ($\theta_2$) = 50°C,

    Barometric reading = 750 mm of mercury, volume of cooling air per second measured at intake ($V_a$) = 30 m$^3$/sec.

    Find the efficiency of the machine. Assume specific heat of air at constant pressure ($C_p$) = 100 J/kg/°C and the volume (V) of 1 kg of air at 0° and a pressure of 760 mm is 0.78 m$^3$. **(AMIE W 1997)**
15. Find the amount of cooling air required per second at the inlet temperature of 25°C for a 2500 kVA alternator working at full load, the efficiency is 96% and the power-factor is 0.87. The temperature of air coming out of machine is 50°C. Assume suitable values for specific heat and density of air. Air pressure is 760 mm of mercury. **(Punjab University 1972)**
16. Explain continuous rating, short time rating and intermittent rating with reference to electrical machines.
17. Why the arrangements are made to limit the temperature rise in any electrical machine?
18. Explain that for a particular frame size, a larger out put can be derived if the machine is short time rated, than that when continuously rated.

# 5

# Construction, Principle and Design Aspects of 3-phase Alternator

## 5.1 INTRODUCTION

When a single conductor or armature coil rotates in a uniform magnetic field with stationary field poles, an alternating voltage is generated. An alternating voltage will also be generated in stationary conductor or armature coil when the field poles rotates around the conductors. So it can be concluded that if there is a relative motion between the armature conductors and the magnetic field flux, there will be a voltage generated in the armature conductors. The sine wave shape of the voltage generated is obtained whether the field is stationary or rotating.

A.C. generators are generally known as alternators. They are also known as synchronous generators, and are universally employed for the generation of three-phase power. Alternators may have either rotating field poles and stationary armature or rotating armature and stationary field poles.

A three-phase synchronous machine is a doubly excited a.c. machine. The field winding is always energised from a d.c. source. The armature winding of synchronous generators supplies a.c. power to external circuit and synchronous motors import a.c. power. The rotor of the synchronous generator is being rotated with the help of the prime-mover. Under steady-state conditions, speed of synchronous machine depends on the frequency of armature currents and the number of field-poles. So a rotating machine that rotates at a speed according to the supply frequency and the number of poles are called synchronous machine and the speed at which they rotate is called as synchronous speed.

## 5.2 CONSTRUCTION OF ALTERNATOR

An alternator consists of two main parts namely, the armature and field system which are placed on the stator and the rotor. The stator is the stationary part and the rotor is the rotating part of the machine. The arrangement of the two main parts of an alternator can be done in two ways. One way is to place the armature winding in the stationary stator and field system on the rotor Fig. 5.1(*b*). Another way is to place armature winding on the rotor and the field system is placed in the stationary stator Fig. 5.1(*a*).

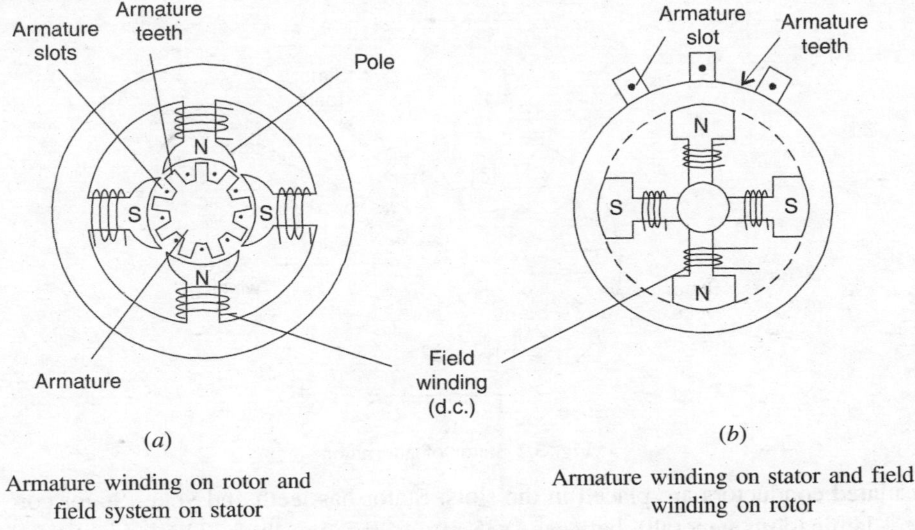


(*a*) Armature winding on rotor and field system on stator

(*b*) Armature winding on stator and field winding on rotor

**Fig. 5.1** Armature and field system.

Normally, the armature winding is placed on the stator and field winding on the rotor. In an alternator the voltage is generated in armature winding and the output is taken from the stator. The rotor is the rotating part of the machine and produces the main field flux.

As mentioned above the armature winding can be placed on the rotor. Rotating field type machines have several advantages over rotating armature type machines:

1. Armature has to be insulated for high voltage (generated voltage may be as high as about 11 kV) for which the alternator is to be designed. A stationary armature can be more easily insulated.
2. The output current can directly be taken from the stator without using slip rings, brushes etc. If rotating armature construction is taken, at least three slip rings would be required. The insulation of high voltage slip rings from shaft will have to be done, which will make rotor more bulky with large centrifugal forces. Larger brushes will be needed for large rated current of armature.
3. D.C. supply is given to field winding. Usually the field voltage is between 100 to 500 volts. Rotating field require only two slip rings to provide direct current. The insulation of these relatively low voltage slip rings from the shaft can be provided easily.
4. The armature winding in the stator can be mechanically braced in better way for high electromagnetic forces due to large short circuit currents.
5. The armature winding in the stator are not subjected to vibration and centrifugal forces. So winding does not get deformed due to operation.
6. Comparatively lighter rotor can be constructed with field winding and can be used for high speed generators.
7. Better cooling can be provided to stationary armature because the armature can be made larger to provide ventilating/cooling ducts. The armature of large alternators are cooled with circulating gas or liquids.

### 5.2.1 Stator Construction

The stator is build up of sheet steel laminations which have slots on their inner periphery as shown in Fig. 5.2.

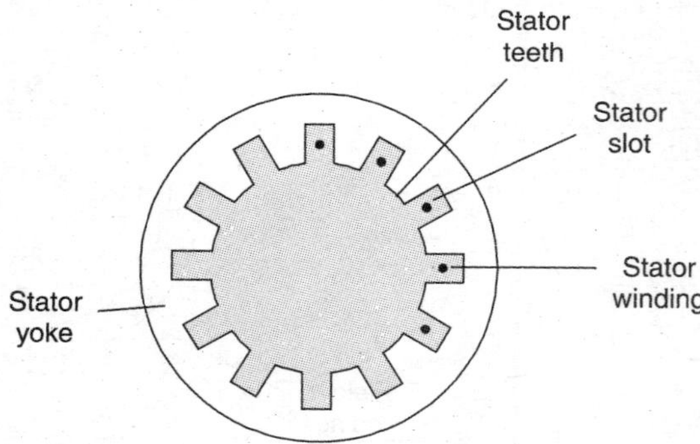


**Fig. 5.2** Stator of alternator.

The insulated conductors are placed in the slots. Stator has teeth and yoke. Stator core is built up of sheet-steel laminations generally between 0.35 mm to 0.5 mm thick. The laminations are punched out carefully to remove shearing and punching strains. They are then coated with an insulating varnish. The insulated punchings are then assembled in the armature frame on keys riveted to the frame or in dove-tailed grooves milled in rivet of the frame. The laminations are generally punched in segments because the stator diameters are usually large for synchronous machines. The number of segments per circle will depend upon the number of slots, the method of punching etc. For large capacity machines, the number of segments per circle should be selected in such a way that the shaft current is zero.

One segment with spot welded tooth support or duct spacers is shown in Fig. 5.3.

The core is clamped between two following rings, which are held in place by a key or by long bolts passing through that frame back of the armature core. The teeth are supported by a finger, or tooth support which extends from the top of the tooth to the inside of the armature core. This tooth support is generally a piece of rolled steel, spot welded to the last lamination.

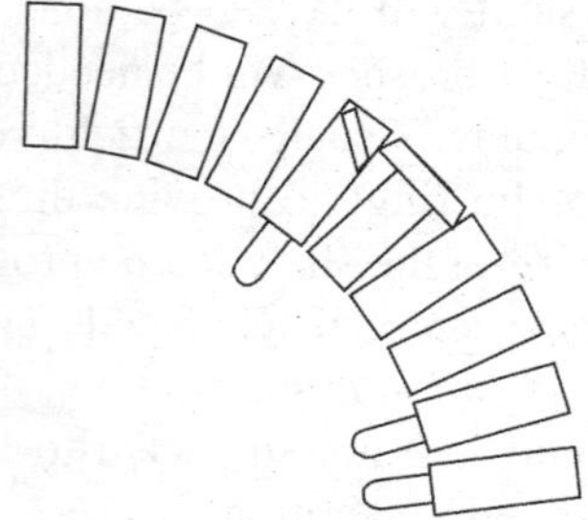

**Fig. 5.3** One segment with spot welded tooth support.

The length of the stator core must be divided into small sections by radial ventilating ducts, to ensure proper ventilation of all parts of the armature. These ducts are usually 1 cm wide for small and medium size machines and 1.25 for large machines. The distance between centres of ducts should be normally 10 cm.

For two layer winding, the armature coils are so-formed that one side may be on the top of one slot and the other side in the bottom of another slot, approximately a pole pitch away. The coils are completely insulated before they are placed into the slots.

For turbogenerators, the armature coil end connections are very long and the coils must be supported to prevent distortion caused by heavy over loads or short-circuits.

For large capacity salient pole machines, the armature coil end connections must be supported in a similar manner.

***Stator Frame:*** The stator frame is made either of cast iron or welded rolled steel. Large ventilating passages are provided in the frame of the stator.

### 5.2.2 Rotor Construction

The rotor is the rotating part of the machine. It produces the main field flux. There are two types of rotor constructions, the cylindrical rotor type and the salient pole rotor type.

***Cylindrical Rotor:*** A cylindrical rotor machine is also called a non-salient pole rotor machine. Cylindrical rotor machine has its rotor so constructed that it forms a smooth cylinder as shown in Fig. 5.4. The construction is such that there are no physical poles such as in the salient pole construction. The d.c. field windings are accommodated in slots in the rotor. The winding is of distributed type. A cylindrical rotor machine has a comparatively small diameter and long axial length. Due to such construction, centrifugal forces are reduced. Thus, they are suitable for high speed machines. There are two or four poles provided in cylindrical rotor alternators. Greater mechanical strength and accurate dynamic balancing can be achieved. The cylindrical rotor machine results in less windage losses and less noisy operation because of uniform airgap. The rotors are manufactured from two pole and four pole cylindrical rotor. Solid steel forging chromium-nickel steel or special chrome-nickel molybdenum steel is used for rotors of large turbo-alternators.

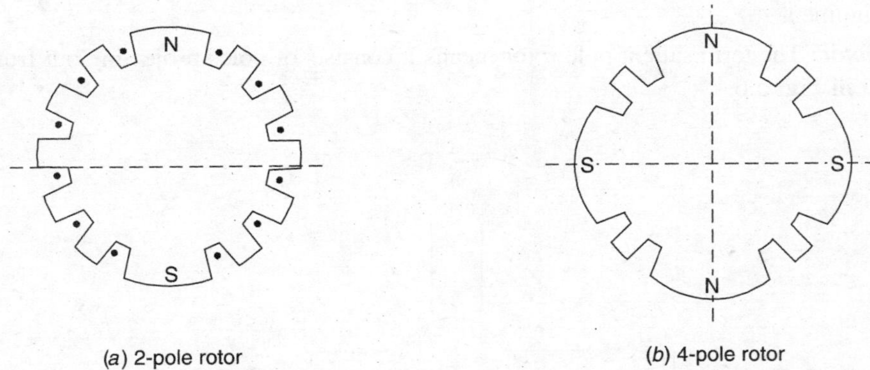


(*a*) 2-pole rotor

(*b*) 4-pole rotor

**Fig. 5.4** Two pole and 4-pole cylindrical rotor.

The rotor of turbo-alternator is slotted to accommodate the field winding.

Normally, two third of the rotor periphery is slotted for accommodating the field winding and one-third is left unslotted to form the pole which is shown in Fig. 5.5. This is done on per pole pitch basis. The total peripheral distance of the rotor is divided into the required number of pole pitches according to the number of poles in the machine. Now, two-third of each pole pitch is slotted to accommodate the field winding and one-third is left unslotted to make pole. Rectangular slots with taper teeth are mode in the rotor and rectangular conductors like flat copper or aluminium strip can be used for the field winding. Due to high speed of the rotor, the overhang of the field winding is secured by steel retaining rings. Forces on the insulation of the field system are quite high so mica, asbestos and other composition insulation materials are normally used in the rotor slots and between the layers of the field winding.

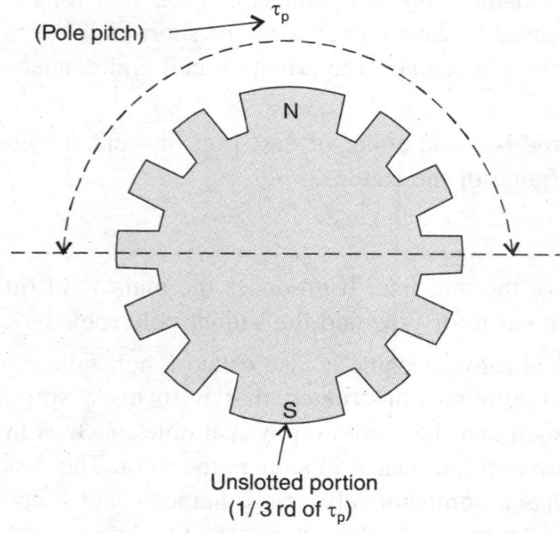


**Fig. 5.5** Pole pitch.

These insulating materials can withstand quite high forces and their temperature limit is also appreciably high and hence they are able to withstand the heat generated in the field core. Heavy non-magnetic metal wedges are placed in the top of the slots to hold the field winding against centrifugal forces. Perfect balancing of the rotor along its length is very essential to obtain the uniform air gap and mechanical alignment.

***Salient Pole Rotor:*** The term salient pole rotor means it consists of poles projecting out from the rotor yoke as shown in Fig. 5.6.

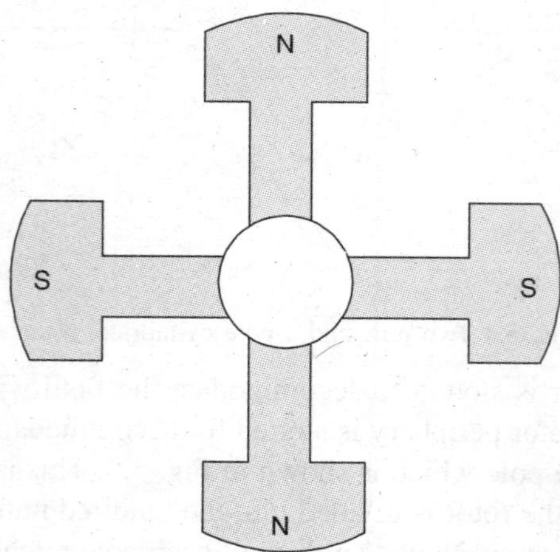


**Fig. 5.6** Salient pole rotor.

The salient poles are made of steel laminations riveted together. The thickness of the sheet used is generally 0.5 mm to 1.0 mm. For low peripheral speeds (upto 30m/second), the poles may be bolted to rotor body but for higher peripheral speed, the poles may be dove-tailed as given in Fig. 5.7.

**Fig. 5.7** Dove-tailed construction.

Figure 5.8 shows a number of pole punchings. The holes near the surface of the pole shoe are for the damper winding. The field coils are placed on the field poles. The individual field pole windings are connected in series to give alternate north and south polarities. The ends of the field windings are connected to a d.c. source through brushes on the slip rings. The slip rings are metal rings mounted on the shaft and insulated from it. They are used to carry current to or from the rotating part of the machine via carbon brushes. Salient pole rotors are used in low medium speed alternators. Salient pole alternators are driven by oil or gap engines and water turbines. A salient pole alternator has comparatively large diameter and a short axial length. The large diameter can accommodate a large number of poles.



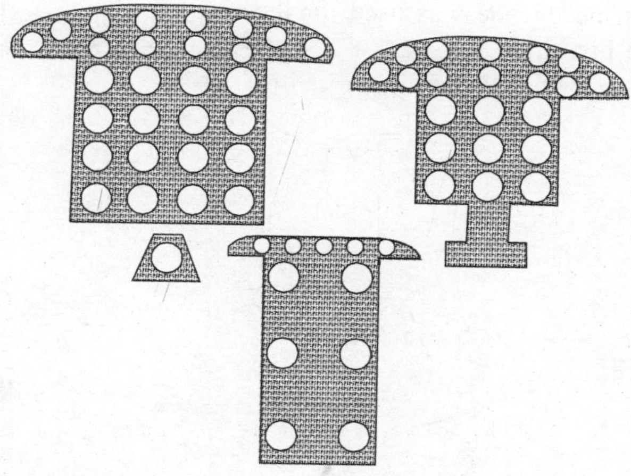

**Fig. 5.8** Field pole punchings.

***Spider:*** The spider on which the field poles are mounted is either of cast iron, cast steel, rolled steel or steel plates. The rim of spider must carry the flux which passes between poles and must therefore have a high permeability besides good mechanical strength. It is difficult to get uniform steel castings, free from flaws, and therefore rolled steel spider rims are preferred. A spider with cast steel hub and arms but with rim built up of steel plate is shown in Fig. 5.9.

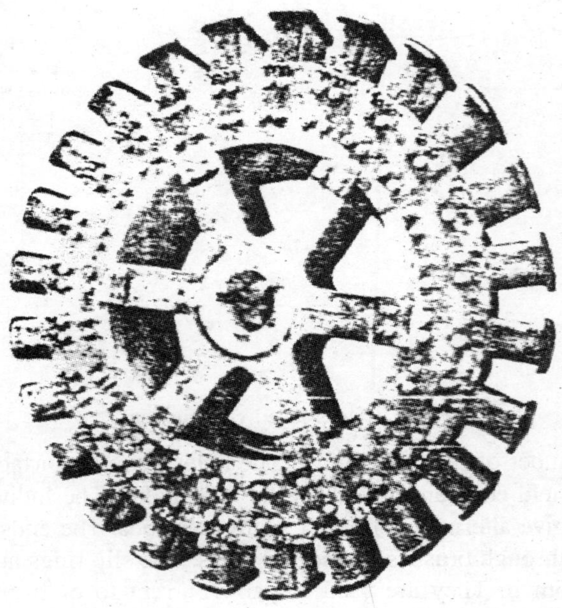

**Fig. 5.9** Rotor spider with laminated rim.

The complete spider for a generator built up of punched steel plates can be made.

For small capacitive generators and motors and with small number of poles, the spider is punched from sheet steel of the same thickness as used for the pole punching. A spider punching and pole punching is shown in the Fig. 5.10.

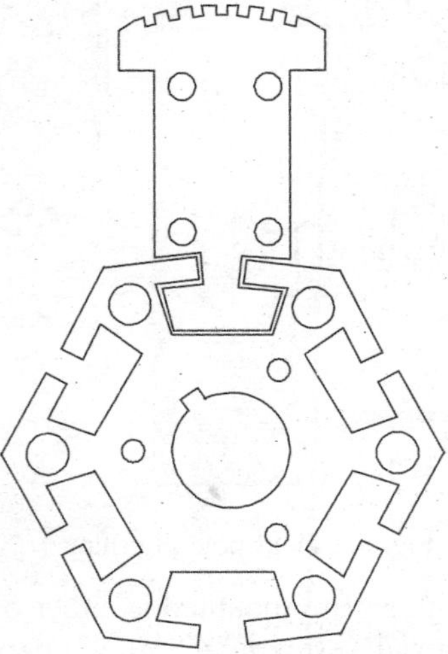

**Fig. 5.10** Spider punching and pole punchings.

## 5.3 MAGNETIC PORTIONS OF THREE-PHASE ALTERNATOR

Let us assume that $\phi$ is the useful flux per pole. Flux emerging out per pole (North pole) are divided in two parts. Each half of the flux ($\phi/2$) produced per pole are being associated to the adjacent opposite poles (South poles). Since reluctance of tooth is quite less as compared to slot so flux tries to pass through the tooth shown in Fig. 5.11. One complete path of the magnetic flux for a pair of pole consists of the following portion of the alternator. The flux path is associated by

(*i*) two length of air gap.
(*ii*) two length of tooth.
(*iii*) one length of stator yoke.
(*iv*) two length of pole shoe.
(*v*) two length of pole core.
(*vi*) one length of pole yoke.

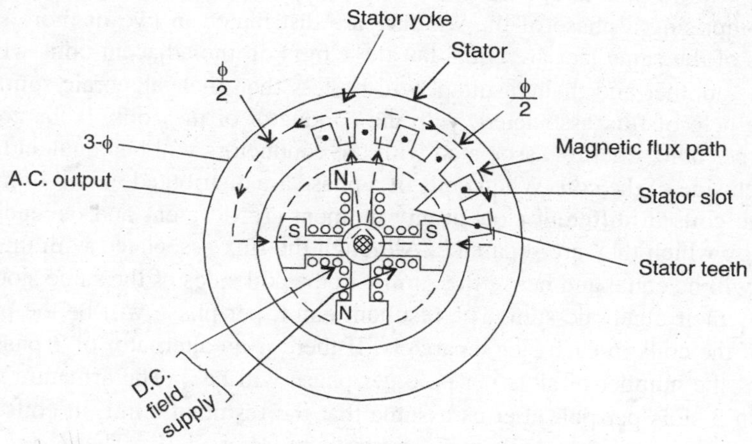


**Fig. 5.11** Magnetic circuit of 4-pole, salient pole alternator.



## 5.4 E.M.F. EQUATION OF AN ALTERNATOR

The r.m.s. value of the generated voltage per phase is given by

$$E_{ph} = 4.44\, f\, \phi_1 N_{ph}$$

where, $f$ = frequency of generated voltage in Hz

$\phi_1$ = the fundamental flux per pole and

$N_{ph}$ = number of turns per phase

The above expression has been derived with the assumptions that

(*i*) Full pitched coils are placed.
(*ii*) All conductors of the phase are concentrated in one stator slot per pole.

But when the winding is distributed in more than one slot the above expression must be multiplied by a factor $k_d$ known as distribution factor (or breadth factor). Since the windings are generally short pitched, the expression must be multiplied by another factor $k_p$ known as pitch factor. Thus the expression for the e.m.f. induced can be written as

$$E_{ph} = 4.44\ k_p k_d\ f\ \phi_1 N_{ph}$$

or

$$E_{ph} = 4.44\ k_w f\ \phi_1 N_{ph}$$

where, $k_w = k_p\ k_d$ and it is called as winding factor.

## 5.5 DISTRIBUTION FACTOR AND PITCH FACTOR

### 5.5.1 Distribution Factor

If the coils of a phase of the winding are placed in one slot per pole, the e.m.f. in the two coil sides are equal and in phase with respect to one another and their resultant will be equal to their algebraic sum. When the coils comprising a phase of the winding are distributed in two or more slots per pole, the e.m.fs. in the coils of the same slot are equal but the e.m.fs. in the adjacent coils will be out of phase with respect to one another and their resultant will be less than their algebraic sum. This can also be explained with the help of flux associated with the conductor of the coils. If the coils of a phase are placed in one slot per pole, the flux associated with the conductors will be equal and equal e.m.fs. will be produced in each side of the coil. When coils of a phase are distributed in two or more slots, the flux associated with the coils in different slots at any moment are different and depends upon the angles between the slots by which they are separated. Although the flux associated with the coils of the same slot in such cases will be equal and hence the e.m.fs. in the coil sides of the same slot will be equal and can be obtained by their algebraic sum. The resultant e.m.f. per phase will be the phasor sum of the resultant e.m.fs. of the coils in each slot separately. If there is an alternator of 3-phase, 4-pole with 36 slots on the stator, the number of slots per pole per phase will be 3. The armature winding per phase has to be placed in 3 slots per pole. Let us assume that the resultant e.m.f. in coils of first slot is $E_1$, resultant e.m.f. in the coils of second slot is $E_2$ and in the third one is $E_3$. These` e.m.fs. of the adjacent slots are displaced by the slot angle (say α degree electrical). The distribution factor, $k_d$, can be defined

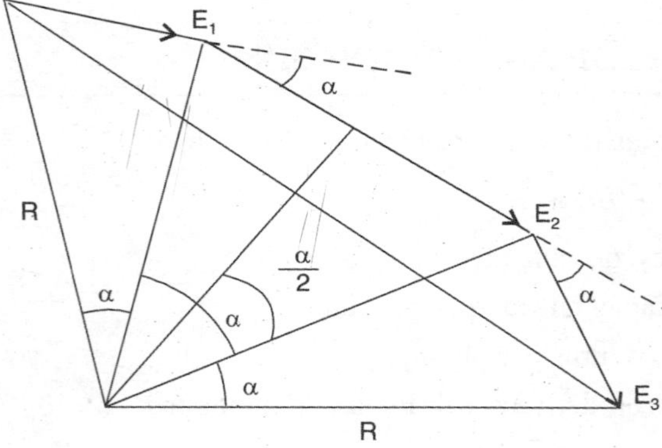


**Fig. 5.12** Calculation of distribution factor.

as the ratio of phasor sum of component e.m.fs. ($E_1$, $E_2$ and $E_3$) to the arithmetic sum of the component e.m.fs. ($E_1$, $E_2$ and $E_3$) as given in Fig. 5.12. Mathematically it can be written as

$$k_d = \frac{\text{Phasor sum of component e.m.fs.}}{\text{Arithmetic sum of component e.m.fs.}}$$

$$k_d = \frac{2\,R\sin(m\alpha/2)}{m\cdot 2R\sin\alpha/2}$$

$$= \frac{\sin(m\alpha/2)}{m\cdot\sin\alpha/2}$$

where, $m$ is the number of slots pole per phase.

### 5.5.2 Pitch Factor

In a full pitch coil, the e.m.fs. in the two coil sides are in phase and therefore, the coil e.m.f. is twice the e.m.f. of each coil side. In a short pitch coil, the e.m.fs. of the two coil sides are not in phase so the resultant e.m.f. (coil e.m.f.) can be obtained by phasor sum of e.m.fs. of each coil side. The pitch factor can be defined as the ratio of phasor sum of coil side e.m.fs. to the arithmetic sum of coil side e.m.fs.. Pitch factor or coil span factor can be written as

$$k_p = \frac{\text{Phasor sum of coil side e.m.fs.}}{\text{Arithmetic sum of coil side e.m.fs.}}$$

Chapter 5

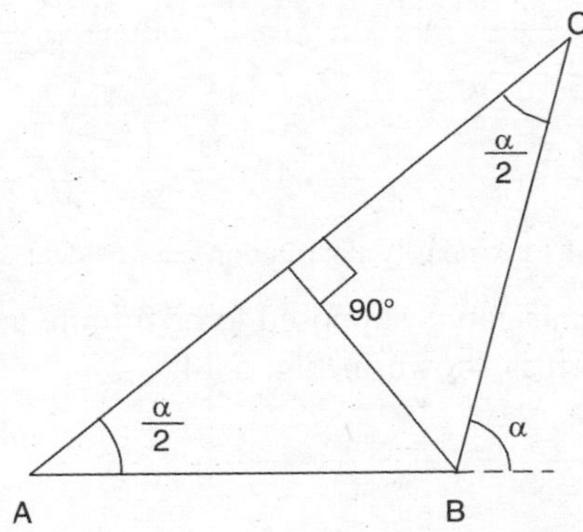


**Fig. 5.13** Calculation of pitch factor.

Figure 5.13 shows the coil side e.m.fs. $AB$ and $BC$ and the resultant coil e.m.f. $AC$ when the coil pitch is short of full pitch by electrical angle α.

Phasor sum of coil side e.m.fs. $= AC = 2\,AB\,\cos\frac{\alpha}{2}$

Arithmetic sum of coil side e.m.fs. $= 2\,AB$

$$k_p = \frac{2\,AB\cos\alpha/2}{2\,AB} = \cos\alpha/2$$

This equation gives the pitch factor for fundamental component of e.m.f.

If the flux density distribution contains harmonics, the pitch factor for the *n*th harmonics is given by

$$k_p = \cos\frac{n\alpha}{2}$$

If $\alpha$ is so much that $\cos\frac{n\alpha}{2} = 0$ or $\frac{n\alpha}{2} = 90°$, the *n*th harmonics will be reduced to zero. The windings could be designed in such a way that the specified harmonics will not be generated (if $\alpha = 60°$, no third harmonic generation). Then fractional pitch windings results in a voltage waveform which resembles a sinusoidal to a better degree than that in a full pitch winding.

## 5.6 FLUX DENSITY DISTRIBUTION CURVE: SALIENT POLE MACHINE

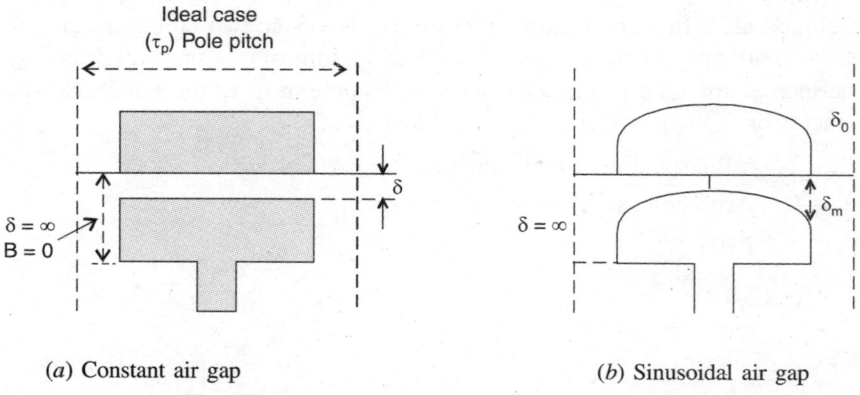


(*a*) Constant air gap (*b*) Sinusoidal air gap

**Fig. 5.14** Flux density distribution curve (Ideal case).

Ideally beyond the pole surface the air gap is supposed to be infinite and hence there should be no flux density beyond the pole surface as it is shown in Fig. 5.14.

$$\text{Flux } \phi = \frac{MMF}{\text{Reluctance}}$$

$$= \frac{MMF}{l/\mu_0 A}$$

$$= \frac{MMF \cdot \mu_0 A}{l}$$

$$\text{Flux density, } B = \frac{\phi}{A} = \frac{MMF \cdot \mu_0}{l}$$

$$B = \frac{\mu_0 \cdot MMF}{\text{Length of air gap}} \qquad ...(5.1)$$

When permeability of iron is supposed to be infinite, no m.m.f. will be required for iron portion whole of m.m.f. is required for air gap of the machine.

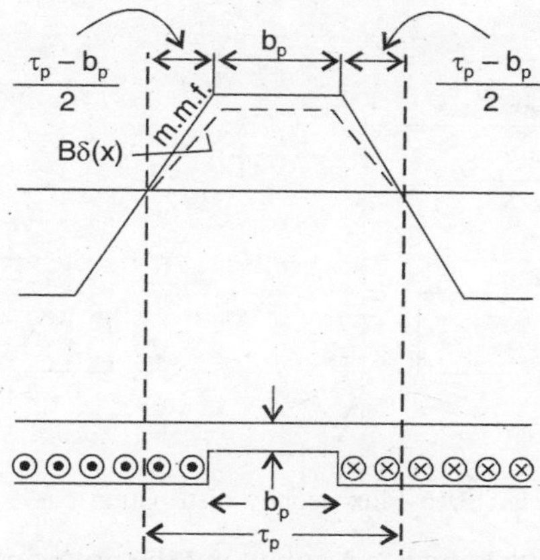


(*a*) m.m.f. and field distribution curve of a cylindrical rotor machine.

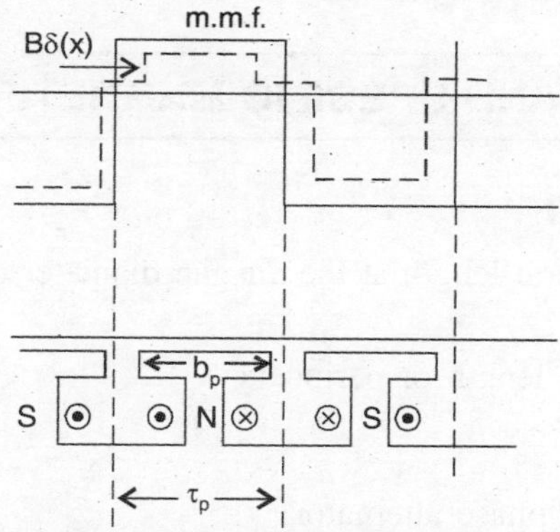


(*b*) m.m.f. and field distribution curve in the case of salient pole synch. machine having cont. air gap.

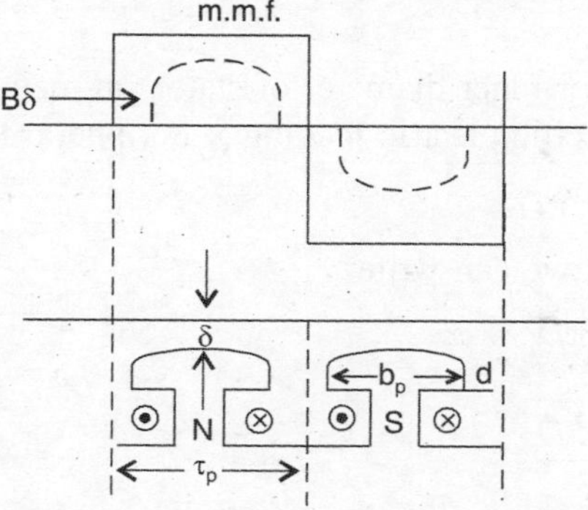


(*c*) m.m.f. and field distribution curve in salient pole synch. machine having sinusoidal air gap.

**Fig. 5.15**

Chapter 5

It can be observed from equation (5.1) that when length of air gap is constant, the flux density is constant. When air gap length varies sinusoidally then as the air gap decreases, the flux density increases and as the air gap increases flux density decreases as shown in Fig. 5.15 for different types of machines.

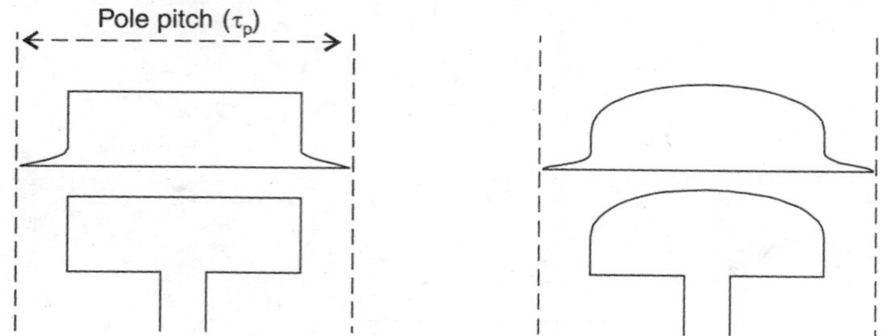


**Fig. 5.16** Flux density distribution curve.

In actual practice the flux do not vanishes just beyond the pole surface. It becomes zero at interpolar point which can be shown in the Fig. 5.16.

## 5.7 SPECIFIC ELECTRIC AND SPECIFIC MAGNETIC LOADINGS

### 5.7.1 Specific Electric Loading

Ampere-conductor per unit peripheral length at the air gap diameter of the machine is called specific electric loading.

$\overline{ac}$ = ampere-conductor/metre length of periphery

Let phase current = $I_{ph}$ ampere

Turns/phase = $N_{ph}$

Total ampere-conductor of a 3-phase alternator

$$= 2 \times 3 N_{ph} I_{ph}$$

$$= 6 N_{ph} I_{ph} \qquad \text{...(5.2)}$$

Let $D$ is the air gap diameter (or inner diameter of stator) in metres.

Total ampere-conductor = Specific electric loading × Peripheral length of the alternator

$$= \overline{ac} \times \pi D \qquad \text{...(5.3)}$$

With equations (5.2) and (5.3), we can write

$$6 N_{ph} \cdot I_{ph} = \overline{ac} \times \pi D$$

or

$$N_{ph} \cdot I_{ph} = \frac{\overline{ac} \times \pi D}{6}$$

Specific electric loading for water wheel alternator is given as

$\overline{ac}$ = 20,000 – 40,000 ampere-conductor per metre of periphery.

Maximum permissible value is 50,000 ampere-conductor/m.

For Turbo-alternator, specific electric loading is given as

$\overline{ac}$ = 50,000 – 75,000 ampere-conductor/metre of periphery.

***Selection of Specific Electric Loading:*** Factors which influence the choice of specific electric loading are being discussed below:

1. *Synchronous reactance:* When ampere-conductor increases the value of leakage reactance and armature reaction increases and as a result a high value of synchronous reactance is there.
2. *Voltage regulation:* A machine with high ampere-conductor means high synchronous reactance and hence it will have a poor inherent voltage regulation.
3. *Short-circuit current:* If $\overline{ac}$ is high, there will be low current under short-circuit condition due to higher synchronous reactance. Many large rating alternators specially turbo-alternators are designed with a high value of $\overline{ac}$ so that short-circuit current can be minimised and alternators can withstand momentary short-circuits without any damage.
4. *Stability limit:* More $\overline{ac}$ will lead to small synchronizing power, low value of steady state stability limit and hence machine is less stable.
5. *Copper loss, efficiency and temperature rise:* Higher copper loss, low efficiency and high temperature rise is observed for machines with high $\overline{ac}$. High value of $\overline{ac}$ can be used in the machines in which better cooling techniques are employed.
6. *Voltage:* Higher value of $\overline{ac}$ can be used for low voltage machines. Since less space is required by the insulation.
7. *Parallel operation:* Nowadays, almost all alternators are connected in parallel with other alternators. The satisfactory parallel operation of alternators depends upon the synchronizing power. If the synchronizing power is higher, the higher is the capability of the system to synchronism. Synchronizing power is inversely proportional to synchronous reactance. Machine designed with high $\overline{ac}$ will not work satisfactorily in parallel, since high $\overline{ac}$ results in high synchronous reactance and hence less synchronizing power.

### 5.7.2 Magnetic Loading: Specific Magnetic Loading $\overline{B}$ or $B_{\delta_1}$

$\overline{B}$ = average value in tesla.

$B_{\delta_1} = \frac{\pi}{2}\overline{B}$, peak value of the fundamental in tesla.

$\phi_1$ = fundamental flux/pole in weber.

$\phi$ = useful flux/pole in weber.

The flux density in different part of stator of alternator and in different portions of salient pole rotor as given in Fig. 5.17 are given below.

1. Flux density in teeth = up to 1.7 tesla

   If the flux density is kept low, the uneconomic use of machine portions are observed and if it is increased very much (above 1.7 tesla), magnetic-saturation will occur in stator teeth.
2. $B_{\delta_1}$ in air gap = 0.8 to 1.0 T for water wheel alternators

   = 0.85 to 1.0 T for turbo alternators.

Chapter 5

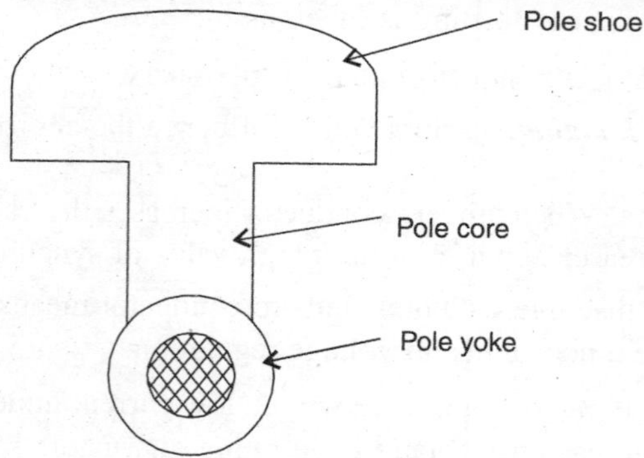


**Fig. 5.17** Different portion of salient pole rotor.

3. $\bar{B}$ in stator yoke or core = 1.3 to 1.5 T

   $\bar{B}$ in pole shoe = 1.8 T

   $\bar{B}$ in pole core or shaft = 1.5 to 1.7 T

   $\bar{B}$ in pole yoke = 1.3 to 1.5 T

***Selection of Specific Magnetic Loading:*** Some of the factors which influence the selection of specific magnetic loading are being given below:

1. *Iron loss, efficiency and temperature rise:* High value of flux density in air gap results in a high value of flux density in stator teeth and yoke giving high iron loss, decrease in efficiency and increase in temperature rise.
2. *Stability:* If flux density is high, the flux per pole is large and therefore a smaller number of turns are required for the armature winding. This results in reduction in the value of synchronous reactance. Therefore high gap density will improve the steady state limit.
3. *Parallel operation:* The higher is the value of gap density the higher is the synchronizing power and hence machines with high gap density will operate satisfactorily in parallel.
4. *Voltage:* For high voltage machines, the space occupied by insulation becomes more and space left for teeth is less. So, a lower value of gap density should be used in high voltage machines.
5. *Short-circuit current:* High gap density results in decrease in the leakage reactance of the machine. Due to this, more current under short-circuit condition will be obtained. Due to high gap density, number of turns required will be less and hence less leakage reactance and less synchronous reactance.

## 5.8 OUTPUT EQUATION AND ITS DERIVATION

### 5.8.1 Output Equation

The output equation is

$$Q = C D^2 L N$$

where, $Q$ = output/rating of machine in kVA

$D$ = inner diameter or air-gap diameter of stator

$L$ = length of machine in metre

$N$ = speed in r.p.m.

$C$ = output coefficient

$= 11.1 \times 10^{-5}\ B_{\delta_1} \cdot \overline{ac}$

$B_{\delta_1}$ = magnetic loading

$\overline{ac}$ = specific electric loading,
ampere-conductor per unit length ($m$) of periphery

### 5.8.2 Derivation of Output Equation

The fundamental expression for the rating of machine ($Q$) can be given as

$$Q = 3V_{ph}\, I_{ph} \times 10^{-3}\ \text{kVA} \qquad \text{(for a 3-phase machine)} \qquad ...(5.4)$$

The basic equation for the phase voltage $V_{ph}$ is

$$V_{ph} = 4.44\ k_w f\, \phi_1 N_{ph}$$

So, $$Q = 3(4.44\, k_w f\, \phi_1 N_{ph}) \times I_{ph} \times 10^{-3}\ \text{kVA} \qquad ...(5.5)$$

If, $P$ is the number of poles and $f$ is the supply frequency then

$$f = \frac{PN}{120}$$

Let winding factor $k_w$ = 0.955 for fractional pitch winding.
Also, $\phi_1$ = fundamental flux/pole

$$= \frac{2}{\pi} B_{\delta_1} \tau_p L$$

where, $\tau_p$ is the pole pitch $\left(= \dfrac{\pi D}{P}\right)$

It is known that $6N_{ph}I_{ph} = \overline{ac} \cdot \pi D$ [From equations (5.2) and (5.3)]

$$N_{ph} \cdot I_{ph} = \frac{\overline{ac} \cdot \pi D}{6}$$

Putting the values of $f$, $k_w$, $\phi_1$, $N_{ph}\, I_{ph}$ in equation (5.5)

$$Q = 3\left[4.44 \times 0.955 \times \frac{PN}{120} \times \frac{2}{\pi} B_{\delta_1} \times \frac{\pi D}{P} \times L \times \frac{\overline{ac} \times \pi D}{6}\right] \times 10^{-3}$$

$$= \frac{4.44 \times 10^{-3}}{120} \times 0.955 \times \pi D^2 L\, \overline{ac}\, B_{\delta_1} N$$

Chapter 5

$$= 11.1 \times 10^{-5} B_{\delta_1} \times \overline{ac}\, D^2 LN$$

$$Q = CD^2 LN \text{ kilovolt ampere} \qquad ...(5.6)$$

This equation is known as output equation. $D$ and $L$ are called as the main dimensions of the machine.

From equation (5.6), we get,

$$D^2 L = \frac{Q}{CN} \text{ (i.e., the volume of the machine).}$$

Then for a specified output, low speed machine becomes costlier since the volume increases. Usual value of output coefficient ($C$) varies from 1.77 to 8.2.

$$C_{min} = 11.1 \times 10^{-5} \times 0.8 \times 20{,}000 = 1.77$$

$$C_{max} = 11.1 \times 10^{-5} \times 1.0 \times 75{,}000 = 8.2$$

## 5.9 ESTIMATION OF MAIN DIMENSION (D AND L)

### 5.9.1 Salient Pole Alternator (Water Wheel Generator)

There are generally two types of construction of salient pole rotor, namely rectangular poles and round poles which is shown in Fig. 5.18. The round core and square shoe are being used for round pole construction.

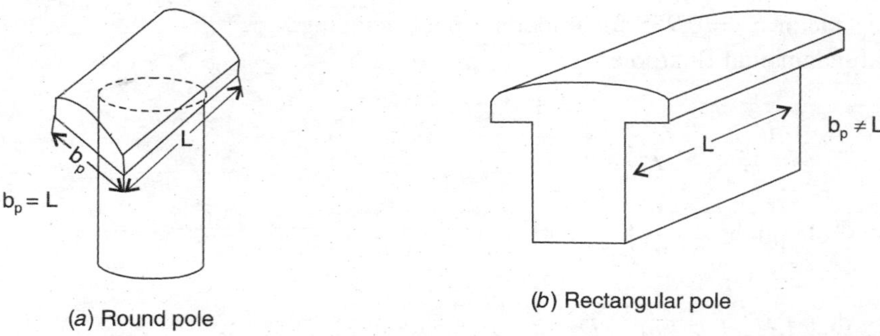


**Fig. 5.18** Round and rectangular poles.

There are several advantages in using round poles as compared to rectangular poles.

(*i*) Since the pole core is round, the length of mean turn of the field winding is smaller, resulting in saving of conductor material and reduced losses in field winding.
(*ii*) Ventilation is better because the field winding is uniformly exposed to air.
(*iii*) Manufacturing of round poles are simple and cheaper.
(*iv*) Since circular field coils are used for round poles so the coils do not get deformed due to centrifugal force while running.

The ratio of pole arc to pole pitch for salient pole machine may be given or it can be selected for round pole machine, as below.

$$\frac{\text{Pole arc}}{\text{Pole pitch}} = \frac{b_p}{\tau_p} = 0.6 \text{ to } 0.7 \qquad \text{...(5.7)}$$

$b_p$ is the pole arc length of the pole shoe and $\tau_p$ is the pole pitch of the machine. It can be represented as in Fig. 5.19. If $\frac{b_p}{\tau_p}$ ratio is selected between 0.6 and 0.7, it means pole arc occupies 60% to 70% space of pole pitch.

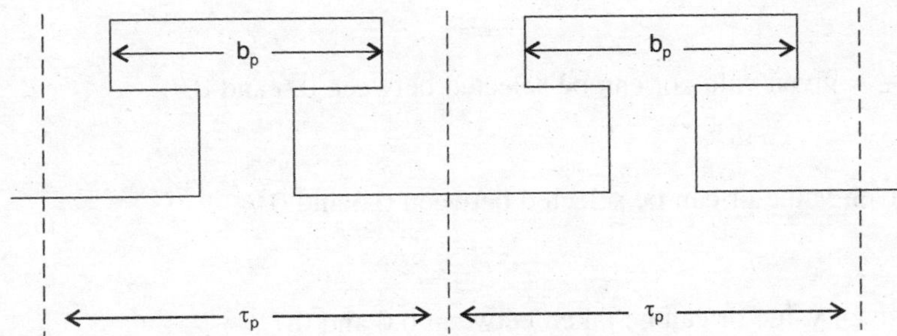


**Fig. 5.19** Representation of pole arc and pole pitch.

If ratio of pole arc to pole pitch is very small (say 0.35), the useful flux in such case will be very small and if the ratio is very high (say 0.91) leakage of flux will be very high. So the optimum value of the ratio is taken and that is lying between 0.6 and 0.7. For very precise design, the ratio can be selected as 0.66 or 0.67.

In case of a round pole machine, since the pole core is round and pole shoe is square and the pole arc may be taken equal to the axial length of the stator, so

$$\frac{L}{\tau_p} = \frac{b_p}{\tau_p} = 0.6 \text{ to } 0.7$$

or

$$\frac{L}{\tau_p} = 0.6 \text{ to } 0.7 \qquad \text{...(5.8)}$$

In case of a rectangular pole, the pole arc length ($b_p$) is different than axial length of machine and $L/\tau_p$ is considered separately and any value can be selected from 1 to 4.

$$\frac{L}{\tau_p} = 1 \text{ to } 4 \qquad \text{...(5.9)}$$

$\frac{b_p}{\tau_p}$ ratio will also exist separately for rectangular pole machine. It may be given in the problem otherwise it can be selected between 0.6 and 0.7.

Chapter 5

### 5.9.2 Steps for Estimation of *D* and *L* of Salient Pole Alternator

If specific type of pole construction is not mentioned the design with round pole construction is usually stated.

*(a) Round pole construction*

(1) $\dfrac{b_p}{\tau_p}$ ratio may be given otherwise select it between 0.6 and 0.7.

(2) $\dfrac{L}{\tau_p} = \dfrac{b_p}{\tau_p}$ for round pole construction.

(3) $\dfrac{L}{\tau_p} = \dfrac{b_p}{\tau_p}$ = given value or can be selected between 0.6 and 0.7.

$\dfrac{L}{\tau_p}$ = given value or can be selected between 0.6 and 0.7.

(4) $\dfrac{L}{\dfrac{\pi D}{p}}$ = given value or can be taken between 0.6 and 0.7. $\left[\because \tau_p = \dfrac{\pi D}{P}\right]$

*i.e.,* $\dfrac{LP}{\pi D}$ = some value ...(5.10)

(5) The output, r.p.m. of the machine may be given or it can be estimated and a suitable value of output coefficient (*C*) can be calculated by selecting $B_{\delta 1}$ and $\overline{ac}$ between specified range for specific machine. Once *Q*, *C* and *N* are known, we have

$D^2L$ = some value ...(5.11)

(6) With the help of equations (5.10) and (5.11), diameter $D_1$ and length *L* of the machine can be calculated.

(7) The feasibility of inner diameter of stator calculated above must be checked. It can be checked with the help of peripheral speed.

Peripheral speed, $V_p = \dfrac{\pi D N}{60}$ metre per second ...(5.12)

Peripheral speed for water wheel generator should be taken as

$V_p$ = 30 to 80 m/sec. (for water wheel generator)

= around 140 m/sec. (for machine with special construction)

Calculate peripheral speed, with equation (5.12). If $V_p$ is in the given range, further design steps can be considered with this diameter. If $V_p$ is not in the range, modify $\overline{ac}$ and $B_{\delta 1}$, in a suitable step and again calculate inner diameter and length of the machine and check for peripheral velocity. If $V_p$ is in the range, the design of machine can be continued with these values of *D* and *L*. If $V_p$ is not again in the range modify $\overline{ac}$ and $B_{\delta 1}$, again and again to get the value of peripheral speed in the range.

**If even by selecting the limiting value of $\overline{ac}$ and $B_{\delta 1}$, peripheral speed is not in the range for round pole construction. The diameter and length can be calculated by considering the rectangular pole construction.**

### *(b) Rectangular pole construction*

(1) A suitable value of ratio of length to pole pitch can be assumed between 1 and 4.

$$\frac{L}{\tau_p} = 1 \text{ to } 4$$

$$\text{or,} \quad \frac{L}{\pi D/P} = \text{some value} \qquad \left\{\tau_p = \frac{\pi D}{P}\right\}$$

$$\text{or,} \quad \frac{L \cdot P}{\pi D} = \text{some value} \qquad \ldots(5.13)$$

Then the number of poles are known.

(2) With the help of equations (5.11) and (5.13), $D$ and $L$ can be calculated.

(3) Check the feasibility of diameter by calculating peripheral speed. If $V_p$ is not in the range, $\overline{ac}$, $B_\delta$ and $\frac{L}{\tau_p}$ ratio can be modified to get $V_p$ in range and finally estimate $D$ and $L$.

The advantage of round pole construction compared with rectangular pole has been discussed in previous section. But the diameter is more for round pole construction for same rating of the machine and keeping other parameters equal.

This can be explained with the help of output equation as

$$Q = CD^2 LN$$

$$\text{or,} \quad D^2 = \frac{Q}{CLN}$$

$$= \frac{Q}{C\frac{L}{\tau_p} N\tau_p}$$

$$= \frac{Q}{C\frac{L}{\tau_p} N\frac{\pi D}{P}}$$

$$\text{or,} \quad D^3 = \frac{QP}{C\frac{L}{\tau_p} N\pi} \qquad \ldots(5.14)$$

In case of Round pole, $\frac{L}{\tau_p} = \frac{b_p}{\tau_p} = 0.6 \text{ to } 0.7$ and for rectangular pole machine, $\frac{L}{\tau_p} = 1 \text{ to } 4$, so, it can be concluded that since $\frac{L}{\tau_p}$ for round pole is always less than that for rectangular pole hence the diameter from equation (5.14) in case of round pole will always be larger as compared to rectangular pole.

### 5.9.3 Main Dimension of Turbo-Alternator

Peripheral speed of turbo-alternator,

$$V_p = 120 \text{ to } 150 \text{ m/sec.}$$

The maximum value of peripheral speed can be taken 175 m.sec. with special construction.

Internal diameter of the stator is quite less as compared to the axial length of the stator. Turbo-alternators are generally designed for high speed (say 3000 r.p.m.). The diameter of the rotor is nearly equal to the inner diameter of stator. Due to centrifugal force, diameter is limited with the consideration of permissible value of peripheral speed. Hence, the internal diameter of the stator is normally calculated on the basis of peripheral speed.

$$\frac{\pi DN}{60} = 120 \text{ to } 150$$

with the help of above equation, diameter $D$ can be calculated.

Taking the maximum value of peripheral speed for a 3000 r.p.m. machine, we get

$$D = \frac{60 \times 175}{\pi \times 3000}$$

$$= \frac{3.5}{\pi}$$

$$\approx 1.1 \text{ m.}$$

For a peripheral speed of 175 m/sec, and with rotor speed of 3000 r.p.m., the rotor diameter can be taken 1.1 m. Thus with limited rotor diameter, the only way to increase the rating of turbogenerator is to increase its gross length. Turbogenerators are characterised by short diameter and long length. As the length of alternator is increased, the mechanical alignment becomes difficult and uniform air gap may not be obtained and hence due to high speed machine may get damaged. Usually, maximum length of turbogenerator is around 5 m.

## 5.10 RUNAWAY SPEED

The speed at which the prime-mover of the generator will run if the total load is suddenly thrown off when generator is working at its rated load is called as runaway speed.

The salient pole alternators are slow speed machines, speed varying from 150 to 600 r.p.m. These machines are driven by hydraulic turbine and are also called as water wheel generators or hydrogenerators. The speed of these machines is mainly dependent upon the water head available, the higher the water head, higher will be the speed. Normally, the hydraulic turbines used depending upon water head available are given below.

(*i*) Pelton wheel — For high water heads, 400 m and above.

(*ii*) Francis turbine — For medium water heads, up to 350 m.
(*iii*) Kaplan turbine — For low water heads, up to 50 m.

Generally, the runaway speed of water wheel turbines are given as —

Pelton wheel — 1.8 times the rated speed.
Francis turbine — 2 to 2.2 times the rated speed.
Kaplan turbine — 2.5 to 2.8 times the rated speed.

Runaway speed of turbogenerators are 1.25 times the rated speed. It is less in turbo-alternators due to fast governor action.

The peripheral speed of hydroalternators at runaway speed will be large compared to that at normal speed. The water wheel generators should be designed to withstand mechanical stresses at runaway speed.

The construction of salient pole alternators are different to as compared to turbo-alternators. The main difference in construction and the design of two types of alternators is due to much difference in speed of rotating parts. The number of poles are large in salient pole alternators, which indirectly results into large diameter of rotor. The diameter of large rating generators may be as high as 15 m. The axial length of alternator is much smaller than the stator diameter. The water wheel generators are characterized by large diameter and short axial length.

## 5.11 EFFECTIVE LENGTH OF ALTERNATOR ($L_e$)

The length of the stator must be provided with small sections of radial ventilating ducts to insure proper ventilation of different parts of the stator. These ducts are normally 1 cm wide for small and medium size machines and 1.25 cm for large machines. The distance between centres of two adjacent ducts should be 10 cm.

The stator is being laminated. The total length of the machine include the length of different section and ventilating ducts as given in Fig. 5.20.

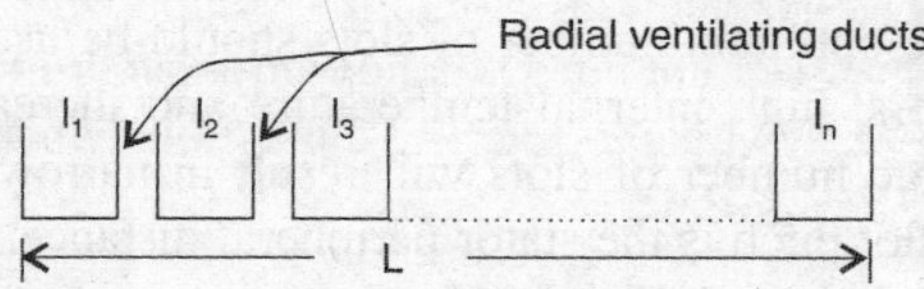


**Fig. 5.20** Radial ventilating ducts.

Let $l_1, l_2, l_3, ..., l_n$ are gross iron section in addition to ventilating ducts and $b_v$ is width of ventilating ducts and $n_v$ is the number of ventilating ducts.

So, total gross iron length

$$l = l_1 + l_2 + l_3 + ... + l_n$$

But actual iron length = $l_i = k_i\, l$

where, $k_i$ = stacking factor

= 0.9 to 0.92 (for 0.5 mm lamination)

= 0.97 to 0.98 (for 1.0 mm lamination)

Chapter 5

L = over all length

$= l + n_v b_v$

The effective length is given by the expression

$$L_e = L - n_v b'_v$$

where, $b'_v$ is the effective width of the ventilating ducts.

Due to effect of fringing of flux, the effective width of the ventilating ducts is different than the physical width of the ducts.

$$b'_v = b_v \frac{5}{5 + \frac{b_v}{\delta}}$$

where δ is air gap length of the machine.

## 5.12 LENGTH OF AIR GAP (δ)

The length of air gap can be calculated for water wheel and turbogenerator from following expression:

$$\frac{\delta}{\tau_p} = 0.01 \text{ to } 0.015 \text{ (for water wheel generator)}$$

$$= 0.02 \text{ to } 0.025 \text{ (for turbogenerator).}$$

## 5.13 SELECTION OF STATOR SLOTS

Some factors are considered for selection of stator slots. The number of slots should be so situated that 3-phase balanced winding is obtained. Since unbalanced winding will lead to generation of space harmonics and overheating. The optimum number of slots should be used. Small number of slots will result in crowding of conductors, high internal temperature and increased leakage flux (reactance). Since conductors are more, large number of slots will result in narrow teeth and hence flux density beyond permissible limit. The slot pitch is the stator peripheral distance from the centre of a stator slot to the centre of adjacent slot as shown in Fig. 5.21.

The normal value of slot pitch depending upon voltage.

$\tau_s$ = stator slot pitch

$\tau_s \leq 25$ mm for low voltage machine up to 1 kV

$\tau_s \leq 40$ mm for voltage up to 6.6 kV

$\leq 60$ mm for voltage up to 15 kV

Sometimes stator slot pitch for large water wheel generator are taken up to 90 mm.

Total number of stator slots

$$S_1 = \frac{\pi D}{\tau_s}$$

Slots per pole per phase, $q_1 = \frac{S_1}{3P}$ for 3-phase machine

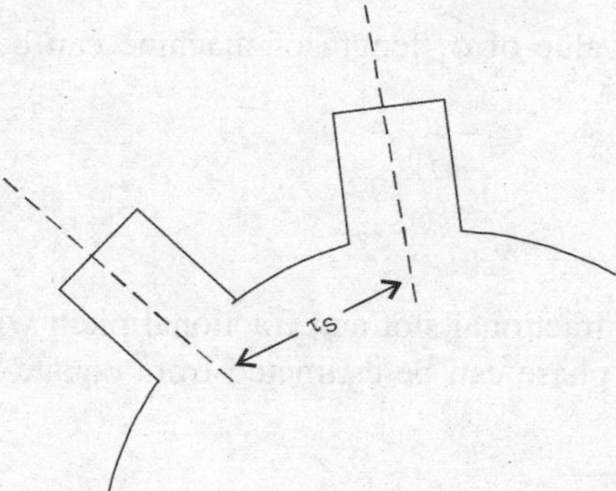

**Fig. 5.21** Slot pitch.

For water wheel generator, the number of slots per pole per phase may be from 2 to 5. The number of slots per pole per phase for turbo-alternator may be much higher (*i.e.*, from 5 to 9) than the salient pole alternator. To reduce the magnitude of higher harmonics in the e.m.f. wave the number of slots per pole per phase may also be taken as fractional number (say $3\frac{1}{2}$). However the total number of slots must satisfy that the slot loading (ampere-conductor per slot) should not be normally exceed 1500.

When the windings have fractional pitch and fractional slot in actual practice. Then we say

$$q_1 = 3\frac{1}{2} = \frac{7}{2} = \frac{x}{y}$$

To have a balanced synchronous winding it is essential that the number of pole pairs must be divisible by *y*.

$$\frac{p}{2} = \text{Number of pole pairs}$$

So,
$$\frac{p}{2}\Big/ y = \text{An integer}$$

## 5.14 ESTIMATION OF TURNS PER PHASE, TOTAL CONDUCTORS, CONDUCTORS PER SLOT AND THEIR CORRECTED VALUES

Let, $Z_{ph}$ = conductor per phase

$Z_T$ = total stator conductor

$N_c$ = conductors per slot

From fundamental equation of phase voltage, we know that

$$V_{ph} = 4.44\, k_w\, \phi_1 f\, N_{ph} \text{ volt} \qquad ...(5.15)$$

where, $$\phi_1 = \frac{2}{\pi} B_{\delta_1} \tau_p \, L_e \qquad ...(5.16)$$

[For estimation of approximate value of $\phi_1$ length of machine can be considered inspite of effective lenth of machine]

Also, $$\tau_p = \frac{\pi D}{P} \qquad ...(5.17)$$

where, $k_w$ = winding factor

= 0.955 (say) for fractional slot and fractional pitch winding.

So, the number of turns per phase can be estimated from equation (5.18) as

$$N_{ph} = \frac{V_{ph}}{4.44 \, k_w \phi_1 f} \qquad ...(5.18)$$

Also, number of conductors per phase, $Z_{ph} = 2N_{ph}$

Total numbers of conductors for a 3-phase machine

$$Z_T = 3 \times 2 \, N_{ph}$$

$$Z_T = 6 \, N_{ph}$$

Conductors per slot, $$N_c = \frac{Z_T}{S_1 \, (\text{Total number of slots})}$$

Conductors per slot may or may not be an integer. We can have only an integer number of conductors in each slot. For a 2-layer winding, the number of conductors in each slot must be even number. Select a suitable integer for $N_c$ and this will give the corrected value of $N_c$ (corrected).

So,

$$Z_T \, (\text{corrected}) = N_c \, (\text{corrected}) \times S_1$$

$$N_{ph} \, (\text{corrected}) = \frac{Z_T}{6} (\text{corrected}) \qquad ...(5.19)$$

Estimation of corrected values of $B_{\delta_1}$, $B_{\delta_m}$, $\phi_1$ and $\phi$ can be calculated with the help of equation (5.18) as

*i.e.*, $$\phi_1 \, (\text{corrected}) = \frac{V_{ph}}{4.44 \, k_w \, N_{ph} \, (\text{corrected}) \times f} \qquad ...(5.20)$$

And from equation (5.16), we get

$$B_{\delta_1} \, (\text{corrected}) = \frac{\phi_1 \, (\text{corrected})}{\frac{2}{\pi} \cdot \tau_p \cdot L_e} \qquad ...(5.21)$$

The factors $k_\delta$ and $k_\phi$ are given below as

$$k_\delta = \frac{B_{\delta_1}}{B_{\delta_m}} \qquad ...(5.22)$$

And $$k_\phi = \frac{\phi_1}{\phi} \quad ...(5.23)$$

The factors $k_\delta$ and $k_\phi$ for different types of machines can be estimated with the help of expressions given by equations (5.24), (5.25), (5.26), (5.27), (5.28), (5.29), (5.30) and (5.31).

$B$ can be defined as the ratio of the maximum flux density in the air gap to the amplitude of the fundamental component of the flux density.

**(i) Cylindrical rotor machine:**

$$B = \frac{B_{\delta_m}}{B_{\delta_1}} = \frac{\pi^2}{8} \frac{1 - \frac{b_p}{\tau_p}}{\sin \frac{\pi}{2}\left(1 - \frac{b_p}{\tau_p}\right)} \quad ...(5.24)$$

**(ii) Salient pole machines with constant air gap in the pole arc region:**

$$B = \frac{B_{\delta_m}}{B_{\delta_1}} = \frac{\pi}{4} \cdot \frac{1}{\sin\left(\frac{\pi}{2} \cdot \frac{b_p}{\tau_p}\right)} \quad ...(5.25)$$

**(iii) Salient pole machines with sinusoidal air gap in the region of the pole arc:**

$$B = \frac{B_{\delta_m}}{B_{\delta_1}} = \frac{1}{\frac{b_p}{\tau_p} + \frac{\sin \pi \frac{b_p}{\tau_p}}{\pi}} \quad ...(5.26)$$

Depending upon the types of machine, $B$ can be estimated from equations (5.24), (5.25) or (5.26), the factor $k_\delta$ can be estimated.

So, $$k_\delta = \frac{1}{B} = \frac{B_{\delta_1}}{B_{\delta_m}} \quad ...(5.27)$$

Let $\theta$ is the ratio of the useful flux per pole to the flux in the air gap corresponding to the fundamental component.

**(i) Cylindrical rotor machine:**

$$\theta = \frac{\phi}{\phi_1}$$

$$= \frac{\pi^3}{32} \frac{1 - \left(\frac{b_p}{\tau_p}\right)^2}{\sin \frac{\pi}{2}\left(1 - \frac{b_p}{\tau_p}\right)} \quad ...(5.28)$$

Chapter 5

**(*ii*) Salient pole machine with constant air gap:**

$$\theta = \frac{\pi^2}{8}\frac{b_p}{\tau_p}\frac{1}{\sin\left(\frac{\pi}{2}\cdot\frac{b_p}{\tau_p}\right)} \qquad ...(5.29)$$

**(*iii*) Salient pole machine with sinusoidal air gap:**

$$\theta = \frac{\pi \sin\left(\frac{\pi}{2}\cdot\frac{b_p}{\tau_p}\right)}{\pi\frac{b_p}{\tau_p} + \sin\pi\frac{b_p}{\tau_p}} \qquad ...(5.30)$$

depending upon the types of machine, $\theta$ can be estimated from equations (5.28), (5.29) or (5.30) and the factor $k_\phi$ can then be estimated as

$$k_\phi = \frac{1}{\theta} = \frac{\phi_1}{\phi} \qquad ...(5.31)$$

$\phi_1$ (corrected) and $B_{\delta_1}$ (corrected) have been estimated with the help of equations (5.20) and (5.21).

Now with the help of equations (5.22) and (5.23), we can calculate $B_{\delta_m}$ (corrected) and $\phi$ (corrected) as below

$$B_{\delta_m}(\text{corrected}) = \frac{B_{\delta_1}(\text{corrected})}{k_\delta}$$

And

$$\phi(\text{corrected}) = \frac{\phi_1(\text{corrected})}{K_\phi}$$

## 5.15 DESIGN OF TEETH AND SLOTS

e.m.f. per conductor can be calculated as

$$E_z = \frac{V_{ph}}{Z_{ph}}$$

and e.m.f. per turn can be given as

$$E_{Turn} = 2E_z = \frac{V_{ph}}{Z_{ph}/2} = \frac{V_{ph}}{N_{ph}}$$

### 5.15.1 Estimation of Size of Teeth and Slots

Parallel sided slots are used. Therefore the teeth are tapered and they become wider towards the bottom of the teeth. The minimum width of the teeth occurs at the top of the teeth (or at the air gap surface). The width is maximum towards the stator yoke side as shown in Fig. 5.22.

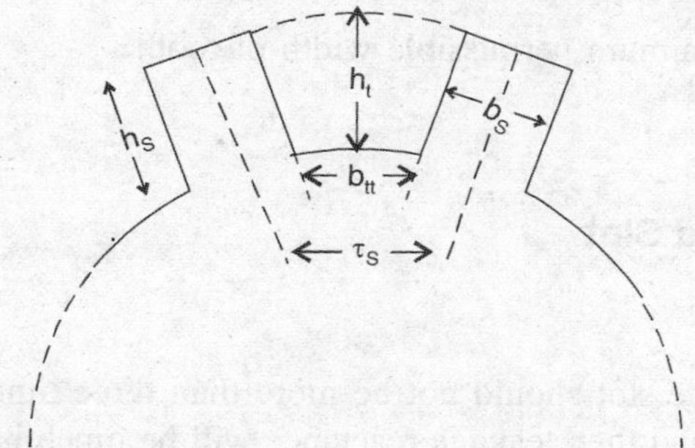


**Fig. 5.22** Teeth and slot dimensions.

Also from the Fig. 5.22, the height of teeth is equal to height of slot *i.e.*,

$$h_s = h_t$$

Since parallel sided slots are used so the width of slots will be equal throughout their height.

Width of slots = $b_s$

Width of teeth at the top = $b_{tt}$

Stator slot pitch has already been estimated and is given by

$$\tau_s = \frac{\pi D}{S_1}$$

and flux in one slot pitch is given by

$$\phi_{\tau_s} = B_{\delta_m(corrected)} \times \tau_s \cdot L_e$$

Iron provide very low reluctance path to flux so it can be assumed that the entire flux per slot pitch passes through the tooth only.

Flux entering or emerging out of a tooth is given by

$$= B_t b_{tt} k_i l$$

where $B_t$ is flux density in teeth.

This flux can be represented by flux per slot pitch.

So, $\quad B_t b_{tt} k_i l = \phi_{\tau s}$

$$\therefore \quad b_{tt} = \frac{\phi_{\tau_s}}{B_t k_i l} = \frac{B_{\delta m(corrected)} \tau_s L_e}{B_t k_i l}$$

where, $k_i$ = stacking factor

$l$ = gross iron length of machine

$B_t$ = 1.6 to 1.8 T

Chapter 5

When the flux density in teeth is above 1.7 tesla magnetic unloading of the tooth takes place, due to saturation of the iron parts.

### 5.15.2 Width of Slot

Maximum permissible width of slot

$$b_s = \tau_s - b_{tt} \text{ (i.e., The minimum permissible width of tooth)}$$

$$b_s = \tau_s - b_{tt}$$

### 5.15.3 Height of Teeth and Slot

$$h_t = h_s = (1 \text{ to } 5)\ b_s$$

Normally, the height of the slot should not be more than three times the width of slot. If this is increased that is deeper slots and then leakage reactance will be much higher. Sometimes deeper slots are made to have a high leakage reactance to limit the short-circuit currents.

## 5.16 PHASE-CURRENT

Phase-current is given by the formula as

$$I_{ph} = \frac{Q \times 10^3}{3\,V_{ph}} \text{ ampere}$$

where $Q$ is output in kVA.

Current in each conductor $I_z = I_{ph}$ but if there are $a$ parallel paths then the conductor current,

$$I_z = \frac{I_{ph}}{a}.$$

## 5.17 SECTIONAL AREA OF STATOR CONDUCTOR AND SLOT INSULATION

Sectional area of conductor is given by

$$F_c = \frac{Iz}{\sigma}$$

where, $\sigma$ is the permissible current density in the armature conductor and it is assumed to be 4 to 5 A/mm$^2$.

The slot has to accommodate the conductor, insulation, support wedges and some space for reactance gap. The size of the slot should be so designed to accommodate the above required things. If the conductors, insulators etc. are not properly being adjusted the size of the slot, conductor size and insulation type can be modified by modifying

(*i*) $b_s =$ (2 to 5) $b_0$.
(*ii*) Permissible current density.
(*iii*) Selecting suitable insulating material.

These modifications should result in proper adjustment of conductor, insulation and support wedges in slots.

A single turn bar winding is used in machines when the armature current per circuit is more than 1500 ampere. Due to large current the conductor cross-section is large. To reduce the eddy current loss in bar windings, the conductors are subdivided into many parts. There are two conductors in slot separated by a separator if a bar winding is used. Each conductor consists of vertical stacks given in Fig. 5.23 of copper laminations (subdivisions of copper conductor) separated by a stack separator. There are wall insulation and insulation strip on conductor side which can be seen in the Fig. 5.24.

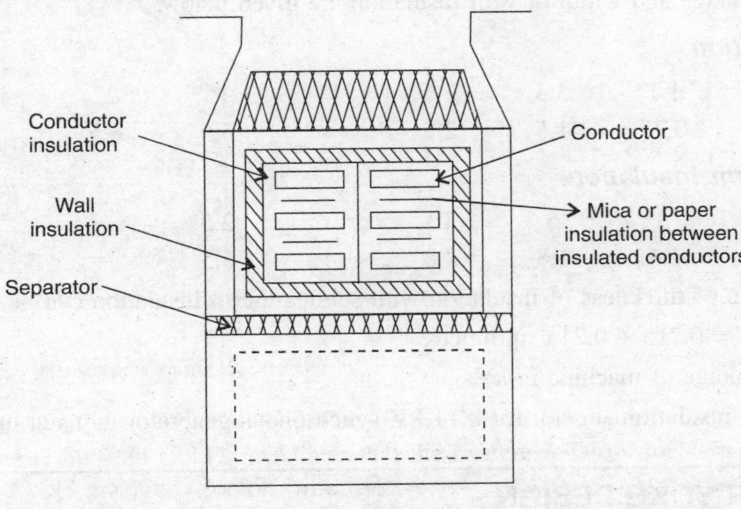


**Fig. 5.23** Filling up to slot.

To reduce the circulating current loss transposition of conductor laminations in the slots are used. Usually, Robell conductors are used. It is a type of transposed conductor. In this transposition each

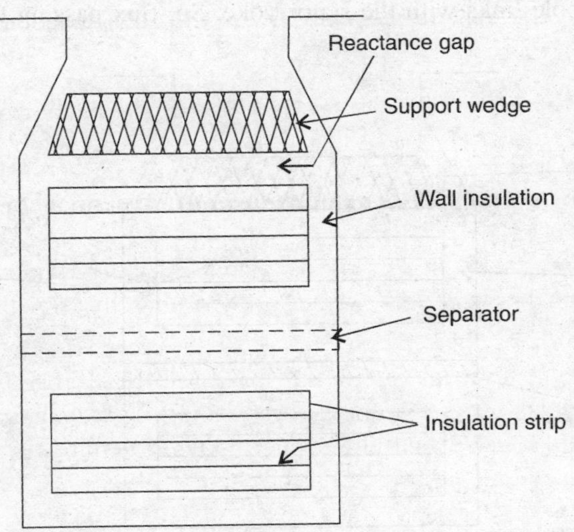


**Fig. 5.24** Strip conductor.

conductor lamination is arranged to move continuously through all position in the path of coil side. So that the leakage reactance of all conductor laminations is equalised and consequently there is no circulating current between the lamination and hence the circulating current is eliminated.

Width of wall insulation for machines depends upon the voltage rating and also upon type of insulating material. Such as for the same voltage rating machine semica-therm occupies less space compared to mica-insulation. So thickness for semica-therm is lesser and heat dissipation is much better.

A table for voltage and width of wall insulation are given below.

***(i) Mica Insulation***

| kV | 0.4 | 1.1 | 3.3 | 6.6 | 11 | 15 | 22 |
|---|---|---|---|---|---|---|---|
| mm | 0.5 | 0.75 | 1.5 | 2.5 | 4 | 5.5 | 6.5 |

***(ii) Semica-therm Insulation***

| kV | 2 | 3 | 6 | 10 | 16 | 2.5 |
|---|---|---|---|---|---|---|
| mm | 1.1 | 1.4 | 1.8 | 2.8 | 4 | 5.5 |

The expression of thickness of insulation with semica-therm insulation can be written as

$$Y = 0.715 + 0.21x \text{ millimetre}$$

where, $x$ is voltage of machine in kV.

The details of insulation in slot for a 11 kV synchronous generator is given in Fig. 5.25.

## 5.18 STATOR YOKE DESIGN

Let $h_y$ is the height of stator yoke as shown in Fig. 5.26.

Half of the flux per pole links with the stator yoke. So, flux passing through the stator yoke $= \frac{\phi}{2}$

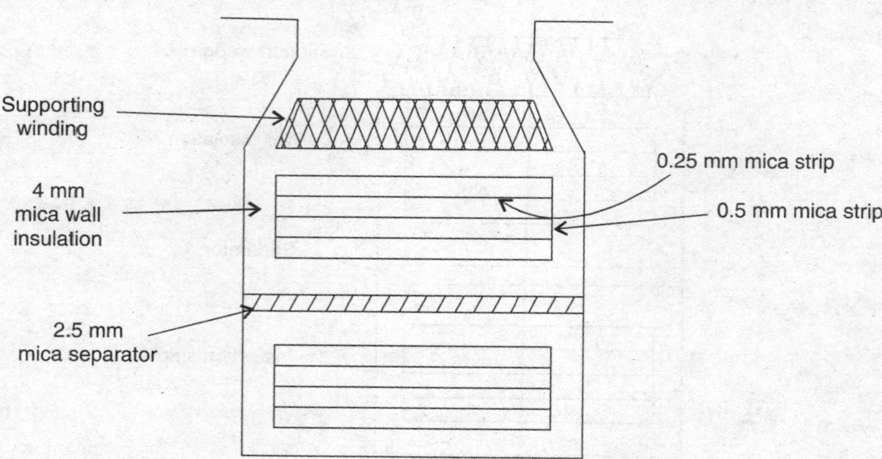


**Fig. 5.25** Details of insulation in slot of 11 kV alternator.

Let

$B_y$ = Flux density in stator yoke

= 1.3 to 1.5 T

Flux through yoke = $B_y \cdot h_y \cdot k_i\, l$

But, here $B_y \cdot h_y \cdot k_i\, l = \phi/2$

$$\therefore \qquad h_y = \frac{\phi}{2 B_y\, k_i\, l} \qquad \text{...(5.32)}$$

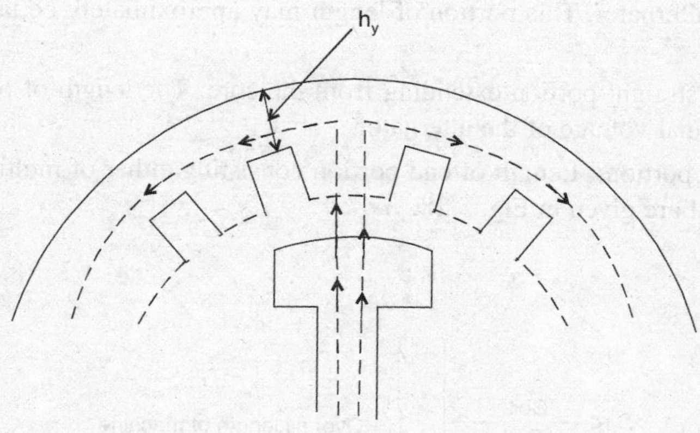


**Fig. 5.26** Stator yoke.

With the help of equation (5.32), the height of stator yoke can be calculated. By calculating height of the stator yoke, the outer diameter of machine can be estimated as shown in Fig. 5.27.

Let $D_0$ = Outer diameter $= D + 2h_t + 2h_y$

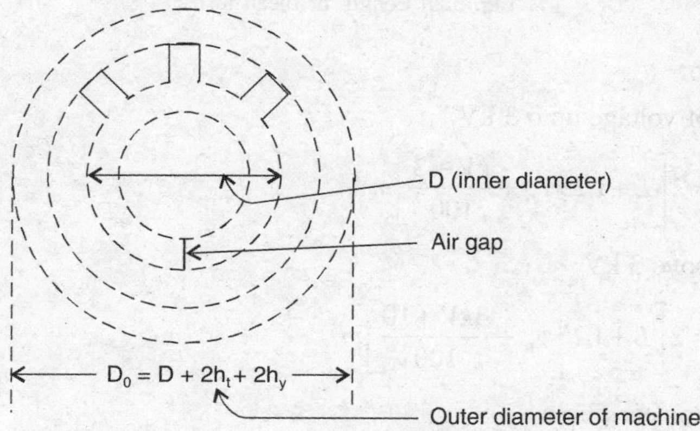


**Fig. 5.27**

## 5.19 STATOR COPPER AND IRON LOSSES, WEIGHT OF IRON AND COPPER

### (a) Stator Winding Resistance and Copper Loss

Mean length of turn for stator winding of synchronous machine consists of following portions:

(i) Length of coil in the slot: Each turn has the length in the slot equal to two times the gross length of stator core.

(ii) Length of overhang portion: The length of overhang portion of the turn will depend upon the pole pitch of alternator. This portion of length may approximately be taken 2.5 times the pole pitch.

(iii) Length of the straight portion extending from the core: The length of portion depends mainly upon the terminal voltage of the alternator.

(iv) Length of end portion: Length of end portion consisting either of multi-turn coil or single bar coil. The detail are given in Fig. 5.28.

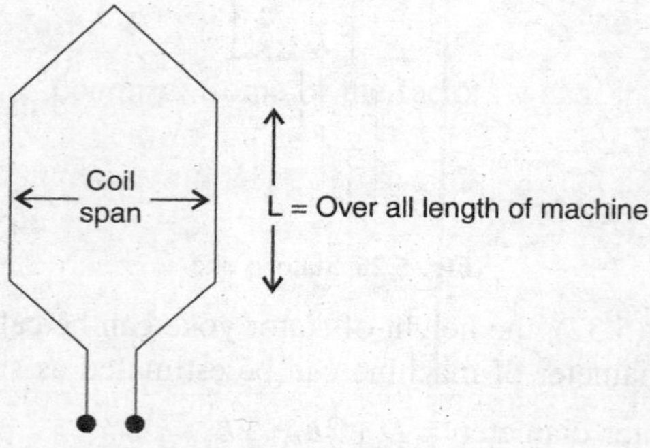


**Fig. 5.28** Length of mean turn.

### Length of mean turn

(i) For machine of voltage upto 3 kV

$$L_{mt} = 2\left[L + 1.25\,\tau_p + \frac{kV + 3}{100}\right] m$$

(ii) For voltage above 3 kV

$$L_{mt} = 2\left[L + 1.25\,\tau_p + \frac{3kV + 10}{100}\right] m$$

Length of conductor per phase

= Number of turns per phase × Mean turn length

$= N_{ph} \times L_{mt}$

Sectional area of conductor $F_c$, has already been estimated.

So, Resistance per phase (at 75°C)

$$R_{dc} = 0.021 \times \frac{N_{ph} \cdot L_{mt}}{F_c}$$

where $L_{mt}$ is length of mean turn in *m*.
and $F_c$ is sectional area of conductor in $mm^2$.

This is the resistance shown by the conductor when d.c. supply is given. When a.c. supply is applied across the conductor, then there is skin effect observed in the conductor. Current tries to flow through the outer surface of the conductor and result in unequal distribution of current in conductor. The effective resistance of the conductor due to skin effect becomes more than its d.c. resistance.

### *Effective stator winding resistance/phase*

$$R_{ph}(75°\text{C}, ac) = (1.15 \text{ to } 1.20)\, R_{dc}$$

### *Stator copper loss*

Stator copper loss for 3-phase machine is given by

$$= 3I_{ph}^2 \cdot R_{ph} \text{ watt}$$

### *(b) Iron Loss, Weight of Iron in Stator*

Let $b_{tb}$ = width of tooth at bottom

$b_{tt}$ = width of tooth at the top.

$$\text{Average width of tooth} = \frac{b_{tt} + b_{tb}}{2}$$

Now, $b_{tt}$ has already been estimated. $b_{tb}$ can be calculated by calculating slot pitch at the bottom of teeth and subtracting slot width from it. Since the slots are parallel sided so width of the slots are same along their height (depth).

It is clear from Fig. 5.29 that

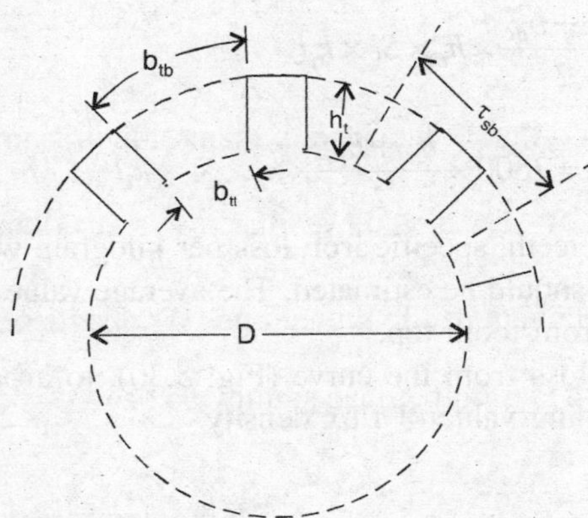


**Fig. 5.29** Slot pitch at tooth bottom.

Chapter 5

Slot pitch at tooth bottom, $\tau_{sb} = \dfrac{\pi[D + 2h_t]}{S_1}$

and $\qquad b_{tb} = \tau_{sb} - b_s$

Area of one tooth $= \dfrac{b_{tt} + b_{tb}}{2} h_t$

Area of all the teeth $= \dfrac{b_{tt} + b_{tb}}{2} \times h_t \times S_1$

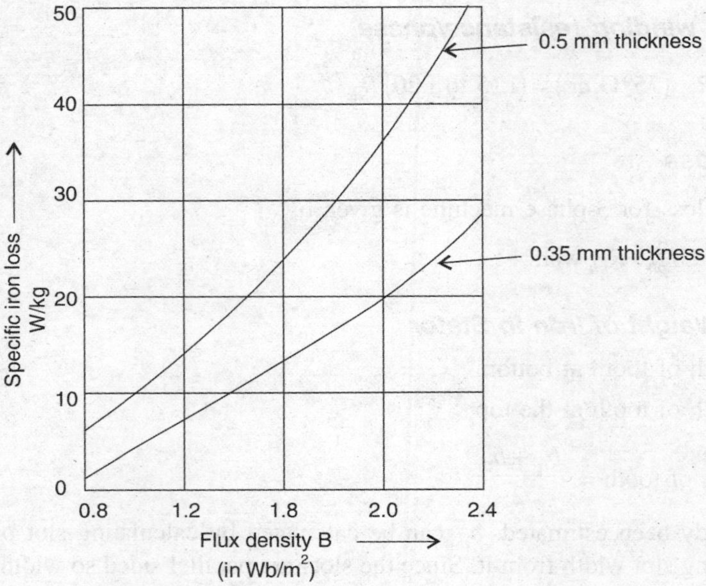


**Fig. 5.30** Specific iron loss for rotating machine.

Volume of all the teeth $= \dfrac{b_{tt} + b_{tb}}{2} \times h_t \times S_1 \times k_i l$

Weight of iron in teeth, $W_{it} = 7600 \times \dfrac{b_{tt} + b_{tb}}{2} \times h_t \times S_1 \times k_i l$

To calculate the iron loss in teeth, specific iron loss per kilogram weight of iron corresponding to average flux density in the teeth should be estimated. The average value of flux density in tooth can be estimated at 1/3rd tooth height from tooth top.

Let $p_{it}$ be the specific iron loss from the curve (Fig. 5.30), to a particular value of thickness of stator lamination and corresponding value of flux density.

Iron loss in teeth = $p_{it} \times W_{it}$

Weight of iron in stator yoke

$W_{iy}$ = 7600 × Volume of iron in stator yoke.

Let $p_{iy}$ is the specific iron loss per kg weight of iron in stator yoke corresponding to average value of flux density in yoke.

Iron loss in yoke = $p_{iy} \cdot W_{iy}$

### (c) Total Iron Loss in Stator

$$W_{is} = \text{Iron loss in teeth + Iron loss in yoke}$$

Weight of iron in stator = Weight of iron in teeth + Weight of iron in yoke

$$= W_{it} + W_{iy}$$

### (d) Weight of Copper in Stator

Volume of copper per phase

$$= \text{Length of conductor per phase} \times \text{Sectional area of conductor}$$

$$= N_{ph} L_{mt}\ F_c \text{ cube metre}$$

Weight of copper in the winding of a 3-phase synchronous machine

$$= 8900 \times 3\, N_{ph} L_{mt} F_c \text{ kilogram}$$

## 5.20 DESIGN OF DAMPER WINDINGS

Damper windings are used for different purposes in synchronous machines. The functions of the damper windings in synchronous generators are:

(*i*) To suppress the inverse rotating field.
(*ii*) To prevent hunting.

The functions of the damper windings in synchronous motors are:

(*i*) To provide starting torque to help in self starting as induction motor and then synchronized to run as synchronous motor.
(*ii*) To develop the damping power when the machine is hunting.

The damper winding is arranged on the rotor of a synchronous machine in the same manner as provided in the case of a squirrel-cage induction motor *i.e.*, special holes are provided in the pole shoes in which bars are inserted and short-circuited at both ends. In the case of turbo-alternators windings which are used to close the slot of the rotor serve as damper bars. A damper winding with low leakage reactance and low resistance almost completely suppress the inverse armature field.

The value of fundamental armature mmf of one phase of a polyphase winding is given by

$$AT = \frac{4\sqrt{2}}{\pi} k_w N I_{ph} \qquad \text{...(5.33)}$$

where, $N$ = number of stator turns per phase under one pole

$I_{ph}$ = phase-current

This pulsating mmf can be resolved into two rotating m.m.fs., one half is called the synchronous m.m.f. and the other half inverse mmf. If the damper winding has to suppress the inverse rotating field,

Chapter 5

the damper winding mmf have approximately the same m.m.f. as that of the inverse rotating field (Negative sequence field). Therefore, the design of damper winding is done on the basis of inverse rotating field for one phase.

Armature m.m.f. (inverse) for one phase is

$$AT_1 = \frac{4\sqrt{2}}{2\pi} k_w \times N \times I_{ph}$$

So, m.m.f. of damper winding, $AT_1 = 0.9\, k_w \times N \times I_{ph}$ ...(5.34)

Total ampere conductors on the stator of 3-phase machine arc

$$= 2 \times 3\, NpI_{ph}$$

$$= 6N\, p\, I_{ph}$$

$$6N\, p\, I_{ph} = \overline{ac} \cdot p \cdot \tau_p$$

or
$$NI_{ph} = \frac{\overline{ac} \cdot \tau_p}{6} \qquad ...(5.35)$$

From equations (5.34) and (5.35), we get

$$AT_1 = 0.9\, k_w \cdot \frac{\overline{ac}\tau_p}{6}$$

$$= 0.15 k_w \cdot \overline{ac}\tau_p$$

Let $F'_D$ is cross-sectional area of damper winding conductors under one pole.

$\sigma_D$ = Permissible value of current density in damper bars.

= 4–5 A/mm²

So, $F'_D \cdot \sigma_D = 0.15 k_w \cdot \overline{ac}\, \tau_p$

or, $$F'_D = \frac{0.15 k_w \cdot \overline{ac}\, \tau_p}{\sigma_D} \qquad ...(5.35a)$$

Now, on putting, $k_w$ = 0.95 for fractional pitch in equation (5.35*a*), we get

$$= 0.143 \cdot \frac{\overline{ac}\tau_p}{\sigma_D}$$

In actual practice, the cross-sectional area provided for damper winding conductor are larger than this.

So,
$$F'_D = 0.2 \cdot \frac{\overline{ac}\, \tau_p}{\sigma_D} \qquad ...(5.36)$$

knowing $\overline{ac}$, $\tau_p$ and $\sigma_D$, $F'_D$ can be calculated by equation (5.36).

### *Number of Damper Bars Used per Pole*

Depending upon the number of bars used, the cross-sectional area of damper winding per pole $F_D'$ is distributed into smaller sections. The damper winding pitch is usually taken as 80% of stator slot pitch.

Let $\tau_{SD}$ = damper winding pitch

$= 0.8\ \tau_s$

$d_D$ = diameter of damper-bars

$Z_D$ = space between two damper bars

$b_P$ = pole arc length

$$\text{Number of damper bars, } n_D = \frac{b_p}{\tau_{SD}}$$

$$= \frac{b_p}{0.8\tau_s}$$

If the values of $b_p$ and $\tau_s$ are known, then $n_D$ can be estimated. The details are given in Fig. 5.31.

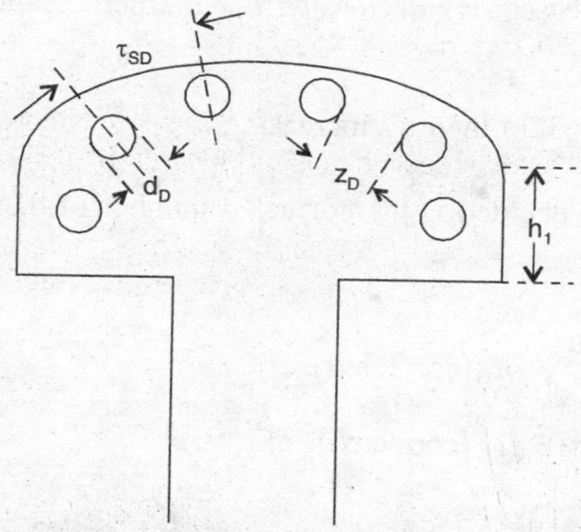


**Fig. 5.31** Design of damper-bars.

Sectional area of one damper bar

$$F_D = \frac{F_D' \text{ (sectional area of damper bars under one pole)}}{n_D \text{ (number of damper bars)}}$$

$$F_D = \frac{\pi}{4} d_D^2$$

or

$$d_D = \sqrt{\frac{4F_D}{\pi}}$$

Figure 5.31 shows the space between damper bars, $Z_D$ and damper winding pitch $\tau_{SD}$.

$$Z_D = \tau_{SD} - d_D$$

The flux density in the pole shoe should be checked since the flux density, $B_D$ in the space $Z_D$ between damper bars should not exceed 1.8 T. The flux entering in or emerging out over a damper winding pitch are

$$\phi_{\tau SD} = B_{\delta_m} \tau_{SD} L_e$$

This flux has to pass through the space $Z_D$, as rest portion of the damper winding pitch is considered non-magnetic.

So, $$\phi_{\tau SD} = B_D Z_D k_i L_p$$

$L_p$ = Length of pole = Length of machine

$$B_D Z_D k_i L_p = B_{\delta_m} \tau_{SD} L_e$$

Flux density in space, $$B_D = \frac{B_{\delta_m} \tau_{SD} L_e}{Z_D k_i L_p}$$

If $B_D > 1.8$ T

Then $\frac{\tau_{SD}}{Z_D}$ should be decreased. By decreasing $n_D$, $Z_D$ increases and $\frac{\tau_{SD}}{Z_D}$ decreases.

If $B_D$ is low (say below 1.65 T) then by increasing $n_D$, $\frac{\tau_{SD}}{Z_D}$ increases.

The number of bars $n_D$ so-decided is the corrected number of damper bars. Thus, $\tau_{sd}$ (corrected) should be obtained.

$$\tau_{SD}(\text{corrected}) = \frac{b_p}{n_D(\text{corrected})}$$

and $$Z_D(\text{corrected}) = \tau_{SD}(\text{corrected}) - d_D$$

$$B_D(\text{corrected}) = B_{\delta_m}\left(\frac{\tau_{SD}}{Z_D}\right)(\text{corrected})\frac{L_e}{k_i L_p}$$

In order to provide short-circuit rings to short the damper bars on both sides of the machine, the length of the damper bars should be

$L_D = 1.1\ L_p$ for short machines, and

$L_D = L_p + 0.1$ metre for larger machines.

End of damper bars are slightly tapered as shown in Fig. 5.32.

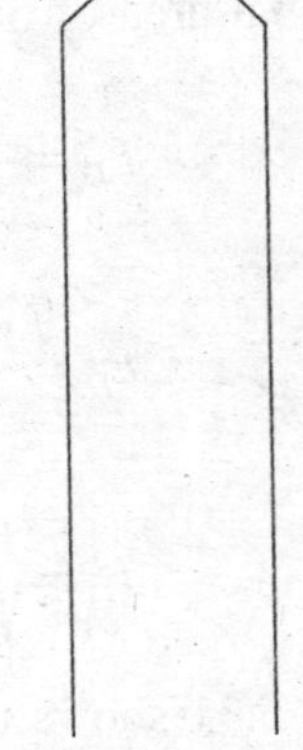

**Fig. 5.32** Tapered end of damper bars.

Sectional area of short-circuiting ring on each side for short-circuiting the damper bars have copper section about 40 to 50 per cent of total copper section in damper winding on a pole.

Sectional area of single ring, $F_{ER} = (0.4 - 0.5) F'_D$

***Height of Pole Shoes Tip:*** The height of pole shoe tip (Fig. 5.33) of the pole shoe is taken as twice of diameter of damper bars so that the damper bars can be satisfactorily accommodated in the pole shoe.

$$h_1 = 2d_D$$

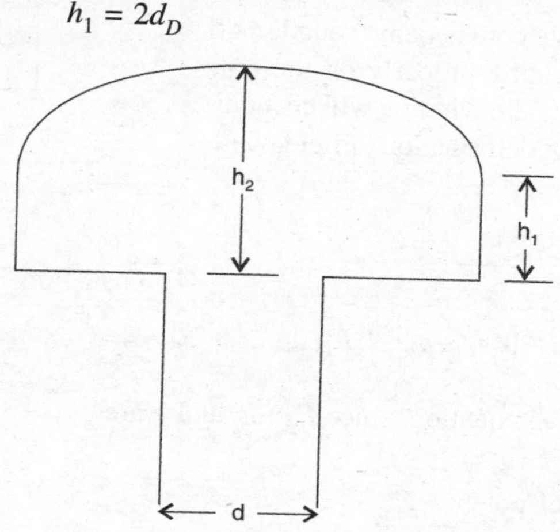


**Fig. 5.33** Pole shoe height.

## 5.21 ROTOR DESIGN OF SALIENT POLE SYNCHRONOUS ALTERNATOR

In rotor designs, the dimensions of the pole, damper winding and the shaft has to be estimated. The flux produced by the poles is higher than the useful flux due to pole leakage flux. The dimensions of pole should be known to calculate the exact value of leakage flux. Since the dimensions are yet not estimated (known) so it is assumed that the pole leakage flux is approximately 15–20 percent of useful flux.

So, $$\phi_p = (1.15 \text{ to } 1.20)\phi$$

***Pole Core Design:*** Cross-sectional area of pole core can be calculated by

$$F_{pc} = \frac{\phi_p}{B_{pc}} \text{ metre-square}$$

where, $B_{pc}$ = flux density in pole core

= 1.3 to 1.5 T

**(i)** ***Round Pole:*** The pole core is round.

Let $d$ = diameter of pole core given in Fig. 5.34.

So, $$F_{pc} = \frac{\pi}{4}d^2$$

or $$d = \sqrt{\frac{4F_{pc}}{\pi}}$$

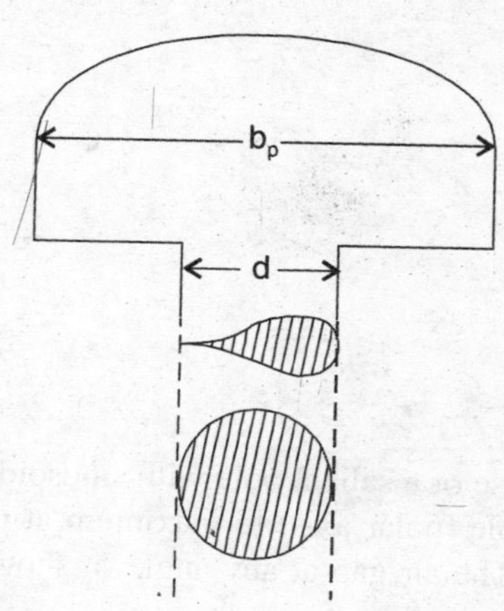


**Fig. 5.34** Diameter of pole core for round pole.

(*ii*) ***Long Pole:*** Rectangular pole core is being used.

Let $d$ = width of pole core as in Fig. 5.35.

The corners of the pole core is being rounded off to place the field winding properly on the pole core. The stress on the field winding will be modified and the chances of deformation of field winding will be reduced.

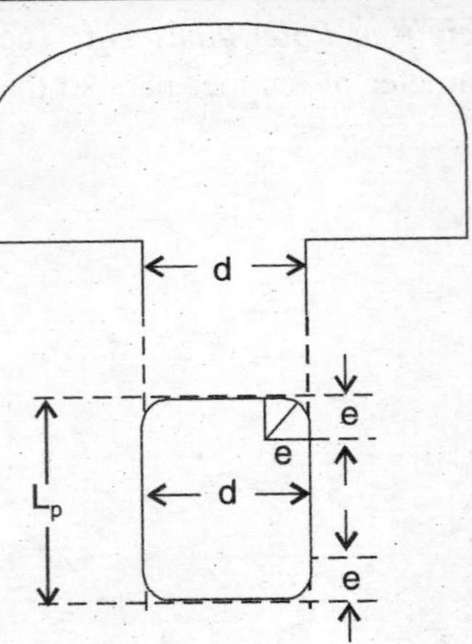


**Fig. 5.35** Width of long pole.

In Fig. 5.35

$$e = 0.05\ L_p$$

$$F_{pc} = \left[L_P \cdot d - \left(4e^2 - \pi e^2\right)\right] K_{ip}$$

$d$ can be calculated from above equation, since $F_{pc}$, $L_p$ and $e$ are known.

$k_{ip}$ = Stacking factor for pole
= 0.98 for 1 mm thickness of lamination
= 0.90 to 0.92 for 0.5 mm thickness of lamination.

### 5.21.1 Pole Profile Drawing of Salient Pole Alternator

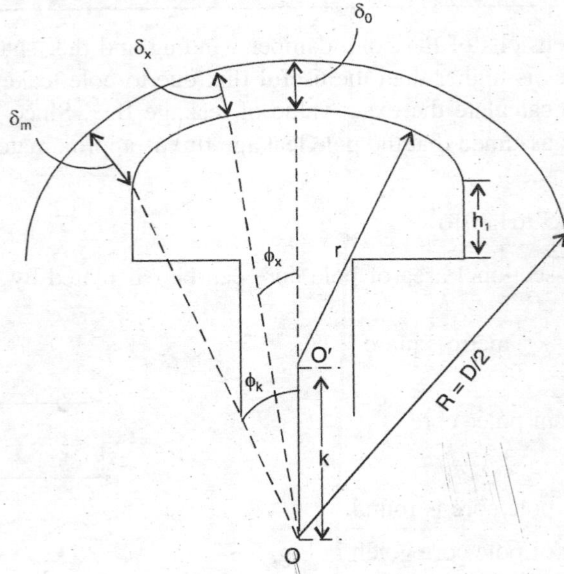


**Fig. 5.36** Pole profile.

In case of a salient pole with sinusoidal air gap machine, the air gap varies from minimum at the centre of pole (polar axis) to maximum at pole shoe tip.

The air gap at any angle $\phi_x$ shown in Fig. 5.36 from poler axis can be given by the

$$\delta_x = \delta_0 + \frac{K}{2r}(K + r)\sin^2 \phi_x \qquad ...(5.37)$$

So, the expression for maximum air gap length $\delta_m$ can be given by

$$\delta_m = \delta_0 + \frac{K}{2r}(K+r)\sin^2\phi_K \quad ...(5.38)$$

where, $\delta_0$ = minimum air gap length of machine

$\delta_x$ = air gap at an angle $\phi_x$

$\delta_m$ = maximum air gap length at any angle $\phi_k$

$R = \frac{D}{2}$, radius of inner diameter with centre O

$r$ = radius of pole arc with centre O′

$K$ = distance between two centres O and O′

$$\phi_k = \frac{\pi\, b_p}{P\, \tau_p} \quad ...(5.39)$$

$$R = K + r + \delta_0 \quad ...(5.40)$$

The ratio between maximum air gap to minimum air gap is normally taken as

$$\delta_m = (1.2 \text{ to } 1.5)\delta_0 \quad ...(5.41)$$

From equations (5.39) and (5.41), the angle $\phi_k$ and $\delta_m$ can be calculated and there by knowing $\phi_k$ and $\delta_m$, the value of $K$ and $r$ can be estimated from equations (5.38) and (5.40).

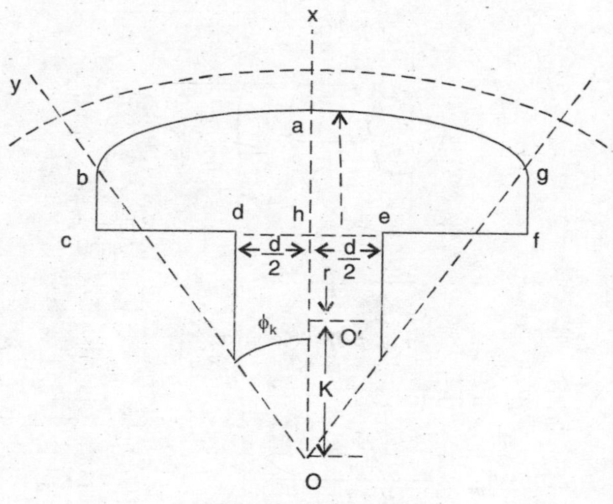


**Fig. 5.37** Pole drawing.

***Pole Drawing:*** Take a point $O$ as centre and draw a circle with radius $R\left(\frac{D}{2}\right)$. Draw a line $OX$. Then draw a line $OY$ at an angle $\phi_k$ from $OX$. Mark a point $O'$ at a distance $k$ on $OX$ from point $O$. Draw circle with $O'$ as centre and $r$ as radius. It will cut the line $OY$ at $b$ and similarly on other side at $g$. Draw lines *bc* and *gf* parallel to *OX* and equal to $2d_D$ (diameter of damper bar) as given in Fig. 5.37. Join the points *c* and *f*. It cut *OX* on *h*. Mark *hd* and *he* equal to *d*/2 (half of width of or diameter of pole core).

Height *ha* will give the height of pole shoe at middle ($h_2$).

### 5.21.2 Height of Pole Yoke

Let height of pole yoke $h_{py}$ (Fig. 5.38)

Useful flux per pole = $\phi$

Due to leakage of flux,

pole flux $\phi_p$ = (1.15 to 1.20) $\phi$. Half of the flux passes through pole yoke. So, the pole yoke will contain flux $\dfrac{\phi_p}{2}$.

$$\frac{\phi_p}{2} = B_{py} \cdot h_{py} \cdot k_i L_p$$

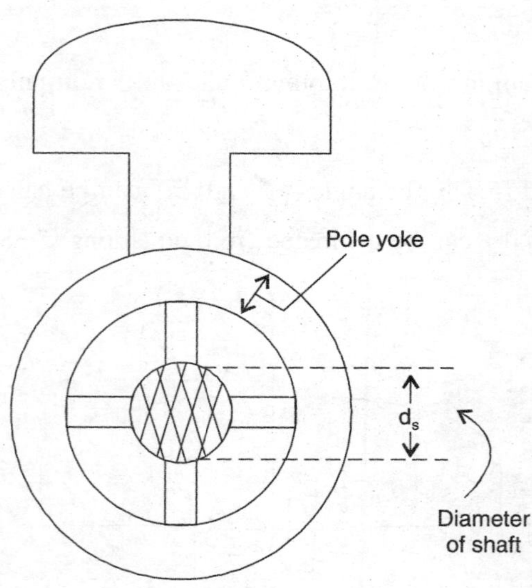


**Fig. 5.38** Pole yoke.

Flux density in pole yoke, $B_{py} = 1.3 \text{ to} 1.5 \text{ T}$

So,
$$h_{py} = \frac{\phi_P}{2 \cdot B_{py} k_i L_p} \qquad ...(5.42)$$

***Diameter of the shaft:*** The diameter of the shaft of the synchronous machine can be given by the expression

$$d_s = k\sqrt{\frac{k_w}{RPM}} \qquad ...(5.43)$$

where, $k = 0.1$ for forged steel.

### 5.21.3 Height of the Pole Core of Salient Pole Machine

Armature m.m.f. for 3-phase per pole, $AT_a = \dfrac{2.7 N_{ph} \cdot I_{ph} \cdot k_w}{P}$

No load field m.m.f., $AT_{f0} = SCR \times AT_a$

Since the dimensions of the pole are not completely known and resistance and leakage reactance of the armature winding are being neglected.

Usual value of $SCR$ = 0.5–0.7 for turbo-alternator

This may be raised to 0.7–1.5 for water wheel generator.

Referring to Fig. 5.39.

Draw $Oa = AT_{fl}$

Draw $ab = AT_a$ at angle $(90 - \phi)$ with $Oa$,

where $\cos \phi$ = P.F. (lagging)

Cut off a.c. such that $ac/ab = k_r$

where, $k_r$ is called the cross-reaction coefficient which depends upon $\dfrac{b_P}{\tau_P}$ ratio given in Fig. 5.40.

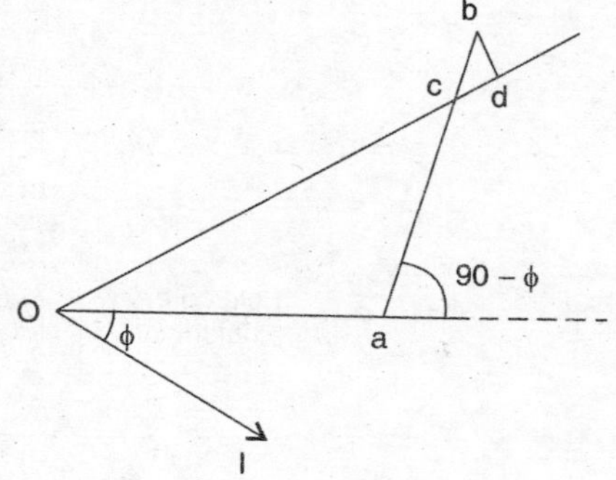


**Fig. 5.39** Phasor diagram.

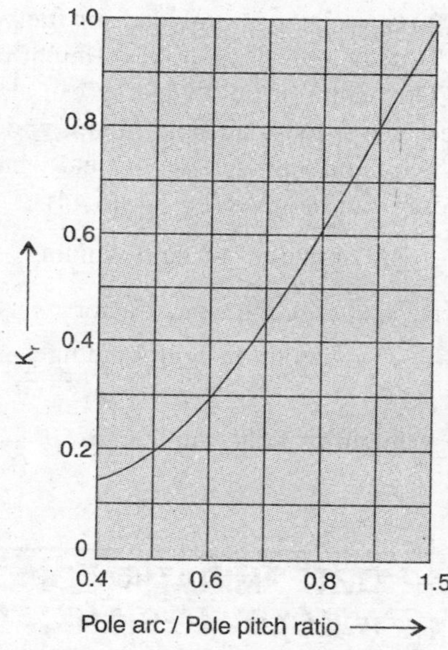


**Fig. 5.40** Cross-reaction coefficient.

Join $Oc$ and extend it. Draw a perpendicular from $b$ on extended $Oc$ line, cutting it at $d$.

Full load field m.m.f. = $Od$ at p.f. $\cos \phi$

Copper area of field winding for one pole $= \dfrac{AT_{fl}}{P \cdot \sigma}$

where, $\sigma$ is the current density of field winding conductor.

Total space required for winding for one pole will be

$$\text{Total space} = \frac{\text{Cu area}}{\text{Space factor}}$$

Space factor = 0.8–0.9 for strip winding. It is used for large and low speed alternators. Bare copper strips are insulated with each other. Width of conductors should not normally exceed 6 mm.

Space factor = 0.45–0.6 for round wire winding.

The height of the field winding $h_f$, can be calculated by dividing the total space of field winding by the depth (or width) of the winding. If $d_f$ is the depth of field winding as shown in Fig. 5.41. Height of field winding, $h_f$, can be calculated by

$$h_f = \frac{\text{Total space}}{d_f}$$

where, $d_f$ can be obtained with the Table 5.2.

Height of pole core $h_{pc} = h_f + 20$ mm.

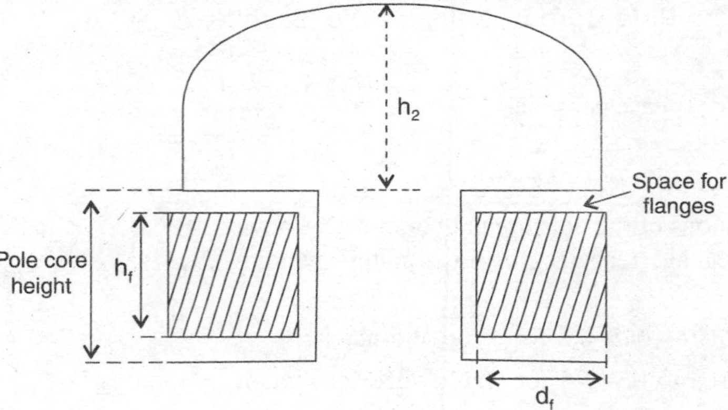


**Fig. 5.41** Height of field winding and pole.

20 mm space is provided for flanges etc. as shown in Fig. 5.41.

Total height of the pole = Height of pole core + $h_2$

The height of field winding can also be calculated with loss per unit surface of field winding ($q_f$), space factor, full load field m.m.f. and depth of the winding.

If loss per unit surface of field winding ($q_f$) in w/m$^2$ is given, then m.m.f. per metre height of field winding can be given by eq. (5.44).

$$MMF/\text{m height of field winding} = 10^4 \times \sqrt{k_s \cdot q_f \cdot d_f} \quad ...(5.44)$$

where, $k_s$ = winding space factor

$d_f$ = depth of field winding

$$\therefore \quad \text{Height of field winding, } h_f = \frac{AT_{fl}}{10^4 \cdot \sqrt{k_s \cdot q_f \cdot d_f}} \quad ...(5.45)$$

## 5.22 DETERMINATION OF OPEN CIRCUIT CHARACTERISTICS (OCC) BY DESIGN DATA

Open circuit characteristic is obtained between terminal voltage at no load and corresponding field m.m.f. per pole *i.e.*, the total field m.m.f. per pole to induce rated voltage in stator winding at no load. Total field m.m.f. is estimated by considering magnetic circuit for pole pair. The sum of m.m.fs. required by different parts of machine in one complete magnetic circuit for a pole pair at no load gives the total field m.m.f. for pole pair. The flux density is calculated or known for different part of machine and thus corresponding m.m.fs. (ampere-turns) per unit length is found with the help of *B-H* or *B*-at curves. Sum of all amperes-turns so-obtained for different parts gives the total ampere-turns required for pole pair. Different portions of the machine in magnetic circuit of one pole pair are given in section 5.3.

### 5.22.1 MMF for Air Gap

#### *Salient Pole Machine*

$\delta$ = actual air gap

$\delta' =$ effective length of air gap

It is given by

$\delta' = k_c \delta$

where, $k_c$ is called Carter's gap coefficient.

By slotting stator, the reluctance offered to flux is more in the region of the slot than in the tooth region. The flux density in the air gap is assumed to be constant but a consideration for the variation of the reluctance of the air gap is made while calculating *AT* for air gap.

$$k_c = \frac{\tau_s}{\tau_s - \gamma\delta} \qquad \text{...(5.46)}$$

where, $$\gamma = \frac{5(b_0/\delta)^2}{5 + \dfrac{b_0}{\delta}}$$

$$\tau_p = \text{ stator slot pitch } = \frac{\pi D}{S_1}$$

$b_0 =$ opening of slot as shown in Fig. 5.46

***Cylindrical Rotor Machine:*** In cylindrical rotor (turbo-alternator) machine both stator and rotor are slotted. Hence, there are air gap coefficient for stator and for rotor also. So, Carter's air gap coefficient for cylindrical rotor machine is given by

$$k_c = k_{c_1} \cdot k_{c_2} \qquad \text{...(5.47)}$$

where $k_{c_1}$ is the Carter's air gap coefficient for stator and $k_{c_2}$ is Carter's air gap coefficient for rotor.

So, from equation (5.46), we get

$$k_{c_1} = \frac{\tau_s}{\tau_s - \gamma_1\delta}, \text{ where } \gamma_1 = \frac{5\left(\dfrac{b_{01}}{\delta}\right)^2}{5 + \dfrac{b_{01}}{\delta}}$$

$b_{01} =$ opening of stator slot.

Also, from equation (5.46), we get

$$k_{c_2} = \frac{\tau_s}{\tau_s - \gamma_2\delta}, \text{ where } \gamma_2 = \frac{5\left(\dfrac{b_{02}}{\delta}\right)^2}{5 + \dfrac{b_{02}}{\delta}}$$

$b_{02} =$ opening of rotor slot.

***(1) MMF for a Pair of Air Gap:***

$$AT_{\delta} = \frac{1}{\mu_0} B_{\delta m} 2\delta' \qquad \text{...(5.48)}$$

where, $\delta'$ is the effective length of air gap.

***(2) MMF for a Pair of Teeth:*** Flux/slot pitch is given by

$$\phi_{\tau_s} = B_{\delta m}\, \tau_s\, l_e$$

This flux passes through tooth so

$$B_{\delta m}\, \tau_s l_e = B_{tt} b_{tt} k_i l$$

$$B_{tt} = \frac{B_{\delta m} \tau_s l_e}{b_{tt} \cdot k_i l}$$

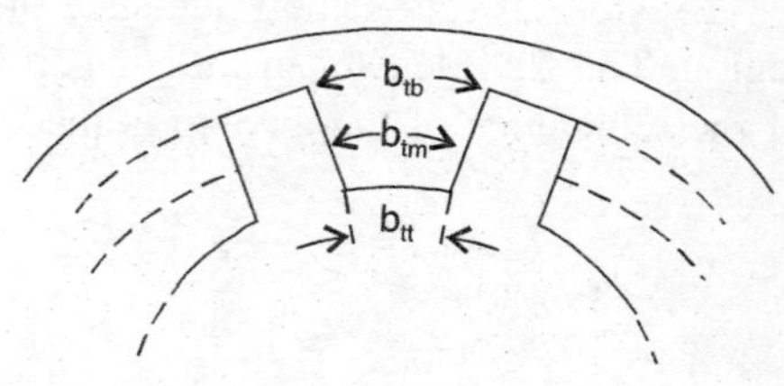


**Fig. 5.42** Tooth m.m.f.

where, $B_{tt}$ is the flux density at tooth top. Corresponding to this flux density, ampere-turn/m ($at_{tt}$) of tooth length for tooth top can be estimated from *B-H* curve.

$at_{tt}$ = corresponding to $B_{tt}$ from *B-H* curve

Flux density at middle of tooth

$$B_{tm} = \frac{B_{\delta m}\, \tau_s\, l_e}{b_{tm} \cdot k_i l}$$

Ampere-turn/m length of tooth at the middle

$at_{tm}$ = corresponding to $B_{tm}$ from *B-H* curve.

Similarly, flux density at bottom of tooth

$$B_{tb} = \frac{B_{\delta m}\, \tau_s l_e}{b_{tb} \cdot k_i l}$$

and ampere-turn per m $at_{tb}$ length of tooth at the bottom

$at_{tb}$ = corresponding to $B_{tb}$ from *B-H* curve.

Width of tooth at the top $b_{tt}$ is given by

$$b_{tt} = \tau_s - b_s = \frac{\pi D}{S} - b_s$$

Width of tooth at the middle $b_{tm}$ is given by

$$b_{tm} = \tau_{sm} - b_s = \frac{\pi(D + h_t)}{S_1} - b_s$$

Similarly, width of tooth at the bottom $b_{tb}$ is given by

$$b_{tb} = \tau_{sb} - b_s = \frac{\pi(D + 2h_t)}{S_1} - b_s$$

The average value of ampere-turn/m length of tooth ($at_{at}$) can be estimated by Simpson's rule as following:

$$at_{at} = \frac{at_{tt} + 4at_{tm} + at_{ab}}{6} \qquad ...(5.49)$$

Total ampere-turn ($AT_{at}$) required for the pair of teeth is given by

$$AT_{at} = at_{at} \times 2h_t \qquad ...(5.50)$$

***(3) MMF for Stator Yoke:*** The height of ($h_y$) stator yoke has already been estimated. In one complete magnetic circuit for a pair of pole the mean length of flux path in stator yoke as shown in Fig. 5.43 is

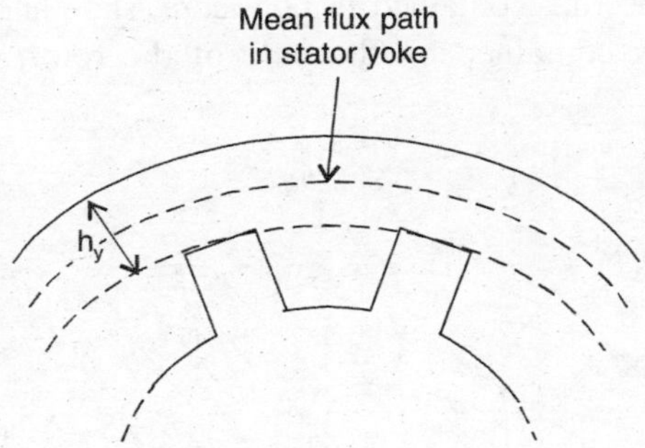


**Fig. 5.43** Mean flux path in stator yoke.

$$l_y = \frac{\pi(D_0 - h_y)}{P}$$

or $$l_y = \frac{\pi(D + 2h_t + h_y)}{P}$$

$D_0$ = outer diameter of stator

Let $at_y$ be the m.m.f. per *m* for corresponding to average flux density by in stator yoke.

Total m.m.f. for stator yoke for one pole pair

$$AT_y = k_y at_y l_y \qquad ...(5.51)$$

where, $k_y$ is the factor obtained from the curve between $\frac{h_y}{\tau_y}$ and $k_y$ (Fig. 5.55).

$h_y$ = height (thickness of stator yoke).

$\tau_y$ = stator yoke mean pole pitch $= \frac{\pi(D + 2h_t + h_y)}{P}$

The m.m.f. which acts upon the air gap, armature teeth and armature yoke is responsible in determining the magnitude of the leakage flux between the pole shoes. This is the equivalent m.m.f.

$$AT_e = AT_\delta + AT_{at} + AT_y \qquad ...(5.52)$$

### 5.22.2 Pole Leakage Flux

Stator slot offers a parallel path for the flux passing through the teeth. Similarly, the interpolar space in the field system offers a parallel path to the flux leaving the pole and entering the stator. A portion (15 to 20% of useful flux) of pole flux do not enter the stator but takes a path from pole shoe to pole shoe, pole core to pole core as shown in Fig. 5.44. Although the reluctance of the interpolar space is larger still the flux which takes a path through the space between the poles cannot be neglected. The flux which enters the stator and complete the magnetic circuit is called as useful flux and this flux induces e.m.f. in the stator winding. The flux which passes through the interpolar space is a waste and is called as pole leakage flux ($\phi_{el}$). So the flux contained in pole core, shoe and yoke is more than the useful flux. To estimate the total flux contained by the pole of the rotor, estimation of leakage flux is necessary.

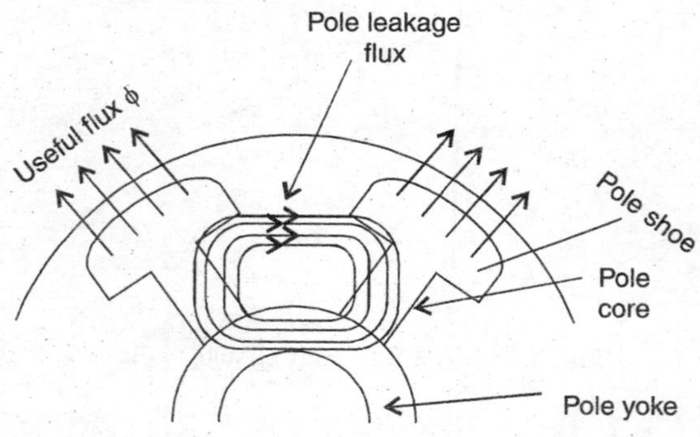


**Fig. 5.44** Pole leakage flux.

***Calculation of Pole Leakage Flux ($\phi_{el}$):*** Pole leakage flux consists of the following leakage flux between the pole pair:

$\phi_i$ = 2 × Leakage flux from pole shoe to pole shoe end or between inner surfaces of pole shoes.

$\phi_{ii}$ = 2 × Leakage flux from pole side to pole side end or between end surfaces of pole shoes.

$\phi_{iii}$ = 2 × Leakage flux from pole core to pole core end or between inside surfaces of pole cores.

$\phi_{iv}$ = 2 × Leakage flux from pole core side to pole core side or between end surfaces of pole cores.

So, pole leakage flux, $\phi_{el}$ will be

$$\phi_{el} = \phi_i + \phi_{ii} + \phi_{iii} + \phi_{iv}$$

$\therefore$ Total flux per pole $\left(\phi_p\right) = \phi_{el}$

The detail of estimation of leakage flux, $\phi_{el}$ is already explained in Chapter 2.

***(4) MMF for a Pair of Pole Core and Shoe ($AT_{pcs}$):*** Flux density in pole core, $B_{pc} = \dfrac{\phi_p}{F_{pc}}$ can be calculated.

where, $F_{pc}$ is cross-sectional area of pole core already estimated.

From *B–H* curve, find out $at_{pc}$ (ampere-turn/m length of pole core) corresponding to $B_{pc}$.

So, $$AT_{pcs} = at_{pcs} \times 2\left(h_{pc} + h_2\right) \quad ...(5.53)$$

***MMF for Pole Yoke:*** $l_{py}$ = length of pole yoke being associate to flux between adjacent poles (Fig. 5.45)

$$= \frac{\pi\left(D - 2\delta - 2h_2 - 2h_{pc} - h_{py}\right)}{p}$$

where, $h_{py}$ is height of pole yoke.

Flux density in the pole yoke can be calculated as below.

$$B_{py} = \frac{\phi_{p/2}}{h_{py}k_i l_p} \text{ tesla}$$

from *B-H* curve, ampere-turn/m length of pole yoke ($at_{py}$) can be found out corresponding to $B_{py}$.

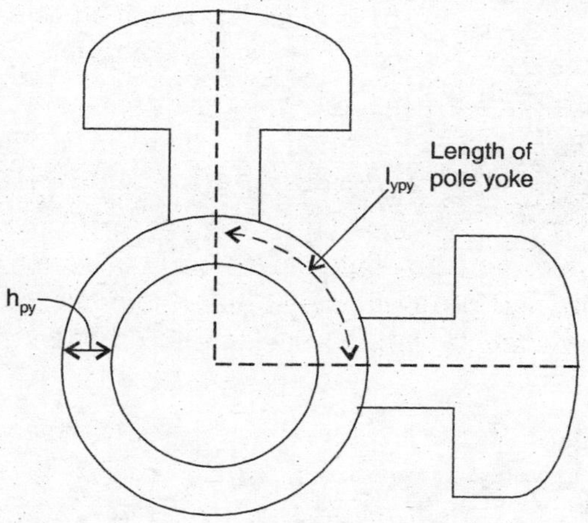


**Fig. 5.45** Length of pole yoke.

Total ampere-turn for pole yoke is given by

$$AT_{py} = at_{py} l_{py} \quad ...(5.54)$$

Total m.m.f. per pole pair to induce rated voltage at no load can be given as

$$AT_{\text{pole pair}} = AT_\delta + AT_{at} + AT_y + AT_{pcs} + AT_{py} \quad ...(5.55)$$

The values of flux density, flux, ampere-turn per metre, and total ampere-turn for different portions of the machine for a magnetic circuit for a pole pair can be calculated and tabulated for ready reference. Although this may not be required in general practice. Only flux which gives terminal voltage at no load and total ampere-turn per pole is generally required to be calculated at different percentage.

For determining the *OCC*, the various values calculated can be tabulated for 20 per cent, 40 per cent, 60 per cent, 80 per cent, 120 per cent as following:

| | 20% | 40% | 60% | 80% | 100% | 120% |
|---|---|---|---|---|---|---|
| $E = \phi_1$ | | | | | | |
| $B_{im}$ | | | | | | |
| $B_{tt}$ | | | | | | |
| $B_{tm}$ | | | | | | |
| $B_{tb}$ | | | | | | |
| $at_{tt}$ | | | | | | |
| $at_{tm}$ | | | | | | |
| $at_{tb}$ | | | | | | |
| $at_{at}$ | | | | | | |
| $B_y$ | | | | | | |
| $at_y$ | | | | | | |
| $\phi_{el}$ | | | | | | |
| $\phi_p$ | | | | | | |
| $B_{pc}$ | | | | | | |
| $B_{py}$ | | | | | | |
| $at_{pc}$ | | | | | | |
| $at_{py}$ | | | | | | |
| $AT_\delta$ | | | | | | |
| $AT_{at}$ | | | | | | |
| $AT_y$ | | | | | | |
| $AT_{pcs}$ | | | | | | |
| $AT_{py}$ | | | | | | |
| $AT_{\text{pole pair}}$ | | | | | | |

$$\frac{AT_{\text{pole pair}}}{2} = AT_{\text{per pole}}$$

The curve drawn between ampere-turn per pole and flux giving terminal voltage at no load gives open circuit characteristics (*OCC*).

## 5.23 STATOR LEAKAGE REACTANCE

$$X_{al} = X_{sl} + X_{tl} + X_{el}$$

where, $X_{sl}$ = slot leakage reactance

$X_{tl}$ = tooth top leakage reactance

$X_{el}$ = end winding leakage reactance

### 5.23.1 Slot Leakage Reactance

$$X_{sl} = 2\pi f\, L_{sl}$$

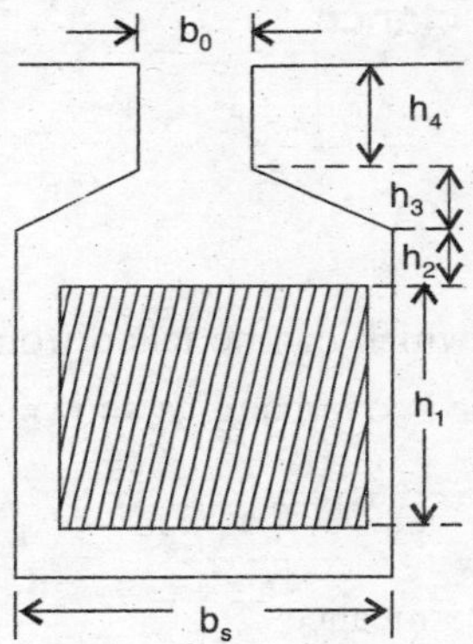


**Fig. 5.46** Slot leakage inductance for single layer winding.

where, $L_{sl}$ is slot leakage inductance for single layer winding (refer Fig. 5.46)

$$L_{sl} = \frac{4\mu_0 N_{ph}^2 L}{P \cdot q_1}\left[\frac{h_1}{3b_s} + \frac{h_2}{b_s} + \frac{2h_3}{b_0 + b_s} + \frac{h_4}{b_0}\right] \text{henry}$$

Slot leakage inductance for double layer winding (refer Fig. 5.47)

$$L_{sl} = \frac{4\mu_0 N_{ph}^2 L}{P \cdot q_1}\left[\frac{2h_1}{3b_s} + \frac{h_2}{4b_s} + \frac{h_3}{b_s} + \frac{2h_4}{b_0 + b_s} + \frac{h_5}{b_0}\right]$$

Depending upon the type of winding whether single or double, the slot leakage inductance, $l_{sl}$ can be estimated and then slot leakage reactance

$X_{sl} = 2\pi f \, L_{sl}$ can be calculated. ...(5.56)

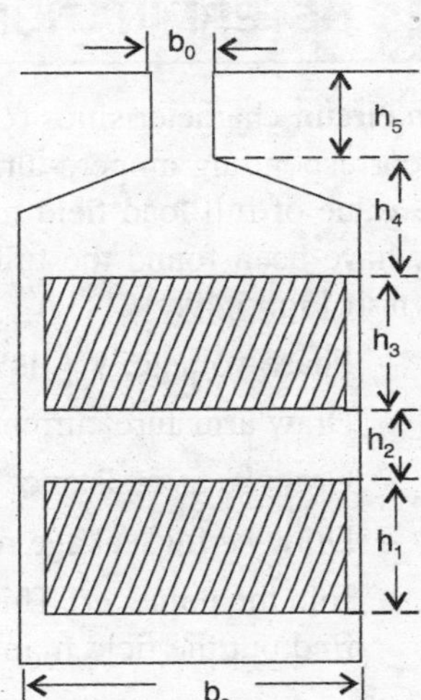


**Fig. 5.47** Slot leakage inductance for double layer winding.

### 5.23.2 Tooth Top Leakage Reactance

$X_{tl} = 2\pi f \, L_{tl}$ ...(5.57)

where, $L_{tl}$ is tooth leakage inductance and given by

$$= \frac{\mu_0 N_{ph}^2}{P \cdot q_1} \cdot L\lambda_{tl}$$

$L$ = overall length of machine.

and
$$\lambda_{tl} = \frac{5\left(\frac{\delta}{b_0}\right)}{5 + 4\left(\frac{\delta}{b_0}\right)} \cdot \frac{b_p}{\tau_p}$$

Now, putting the value of $\lambda_{tl}$, $L_{tl}$ can be calculated and then $X_{tl}$.

### 5.23.3 End Winding Leakage Reactance

End Winding (overhang) permeance

$$L_0\,\lambda_{el} = \frac{\mu_0\,K\,\tau_p^2}{\pi\,\tau_s} \qquad ...(5.58)$$

where, $K$ = Slot leakage factor which can be taken from Fig. 6.20 (*a*)

$L_0$ = Length of conductor is overhang as shown in Fig. 6.20 (*b*)

$\tau_s$ = slot pitch

$\tau_p$ = pole pitch

$\lambda_{el}$ = specific overhang permeance

End winding (overhang) leakage reactance per phase can be calculated by

$$X_{el} = 8\pi f \frac{N_{ph}^2}{P \cdot q_1} \cdot L_0 \lambda_{el}$$

Stator leakage reactance

$$X_{al} = X_{sl} + X_{tl} + X_{el} \qquad ...(5.59)$$

There is voltage drop in the stator winding for phase current $I_{ph}$ and leakage reactance $X_{al}$.

So, Leakage drop = $I_{ph} \times X_{al}$ ...(5.60)

PU leakage reactance, $X = \dfrac{I_{ph} X_{al}}{E_{ph}}$ ...(5.61)

## 5.24 DETERMINATION OF FULL LOAD FIELD MMF

Open circuit characterisitics (*OCC*) as shown in Fig. 5.48 gives the relation between voltage at no load and corresponding ampere-turn per pole. To determine the approximate height of pole core, approximate value of full load field m.m.f. was estimated earlier. Since leakage reactance drop per phase and *OCC* have been found the full load field m.m.f. ($AT_{fl}$) can be determined by the process as below as shown in Fig. 5.49.

- Phase voltage, $V_{ph}$, is drawn taking suitable scale by line – *ab*
- Draw armature current/phase. $I_{ph}$ at an angle $\phi$ w.r.t. $V_{ph}$ ($\cos\phi$ = p.f. lapping) – *aI*
- Draw *bc* as resistance drop per phase ($I_{ph} \times R_{ph}$) parallel to *aI*.
- Draw *ce* for leakage reactance drop per phase ($I_{ph} \times X_{al}$) perpendicular to *aI*.
- Join the points *ae*. This gives the generated voltage $E_g$. Corresponding to generated voltage $E_g$, find out the field m.m.f. from *OCC* curve given in Fig. 5.48.

Let this is $AT_g$.

- Mark point *d* so that $ad = AT_g$
- Draw field m.m.f. (*dg*) equivalent to armature m.m.f. per pole at full load perpendicular to *al* at *d*.

Field m.m.f. equivalent to armature m.m.f. per pole $= \dfrac{2.7 I_{ph} N_{ph} k_w \rho_d}{p}$ ...(5.62)

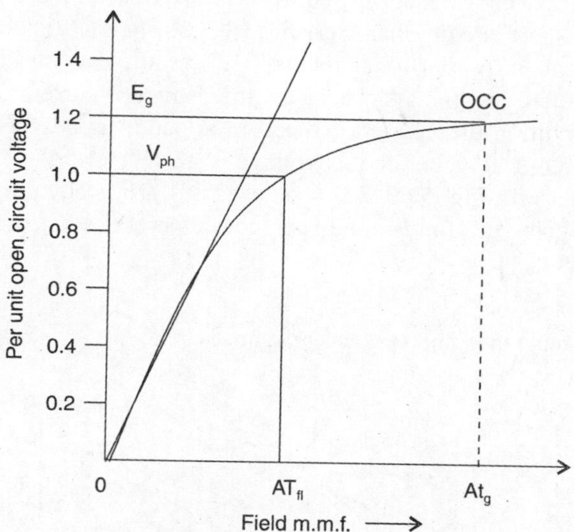


**Fig. 5.48** Open circuit characteristics

where, $\rho_d = \dfrac{\alpha + \sin\alpha}{4\sin\dfrac{\alpha}{2}}$

where, $\alpha$ is angle span of pole arc and is equal to $\dfrac{b_p}{\tau_p} \cdot \pi$

Generally the value of $\rho_d$ is taken as 0.85.

$\rho_d$ for cylindrical rotor machine $= \dfrac{\pi^2 b_p/\tau_P}{8\sin\left(b_p/\tau_p \cdot \pi/2\right)}$

– Mark a point $f$ on $dg$ so that

$$\frac{df}{dg} = k_r$$

where, $k_r$ is the cross-reaction coefficient and depends upon the pole arc to pole pitch ratio and it can be found out from Fig. 5.40.

So, $df = k_r \cdot dg$

– Join $af$ and extend it.

– Draw a perpendicular from point $g$ on extended portion of $af$. If perpendicular cut at point $j$, $aj$ = full load field m.m.f. ($AT_{fl}$) (Fig. 5.49)

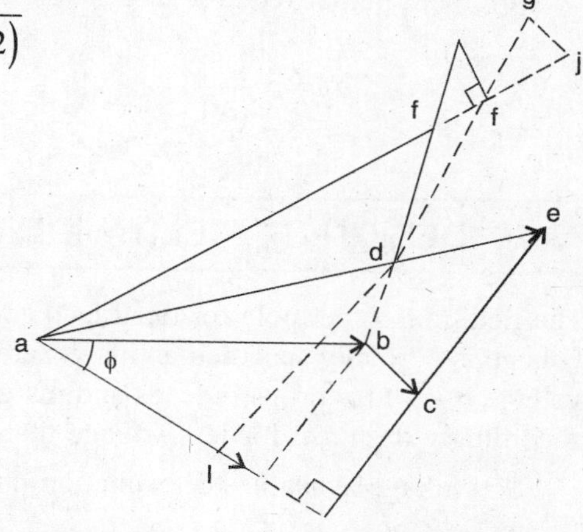


**Fig. 5.49** Determination of full load field m.m.f.

## 5.25 SHORT-CIRCUIT CHARACTERISTICS

The characteristics shown between the field m.m.f. and the armature current when the armature is short-circuited is known as short circuit. characteristics (SCC). The *SCC* is obtained straight line since armature short-circuit current is proportional to the field m.m.f. over a wide range. The armature m.m.f. has demagnetising effect on the field because the armature current is in quadrature with the voltage. In short-circuit condition the voltage induced in armature is nearly equal to $I_{ph}\ X_{al}$ (leakage reactance drop). The field m.m.f. to generate this voltage is *AB* (Fig. 5.50). The field m.m.f. equivalent to armature m.m.f. = $\rho_d \times AT_a$. In Fig. 5.50, $BD = \rho_d \times AT_a$. If *DE* is equal to rated arm current, point *E* should be on *SCC*. *AD* gives the field m.m.f. per pole to produce rated arm current under short-circuit condition.

Estimation of SCR:

(*i*) Calculate leakage reactance and leakage reactance drop.

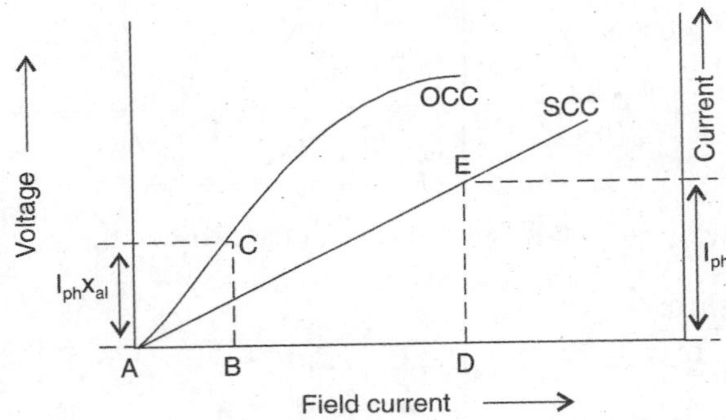


**Fig. 5.50** OCC and SCC.

(*ii*) Estimate *MMF* corresponding to leakage reactance drop from *OCC*.

(*iii*) Find out field m.m.f. equivalent to arm m.m.f. = $\rho_d \times AT_a$

(*iv*) Full load field m.m.f. = (*ii* + *iii*)

(*v*) Field m.m.f. required to generate rated voltage at no load.

$$\therefore \qquad SCR = \frac{(v)}{(iv)}$$

## 5.26 DESIGN OF FIELD WINDING OF SALIENT POLE ALTERNATOR

The field m.m.f. per pole for rated load and power factor condition has been estimated. The excitation voltage is normally specified and is generally 50 to 500 volt d.c. ($V_{dc}$). 15 to 20 per cent of excitation voltage is kept for drops in leads and margins for variation in the field excitation. The field winding is accordingly designed for less voltage than the exciter voltage.

So, voltage available for excitation purpose = $(0.80–0.85)V_{dc}$

Thus $(0.80–0.85)V_{dc} = I_{fd} \times R_{fd}$ volt ...(5.63) where, $I_{fd}$ = field current,

$R_{fd}$ = resistance of the field winding at 75°C in ohm.

$$R_{fd} = 0.021 \times 10^{-6} \frac{L_{mft} N_{ft}}{F_{fd}} \qquad ...(5.64)$$

where, $F_{fd}$ = sectional area of field winding conductor in $m^2$

$L_{mft}$ = mean length of field turn (Fig. 5.51) in m

$= 2L_m + \pi(d + 0.01 + d_f)$

$L_m = 0.9\, L_p$

$d$ = width/diameter of pole core

$d_f$ = depth of field winding given in Fig. (5.51)

$N_{ft}$ = total no. of field turns

$$I_{fd} = \delta_{fd} \,.\, F_{fd} \qquad ...(5.64a)$$

$\sigma_{fd}$ = current density permissible for field conductor

= 3 to 4 amp/mm$^2$

Putting the value of $I_f$ and $R_{fd}$ from equation (5.64 & 5.64(*a*)) in equation (5.63), we get



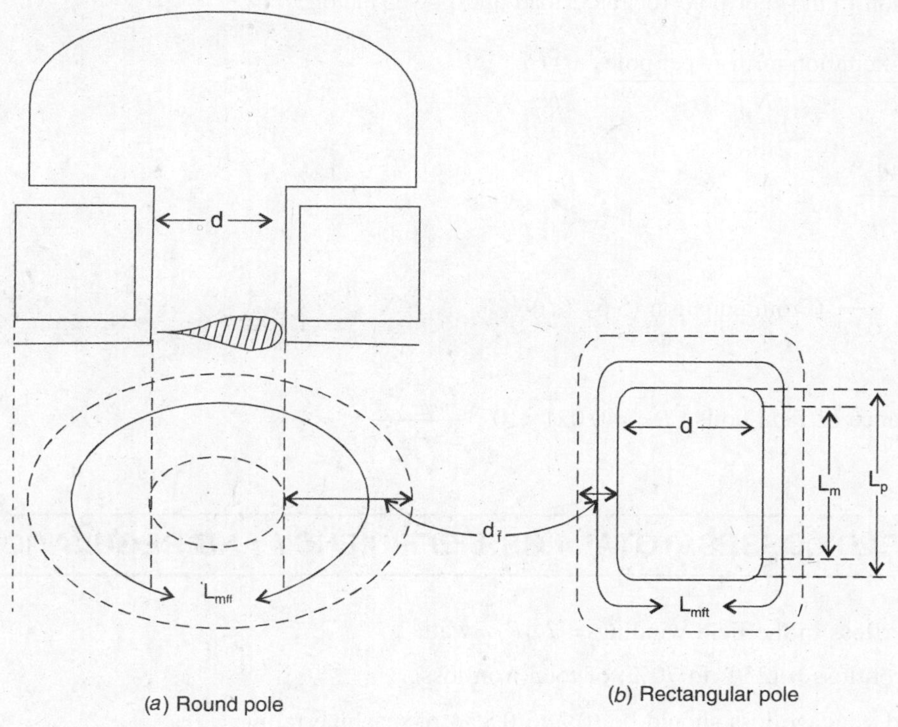


**Fig. 5.51** Depth of field winding.

$$(0.80\text{–}0.85)\,V_{dc} = \sigma_{fd} \cdot F_{fd} \times 0.021 \times 10^{-6}\,\frac{L_{mft}\,N_{ft}}{F_{fd}}$$

$$= 0.021 \times 10^{-6}\,L_{mft} \cdot N_{ft} \cdot \sigma_{fd}$$

$$\therefore \quad N_{ft} = \frac{(0.80 - 0.85)\,V_{dc}}{0.021 \times 10^{-6}\,L_{mft}\sigma_{fd}} \qquad \text{...(5.65)}$$

So, $N_{ft}$ can be estimated with the help of equation (5.65)

$$\text{Number of field turns per pole} = \frac{\text{Total number of field turns}}{\text{Number of poles}}$$

$$\text{Field turns/pole, } N_p = \frac{N_{ft}}{p}$$

Number of field turns per pole may not be an integer. It may be made an integer by modifying the current density. Further the height and the depth of the field winding can be modified to accommodate the field winding and also to keep the temperature rise within the limit.

Excitation m.m.f. per pole for rated load and power factor, $AT_{fl} = I_{fd} \cdot N_p$

$$\therefore \quad I_{fd} = \frac{\text{Excitation m.m.f. per pole}}{N_p} = \frac{AT_{fl}}{N_p}$$

$$\Rightarrow \quad I_{fd} = \frac{AT_{fl}}{N_p}$$

Also, $F_{fd} = \dfrac{I_{fd}}{\sigma_{fd}}$ (From equation (5.64 (*a*))

$$\text{Total resistance of field coils, } R_{fd} = 0.021 \times 10^{-6}\,\frac{L_{mft}\,N_{ft}}{F_{fd}} \qquad \text{...(5.66)}$$

## 5.27 FIELD LOSSES, TOTAL LOSS, EFFICIENCY AND REGULATION

Total copper loss in the field winding = $I_{fd}^2 R_{fd}$ watt.

Pole face iron loss = (25% to 70%) of total iron loss.

Friction and windage loss should be (0.2 to 0.8%) of machine rating.

Friction and windage loss with hydrogen cooling should be (0.3 to 0.4%) of m/c rating.

Taking a drop of 1.0 V at each brush and brush contact loss can be estimated

$$= 2 \times 1.0 \times I_{fd} \text{ watt}$$

Field copper loss + Brush constant loss = ? watt

Stray load loss is taken as 20% of total Cu loss in stator.

Total loss in the machine can be estimated.

Total loss = Copper loss in stator and rotor
+ Iron loss in stator and rotor
+ Stray load loss
+ Friction and windage loss

$$\text{Efficiency} = \frac{\text{Output}}{\text{Input}}$$

$$= \frac{\text{Rated output}}{\text{Rated output} + \text{Total losses}}$$

$$= \frac{KVA \times \text{power factor}}{KVA \times \text{Power factor} + \text{Total losses}} \qquad ...(5.67)$$

***Voltage Regulation:*** The voltage regulation of a synchronous generator is the rise in the voltage at the terminals when the load is reduced from full load rated value to zero and speed and field current remaining constant.

$$\text{Per unit voltage regulation} = \frac{\Delta |E_g| - |V_{ph}|}{|V_{ph}|}$$

where, $|E_g|$ = magnitude of generated voltage per phase,

and $|V_{ph}|$ = magnitude of rated voltage per phase.

The voltage generated ($E_g$) per unit corresponding to full load field m.m.f. estimated.

$$\% \text{ Regulation} = \frac{E_g - V_{ph}}{V_{ph}} \times 100 \qquad ...(5.68)$$

## 5.28 TEMPERATURE RISE IN ALTERNATOR

### 5.28.1 Temperature Rise of Field Coils

$$\text{Voltage applied across each field coil} = \frac{(0.80\text{–}0.85)\, V_{dc}}{P} = V_f$$

Resistance of each field coil, $R_f = 0.021 \times \dfrac{L_{mft} \times N_p}{F_{fd}} \times 10^{-6}$

Actual value of field current, $I_f = \dfrac{V_f}{R_f}$

m.m.f. provided, $AT_{fl} = I_f N_p$

Loss in each field coil, $Q_{fd} = I_f^2 R_f$

Area of dissipating surface, $S_f = 2 L_{mft} \left(h_f + d_t\right)$

where $h_f$ = height of field winding

and $d_f$ = depth of field winding

Cooling coefficient, $C_{fd}$, can be estimated with the help of Table 5.1 by selecting a suitable value at calculated peripheral speed in m/sec.

$$\text{Temperature rise, } \theta = \frac{Q_{fd} C_{fd}}{S_f} \qquad \text{...(5.69)}$$

The temperature should be within the specified limit depending upon the insulating materials used. If the temperature is more the depth of winding should be modified (increase) to reduce the temperature suitably.

### 5.28.2 Temperature Rise of Stator

Stator temperature rise depends upon the sum of all losses to be dissipated from the stator surface and the total loss dissipated from the stator surface per °C temperature. It can be written as

$$\text{Stator temperature, } \theta = \frac{\text{Total loss to be dissipated from stator surface in watt}}{\text{Total loss dissipated from stator surface in watt per °C.}}$$

Total loss to be dissipated from different portions of stator and losses dissipated from different portions of the stator can be calculated as below. The losses are dissipated from outer cylindrical surface of stator, end surfaces of stator, inner cylindrical surface of stator and through ducts.

***Loss Dissipated from Outer Cylindrical Surface of Stator and End Surfaces:*** Outer cylindrical surface of stator $= \pi D_0 L$

Area of end surfaces $= 2\dfrac{\pi}{4}\left(D_0^2 - D^2\right)$

Cooling coefficient for outer surfaces can be calculated from Table 5.1 with peripheral velocity ($V_c = V_p$).

Cooling coefficient for end surfaces can also be selected from Table 5.1 ($V_c = 0$).

**Table 5.1** Cooling coefficient

| *Portion of machine* | $C_{fd}$ | $V_c$ |
|---|---|---|
| Cylindrical surface of stator and rotor | $\dfrac{0.03 \text{ to } 0.05}{1+0.1V_c}$ | Relative peripheral speed |
| End of stator surfaces | 0.02 to 0.04 | Zero |
| Cylindrical surface of d.c. armature | $\dfrac{0.015 \text{ to } 0.035}{1+0.1V_c}$ | Armature peripheral speed |
| Stationary field coils | $\dfrac{0.14 \text{ to } 0.16}{1+0.1V_c}$ | Armature peripheral speed |
| Rotating field coil | $\dfrac{0.08 \text{ to } 0.12}{1+0.1V_c}$ | Armature peripheral speed |
| Ventilating ducts | $\dfrac{0.08 \text{ to } 0.2}{V_c}$ | Air velocity in ducts |
| Commutative | $\dfrac{0.015 \text{ to } 0.025}{1+0.1V_c}$ | Commutator peripheral speed |



$$\text{Loss dissipated from outer surface (watt/°C)} = \frac{\text{Outer cylindrical surface}}{\text{Cooling coefficient for outer surface}}$$

$$\text{Loss dissipated from end surfaces (watt/°C)} = \frac{\text{Area of end surfaces}}{\text{Cooling coefficient for end surfaces}}$$

So, total loss dissipated from outer and end surfaces can be found out.

***Loss Dissipated by Inner Cylindrical Surface***

Inner cylindrical surface = $\pi DL$

Cooling coefficient can be calculated from Table (5.1), when $V_c = V_p$

$$\text{Loss dissipated from inner surface (watt/°C)} = \frac{\text{Inner cylindrical surface}}{\text{Cooling coefficient}}$$

***Losses Dissipated by Ducts***

$$\text{Area of ducts } = \frac{\pi}{4}\left(D_0^2 - D^2\right)n_v$$

where, the velocity for ducts, $V_c = 0.1\ V_p$.

$$\text{Loss dissipated from ducts} = \frac{\text{Area of ducts}}{\text{Cooling coefficient}}$$

**Table 5.2** Depth of field winding.

| *Pole pitch (m)* | *Winding depth (mm)* |
|---|---|
| 0.10 | 30 |
| 0.20 | 35 |
| 0.30 | 40 |
| 0.40 | 45 |
| 0.50 | 50 |
| 0.55 and above | 55 |

***Estimate Total Loss Dissipated in Watt/°C:*** Copper loss in slot portion of conductors

$$= k_{e(\text{average})} \times I_{ph}^2 \times \text{ D.C.}$$

resistance of stator conductor in slot portion.

where, $K_{e(average)}$ is the average eddy current loss factor and can be expressed as

$$= 1 + (\alpha h')^4 \frac{M^2}{9}$$

$$\alpha = \sqrt{\frac{\text{Copper width in the slot}}{\text{Slot width}}}$$

$h'$ = height of conductor in the slot,

$M$ = number of conductor layers in the slot,

So, D.C. resistance of conductors in the slot portion per phase can be estimated as

$$= 0.021 \times \frac{N_{ph} \cdot 2L}{F_c}$$

where, $L$ is in m and $F_c$ is in mm$^2$.

***Estimate Copper Loss in Slot Portion of Conductors:*** Iron loss in stator has already been estimated.

The total loss to be dissipated from stator surface = Copper loss in slot portion of conductor + Iron loss in stator.

$$\text{So, stator temperature rise, } \theta = \frac{\text{Total loss to be dissipated from stator surface (in watt)}}{\text{Total loss dissipated from stator surface (watt/°C)}} \quad ...(5.70)$$

## 5.29 ROTOR DESIGN FOR TURBO-ALTERNATOR

The rotor of turbo-alternators are cylindrical. Slots are provided on rotor to accomodate the field winding. The field winding is not concentric but it is distributed in different slots of the rotor as shown in Fig. 5.52. 4 to 8 slots per pole are selected for turbo-alternator.

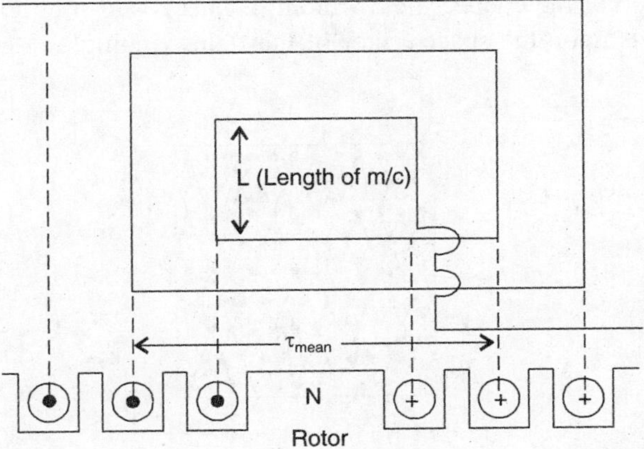


**Fig. 5.52** Field winding of turbo-alternator.

Full load field m.m.f. ($AT_{fl}$) can be expressed as

$$AT_{fl} = 2AT_a$$

where, $AT_a = \dfrac{2.7\,N_{ph} \cdot I_{ph} \cdot k_w}{P}$, knowing $AT_a$, $AT_{fl}$ can be estimated.

Since 15–20% of the excited voltage is kept reserved, voltage across each field coil can be estimated by

$$E_f = \frac{(0.80\text{–}0.85)\,V_{dc}}{P}$$

$$E_f = I_f R_f = I_f \times 0.021 \frac{N_f\, L_{mft}}{F_{fd}}$$

$$= 0.021 \times \frac{I_f (N_f) L_{mft}}{F_{fd}} = 0.021 \times \frac{AT_{fl} \cdot L_{mft}}{F_{fd}}$$

where, $N_f$ = number of field winding turns per pole

$L_{mft}$ = mean length of turns in metre

$= 2[L + (1.5\text{-}2.5)\,\tau_{mean}]$

$\tau_{mean}$ = mean length of coil span

So, $F_{fd}$ can be estimated.

Copper section required in the field winding per pole

$$F_{tp} = N_f \times F_{fd}$$

Slot section required for the slots per pole will depend upon the copper section required for the field winding and the space required for the insulation, supports etc. The cross-section of slot will be more than the cross-section of the copper (field winding). This is shown in Fig. 5.53. The actual cross-section of the slot depends upon the space factor of the field winding.

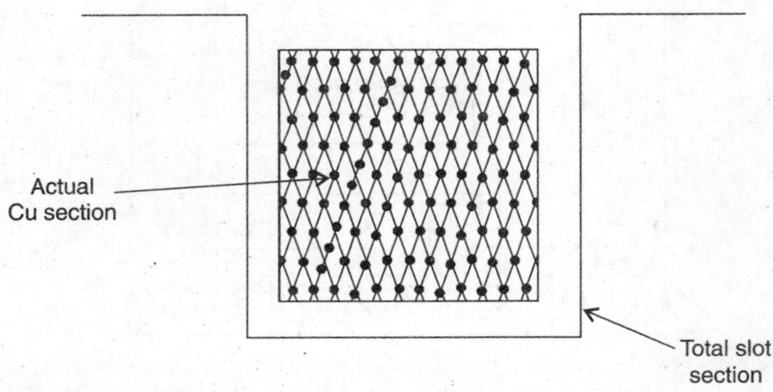


**Fig. 5.53** Rotor slot.

$$\text{Actual section required by the slots per pole} = \frac{F_{tp}}{k_f}$$

where, $k_f$ is the winding space factor, having value between 0.6 and 0.7

Select suitable number of slots between 4 to 8.

$$\text{Slot section} = \frac{\text{Actual section by the slots per pole}}{4\text{–}8} \qquad \text{...(5.71)}$$

**Fig. 5.54** Height and width of rotor slot.

Let $h_r$ = height of rotor slot

$b_r$ = width of rotor slot

Then, $h_r \times b_r$ = Slot section, ...(5.72)

which is shown in Fig. 5.54.

The ratio of height to width of rotor slot may be considered as

$$\frac{h_r}{b_r} = 2 \text{ to } 3 \qquad \text{...(5.73)}$$

So, $h_r$ and $b_r$ can be estimated, from equations (5.72) and (5.73).

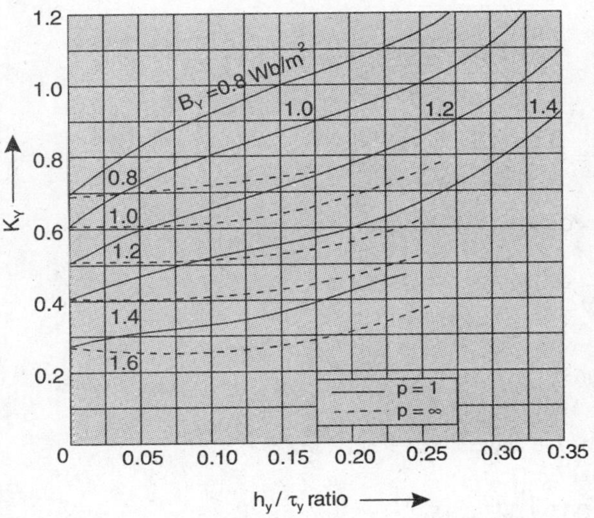


**Fig. 5.55** Curve for $K_Y$.

## SOLVED PROBLEMS ON ALTERNATOR

**Q. 1** Estimate the main dimensions of 1800 kVA, 50 Hz, 3-phase, 187.5 rpm, water wheel generator. The specific magnetic loading is 0.8 Wb/m² and the specific electric loading is 26000 A-conductor/m. Pole arc to pole pitch ratio is 0.66. Calculate peripheral speed also and justify the estimation of *D* and *L*.

**Solution:**

Output equation, $Q = CD^2LN$

where, $C = 11.1 \times 10^{-5}\ B_{\delta_1} \cdot \overline{ac}$

$$B_{\delta_1} = 0.80 \text{ Wb/m}^2$$

$$\overline{ac} = 26000 \text{ A-conductor/m}$$

$$\therefore \quad C = 2.3$$

For round pole construction,

$$\frac{b_p}{\tau_p} = \frac{L}{\tau_p}$$

$$\therefore \quad \frac{L}{\tau_p} = 0.66 \text{ (given)} \qquad \ldots(i)$$

$$\tau_p = \frac{\pi D}{P} = \frac{3.14 \times D}{P}$$

Number of poles, $P = \dfrac{120 \times 50}{187.5} = 32$

$$\tau_p = \frac{3.14 \times D}{32}$$

Putting this value in equation (*i*) we get,

$$\frac{L \times 32}{3.14 \times D} = 0.66$$

$$\therefore \qquad \frac{L}{D} = 0.065 \qquad \text{...(ii)}$$

From output equation, we have

$$D^2L = 4.17$$

So, $D^2 \times (0.065\ D) = 4.17$

or $D^3 = \dfrac{4.17}{0.065} = 64.15$

or $D = 4.003$ m

Now putting the value of $D$ from above in equation (*ii*), we get

$$L = 0.065 \times 4.003 \text{ m} = 0.26 \text{ m}$$

Peripheral speed, $V_p = \dfrac{\pi DN}{60}$

$$= \frac{3.14 \times 4.003 \times 187.5}{60}$$

$$= 39.28 \text{ m/sec.}$$

This is within the range of peripheral velocity *i.e.*, between 30 and 80 m/sec. Hence, the calculated value of $D$ and $L$ are justified.

**Q. 2** Calculate the main dimension of 3-phase, 12MVA, 50Hz, 6.6 kV, 2-pole turbogenerator. Specific magnetic loading is 0.88 Wb/m$^2$ and ampere-conductor loading per metre length of periphery is 52,000. Peripheral speed is 120m/s. Air gap length is 12 mm. Rotor diameter is considered for peripheral velocity.

**Solution:**

Let $D_r$ = Rotor diameter

Given, $\dfrac{\pi D_r N}{60} = 120$

Now, $f = \dfrac{PN}{120}$,

so, $N = \dfrac{120 \times 50}{2} = 3000$ r.p.m.

$$\therefore \quad D_r = \frac{120 \times 60}{\pi \times 3000} = 0.764\,\text{m}$$

Inner or air gap diameter, $D = D_r + 2\delta$

$= 0.764 + 2(12 \times 10^{-3})$

$D = 0.788$ m

Given, $B_{\delta_1} = 0.88$ Wb/m²

$\overline{ac} = 52{,}000$ A-conductor/m length of periphery

$Q$ = Rating in kVA = 12,000 kVA

Output equation, $Q = CD^2LN$

$12{,}000 = 11.1 \times 10^{-5} B_{\delta_1} \cdot \overline{ac} \cdot D^2 LN$

$$\Rightarrow \quad L = \frac{12{,}000}{11.1 \times 10^{-5} \times 0.88 \times 52{,}000 \times (0.788)^2 \times 3000}$$

$L = 1.268$ m

So, diameter, $D = 0.788$ m

and length, $L = 1.268$ m

which are well in limits.

**Q. 3** Estimate the main dimensions of 8 MVA, 6.6 kV, 50 Hz, 3-phase, 2 pole cylindrical rotor generator with following specifications:

Specific magnetic loading = 0.92

Specific electric loading = 50,000

Rotor peripheral velocity = 126 m/s

Air-gap length is 17 mm.

**Solution:**

$$N_s = \frac{120f}{p} = \frac{120 \times 50}{2} = 3000 \text{ r.p.m.}$$

$$= 50 \text{ r.p.s.}$$

Let, $D_r$ = diameter of rotor,

$V_p$ = 126 m/sec. (given)

Now, $\dfrac{\pi D_r N}{60} = 126$

Chapter 5

$$\Rightarrow \quad D_r = \frac{126 \times 60}{3.14 \times 50} = 0.8025 \text{ m}$$

Inner diameter or air-gap diameter is given by

$$D = D_r + 2$$
$$= 0.8025 + 2 \times 17 \times 10^{-3}$$
$$= 0.8365 \text{ m}$$

Output equation is

$Q = CD^2LN$, Take Q in kVA *i.e.*, $8 \times 10^3$ kVA

or $\quad 8 \times 10^3 = 11.1 \times 10^{-5} \times B_{\delta_1} \times \overline{ac} \times D^2 \cdot LN$

or $\quad = 11.1 \times 10^{-5} \times 0.92 \times 50{,}000 \times (0.836)^2 \times L \times 3000$

$$\Rightarrow \quad L = \frac{8 \times 10^3}{107.1 \times 10^2}$$

or $\quad L = 0.746$ m

So, $\quad D = 0.836$ m

**Q. 4** A 1250 kVA, 3.3 kV, 50 Hz, 300 rpm, 3-phase star connected alternator has 180 slots and 5 conductors per slot having single layer winding with full pitch coils and one circuit per phase. Determine the specific electric and specific magnetic loadings if the air gap diameter of alternator is 2.2 m and axial length is 0.5 m. Assume phase spread of 60°.

**Solution:**

Total number of conductors = Number of slots × Conductors per slot
= 180 × 5 = 900

Turns per phase, $N_{ph} = \frac{900}{2 \times 3} = 150$

Synchronous speed, $N_s$ = 300 r.p.m.

$$N_s = \frac{300}{60} = 5 \text{ r.p.s.}$$

Number of poles $= \frac{120 \times 50}{300} = 20$

Number of slots per pole per phase, $q_1 = \frac{180}{3 \times 20} = 3$

Distribution factor, $k_d = 0.96$ for $q_1 = m = 3$

Coil has full pitch, so pitch factor, $k_p = 1$

So, winding factor $k_w = k_p \times k_d = 0.96 \times 1 = 0.96$

$$\text{Voltage per phase, } V_{ph} = \frac{3300}{\sqrt{3}} = 1910\,\text{V}$$

$$\therefore \quad \text{Flux per pole, } \phi = \frac{1910}{4.44 \times 50 \times 150 \times 0.96}$$

$$= 59.8 \times 10^{-3}\ \text{Wb.}$$

$$\text{Pole pitch, } \tau_p = \frac{\pi D}{P} = \frac{\pi \times 2.2}{20} = 0.3454\,\text{m}$$

Area under one pole through which flux is contained = 0.3454 × 0.5 = 0.172 m$^2$

$$\text{Specific magnetic loading, } \overline{B} = \frac{59.8 \times 10^{-3}}{0.172} = 0.347\,\text{Wb/m}^2$$

$$\text{Current per phase, } I_{ph} = \frac{1250}{\sqrt{3} \times 3.3}$$

$$I_{ph} = 218.7\ \text{A}$$

Since there is one circuit per phase

∴ Conductor current = $I_{ph}$ = 218.7 A

So, specific electric loading

$$\overline{ac} = \frac{6 I_{ph} \cdot N_{ph}}{\pi \times D}$$

$$= \frac{6 \times 218.7 \times 150}{3.14 \times 2.2} = 28493\ \text{A/m}.$$

**Q. 5** Calculate flux per pole, length and width of pole and height of winding of a 2000 kVA, 200 r.p.m. 3-phase, 50 cycle/sec., 2000 V star connected salient pole alternator with long pole construction with the following details:

| | |
|---|---|
| Inner diameter | = 2m |
| Length | = 0.5 m |
| Slots per pole per phase | = 3.5 |
| Conductors per slot | = 2 |
| Leakage factor | = 1.2 |
| Winding factor | = 0.95 |
| Flux density in pole, $B_p$ | = 1.5 Wb/m$^2$ |
| The field winding depth | = 0.03 m |

Field m.m.f. is taken twice of armature m.m.f. and field winding space factor = 0.8

Field winding current density = 4 A/mm$^2$

Chapter 5

**Solution:**

Number of poles $= \dfrac{120 \times 50}{200} = 30$

Total number of slots = 3 × 3.5 × 30 = 315

Total conductors = 315 × 2 = 630

Number of turns per phase, $N_{ph} = \dfrac{630}{3 \times 2} = 105$

Voltage per phase, $V_{ph} = \dfrac{2000}{\sqrt{3}} = 1154.7\,\text{V}$

$$V_{ph} = 4.44\, f\, \phi\, N_{ph} \times k_w$$

Useful flux per pole, $\phi = \dfrac{1154.7}{4.44 \times 50 \times 0.95 \times 105} = 0.052\text{ T}$

Flux in pole = Leakage factor × Useful flux per pole

$$\phi_p = 1.2 \times 0.052$$

$$= 0.062$$

Flux per pole, $\phi_p = B_p \times \text{Width of pole} \times L_p$

$L_p$ = Length of pole = Length of m/c = 0.5 m

∴ Width of pole $= \dfrac{0.062}{1.5 \times 0.5} = 0.082\text{ m}$

Current per phase, $I_{ph} = \dfrac{2000 \times 1000}{\sqrt{3} \times 2000} = 577.37\text{ A}$

Armature m.m.f. per pole

$$AT_a = \frac{2.7 \times 577.37 \times 105 \times 0.95}{30}$$

$$= 5183\text{ A}$$

Field m.m.f. per pole = 2 × 5183 = 10366.0

Area of field winding, $h_f \times d_f = \dfrac{AT_{fl}}{\text{Space factor} \times \text{Current density}}$

$$= \frac{10366}{0.8 \times 4 \times 10^6} = 3239 \times 10^{-6}\text{ m}^2$$

But, $d_f = 0.03$ m, (given)

So, $h_f = \dfrac{3239 \times 10^{-6}}{0.03} = 0.1$ m.

**Q. 6** Calculate the ampere-conductor per metre length of periphery and total number of conductors for a 1000 kVA, 3000 volt, 50 Hz, 3-phase star connected, 20-pole alternator, inner diameter of 1.9 m and length of 0.32 m. Maximum value of flux density is 0.9 T and winding factor is 0.96 and single circuit.

**Solution:**

$$\text{Speed, } N = \frac{120 \times 50}{20} = 300 \text{ r.p.m.}$$

We have $Q = CD^2LN$

$$\text{or} \quad 1000 = 11.1 \times 10^{-5} \times B_{\delta_1} \cdot \overline{ac} \times D^2 LN$$

$$= 11.1 \times 10^{-5} \times 0.9 \times \overline{ac} \times (1.9)^2 \times 0.32 \times 300$$

$$= 3462.13 \times 10^{-3} \times \overline{ac}$$

$$\text{or} \quad \overline{ac} = \frac{1000}{3462.13 \times 10^{-3}} = 0.28883 \times 10^5$$

$$= 28883 \text{ A/m}.$$

Let conductor current is $I_z$, total conductor = $Z$

$$\therefore \quad I_z \cdot Z = \overline{ac} \cdot \pi D$$

$$\text{since} \quad I_z = \frac{1000 \times 10^3}{\sqrt{3} \times 3000} = 192.45 \text{ A}$$

$$\text{So,} \quad Z = \frac{28883 \times 3.14 \times 1.9}{192.45}$$

$$= 895$$

**Q. 7** Estimate suitable air gap diameter and length of a 10 MVA, 11 kV, 8-pole 3-phase, 50 Hz star-connected synchronous generator. Maximum air-gap flux density is 0.92 tesla. The ampere per conductor per metre length of periphery is varying from 20,000 to 40,000 A/m. The peripheral velocity should not be more than 80 m/sec. Suggest the type of pole to be constructed. Pole arc to pole pitch ratio is 0.66.

**Solution:**

$$\text{Speed} = \frac{120 \times 50}{8} = 750 \text{ r.p.m.}$$

$$B_{\delta_1} = 0.92 \text{ Wb/m}^2$$

Let us consider $\overline{ac}$ is 30,000

$\overline{ac} = 30{,}000$ A/m

We have output equation

$Q = CD^2 LN$ Take Q in kVA, *i.e.*, $10 \times 10^3$ kVA.

$$10 \times 10^3 = 11.1 \times 10^{-5} \times B_{\delta_1} \cdot \overline{ac} \times D^2 \times L \times 750$$

or $$D^2 L = \frac{10 \times 10^3}{11.1 \times 10^{-5} \times 0.92 \times 30{,}000 \times 750}$$

$$= \frac{10 \times 10^3}{2.29 \times 10^3} = \frac{10}{2.29} = 4.36$$

So, $$D^2L = 4.36 \text{ m}^3 \qquad ...(i)$$

Let us start with round pole construction. For round pole construction,

$$\frac{b_p}{\tau_p} = \frac{L}{\tau_p} = 0.66$$

$\therefore$ $$\frac{L}{\tau_p} = 0.66 \text{ or } L = 0.66 \times \frac{\pi D}{P} = 0.259\ D$$

Putting the value of L in equation (*i*), we get

$$D^3 = \frac{4.36}{0.259} = 16.83$$

$\therefore$ $$D = 2.56 \text{ m.}$$

Let us check the feasibility of diameter by calculating the peripheral speed for this diameter peripheral speed, $V_p = \frac{\pi DN}{60} = \frac{3.14 \times 2.56 \times 750}{60}$

$$= 100.59$$

So, peripheral speed is much more than the maximum limit of 80 m/sec.

Modifying the value of $\overline{ac}$ to its maximum limit.

Let, $\overline{ac} = 40{,}000$

$$Q = CD^2 LN.$$

or $$10 \times 10^3 = 11.1 \times 10^{-5} \times 0.92 \times 40{,}000 \times D^2L \times 750$$

or $$D^2 L = \frac{10 \times 10^3}{3.06 \times 10^3} = 3.26$$

So, $$D^2L = 3.26 \text{ m}^3 \qquad ...(ii)$$

For round pole construction $\frac{L}{\tau_p} = 0.66$ or $L = 0.259\,D$

Putting this value of $L$ in equation (*ii*)

We get, $$D^3 = \frac{3.26}{0.259} = 12.60 \text{ m}^3$$

or $$D = 2.32 \text{ m}$$

Peripheral speed for this diameter, $$V_p = \frac{3.14 \times 2.32 \times 750}{60}$$

$$= 91.37 \text{ m/sec.}$$

This is also above the maximum permissible limit of 80 m/sec.

We have already considered the maximum value of $\overline{ac}$ and the value of $B_{\delta_1}$ is given as 0.92. So, the only alternative to reduce the diameter and hence to bring down the peripheral speed within the limit is to adopt long pole or rectangular pole construction.

For long pole construction, $\frac{L}{\tau_p} = 1$ to 5

Now the $\overline{ac}$ can be taken between 20,000 to 40,000 and $\frac{L}{\tau_p}$ may also be selected between 1 to 5.

Taking, $$\overline{ac} = 34{,}000 \text{ and } \frac{L}{\tau_p} = 3.5$$

So, $$\frac{L}{\tau_p} = 3.5$$

or $$L = 3.5 \times \frac{\pi D}{P} = \frac{3.5 \times 3.14}{8} \times D = 1.37\,D \qquad ...(iii)$$

From output equation, we get

$$D^2 L = \frac{Q}{CN} = \frac{10 \times 10^3}{11.1 \times 10^{-5} \times 0.92 \times 34{,}000 \times 750}$$

$$= \frac{10 \times 10^3}{2.60 \times 10^3} = \frac{10.0}{2.60} = 3.84$$

From equation (*iii*)

$$L = 1.37\,D$$

Chapter 5

So, $D^3 = \frac{3.84}{1.37} = 2.80$

or $D = 1.41$ m

Peripheral velocity, $V_p = \frac{\pi DN}{60} = \frac{1.41 \times 3.14 \times 750}{60}$

$= 55$ m/sec.

So, this diameter can be taken for alternative construction. From equation (*iii*),

$L = 1.37 \times 1.41 = 1.93$ m

So, $D = 1.41$ m

$L = 1.93$ m

Above details are obtained with rectangular or long pole construction. When we adopted round pole construction, we could not get the diameter for which the peripheral speed could be within limit by selecting the maximum limit of $\overline{ac}$. So, we had to adopt the long pole construction.

**Q. 8** Estimate the number of damper bars per pole, area of each damper bar, diameter of each bar and length of damper bar for the alternator designed in Q. No. 7 with following other details. Stator slot pitch is 26 mm. Current density in damper bars is 3.5 A/mm².

**Solution:**

We have, $D = 1.41$ m, $L = 1.93$ m, $\frac{b_p}{\tau_p} = 0.66$

Number of poles = 8, $B_{\delta_1} = 0.92$, $\overline{ac} = 34{,}000$

Pole pitch, $\tau_p = \frac{\pi D}{P} = \frac{3.14 \times 1.41}{8} = 0.55$ m

Cross-sectional area of damper bars per pole (or section of winding conductors under one pole)

$$F'_D = 0.2 \frac{\overline{ac} \cdot \tau_p}{\delta_d}$$

$\delta_d$ = current density in damper winding conductor

$= 3.5$ A/mm²

$$\therefore \quad F'_D = \frac{0.2 \times 34{,}000 \times 0.55}{3.5} = 1075 \text{ mm}^2$$

Damper winding pitch

$\tau_{SD} = 0.8 \times$ Stator slot pitch

$= 0.8 \times 26 = 20.8$ mm

Pole arc, $b_p = 0.66 \times \tau_p = 0.66 \times \frac{\pi D}{P}$

$= 0.36$ m

Number of bars on one pole

$$n_D = \frac{b_p}{\tau_{SD}} = \frac{0.36 \times 1000}{20.0}$$

$= 18$

Cross-sectional area of each bar

$$F_D = \frac{F'_D}{n_d} = \frac{1075}{18} = 59.7 \text{ mm}^2$$

$$\approx 59 \text{ mm}^2$$

For diameter of each bar, $F_D = \frac{\pi}{4} d_D^2$

$$\text{diameter, } d_D = \sqrt{\frac{4 \cdot F_D}{\pi}} = \sqrt{\frac{4 \times 59}{3.14}}$$

$= 8.66$ mm

Length of damper bars are taken slightly greater than the length of machine.

Length of bar, $L_D = 1.1 \times L$

$= 1.1 \times 1.93 = 2.12$ m.

**Q. 9** Estimate the main dimensions and design as far as possible a 150 MVA, 3-phase, 11 kV, star connected, 24-pole, 50 Hz water wheel generator. Make suitable assumptions.

**Solution:**

Let, $\overline{ac} = 36{,}000$ A/m, $\frac{b_p}{\tau_p} = 0.68$

$B_{\delta_1} = 0.88$

Output coefficient, $C = 11.1 \times 10^{-5} \times B_{\delta_1} \times \overline{ac}$

$= 11.1 \times 10^{-5} \times 0.88 \times 36{,}000$

$= 3.5$

Speed, $N = \frac{120 \times 50}{24} = 250$ r.p.m.

Output equation, $Q = CD^2 LN$

Chapter 5

or $$D^2 L = \frac{Q}{CN} = \frac{15 \times 10^3}{3.5 \times 250}$$

or $$D^2 L = 17.1 \quad ...(i)$$

Round pole construction can be considered.

So, $$\frac{b_p}{\tau_p} = \frac{L}{\tau_p} = 0.68$$

or $$L = 0.68 \times \frac{\pi D}{P} = 0.088\ D \quad ...(ii)$$

From equations (*i*) and (*ii*), we get

$$D^3 = \frac{17.1}{0.088}$$

$\Rightarrow$ $$D = 5.79 \text{ m}$$

and $$L = 0.50 \text{ m}$$

Peripheral speed, $$V_p = \frac{\pi DN}{60}$$

$$= 75 \text{ m/sec.}$$

This is within the limit.

So,

$$D = 5.79$$

and $$L = 0.50$$

## Overall Design of Alternator

Design a 3-phase alternator with following specifications:

| Rating | Voltage | No. of Poles | $\frac{\text{Pole arc}}{\text{Pole pitch}}$ | Freq. | Air gap | Type |
|---|---|---|---|---|---|---|
| 38 MVA | 11 kV | 12 | 0.67 | 60 | Sinusoidal | Water wheel |

Estimate efficiency and percentage regulation.

## Solution:

Estimating $D$ (inner diameter) and $L$ (length):

Starting with round pole construction

$$\frac{b_p}{\tau_p} = \frac{L}{\tau_p} = 0.67$$

$$N = \frac{120f}{P} = \frac{120 \times 60}{12} = 600 \text{ r.p.m.}$$

For water wheel, $B_{\delta_1} = 0.8 - 1.0$

$$\overline{ac} = 20{,}000 - 40{,}000$$

$$\frac{L}{\tau_p} = \frac{L}{\pi D/p} = 0.67 \Rightarrow \frac{L \times 12}{\pi D} = 0.67 \Rightarrow L = 0.1754\,D \qquad ...(i)$$

$$C = 11.1 \times 10^{-5} \times B_{\delta_1} \times \overline{ac} = 11.1 \times 10^{-5} \times 0.8 \times 40{,}000 = 3.55$$

$$D^2 L = \frac{Q}{CN} \Rightarrow D^2 L = \frac{38 \times 1000}{3.55 \times 600} \Rightarrow D^2 L = 17.8 \qquad ...(ii)$$

From equations (*i*) and (*ii*)

$$D^3 = \frac{17.8}{0.1754} \Rightarrow D = 4.66$$

Peripheral velocity, $V_P = \dfrac{\pi DN}{60} = \dfrac{\pi \times 4.66 \times 600}{60} = 146.6$

Since the velocity is exceeding the range (30–80 m/sec). Though the maximum value of $\overline{ac}$ has been taken. So, we have to look for LONG POLE Construction.

### Long Pole

$\dfrac{L}{\tau_p} = 1\text{–}5$ (up to 3.5 the peripheral velocity was exceeding).

Taking,

$$\frac{L}{\tau_p} = 4.5 \Rightarrow \frac{L}{\pi D/p} = 4.5 \Rightarrow L = 1.178\,D \qquad ...(iii)$$

From equation (*ii*),

$D^2 L = 17.8$

Now, putting the value of $L$ from equation (*iii*) in above equation, we get

$$\therefore \quad D^3 = \frac{17.8}{1.178} = 15.109$$

$D = 2.47$ m.

Chapter 5

$\therefore$ Peripheral velocity, $V_P = \dfrac{\pi \times 2.47 \times 600}{60} = 77.59$ m/sec.

This is in range.

So, $\boxed{D = 2.47 \text{ m}}$

and $L = 1.178 \times 2.47 = 2.9$ m

$\boxed{L = 2.9 \text{ m}}$

$$\left.\begin{aligned} &\frac{L}{\tau_p} = 4.5 \\ &B_{\delta_1} = 0.8 \\ &\overline{ac} = 40{,}000 \\ &\text{Long pole m/c} \end{aligned}\right\}$$

## Calculation of Effective Length, Gross Iron Length, Actual Length of iron

Number of ventilating ducts will be,

$n_v = \left(\dfrac{L}{11} - 1\right)$, where $L$ is in cm.

$$\Rightarrow \; n_v = \left(\frac{290}{11} - 1\right) = 25.36 \approx 25$$

$$b_v' = b_v \cdot \frac{5}{5 + \dfrac{b_v}{\delta}}$$

where, $b_v$ = width of ventilating ducts,

$b_v'$ = effective width of ventilating ducts

$$\therefore \; b_v' = 1 \times \frac{5}{5 + \dfrac{1}{0.8377}}$$

$= 0.807$ cm

$$\frac{\delta}{\tau_p} = (0.01 - 0.015) \text{ for water wheel}$$

$$\frac{\delta}{\tau_p} = 0.013$$

$$\frac{L}{\tau_p} = 4.5 \Rightarrow \tau_p = \frac{290}{4.5}$$

$$\therefore \; \delta = 0.013 \times \frac{290}{4.5}$$

$= 0.8377$ cm

Effective length $(l_e) = L - n_v b_v'$

$= 290 - 25 \times 0.807$

$= 269.8$ cm

$= 2.698$ m

Gross iron length,

$l = L - n_v b_v = 290 - (25 \times 1) = 265$ cm $= 2.65$ m

Actual length of iron,

$l_i = k_i l = (2.65 \times 0.91)$

$= 2.411$ m [$k_i$, 0.9 to 0.92 for 0.35 mm thick insulation].

$l_e = 2.698$

$l = 2.65$

$l_i = 2.411$

### Number of slots: ($S_1$)

$\tau_{sg_1}$ = Slot pitch between 40 mm and 60 mm for voltage up to 15 kV.

Taking $\tau_{sg_1}$ = 50 mm

$$\tau_{sg_1} = \frac{\pi D}{S_1} \Rightarrow S_1 = \frac{\pi D}{50} = 155.19 \approx 155$$

$$\text{Slot / pole / phase} = q_1 = \frac{155.19}{3 \times 12} = 4.3 \approx 4$$

$q_{1\,(corrected)} = 4$

$S_{1\,(corrected)} = 3 \times 12 \times 4 = 144.$

$$\text{Actual } \tau_{sg_1} = \frac{\pi D}{S_{1\,(corrected)}} = \frac{\pi \times 2470}{144} = 53.88 \text{ mm}$$

$\therefore \quad \tau_{sg_1} = 53.88$ mm.

### Estimation of Corrected Value of $N_{ph}$, $Z_{ph}$, $Z_T$, $N_c$, $B_{\delta_1}$, $B_{\delta m}$, $\phi_1$, $\phi_2$.

We know,

$$V_{ph} = 4.44\, k_w \cdot \phi_1 \cdot N_{ph} \cdot f$$

$$\Rightarrow N_{ph} = \frac{V_{ph}}{4.44\, k_w \cdot \phi_1 \cdot f} \qquad \text{...(iv)}$$

where, $\phi_1$ is the fundamental flux and can be expressed as

$$\phi_1 = \frac{2}{\pi} B_{\delta_1} \tau_p l_e \text{ where, } \tau_p = \frac{\pi D}{P} = \frac{\pi \times 2.47}{12} = 0.6466 \text{ m} \qquad \text{...(v)}$$

$$\therefore \quad \phi_1 = \frac{2}{\pi} \times 0.8 \times 0.6466 \times 2.698 = 0.888 \text{ weber} \qquad \left\{ \begin{array}{l} \because B_{\delta_1} = .8 \\ l_e = 2.698 \end{array} \right.$$

$$k_{d_1} = \frac{\sin q_1 \dfrac{\alpha}{2}}{q \cdot \sin \dfrac{\alpha}{2}} \qquad \left\{ \begin{array}{l} \because q_1 = 4, \text{ and} \\ \alpha = \text{slot angle} = \dfrac{360}{S_1} \times \dfrac{P}{2} \end{array} \right.$$

$$\Rightarrow \quad k_{d_1} = \frac{\sin 4 \cdot \frac{15}{2}}{4 \cdot \sin \frac{15}{2}} = 0.957$$

$$V_{ph} = \frac{11}{\sqrt{3}} = 6.3 \text{ kV}$$

Now, putting values in equation (*iv*), we get

$$N_{ph} = \frac{6.35 \times 1000}{4.44 \times 0.957 \times 0.888 \times 60} = 28.04 \approx 28$$

Number of conductor/phase, $Z_{ph} = 2 \times 28 = 56$

Total number of conductors, $Z_T = 56 \times 3 = 168$

Conductor per slot, $Z_c = \frac{Z_T}{S_1} = \frac{168}{144} = 1.166 \approx 1$

$$Z_{C(corr)} = 1$$

$$\therefore \quad Z_{T(corr)} = 1 \times S_1 = 144$$

$$N_{ph(corr)} = \frac{Z_{T(corr)}}{6} = \frac{144}{6} = 24$$

$$Z_{ph(corr)} = 24 \times 2 = 48$$

$$\phi_{1(corr)} = \frac{V_{ph}}{4.44 \times k_{d_1} \times N_{ph(corr)} \times f} = 1.037 \text{ weber}$$

$$B_{\delta_{1(corr)}} = \frac{\phi_{1(corr)}}{\frac{2}{\pi} \cdot \tau_p \cdot l_e} = 0.934 \text{ Wb/m}^2$$

$$k_\delta = \frac{B_{\delta_1(corr)}}{B_{\delta_m}} \quad \text{...(vi)}$$

$$k_\phi = \frac{\phi_{1(corr)}}{\phi} \quad \text{...(vii)}$$

## For Salient Pole with Sinusoidal Air Gap

$$\beta = \frac{1}{\dfrac{b_p}{\tau_p} + \dfrac{\sin(180° \cdot b_p/\tau_p)}{\pi}} \qquad \left[\because \frac{b_p}{\tau_p} = 0.67\right]$$

$$= 1.059$$

$$k_\delta = \frac{1}{\beta} = 0.943$$

From equation (*vi*), we get,

$$B_{\delta_m} = \frac{B_{\delta_1(corr)}}{k_\delta} = \frac{0.934}{0.943} = 0.989 \text{ Wb/m}^2$$

$$\frac{1}{k_\phi} = \frac{\pi \sin(90° \cdot b_p/\tau_p)}{\pi b_p/\tau_p + \sin(180° \cdot b_p/\tau_p)} = 0.9198$$

$$k_\phi = \frac{\phi_{1(corr)}}{\phi} \Rightarrow \phi = \frac{\phi_{1(corr)}}{k_\phi} = 1.037 \times 0.919$$

$$= 0.95 \text{ Wb}$$

$$\left.\begin{aligned} N_{ph(corr)} &= 24 \\ Z_{ph(corr)} &= 48 \\ Z_{T(corr)} &= 144 \\ \phi_{1(corr)} &= 1.037\ Wb \\ B_{\delta_1(corr)} &= 0.934 \text{ Wb-m}^2 \\ B_{\delta_m(corr)} &= 0.989 \text{ Wb-m}^2 \\ \phi_{(corr)} &= 0.95 \text{ Wb} \\ Z_{(corr)} &= 1 \end{aligned}\right\}$$

## Teeth Design

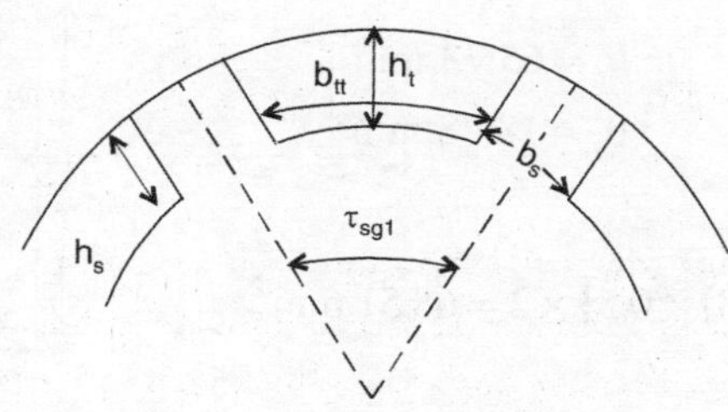

$\tau_{sg_1}$ = stator slot pitch

$b_{tt}$ = width of teeth at the top

$b_s$ = width of slot

$E_z$ = E.m.f. per conductor

$E_T$ = E.m.f. per turn

$$E_z = \frac{V_{ph}}{Z_{ph}} = \frac{6.35 \times 10^3}{48} = 132.29 \text{ volts}$$

$$\therefore \quad E_T = 2E_z = 264.58 \text{ volts}$$

$$b_{tt} = \frac{B_{\delta_m(corr)} \cdot \tau_{sg_1} \cdot l_e}{B_t \, k_i l}$$

$= 35.09$ mm

$$b_s = \tau_{sg_1} - b_{tt} = 53.88 - 35.09$$

$= 18.78$ mm

$\left\{ \begin{array}{l} B_t = 1.6\text{–}1.8 \text{ T} \\ \text{taking } B_t = 1.7 \\ \tau_{sg_1} = 53.88,\ B_{\delta_m(corr)} = 0.989 \end{array} \right.$

$$h_t = h_s = (2 \text{ to } 5) \cdot b_s, \text{ taking } h_t = 3.5\, b_s = (3.5 \times 18.78)$$

$= 65.73$ mm

$$\text{Phase current, } I_{ph} = \frac{Q \times 10^3}{3V_{ph}} = 1994.7 \text{ ampere}$$

$$\text{Sectional area of conductor, } F_c = \frac{I_{ph}}{\sigma}$$

$$\sigma = 4 \text{ to } 5 \text{ A/mm}^2, \text{ taking } \sigma = 4.5 \text{ A/mm}^2$$

$$\therefore \quad F_c = \frac{1994.7}{4.5} = 443.2 \text{ mm}^2$$

Considering, one conductor/slot.

Area of insulation for 4-wall sides,

$$2(4 \times h_t) + \{(b_s - 8) \times 4\}2$$

$$= 611.84 \text{ mm}^2 \qquad \left[ \because h_t = 65.73 \text{ mm},\ b_s = 18.78 \text{ mm} \right]$$

Area of conductor insulation

$$\{(b_s - 8) \times 0.5\} \times 2 + \{(h_t - 8) \times 0.5\} \times 2 = 68.51 \text{ mm}^2$$

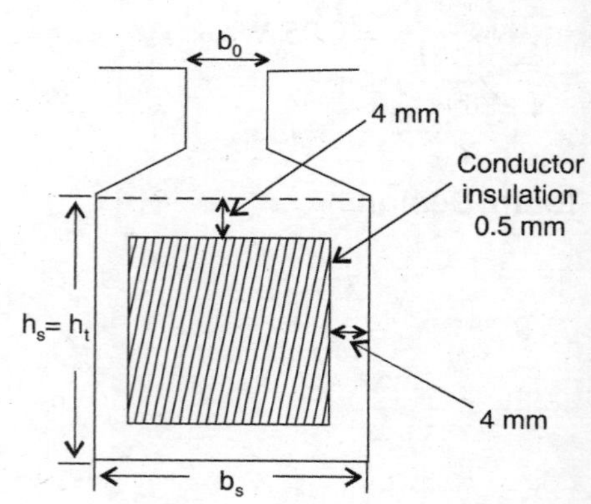

Cross-sectional area of conductor and insulation is,

$= F_c$ + Total insulation area

$= (443.2 + 611.84 + 68.51)\ \text{mm}^2$

$= 1123.55\ \text{mm}^2$

$b_s \times h_s = (65.73 \times 18.78) = 1234.4\ \text{mm}^2$

We see, $b_s \times h_s$ is more than the cross-sectional area of conductor and insulation (near about 10% more). So, $h_s = 3.5\ b_s$ chosen is correct.

**Stator Yoke Design**

$$h_y = \frac{\phi}{2 \cdot B_y \cdot k_i l}$$

where, $B_y$ = 1.36 to 1.5T; taking, $B_y$ = 1.4 T

$\therefore \quad h_y = 140.7$ mm

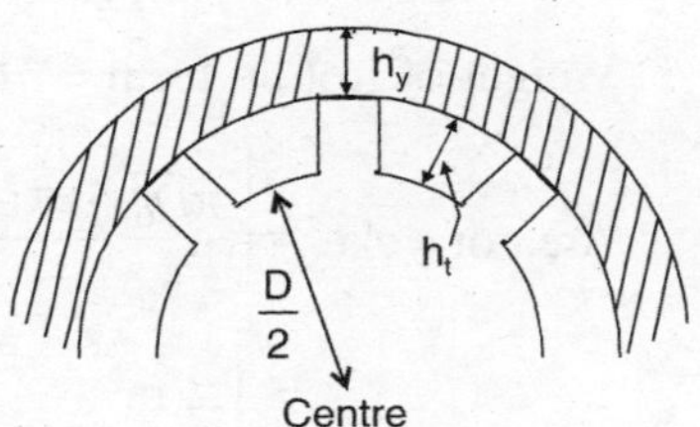


**Outer Diameter of Alternator ($D_0$) →**

$D_0 = D + 2h_t + 2h_y$

$= 2.882$ m

**Estimation of Resistance of Stator Winding, Losses and Weight of Cu and Iron**

Since m/c is above 3 kV rating:

***Length of mean turn***

$$L_{mt} = 2\left[L + 1.25\tau_p + \frac{3 \times \text{kV (phase voltage)} + 10}{100}\right] \text{metre}$$

$= 5.26$ m

Length of conductor or winding per phase

$$= N_{ph} \cdot L_{mt} = 126.24 \text{ m}$$

Cross-sectional area of conductor, $F_c = 443.2\ \text{mm}^2$

Resistance per phase, (at 75°C)

$$R_{dc} = 0.021 \frac{N_{ph} \cdot L_{mt}}{F_c} = 5.98 \times 10^{-3}\ \Omega$$

$$R_{ac} = 1.4 \times 5.98 \times 10^{-3} = 8.37 \times 10^{-3}\ \Omega$$

Stator Cu loss = $3I_{ph}^2 R_{ac}$ = 0.099 MW = 99,000 W

Total volume of Cu, = $3 \times N_{ph} \times L_{mt} \times F_c = 0.167\ \text{m}^3$

Weight of Cu = 89,000 × 0.167 = 1486.8 kg

Chapter 5

***For iron loss***

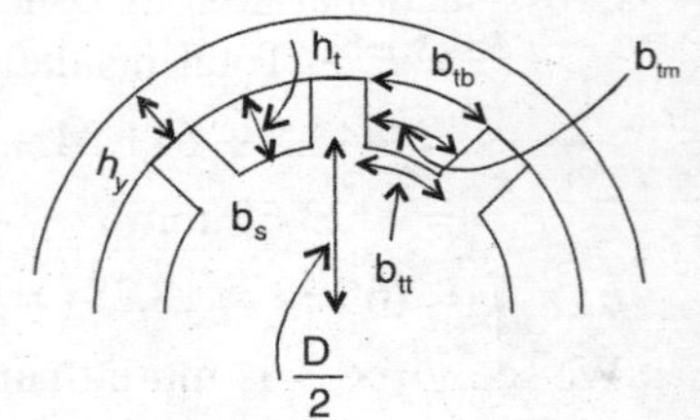


$$b_{tt} = 35 \text{ mm}$$

$$b_{tb} = \frac{\pi(D+2h_t)}{S_1} - b_s$$

$$= 37.97 \text{ mm}$$

Average width of tooth at middle, $b_{tm} = \frac{b_{tt}+b_{tb}}{2} = 36.48$ mm

Area of teeth $= 36.48 \times h_t = 36.48 \times 65.73 = 2398.3 \text{ mm}^2$

Volume of total teeth = Area × $S_1$ × $L_i$

$= 0.83 \text{ m}^3$

Weight of iron in teeth = 7600 × 0.83 = 6308 kg

$$\text{Area of yoke} = \left[\frac{\pi D_0^2}{4} - \frac{\pi(D+2h_t)^2}{4}\right]$$

$= 1.2 \text{ m}^2$

Volume of yoke = 1.2 × 2.411 = 2.9 $\text{m}^3$

Weight of iron in yoke = 7600 × 2.9 kg

= 22040 kg

In yoke for $B_y$ = 1.4, core loss = 8 watt/kg (from specific iron loss curve, Fig. 5.30)

∴ Total core loss in yoke = 22,040 × 8

= 176,320 watt

In teeth for $B_t$ = 1.7, core loss = 13 watt/kg

∴ Total core loss in teeth = 13 × 6308 = 82,004 watt

Total core loss = 258,324 watt

## Estimate the Dimension of Pole Core

Flux at pole side of m/c ($\phi_p$) = 1.17 ×$\phi_{(corr)}$

= 1.111 weber

Cross-sectional area of pole core,

$$F_{pc} = \frac{\phi_p}{B_{pc}} = \frac{1.111}{1.6} = 0.694 \text{ m}^2 \qquad \begin{cases} B_{pc} = 1.5 \text{ to } 1.7 \text{ T} \\ \text{Taking, } B_{pc} = 1.6 \end{cases}$$

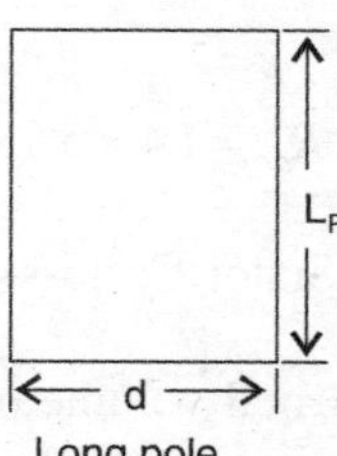

Long pole

$$e = 0.05L_p = 0.05\,L = 0.05 \times 2.9 = 0.145 \text{ m}$$

Since rounding the corners are done, so

$$F_{pc} = \left[L_p \cdot d - \left(4e^2 - \pi e^2\right)\right] k_i$$

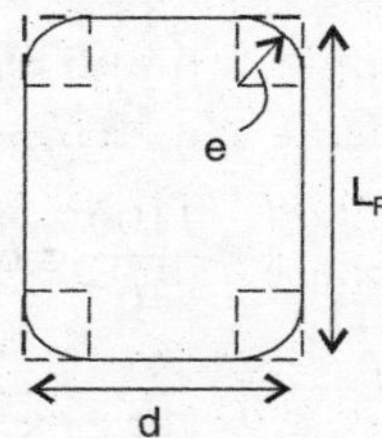


$$\Rightarrow \; d = \frac{F_{pc} + \left(4e^2 - \pi e^2\right) k_i}{k_i L_p}$$

$$= 0.2692 \text{ m}$$

$\therefore$ Width of pole core,

$d = 0.2692$ m

## Damper Winding Design

$\tau_{SD}$ = damper per pitch = 80% of $\tau_{sg}$
$d_D$ = diameter of damper bar
$Z_D$ = spacing between damper bar
$n_d$ = number of bars
$F'_D$ = sectional area of bars (under one pole)

$$F'_D = 0.20 \times 0.957 \times \frac{\overline{ac} \cdot \tau_p}{\delta_D} = 1100 \text{ mm}^2 \qquad \left[\because S_D = 4.5 \text{ A/mm}^2 = 4.5 \times 10^6 \text{ A/m}^2\right]$$

$$\tau_{SD} = 0.8 \times 53.88 = 43.1 \text{ mm.}$$

$$n_d = \frac{b_p}{\tau_{SD}} = \frac{433.2}{43.1} = 10.05 \approx 10 \qquad \left[\because \frac{b_p}{\tau_p} = 0.67 \Rightarrow b_p = 433.2 \text{ mm}\right]$$

Sectional area of one damper bar,

$$F_D = \frac{F'_D}{n_d} = 110 \text{ mm}^2$$

$$d_D = \sqrt{\frac{4F_D}{\pi}} = \sqrt{\frac{4 \times 110}{\pi}} = 11.83 \text{ mm}$$

$$Z_D = \tau_{SD} - d_D = 31.26 \text{ mm}$$

$B_D$ = Flux density in space $Z_D$

$$B_D = \frac{B_{\delta_m} \cdot \tau_{SD} \cdot l_e}{Z_D \cdot k_i \cdot L_P} = 1.39 \text{ T}$$

Since $B_D$ is less than 1.6 T, so, increase $n_d$ and recalculate.
Say,
$n_d = 16$,

$$\tau_{SD(corr)} = \frac{433.2}{16} = 27.075 \text{ mm}$$

$$F_{D(corr)} = \frac{1100}{16} = 68.75 \text{ mm}^2$$

$$d_{D(corr)} = \sqrt{\frac{4 \times 68.75}{\pi}} = 9.35 \text{ mm}$$

$$Z_{D(corr)} = 27.075 - 9.35 = 17.71 \text{ mm}$$

$$B_{D(corr)} = \frac{0.989 \times 0.627 \times 2.698}{0.017 \times 0.91 \times 2.9} = 1.605 \text{ T. It is in range.}$$

$$h_1 = 2d_D = (2 \times 9.35) = 18.7 \text{ mm}$$

$$F'_D = 1100 \text{ mm}^2/\text{pole}$$

## Design of Field Winding

Length of mean turn of field winding

$$L_{mfd} = 2L_m + \pi(d + 0.01 + d_f)$$

where, $d$ = width of core, and

$d_f$ = depth of core = 55 mm for (above 0.55 pole pitch in metre from Table 5.2)

$$L_m = 0.9L = 2.61 \text{ m}$$

$$L_{mfd} = 7.2 \text{ m}$$

$$N_{fd} = \frac{(0.8 \text{ to } 0.85)\, V_{dc}}{0.021\, L_{mfd}\, \delta_{fd}} \qquad \begin{bmatrix} \because V_{dc} = 200 \text{ to } 600\text{V} \\ \delta_{fd} = 3.5 \text{ A/mm}^2 \end{bmatrix}$$

$$= \frac{0.84 \times 400}{0.21 \times 7.2 \times 3.5}$$

$$= 634.9 \approx 635$$

$$N_p = \text{Number of field turn/pole} = \frac{N_{fd}}{P} = \frac{635}{12} = 52.9 \approx 53$$

$$I_{fd} = \frac{2 \times 2.7 \times I_{ph} \times N_{ph} \times k_{d_1}}{P \cdot N_p} = 388.98 \text{ ampere}$$

Now, sectional area of one field condcutor $F_{fd} = \dfrac{I_{fd}}{\delta_{fd}} = \dfrac{388.98}{3.5} = 111.1 \text{ mm}^2$.

Total area by field winding/pole = 111.1 × 2 × 53

= 11776.6 $\text{mm}^2$/pole

Total space required by the field winding per pole.

$$= \frac{\text{Cu area}}{\text{Space factor}} = \frac{11776}{0.85} = 13854.8 \text{ mm}^2$$

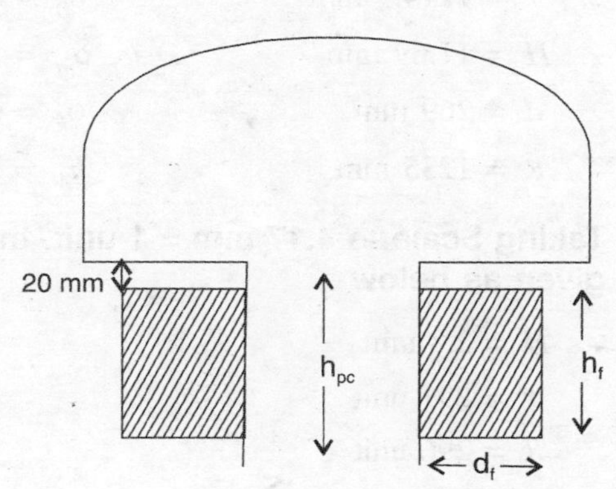


Let, $h_f$ = height of the field winding

$N_{fd(corr)}$ = 53 × 12 = 636.

$d_f$ = depth of the field winding

$h_f \times d_f$ = 13854.8 $\text{mm}^2$, $d_f$ = 55 mm

∴ $h_f$ = 251.9 mm

$h_{pc}$ = height of pole core = 20 + 251.9

= 271.9 mm

## Design of Pole Profile

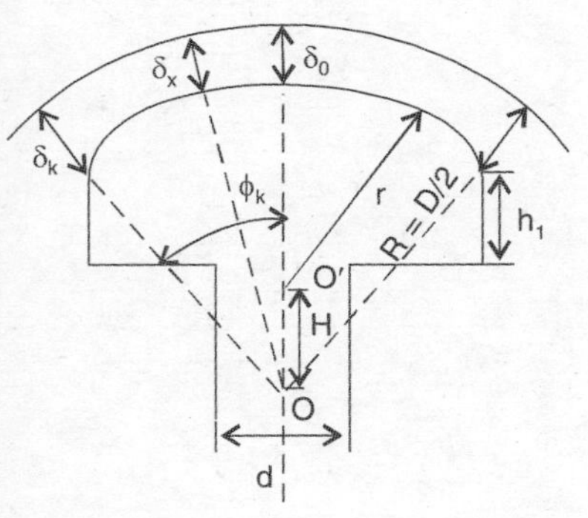


$O$ = centre of the inner periphery of stator
$O'$ = centre of pole arc.

$$\delta_m = \delta_k = \delta_0 + \frac{H}{2r}(H + r)\sin^2 \phi_k \quad \text{...(viii)}$$

$$\phi_k = \frac{180°}{2} \times 0.67 = 60.3°$$

$$R = \frac{D}{2} = \frac{2.47}{2} = 1.235 \text{ m} = 1235 \text{ mm}$$

$$R = H + r + \delta \quad \text{...(ix)}$$

$$\delta_m = 1.4\delta_0 = 11.76 \text{ mm} \quad \text{...(x)}$$

$$\frac{\delta_0}{\tau_p} = 0.013, \qquad \delta_0 = 8.4 \text{ mm}$$

From equation (*ix*), $H + r$ = (1235 – 8.4) = 1226.6 mm ...(*xi*)

Now, on putting the values of $\delta_0$, $H + r$, $\phi_k$, $\delta_m$ in equation (*viii*), we get

$$11.76 = 8.4 + \frac{H}{2r}(1226.6) \times 0.754$$

$$\Rightarrow \quad \frac{H}{r} = 9.6299 \times 10^{-3}$$

From equation (*xi*), $H + r = 1226.6$

$\therefore \quad r + (9.629 \times 10^{-3})r = 1226.6$

$\therefore \quad r = 1214.9$ mm $\qquad \delta_0 = 8.4$ mm

$H = 11.69$ mm $\qquad \delta_m = 11.76$ mm

$d = 269$ mm $\qquad \phi_k = 60.3°$

$R = 1235$ mm $\qquad h_1 = 18.7$ mm

**Taking Scale → 4.17 mm = 1 unit, the pole profile can be drawn on graph with units given as below**

$\therefore \quad H = 2.8$ unit

$r = 291$ unit

$R = 296$ unit

$h_1 = 5$ unit

$d = 65$ unit

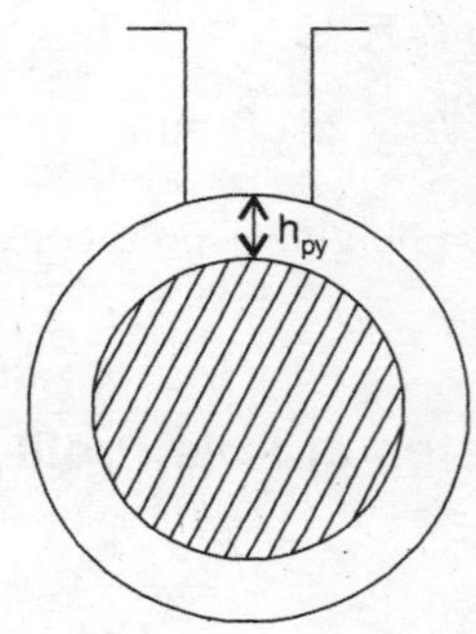


### Height of Pole Yoke

$$h_{py} = \frac{\phi_p}{2 \times B_{py} \times k_i \times L_p}$$

$$= \frac{1.17 \times 0.95}{2 \times 1.4 \times 0.91 \times 2.9}$$

$$= 150.4 \text{ mm} \qquad \left[B_{py} = 1.4\,\text{T},\ \phi_p = 1.17 \times 0.95\right]$$

### Copper (*Cu*) Loss in Field Winding

$$R_{fd} = \frac{0.021\, L_{mfd}\, N_{fd}}{F_{fd}} = \frac{0.021 \times 7.2 \times 636}{111.1}$$

$$= 0.86\,\Omega$$

$Cu$ loss $= I_{fd}^2 \times R_{fd}$

$= 129,467$ watt

Iron loss in rotor = 35% of Iron loss in stator

= 90413.4 watt

m/c rating, $38 \times 10^6 \times 0.8$

$= 30.4 \times 10^6$ watt (say, p.f. = 0.8)

Friction and windage loss $= 0.002 \times 30.4 \times 10^6$

$= 60{,}800$ watt [$\because$ 0.2% = 0.002]

## Total Loss in Alternator

Stator + Rotor copper loss = 228,467.6 watt

Stator + Rotor core loss = 348,737.4 watt

Friction and windage loss = 60,800 watt

Total loss → 638,004 watt

$$\% \text{ age loss} = \frac{\text{Total loss}}{m/c \text{ rating}} = 2.09$$

$$\text{Efficiency, } \eta = \frac{29{,}762}{30{,}400}$$

$= 97.9\%$

## Weight of Cu in Field Winding

$= 8900 \times$ Volume of Cu in field winding

$$= 8900 \times \left[111.1 \times 2 \times 10^{-6} \times 7.2 \times 636\right]$$

$= 9055$ kg

## Weight of Cu in Damper Winding

8900 [Volume of damper bar + Volume of end rings]

But, volume of damper bar $= F_{D(corr)} \times n_{d(corr)} \times L_D$

$= 0.0035\ m^3$

## Volume of End Rings

End ring in each pole is taken.

$$\text{Volume for one pole} = \frac{F'_D}{2} \times b_p = \frac{1100}{2} \times 433$$

$$\therefore \text{ Total volume } = \frac{1100 \times 433 \times 12}{2} = 0.0028\,m^3$$

Weight of Cu in damper winding,

$= 8900\,[0.0035 + 0.0028]$

$= 56.5$ kg

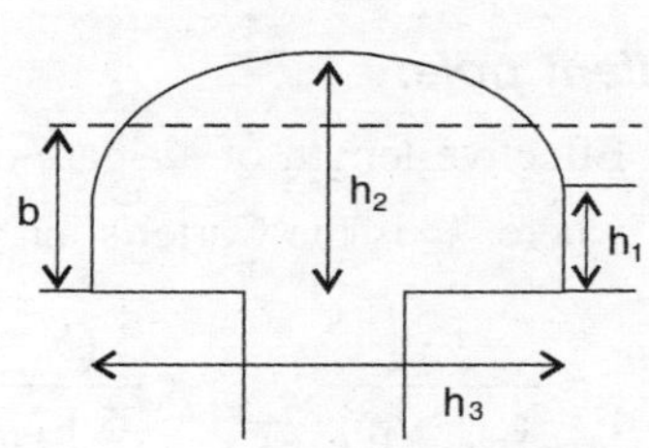

Total weight of Cu in rotor side = (9055.9 + 56.5) kg

= 9112.4 kg

From pole profile, $h_2$ = 157 units

= 654 mm

and $h_1$ = 18.7 mm

$$\therefore\ b = \frac{h_1 + h_2}{2} = 336.6 \text{ mm}$$

$h_3$ = 500 units = 2085 mm

Volume of iron in pole shoe = $b \times h_3 \times L_p$ = (2.08 × .3366 × 2.9) m$^3$

= 2.03 m$^3$

Volume of iron in pole core = $d \times h_{pc} \times L_p$

= 0.2693 × 0.271 × 2.9 m$^3$

= 0.211 m$^3$

Volume of iron in pole yoke

$$\left[\pi\left(D/2 - \delta_0 - h_2 - h_{pc}\right)^2 - \pi\left(D/2 - \delta_0 - h_2 - h_{pc} - h_{py}\right)^2\right]L_P$$

= 0.062 m$^3$

Weight of iron in pole

= 7600 [2.03 + 0.62 + 0.211]

= 21,743.6 kg

Diameter of shaft ($d_s$) $= k \cdot \sqrt{\dfrac{k_w}{\text{rpm}}}$, where, $k = 0.1$

= 0.71 m

$$\text{Weight of shaft } = \left[\frac{\pi(0.71)^2}{4} \times 2.9\right] \times 7600$$

= 8726 kg.

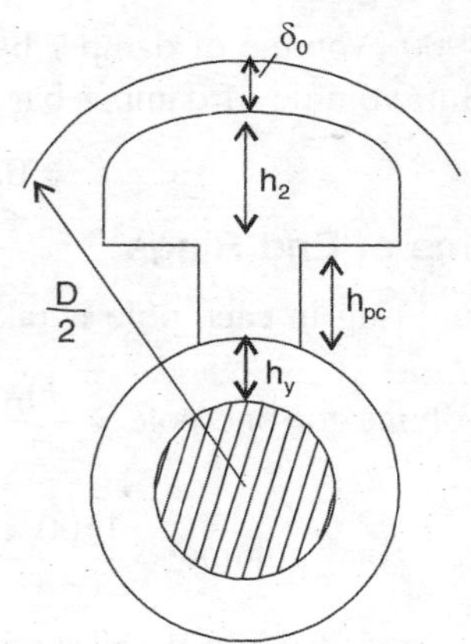


## Determination of OCC

### *Salient pole:*

Effective length of air gap, $\delta' = k_c\, \delta_0$

where, $k_c$ is the Carter's air gap coefficient.

$$= \frac{\tau_{sg_1}}{\tau_{sg_1} - \gamma\delta_0}, \quad \gamma = \frac{5\,(b_0/\delta_0)^2}{5 + b_0/\delta_0}$$

$$= \frac{5\,(3/8.4)^2}{5 + 3/8.4} = 0.115 \qquad \left[\begin{array}{l} \because\ b_0 = 3 \text{ mm (opening of slot)}, \\ \delta_0 = 8.4 \text{ mm}, \\ \tau_{sg_1} = 53.88 \text{ mm} \end{array}\right]$$

$\therefore\ k_c = 1.018$

$\delta' = k_c \times \delta_0 = 8.55$ mm

**Magnetic ckt. for One Pole Pair**

***1. MMF for a Pair of Air Gap***

Total amp-turn in the air gap of alternator:

$$AT_\delta = \frac{1}{\mu_0} B\delta_m 2\delta'$$

$= 13{,}458$

***2. MMF for a Pair of Teeth***

$$B_{tt} = \frac{B_{\delta m}\,\tau_{sg1} l_e}{k_i l\, b_{tt}} = 1.7 \text{ T}$$

By *B-H* curve, $at_{tt} = 6000$ AT/m

$$B_{tm} = \frac{B_{\delta m}\tau_{sg1} \cdot l_e}{k_i l \cdot b_{tm}} = 1.63 \text{ T}$$

$\therefore\ at_{tm} = 3000$ AT/m

$$B_{tb} = \frac{B_{\delta m}\tau_{sg1} \cdot l_e}{b_{tb} \cdot k_i l} = 1.57 \text{ T}$$

$\therefore\ at_{tb} = 2500$ AT/m

Average value of ampere-turn/m length of teeth

$$at_{at} = \frac{at_{tt} + 4at_{tm} + at_{tb}}{6} = 3416.6 \text{ AT/m}$$

Total ampere-turn in pair of teeth,

$AT_{at} = at_{at} \times 2h_t = 448.25$ AT

***3. MMF for Stator Yoke***

$$B_y = \frac{\phi/2}{h_y \cdot k_i l} = 1.4 \text{ T}$$

$\therefore\ at_y = 1000$ AT/m

Chapter 5

Length through which the flux travels,

$$\tau_y = L_y = \frac{\pi(D_0 - h_y)}{P} = 0.71\,\text{m}$$

Total ampere-turn in the yoke:

$$AT_y = k_y \cdot at_y \cdot L_y$$

$$= 0.59 \times 1000 \times 0.71$$

$$= 418.9 \text{ AT}$$

$\because \frac{h_y}{\tau_y} = \frac{0.14}{0.71} = 0.195$

taking $P = 1$,

from Fig. 6.59, $k_y = 0.59$.

Equivalent ampere-turn required for air gap, teeth and yoke:

$$AT_e = AT_\delta + AT_{at} + AT_y$$

$$= 13{,}458 + 448.95 + 418.9$$

$$= 14{,}325.85 \text{ AT}$$

So, with the help of $AT_e$ leakage flux can be calculated.

### *Leakage Flux Calculation*

$\phi_1$ = 2 × Leakage flux between inner surfaces of pole shoes.

$$a_{ps} = \frac{\pi(2b + 2h_{pc} + 2h_{py} + d_s) - (P \cdot b_p)}{P}$$

$$= 0.152 \text{ m}$$

$$\therefore \quad \phi_1 = 2 \times AT_c \times \frac{\mu_0 b \cdot l_p}{a_{ps}}$$

$$= 0.23 \text{ Wb}$$

$\phi_2$ = 2 × Leakage flux from pole side to pole side end

$$= 4\,AT_c \cdot \frac{\mu_0 b}{\pi}\left[l_n\left(1 + \frac{b_p \pi}{2a_{ps}}\right)\right]$$

$$= 0.013 \text{ Wb}$$

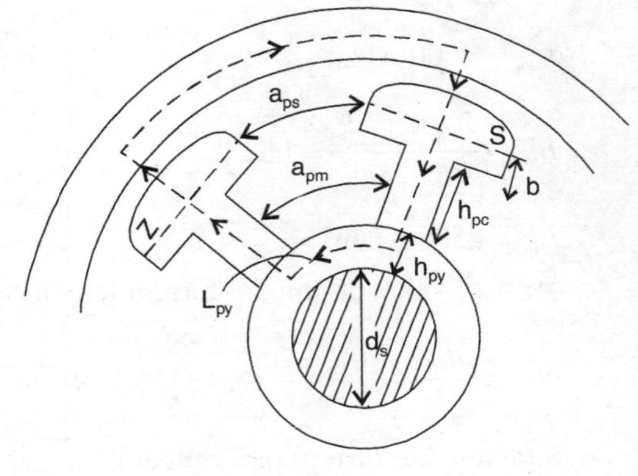


$\phi_3$ = 2 × Flux leakage between end surfaces of pole core

$$= AT_e \cdot \frac{\mu_0 h_{pc} \cdot L_p}{a_{pcm}}$$

$$a_{pcm} = \frac{\pi(h_{pc} + 2h_{py} + d_s) - P \cdot d}{12} = 0.066 \text{ m}$$

$\therefore \quad \phi_3 = 0.21$ Wb

$\phi_4 = 2 \times$ Leakage between end surfaces of pole core

$$= 2AT_e \times \mu_0 \times h_{pc} \times l_n \left(1 + \frac{b_p \cdot \pi}{2a_{pcm}}\right)$$

$$= 0.023 \text{ Wb}$$

**Pole Flux Leakage**

$$\phi_{e1} = \phi_1 + \phi_2 + \phi_3 + \phi_4 = 0.476$$

Total flux per pole $(\phi_p) = \phi + \phi_{e1} = 1.41$ Wb

### *1. Flux Density in Pole Core*

$$B_{pc} = \frac{\phi_p}{F_{pc}} \qquad \left\{ \because F_{pc} = \left[\left(d \times L_p\right) - 4e^2 + \pi e^2\right] \right.$$

$$= 1.85 \text{ T} \qquad = 0.76 \text{ m}$$

$\therefore \quad at_{pc} = 14{,}000$ AT/m

$$AT_{pcs} = at_{pc} \times 2\left(h_{pc} + b\right) = 16996 \text{ AT}$$

### *2. MMF required for Pole Yoke*

$$B_{py} = \frac{\phi_p/2}{h_{py} \cdot k_i \cdot L_p} = 1.8 \text{ T}$$

$\therefore \; at_{py} = 12000$ AT/m $\qquad \left[\because \; L_{py} = \dfrac{\pi\left(D - 2\delta_0 - 2h_z - 2h_{pc} - h_{pc}\right)}{P} = 0.118\text{m}\right]$

$$AT_{py} = at_{py} \cdot L_{py}$$

$$= 12{,}000 \times 0.118$$

$$= 1416 \text{ AT}$$

Total *MMF* per pole pair to induce voltage at no load:

$$AT_{pole\ pair} = AT_\delta + AT_{at} + AT_y + AT_{pcs} + AT_{py}$$

$$= 32737 \text{ AT}$$

$$AT_{per\ pole} = 16368 \text{ AT}$$

Chapter 5

Rated voltage = 100%. So, above values are for 100% voltage:

| Voltage ⟶ | 40% | 60% | 80% | 100% | 120% |
|---|---|---|---|---|---|
| $\phi$ ⟶ | 0.38 | 0.57 | 0.76 | 0.95 | 1.14 |
| $B_{\delta m}$ ⟶ | 0.395 | 0.59 | 0.79 | 0.989 | 1.186 |
| $B_{tt}$ ⟶ | 0.68 | 1.02 | 1.36 | 1.7 | 2.04 |
| $B_{tm}$ ⟶ | 0.65 | 0.978 | 1.3 | 1.63 | 1.956 |
| $B_{tb}$ ⟶ | 0.628 | 0.94 | 1.256 | 1.57 | 1.884 |
| $at_{tt}$ ⟶ | 2400 | 3600 | 4800 | 6000 | 7200 |
| $at_{tm}$ ⟶ | 1200 | 1800 | 2400 | 3000 | 3600 |
| $at_{tb}$ ⟶ | 1000 | 1500 | 2000 | 2500 | 3000 |
| $at_{at}$ ⟶ | 1366.4 | 2049 | 2732 | 3416 | 4099 |
| $B_y$ ⟶ | 0.56 | 0.84 | 1.12 | 1.4 | 1.68 |
| $at_y$ ⟶ | 400 | 600 | 800 | 1000 | 1200 |
| $\phi_{e_1}$ ⟶ | 0.19 | 0.285 | 0.38 | 0.476 | 0.57 |
| $\phi_p$ ⟶ | 0.56 | 0.84 | 1.128 | 1.41 | 1.69 |
| $B_{pc}$ ⟶ | 0.74 | 1.11 | 1.48 | 1.85 | 2.22 |
| $B_{py}$ ⟶ | 0.72 | 1.08 | 1.44 | 1.8 | 2.16 |
| $at_{pc}$ ⟶ | 5600 | 8400 | 11,200 | 14,000 | 16,800 |
| $at_{py}$ ⟶ | 4800 | 7200 | 9600 | 12,000 | 14,400 |
| $AT_\delta$ ⟶ | 5383 | 8074 | 10,766 | 13,458 | 16,149 |
| $AT_{at}$ ⟶ | 179.5 | 269.3 | 359 | 448.95 | 538.7 |
| $AT_y$ ⟶ | 167.5 | 251 | 335 | 418.9 | 502 |
| $AT_{pcs}$ ⟶ | 6798 | 10,197 | 13,596 | 16,996 | 20,395 |
| $AT_{py}$ ⟶ | 566 | 849.6 | 1132.8 | 1416 | 1699 |
| $AT_{\text{pole pair}}$ ⟶ | 13,094 | 19,642 | 26,189 | 32,737 | 39,284 |
| $AT_{\text{per pole}}$ ⟶ | 6547 | 9820 | 13,024 | 16,368 | 19,641 |

## Leakage Reactance

### 1. *Slot Leakage Reactance for Single Layer Winding*

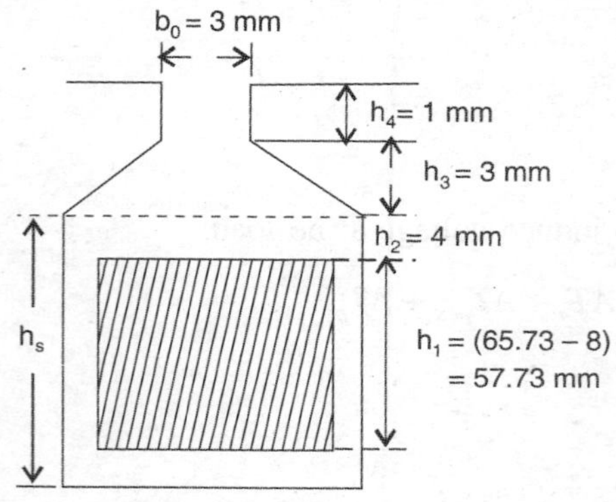

Slot leakage inductance,

$$L_{sl} = 4\mu_0 \frac{N_{ph}^2}{p \cdot q} l_e \left[ \frac{h_1}{3b_s} + \frac{h_2}{b_s} + \frac{2h_3}{b_0 + b_s} + \frac{h_4}{b_0} \right]$$

$$= 2.98 \times 10^{-4} \text{ H}$$

$$\therefore \quad X_{sl} = 2\pi f L_{sl} = 0.112 \ \Omega$$

## 2. Tooth Leakage Reactance ($X_{tl}$)

$$L_{tl} = \frac{\mu_0 N_{ph}^2}{P \cdot q} \cdot L_e \cdot \lambda_{tl}$$

$$= 2.35 \times 10^{-5} \text{ H}$$

$$\because \quad \lambda_{tl} = \frac{5\left(\frac{\delta_0}{b_0}\right)}{5 + 4\left(\frac{\delta_0}{b_0}\right)} \times \frac{b_p}{\tau_p}$$

$$= 0.579$$

$$X_{tl} = 2\pi f L_{tl}$$

$$= 8.88 \times 10^{-3} \Omega$$

## 3. End turn leakage, ($X_{el}$)

$$L_{el} = \frac{\mu_0 N_{ph}^2}{P} \cdot L_w \cdot \lambda_{e_1}$$

$$= 1.71 \times 10^{-6} \text{ H}$$

$$\because \quad L_w = \frac{l_{e1}}{2} + l_{e_2}$$

$$\text{where, } l_{e_1} = \frac{1}{2} \cdot \frac{\tau_p \cdot b_s}{\sqrt{\tau_p^2 - b_s^2}} = 9.34 \times 10^{-3} \text{ m}$$

and $l_{e_2} = 9 \text{ cm} = 0.09 \text{ m}$

$\lambda_{e_1} = 0.3$ (for stator and 1-layer)

$$\therefore \quad L_w = 0.0946 \text{ m}$$

$$\therefore X_{el} = 2\pi f L_{el}$$

$$= 6.45 \times 10^{-4} \ \Omega$$

$$\therefore X_{al} = X_{sl} + X_{tl} + X_{el} = 0.1215 \Omega$$

Leakage drop = 1994 × 0.121 = 239 V

$$\% \text{ regulation} = \frac{239 \times 100}{6300} = 3.79$$

# DESIGN SHEET

### Rating

1. Full load kVA $\longrightarrow$ $38 \times 10^3$

2. Full load power ⟶ $30.4 \times 10^6$ watt
3. Line voltage ⟶ 11kV
4. Phase voltage ⟶ 6.35 kV
5. Frequency ⟶ 60 Hz
6. Power factor (P.f.) ⟶ 0.8
7. Speed (rpm) ⟶ 600
8. Number of Poles ⟶ 12

### *Main Dimensions*

1. Specific magnetic loading ⟶ 0.8
2. Specific electric loading ⟶ 40,000
3. Inner diameter of stator ⟶ 2.47 m
4. Length of m/c ⟶ 2.9 m
5. Gross iron length ⟶ 2.65 m
6. Net iron length ⟶ 2.411 m
7. Pole pitch ⟶ 644.4 mm
8. Current/phase ⟶ 1994.7 ampere
9. $B_{\delta m\ (corr)}$ ⟶ 0.989 T
10. $\phi_{(corr)}$ ⟶ 0.95 Wb

### *Stator*

1. Number of slots ⟶ 144
2. Total number of conductors ⟶ 144
3. Number of turns/phase ⟶ 24
4. Number of conductor/slot ⟶ 1
5. Slot pitch ⟶ 53.88 mm
6. Width of teeth
   - on top ⟶ 35.09 mm
   - on middle ⟶ 36.48 mm
   - on bottom ⟶ 37.97 mm
7. Width of slot ⟶ 18.78 mm
8. Height of teeth = height of slot ⟶ 65.73 mm
9. Sectional area of one conductor ⟶ 443.2 mm$^2$
10. Length of mean turn ⟶ 5.26 m
11. Resistance/phase ⟶ $8.37 \times 10^{-3}\ \Omega$
12. Stator (Cu loss + Iron loss) ⟶ 357,324 watt
13. Outer diameter ⟶ 2.88 m
14. Height of stator yoke ⟶ 140.7 mm
15. $B_y$ ⟶ 1.4 T
16. Length of flux path, $L_y$ ⟶ 0.71 m
17. Arm yoke *MMF* ⟶ 418.9 AT

**Rotor**

| | | |
|---|---|---|
| 1. Type | → | Salient pole (long pole) |
| 2. Field *MMF* | → | 16,368 AT/pole |
| 3. Number of slots<br>4. Rotor slot pitch<br>5. Slot width<br>6. Slot height | → | For turbo |
| 7. Number of field turns/pole | → | 53 |
| 8. Number of damper bars | → | 16 |
| 9. Sectional area of damper bars | → | 68.75 mm$^2$ |
| 10. Rotor Cu loss | → | 129,467 watt |
| 11. Rotor winding resistance | → | .86 Ω |
| 12. Field current | → | 388.98 ampere |
| – Efficiency | → | 97.9% |
| – Total leakage reactance ($X_{al}$) | → | 0.1215 Ω |
| – Leakage drop | → | 239 V |
| – % regulation | → | 3.79 |

## PROBLEMS WITH SHORT ANSWERS

**Q. 1 Explain, why the inherent voltage regulation of a synchronous machine with low value of short-circuit ratio is poor?**

**Ans.** A low value of *SCR* means that the synchronous reactance has a large value, synchronous machines with low value of *SCR* thus have greater changes in voltage under fluctuations of load *i.e.*, the inherent voltage regulation of the machine is poor.

**Q. 2 Explain, why a machine with a large air gap has a higher synchronising power?**

**Ans.** The length of air gap greatly influences the performance of a synchronous machine. A large air gap offers a large reluctance to the path of flux produced by the armature m.m.f. and thus reduces the effect of armature reaction. This results in a small value of synchronous reactance and a high value of *SCR*. Thus a machine with a large air gap has a higher synchronising power which makes the machine less sensitive to load variations.

**Q. 3 Explain, why a machine designed with a higher flux density has better stability?**

**Ans.** The maximum power which a synchronous machine can deliver under steady state conditions is $P_{max} = EV/X_s$ where $E$ is the excitation voltage, $V$ is the terminal voltage and $X_s$ is the synchronous reactance. Therefore the maximum power or the steady state stability limit of a machine is inversely proportional to its synchronous reactance. If a high value of gap density is used, the flux per pole is large and therefore a smaller number of turns are required for the armature winding. This results in reduction of the value of synchronous reactance. Therefore, the use of a high gap density improves the steady state stability limit. This is also explained in the book while solution of flux density is discussed.

Chapter 5

**Q. 4 Explain, why a machine designed with higher specific electric loading has a poor voltage regulation?**

**Ans.** The value of specific electric loading affects the leakage reactance and armature reaction in the machine. A high value of (ampere-conductor/m) leads to high values of leakage reactance and armature reaction and consequently a high value of synchronous reactance. Therefore a machine designed with a high value of *ac* will have poor inherent voltage regulation.

**Q. 5 Define *SCR* and explain its effects on machine performance.**

**Ans.** The Short-circuit Ratio (*SCR*) of a synchronous machine is defined as the ratio of field current required to produce rated voltage on open circuit to field current required to circulate rated current at short-circuit.

Effects of *SCR* on Machine Performance:

(*i*) *Voltage Regulation*: Synchronous machines with low *i.e.*, value of *SCR* have greater changes in voltage under fluctuations of load *i.e.*, the inherent regulation of the machine is poor.

(*ii*) *Stability*: A machine with a low value of *SCR* (*i.e.*, high value of $X_d$) has a lower stability limit as the maximum power output of machine is inversely proportional to $X_d$.

(*iii*) *Parallel Operations*: Machine with a low value of *SCR* are not suitable to operate in parallel because a high value of $X_d$ gives a small synchronizing power.

(*iv*) *Short-Circuit Current*: A machine with small value of *SCR* has smaller value of current under short-circuit conditions due to large value of synchronous reactance.

(*v*) *Self Excitation*: Synchronous machines feeding long transmission lines should not be designed with a small short-circuit ratio since this would lead to large voltage on open circuit produced by self excitation due to large capacitive currents drawn by the transmission lines. A machine with higher value of *SCR* is costlier to build. Present trend is to design the alternator (specially turbo-alternator) with a low value of *SCR*.

**Q. 6 Explain the difference between dynamo-grade steel and transformer-grade steel.**

**Ans.** Dynamo-grade steel is a steel with low silicon content and therefore possess good permeability at high flux densities, high ductility but it has high iron losses. In rotating electrical machines, it is desirable that iron parts should work satisfactorily at higher values of flux density in order to achieve a higher output to weight ratio. The low silicon steels have about 0.5% silicon or less.

Sheet steels possessing higher silicon content of about 4–5% are called transformer-grade steels. These are mainly used in transformers as the magnetizing current is not of much importance in transformers. This steel is also termed as high resistance steel.

**Q. 7 Why turbogenerators are designed for smaller diameter and larger length while water wheel generators have large diameter and short axial length?**

**Ans.** Turbo-alternators are high speed machines since they are driven by steam turbines, which have better efficiency at high speed. Hence, if it is designed with large diameter will have higher peripheral speed which may cause mechanical damages, hence to provide a given output diameter is reduced and length is increased. Water wheel generators are low speed machines and hence have large number of poles. To accommodate large number of poles diameter required is larger which does not have the peripheral speed high because the rated speed of generator is low. Due to low speed the centrifugal force is less.

**Q. 8 Why are the pole shoes chamfered for salient pole synchronous machines?**

**Ans.** The function of pole shoes in a salient pole machine is to provide uniform distribution of flux in the polar region. As the pole body is of smaller cross-section, if unchamfered pole shoes are provided the space covered by pole shoes will be less and hence the flux will not be uniformly distributed in the air gap of machine. Thus by chamfering flux is distributed almost uniformly in the machine.

## UNSOLVED PROBLEMS

**1.** Obtain the main dimensions of the rotor of a 50 MVA, 2-pole, 50 Hz, synchronous generator. The peripheral speed is limited to approximately 160 m/sec. Take an electric loading of 65,000 A/m and a mean gap density of 0.575 Wb/m$^2$. Assume gap length of 25 mm.

Explain why special consideration for diameter and length is required in design of turbo-alternator?

**(DDU, GKP University, 1994)**

**2.** Why the damper windings are provided in synchronous machines? Derive the expressions in connection with design of damper winding of synchronous machine.

**(DDU, GKP University, 1994)**

**3.** For a 750 kVA, 2.2 kV, 375 r.p.m., 50 Hz, 3-phase, star-connected salient pole alternator, calculate:

(*a*) Main dimensions of stator frame.

(*b*) Number of stator turns per phase.

Given:

Specific magnetic loading = 0.57 tesla
Specific electric loading = 27500 ampere-conductor/metre
Winding factor for full pitch coil = 0.955

**(DDU, GKP, University, 1999)**

**4.** Discuss various factors, which contribute to the rejection of the salient pole field system for high speed alternators and the adoption of forged steel cylindrical rotor construction with parallel slots.

Calculate:

(*i*) Core length.
(*ii*) The bore.
(*iii*) The number of slots.
(*iv*) The number of turns per phase.

For the stator of an alternator developing 1500 kVA, 2200 volts, 3-phase, star connected, 50 Hz at 600 rpm. Ampere-conductors per metres of periphery may be assumed 31,000 and flux density of 0.54 T.

**(DDU, GKP, University, 2001)**

**5.** Design the stator frame for a 500 kVA, 6600 V, 50 Hz, 12-pole, star connected, 3-phase salient pole alternator, giving the following informations:

(*a*) Internal diameter and gross length of stator frame.

(*b*) Number of stator conductors.

(*c*) Number of stator slots.

Take:

Specific magnetic loading = 0.56 tesla

Specific electric loading = 26,000 amp-conductor per metre.

Assume other data needed.

**(DDU, GKP, University, 1996)**

6. What are the usual limits of specific magnetic and specific electric loading for slow speed water wheel salient pole alternators and for high speed turbogenerators. Discuss the factors on which these are basically dependent upon.
7. Discuss the following:
   (*i*) Hydrogen cooled turbogenerators.
   (*ii*) Discuss the steps that are followed in the design of salient pole alternators to ensure a good wave forms for the e.m.f.
   (*iii*) How does the value of short-circuit ratio affect the design of alternators?
   (*iv*) Leakage reactance in an alternator having single layer winding.
8. Derive the output equation for a 3-phase alternator in terms of specific magnetic and electric loadings. What are the usual limits for these quantities in water wheel and turbogenerators? Discuss why high speed alternators have very long armature whereas low speed water wheel alternators have very large diameters?
9. Explain
   (1) Various types of insulation used in electromagnetic apparatus.
   (2) Leakage reactance in an alternator having double layer winding.
   (3) Need for bracing the end connections of turboalternators.
10. Derive from fundamental principles an expression for the output coefficient of a 3-phase alternator in terms of specific magnetic and electric loadings. What are the usual limits for these quantities in water wheel and turbogenerators?
11. Find the main dimensions of a 25,000 kVA, 187.5 rpm, 50 Hz, 3-phase, 3000 volts salient pole synchronous generator. The specific magnetic loading is 0.6 Wb/m$^2$ and the specific electric loading is 34,000 ampere-conductor per meter of periphery. Use round pole and assume suitable data.

    Specify the type of pole construction if the runaway speed is about 2 times the normal speed.

    **(DDU, GKP, Univ. 1995)**
12. Determine for a 12.5 MVA, 6.6 kV, 20-pole 50 Hz, 3-phase alternator suitable values for diameter, length, number of stator slots and stator conductors. Assume suitable data required.
13. Define and explain clearly what do you understand by 'short-circuit ratio (*SCR*)' of a synchronous generator? If the synchronous generator is redesigned with a higher value of *SCR*, compare the stability and voltage regulation of the redesigned machine with those of the former machine.
14. Estimate the main dimensions and design as far as possible a 3-phase, 2000 kVA, 0.85 power factor, 3.3 kV, 50 Hz, 600 r.p.m. water wheel generator. Make suitable assumptions.
15. Explain and justify why for the same rating of alternator to be designed the diameter of water wheel generator with round pole construction is higher than the diameter of the water wheel generator with long pole construction? Compare the two types of construction.

16. Explain the method of determination of full load field m.m.f. for a salient pole synchronous generator.

17. Explain
   (*i*) Ventilation of electrical machines.
   (*ii*) Effects of cooling of synchronous machines.
   (*iii*) Synchronous reactance and how does it effect the design of alternator?
   (*iv*) Field winding design for salient pole alternator.

18. Calculate the main dimensions for a 10 MVA, 11 kV, 50 Hz, 3-phase, 2-pole, turbo-alternator based on the following information:

| | | |
|---|---|---|
| Specific magnetic loading | = | 0.63 tesla |
| Specific electric loading | = | 48000 amp-conductor per metre |
| Peripheral speed | = | 120 m/sec. |
| Length of air gap | = | 2.0 cm |
| Stator winding factor | = | 0.955 |

**(DDU, GKP, University, 1998)**

19. A 200 kVA, 3-phase, 50 Hz, 2.2 kV, 375 rpm alternator is to be designed with specific electric loading of 23,000 ampere-conductor per metre of the periphery. Flux density 0.5 tesla with an air gap of 2.0 cm. Assume that the stator winding is uniformly distributed with a 60° phase spread and with full pitch winding.

   Determine approximately the diameter and length of the stator, total number of slots, conductors per slot and the size of the slot. The peripheral speed must not exceed 30 m/sec.

**(DDU, GKP, University, 1999)**

20. Explain
   (*i*) How the oscillations are minimized in an alternator by providing damper winding? Derive the expressions to estimate the number of damper bars to be provided in an alternator.
   (*ii*) Field winding design of cylindrical rotor alternator.

21. Why is the field structure usually made a rotating part in a 3-phase synchronous alternator?

# 6

# Construction and Design Steps of Induction Motors

## 6.1 INTRODUCTION

Induction motors are widely used in the industry and in other places. There are several advantages associated with the induction motors due to which it has become popular. The advantages of 3-phase induction motor are:

(*i*) It is very simple, robust, rugged and capable of withstanding rough use.
(*ii*) It is cheaper and reliable.
(*iii*) Maintenance cost is low.
(*iv*) Losses are small so high efficiency.
(*v*) It is trouble free motor.
(*vi*) It is self starting since it has self starting torque.
(*vii*) Its power factor is reasonably good at full load.
(*viii*) It is equivalent to a transformer whose secondary is capable of rotating with respect to the primary. Usually the stator is primary while the rotor is secondary.

But there are several disadvantages associated with it and they are:

(*i*) Its speed cannot be varied without compromising with efficiency.
(*ii*) Its speed decreases with increase in load.
(*iii*) Its starting torque is less than a d.c motor.

This is called an induction motor since only the stator winding is connected to the voltage source which produces the field rotating at synchronous speed in air gap of motor and is called as rotating magnetic field (r.m.f.) and this induces e.m.f. in the rotor winding (at standstill) and the e.m.f. in the rotor produces current in rotor. This is explained with Faraday's law of electromagnetic induction. Now, there is a field set-up by the rotor current and which opposes the cause by which it is produced (it is explained by Lenz's Law) *i.e.*, rotating magnetic field. So in the process of opposition rotor tries to minimize the difference of speed between rotor and r.m.f. Since the rotor is at standstill and r.m.f. is rotating at synchronous speed, so rotor starts running in the direction of r.m.f. to minimize the speed difference. Since there is no electrical connection to the rotor from the 3-$\phi$ supply or from any other d.c. supply and current in the rotor winding are set-up entirely by the effect of electromagnetic induction from the stator hence it is known as induction motor.

## 6.2 CONSTRUCTION

An induction motor consists of following parts:

(*i*) Stator
(*ii*) Rotor
(*iii*) Stator winding
(*iv*) Rotor winding
(*v*) Slip rings
(*vi*) Shafts and bearings
(*vii*) Frame

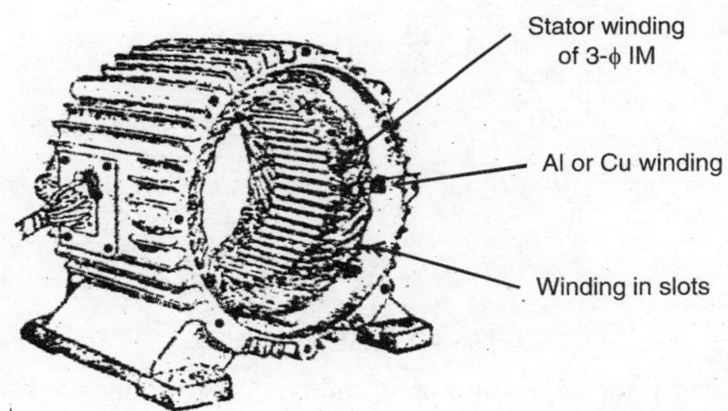


**Fig 6.1** Stator of a 3-$\phi$ induction motor.



### 6.2.1 Stator

It is an outer stationary hollow cylindrical structure made of laminated sheet steel of dynamo grade laminations of either 0.35 mm or 0.5 mm thick. These laminations are insulated with each other by paper or varnish. The laminations are slotted on their inner periphery with open or semi-closed and usually rectangular shape as shown in Fig. 6.1. The laminations are punched in one piece for small machines. The laminated core is welded at several places around the outer cylindrical surface as shown in Fig. 6.2.

### 6.2.2 Rotor

Rotor is an inner cylindrical core, built of laminated sheet steel strips with slots at its outer periphery. Rotor laminations are mounted in one piece for small machines. The laminations are segmented and built up on a fabricated-spider in a longer machine, as shown in Fig. 6.3. Axial and radial ventilating ducts are provided for rotor as well as for stator if the length of machine is large. The slots of the rotor are not always parallel to the slots on the stator but they are given a twist known as skew. The rotor is then known as skewed rotor and this reduces the noise, eliminates cogging, magnetic locking of the rotor and also increases the starting torque. The windings are placed in the rotor slots.

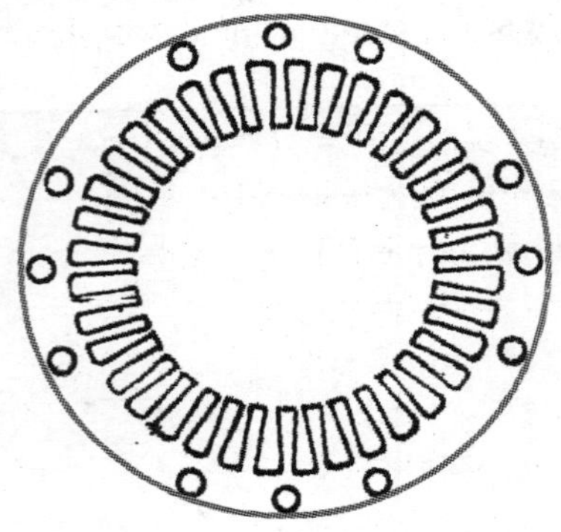

**Fig 6.2** Stator lamination for small machine.

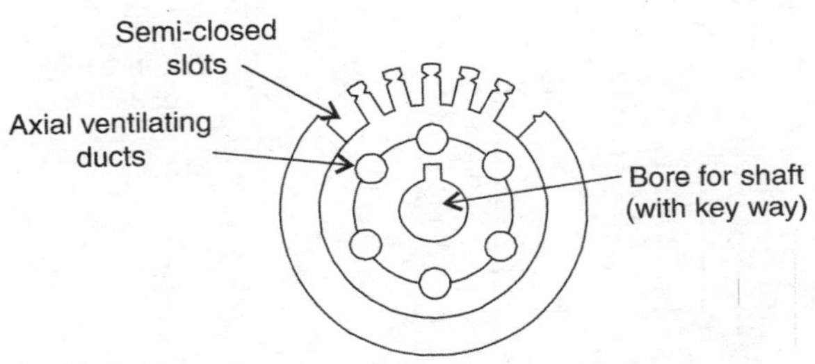


**Fig. 6.3** Rotor stumping.

There are two types of rotor depending upon the type of winding used.

***(a) Squirrel Cage Rotor:*** It consists of a slotted laminated rotor core of sheet steel with skewed slots on its outer periphery as shown in Fig. 6.4.

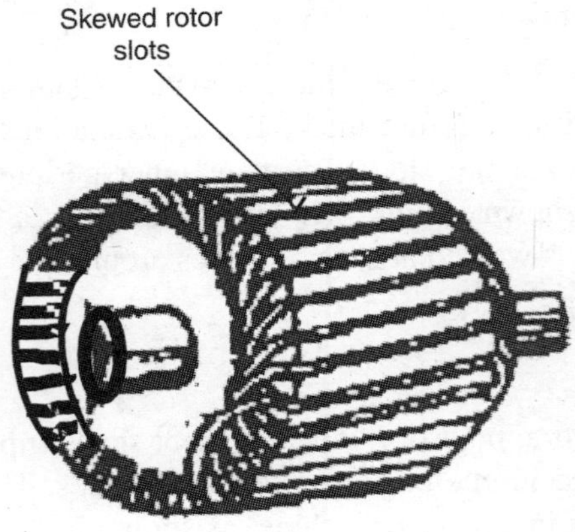


**Fig . 6.4** Squirrel cage rotor.

Most of the 3-$\phi$ induction motors are of squirrel cage type because its rotor is simple and rugged in construction.

***(b) Slipring (or Wound) Rotor:*** It consists of windings in the rotor slots, usually connected in star. The three ends of the winding which are free are taken out and connected to three sliprings mounted on the shaft. On each slipring a brush is there in contact with it. The three brushes are further connected to a 3-phase star-connected rheostat. This arrangement makes possible to include an addition resistance in each phase of rotor at the time of starting to achieve higher starting torque. A slipring rotor is shown in Fig. 6.5.

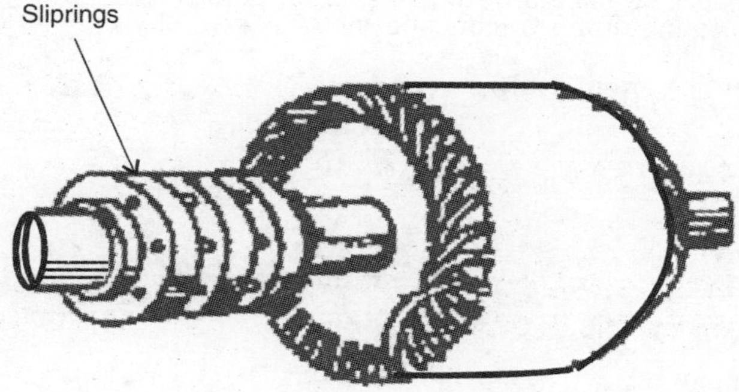


**Fig. 6.5** Slipring rotor.

### 6.2.3 Stator Winding

There are two types of winding usually used in induction motors are single layer mush winding or double layer lap winding. For small motors mush winding is suitable. For medium type motors double layer lap winding with diamond shaped coils and for large motors single layer with concentric coils are being used. In open slots, normally, double layer windings are used and in semi-closed slots mush winding is used.

### 6.2.4 Rotor Winding

There are two types of rotor windings as squirrel cage and wound (or slipring) rotor. It has been already discussed. In cage type rotor, each slot has one bar of copper or aluminium. These bars are short-circuited on both ends of rotor by copper or aluminium ring called end ring. The rotor bars and end rings take the shape of cage and so called cage rotor. Die-cast rotors are being used since it has no joints so it has good mechanical strength with good running performance. Slipring induction motors are used where speed control and high starting torque are needed. This is achieved by varying the rotor resistance through slipring of the rotor of the motor.

### 6.2.5 Sliprings

Three sliprings are needed, as discussed earlier, in a slipring rotor as shown in Fig. 6.5. Sliprings are made of brass or phosphorus-bronze placed on reinforced thermosetting resin in mild steel hub.

### 6.2.6 Shafts and Bearings

The air gap length in an induction motor is less, so any irregularity in air gap length will cause unbalanced magnetic pull. Therefore short and well fitted shaft with good mechanical strength are used. To achieve accurate centering of the rotor, ball or roller bearings are used.

### 6.2.7 Frame

It is casted or fabricated in which stator core is assembled. It encloses the stator and windings, covers the different part of machine. The casted frames are normally used for small machines. For medium and large motors the frames are fabricated.

## 6.3 OUTPUT EQUATION

The kilowatt (kW) rating of a 3-$\phi$ induction motor is given by

$$Q = 3V_{ph} I_{ph} \eta \cos\phi \times 10^{-3} \quad \text{(kilowatt)}$$

$$= 3\left[4.44 k_w \, f\phi \, N_{ph} I_{ph}\right] \eta \cos\phi \times 10^{-3} \qquad \text{...(6.1)}$$

Frequency, $f = \dfrac{PN}{120}$

$P =$ No. of poles

$N =$ r.p.m

Average value of magnetic loading is $\overline{B}$, then

$$\phi = \overline{B}\tau_P \cdot L$$

where, $L$ is the length of machine, and $\tau_p$ is pole pitch.

Total ampere-conductor of the machine

$$3 \times 2 N_{ph} I_{ph} = \overline{ac} \cdot \pi D$$

So, $$N_{ph} I_{ph} = \frac{\overline{ac} \cdot \pi D}{6}$$

where, $D$ is the inner diameter of stator.

Now, on putting the values of $f$, $\phi$ and $N_{ph} I_{ph}$, in equation (6.1), we get

$$Q = 3\left[4.44 k_w \frac{PN}{120} \cdot \overline{B}\tau_P \cdot L \cdot \overline{ac} \cdot \frac{\pi D}{6}\right] \cdot \eta \cos\phi \times 10^{-3}$$

$$= 18.3 \times 10^{-5} k_w \overline{B}\, \overline{ac}\, \eta \cos\phi D^2 LN$$

$$\therefore \qquad Q = CD^2 LN \qquad \text{...(6.2)}$$

where, $C$ is output coefficient and

$$C = 18.3 \times 10^{-5} k_w \overline{B}\overline{ac}\, \eta \cos\phi$$

$$= 18.3 \times 10^{-5} \times 0.955 \times \frac{2}{\pi} B_{\delta_1} \cdot \overline{ac} \cdot \eta \cos\phi$$

The winding factor $k_w = 0.955$ for fractional pitch distributed winding and $B_{\delta_1} = \dfrac{\pi}{2}\overline{B}$.

hence, $C = 11.1 \times 10^{-5} B_{\delta_1} \overline{ac} \eta \cos\phi$

Equation (6.2) is the output equation. $D$ and $L$ are the main dimensions of machine.

## 6.4 SELECTION OF FLUX DENSITY $(\overline{B})$ AND SPECIFIC ELECTRIC LOADING

***Selection of $\overline{B}$:*** Usual value is taken between 0.3 and 0.6 T. The value of magnetic loading ($\overline{B}$) has the effect on following:

(*i*) *Overload Capacity:* $V_{ph} = 4.44\, k_w f \phi N_{ph}$. For the same voltage if $\overline{B}$ is small, flux will be less and number of turns required should be more resulting in large leakage reactance. The circle diagram of machine has low diameter for large leakage reactance. This means that maximum output is less so the overload capacity will be small. To have large overload capacity, the flux density $\overline{B}$ should be high.

(*ii*) *Power Factor:* If the flux density $\overline{B}$ is higher, machine will draw large magnetizing current and so power factor will be poor. To get good power factor magnetic loading should be low.

(*iii*) *Cost and Size:* If $\overline{B}$ is high, diameter of machine will be less and hence less costlier.

(*iv*) *Temperature Rise:* If $\overline{B}$ is high, iron loss will be more and more temperature rise in the machine.

(*v*) *Iron Loss, Efficiency, Magnetizing Current:* If $\overline{B}$ is high, iron loss will be high, machine will have large magnetizing current and efficiency will be low.

### Selection of Ampere-conductor Loading Per Metre Length of Periphery

10,000–17,500 ampere-conductor/metre, for small m/c up to 10 kW.

20,000–30,000 ampere-conductor/metre, for medium m/c up to 100 kW.

30,000–40,000 ampere-conductor/metre, for m/c, above 100 kW.

(*i*) *Overload Capacity:* Ampere-conductor loading of the machine should be low. Low $\overline{ac}$ means less number of turns per phase and hence low leakage reactance and large diameter of machine.

(*ii*) *Temperature Rise:* $\overline{ac}$ should be low since a large $\overline{ac}$ means more copper and high loss so more temperature rise.

(*iii*) *Copper Loss:* $\overline{ac}$ should be low to keep low copper loss.

(*iv*) *Size and Cost of m/c:* High value of $\overline{ac}$ means high leakage reactance and low diameter of machine and hence smaller m/c and also cheaper.

## 6.5 ESTIMATION OF MAIN DIMENSIONS

Output equation of a 3-$\phi$ induction motor,

$$Q = CD^2 LN$$

or $$D^2L = \frac{Q}{CN} \qquad ...(6.3)$$

Length to pole pitch ratio $(L/\tau_P)$ may be selected based on

$$\frac{L}{\tau_P} = 1 \text{ for square pole, overall good design}$$

$= 1$ to $1.25$ for good power factor

$= 1.5$ for high efficiency

$= 1.5$ to $2$ for cheaper m/c ....(6.4)

And pole pitch, $\tau_P = \dfrac{\pi D}{P}$

With equations (6.3) and (6.4) main dimensions $D$ and $L$ can be calculated.

High values of $\dfrac{L}{\tau_P}$ results in small diameter of machine. If the motor is of low rating, motor may not be able to accommodate the slots for larger value of $(L/\tau_P)$. $\dfrac{L}{\tau_P}$ ratio can be selected up to 0.65 depending upon the (lower) rating and the size required.

The peripheral speed ($V_p$) can be estimated as

$$V_p = \frac{\pi DN}{60}$$

where, $D$ is the inner diameter of stator.

$V_p = 30–60$ m/sec.

$V_p$ can be taken up to 75 m/sec for the *m/c*s with special construction. $V_p$ should normally be around lower portion of the peripheral speed range *i.e.*, up to 40 m/s.

## 6.6 EFFECTIVE LENGTH OF MACHINE ($L_e$)

If the length of the motor exceeds to a certain value (normally 10 cm), radial ventilating ducts along the length of the motor should be provided. Due to ventilating ducts the fringing of flux occurs and hence there is need to estimate the effective length of the motor.

Let $b_v$ = width of ventilating ducts shown in Fig. 6.6.

= 1 cm at each interval of 8 cm *m/c* length.

$n_v$ = number of ventilating ducts.

$b'_v$ = effective width of ventilating ducts.

Gross iron length, $l = l_1 + l_2 + l_3 + \cdots + l_n$

Actual iron length in gross iron length (which includes iron length and the insulation length required in lamination) depends upon thickness of lamination (stacking factor).

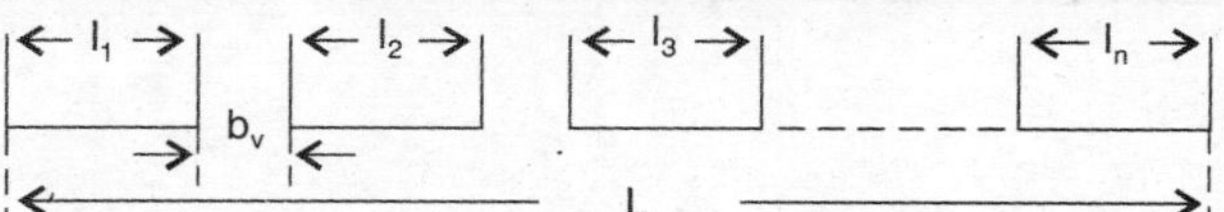


**Fig. 6.6** Effective length of m/c.

Actual iron length, $l_i = k_i l$

where, $k_i$ = stacking factor

= 0.96–0.98 for 0.50 mm thickness of lamination

= 0.90–0.92 for 0.35 mm thickness of lamination

Overall length, $L = l + n_v b_v$

Effective length, $L_e = L - n_v b'_v$

where, $$b'_v = b_v \cdot \frac{5}{5 + \dfrac{b_v}{\delta}}$$

$\delta$ is length of air gap of motor.

$= \left(0.2 + 2\sqrt{DL}\right)$ millimetre where, $D$ and $L$ are in metres.

## 6.7 ESTIMATION OF TURNS PER PHASE ($N_{ph}$), TOTAL CONDUCTORS ($Z_T$) AND CONDUCTORS PER SLOT ($N_C$) ETC. AND THEIR CORRECTED VALUES

$$V_{ph} = 4.44\, k_w \cdot f \cdot \phi_1 \cdot N_{ph} \qquad ...(6.5)$$

For a given machine,

$$N_{ph} = \frac{V_{ph}}{4.44\, k_w f \phi_1} \qquad ...(6.6)$$

where, $$\phi_1 = \overline{B} \tau_P L_e \qquad ...(6.7)$$

For approximate estimation length of machine ($L$) can be considered in place of $L_e$ to calculate $\phi_1$.

$$f = \frac{PN}{120} \text{ and } \tau_P = \frac{\pi D}{P}$$

Knowing the value of $N_{ph}$, conductors per phase, $Z_{ph}$ can be estimated as

$$Z_{ph} = 2 \cdot N_{ph}$$

$\therefore$ Total conductors in machine, $Z_T = 3 \times 2 N_{ph} = 6 N_{ph}$

Chapter 6

Conductors per slot, $N_C = \dfrac{Z_T}{S_1}$ ...(6.8)

where, $S_1$ = total number of slots in stator

$= 3 \cdot P \cdot q_1$

$q_1$ = number of slots per pole per phase

= may be taken from 3 to 10.

From equation (6.8), $N_c$ can be calculated. This may be fractional. It should be made of an integer and it should also be divisible by 2 for double layer winding. A suitable value of $N_c$ is estimated (corrected value, $N_{corr}$).

Calculate, $Z_{T\,(corr)} = S_1\, N_{C\,(corr)}$

$$N_{ph(corr)} = \frac{Z_{T\,(corr)}}{6}$$

Knowing $N_{ph\,(corr)}$, from equation (6.5), then corrected value of $\phi_1$ ($\phi_{1\,corr}$) can be calculated.

$$\phi_{1\,(corr)} = \frac{V_{ph}}{4.44\, k_w f\, N_{ph(corr)}}$$

Knowing $\phi_{1\,(corr)}$, $\overline{B}_{(corr)}$ can be calcualted from equation (6.7).

$$\overline{B}_{(corr)} = \frac{\phi_{1\,(corr)}}{\tau_p L_e}$$

***Check for Ampere-conductor Per Slot:*** If $q_1$ is higher, ampere-conductor will be higher and leakage reactance will be higher.

Slot pitch, $\tau_s = \dfrac{\pi D}{S_1}$ or $S_1 = \dfrac{\pi D}{\tau_s}$

Usually, $\tau_s = 10-20$ for m/cs having semi-closed slots,

$= 15-30$ for m/cs having open slots.

A suitable number of slot per pole per phase ($q_1$) and total slots ($S_1$) and hence the number of conductors per slot ($N_c$) can be estimated $\left(\dfrac{Z_T}{S_1}\right)$ to keep the slot ampere-conductor ($N_c \cdot I_{ph}$) within limit (around 500).

***Check for Ampere-conductor Per Slot:*** According to slot pitch, total slots should be between $\dfrac{\pi D}{30 \times 10^{-3}}$ and $\dfrac{\pi D}{10 \times 10^{-3}}$.

– $Z_T$ has been estimated

- Select $q_1$
- find out $S_1$ for $q_1$ selected. It should lie within the above range.
- Estimate conductors per slot, $N_c = \dfrac{Z_T}{S_1}$
- Calculate, $I_{ph} = \dfrac{Q \times 10^3}{3V_{ph} \cdot \eta \cdot \cos\phi}$
- Find $(I_{ph} \cdot N_c)$. It should be below 500.
- If $I_{ph} \cdot N_c > 500$

Select next higher value of $q_1$ and calculate $S_1$, and accordingly estimate, $N_c = \dfrac{Z_T}{\text{New value of } S_1}$.

Find out $I_{ph} \cdot N_c$ till it becomes $\leq 500$.

Finally $q_1$, $S_1$ and corresponding, $N_c$ can be estimated.

## 6.8 AIR GAP LENGTH OF MOTOR

Length of air gap of the machine can be calculated by expression

$$\delta = 0.2 + 2\sqrt{DL} \text{ millimetre}$$

where, $D$ and $L$ are taken in meters.

## 6.9 CROSS-SECTIONAL AREA OF CONDUCTOR

Phase current, $I_{ph} = \dfrac{Q \times 10^3}{3V_{ph} \cdot \eta \cdot \cos\phi}$ ampere

Cross-sectional area of each stator conductor, $F_c = \dfrac{I_{ph}}{\text{Current density } (\sigma)}$

$$\sigma = 3\text{-}4 \text{ ampere / mm}^2$$

Stator conductors may be

(*i*) Double cotton covered (D.C.C.) wires.

(*ii*) Super enamelled wires.

(*iii*) Bar or strip conductors.

Round wires are selected for low current rating m/c and bar or strip conductors are used for higher current rating. It is difficult to do winding with the conductors having large diameter. Also higher space factor is obtained for strip type of conductors.

The size (dia) of round conductors depends upon the insulation to be provided as given in Fig. 6.7. Similarly, the size of bar or strip conductors depends upon width and thickness of conductors. The size of conductors in each case can be selected from Indian cable tables or from any other standard table (Table A.1).

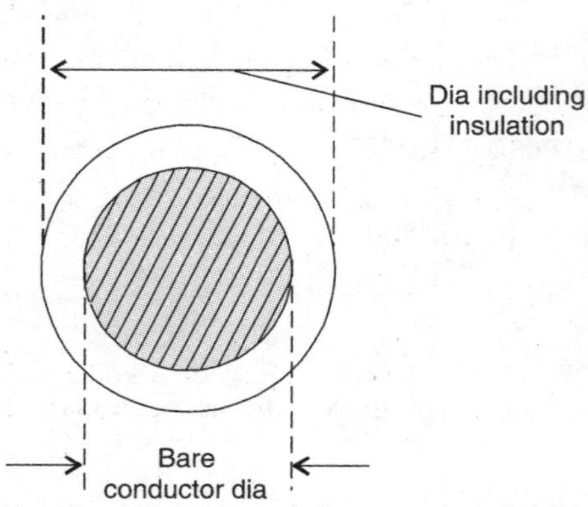


**Fig. 6.7** Conductor insulation.

## 6.10 STATOR SLOT DESIGN

Different shapes of the slot are discussed earlier. Semi-closed slots are being used to optimize the leakage flux. Details of stator slot insulation of 400V induction motor is given in Fig. 6.8.

Copper area per slot = $F_c$

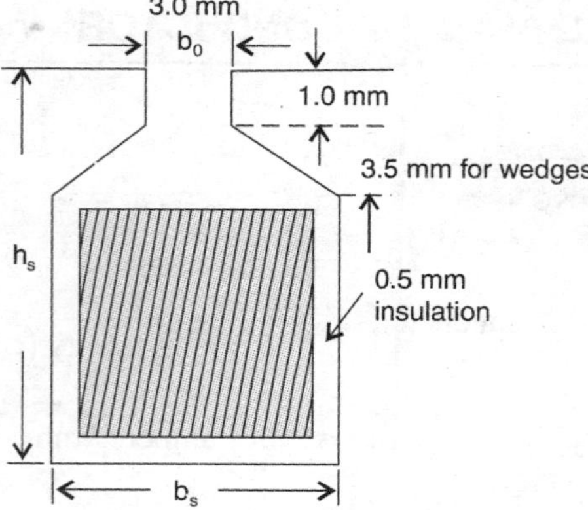


**Fig. 6.8** Details of a stator slot insulation of 400 V induction motor.

Space required by each slot on area of slot,

$$h_s \times b_s = \frac{F_c}{k_f} \quad \text{...(6.9)}$$

where, $k_f$ = winding space factor

= 0.3 to 0.45 for round wires

= 0.45 to 0.6 for strip winding

$$h_s = (2 \text{ to } 5)\ b_s \qquad ...(6.10)$$

From equations (6.9) and (6.10) the height ($h_s$) and width of slot ($b_s$) can be calculated. The conductor and insulation can be accommodated in the slot as in Fig. 6.9.

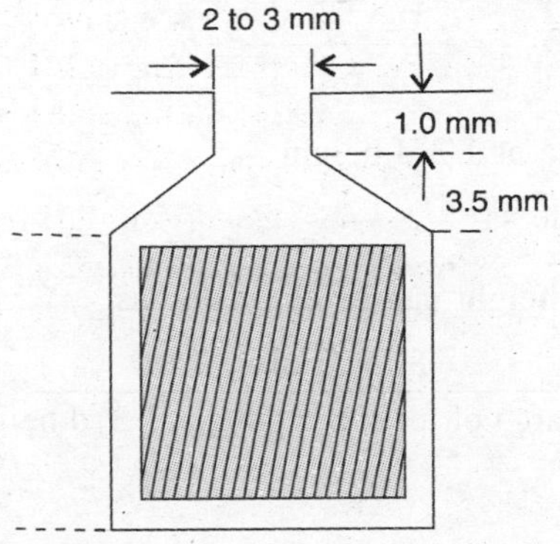


**Fig. 6.9** Details of slot insulation.

## 6.11 TOOTH DESIGN

In induction motor, the permissible value of flux density in stator teeth is 1.6 T at 1/3rd height of the tooth height from the air gap end given in Fig. 6.10.

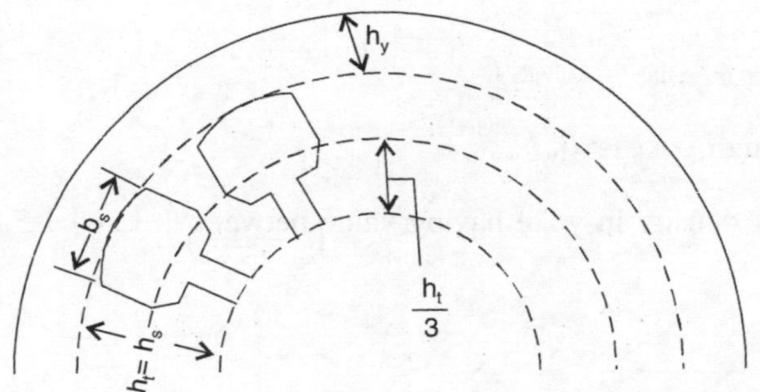


**Fig. 6.10** Tooth design.

Let $h_t$ is the height of teeth.

$$h_t = h_s \text{ (height of slot)}$$

$$\text{Diameter at 1/3rd height } = D + \frac{2}{3} h_t$$

Chapter 6

Slot pitch at 1/3rd tooth height,

$$\tau_{S^{1/3}} = \frac{\pi\left(D + \frac{2}{3}h_t\right)}{S_1}$$

Width of tooth at 1/3rd tooth height,

$$b_{t^{1/3}} = \frac{\pi\left(D + \frac{2}{3}h_t\right)}{S_1} - b_S$$

Sectional area of teeth per pole at 1/3rd height

$= \text{No. of teeth/pole} \times b_{t^{1/3}} \times k_i\, l$

Flux density in tooth at 1/3rd height can be estimated as

$$B_{t^{1/3}} = \frac{\text{Flux / pole}}{\text{Sectional area of teeth per pole at } 1/3 \text{ rd height}}$$

$B_{t^{1/3}}$ should not exceed 1.6 T.

If $B_{t^{1/3}} > 1.6\,\text{T}$, decrease $q_1$.

Area of teeth will increase. In case the slots are more tapered Simpsons rule must be applied.

## 6.12 HEIGHT OF STATOR YOKE

Flux through yoke $= \dfrac{\text{Flux per pole}}{2} = \dfrac{\phi}{2}$

Sectional area of stator yoke $= h_y \cdot k_i\, l$

Flux through yoke $= B_y\, h_y \cdot k_i\, l$

where, $B_y$ is the flux density in yoke having value between 1.3 and 1.5 T.

So, $\dfrac{\phi}{2} = B_y \cdot h_y \cdot k_i\, l$

or $h_y = \dfrac{\phi}{2 \cdot B_y k_i\, l}$

## 6.13 OUTER DIAMETER

Outer diameter of the motor is shown in Fig. 6.11 is equal to

$D_0 = D + 2h_t + 2h_y$

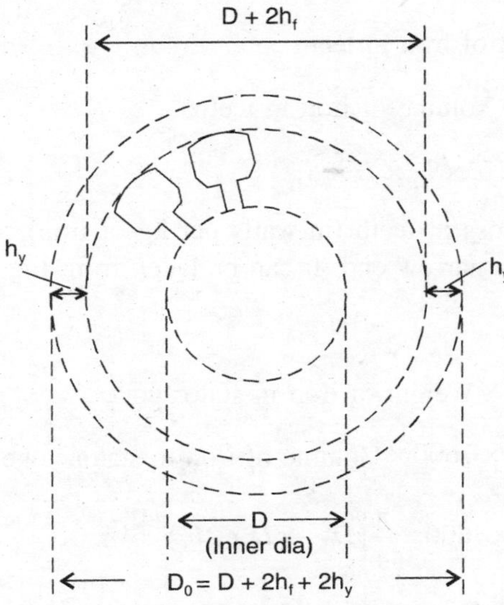


**Fig. 6.11** Outer diameter.

## 6.14 STATOR WINDING RESISTANCE, COPPER LOSS, IRON LOSS, STATOR WEIGHT ETC.

### Length of Mean Turn

$$L_{mt} = 2\left[L + 1.15\ \tau_P + 0.08\right] \text{ metre}$$

Length of stator winding conductor per phase,

$$= L_{mt} \cdot N_{ph} \text{ metre}$$

$$R_{dc\ (75°\text{C})} \text{ per phase } = 0.021 \frac{L_{mt} \cdot N_{ph}}{F_c}$$

where, $F_c$ is the sectional area in mm$^2$ and $L_{mt}$ is in m.

A.C. resistance per phase will be more than $R_{dc}$ due to skin effect and it is usually taken as $R_{ac} = R_{ph} = (1.15 \text{ to } 1.20)\, R_{dc}$

Stator copper loss for a 3-ϕ motor is $= 3I_{ph}^2\, R_{ph}$ watt

Volume of Cu in stator winding $= 3F_c \cdot L_{mt} \cdot N_{ph}$ cubic metre

Weight of copper in stator winding $= 8900 \times 3F_c \cdot L_{mt} \cdot N_{ph}$ kilogram.

Chapter 6

### Stator Iron Loss

(*i*) Iron loss in stator teeth,

$$= p_{it} \times \text{Weight of iron in teeth}$$

$$= p_{it} \times 7600 \times \text{Volume of iron in teeth}$$

$$= p_{it} \times 7600 \times b_t \times h_t \times L \times S_1$$

where, $p_{it}$ is the specific iron loss in teeth (in watts per kg of iron) corresponding to flux density in teeth at 1/3rd tooth height from narrow end. It can be taken from Fig. 5.30.

(*ii*) Iron loss in stator yoke,

$$= p_{iy} \times \text{Weight of iron in stator yoke}$$

$$= p_{iy} \times 7600 \times \text{Volume of iron in stator yoke}$$

$$= p_{iy} \times 7600 \times \frac{\pi}{4}\left[D_0^2 - (D + 2h_t)^2\right]$$

where, $p_{iy}$ is the specific iron loss in yoke in watts per kg of iron corresponding to the flux density in yoke.

Weight of stator = Weight iron in stator + Weight of copper in stator + Weight of insulation, fixtures etc. (15–20% of total stator iron and copper weight).

## 6.15 ROTOR DESIGN

### 6.15.1 Rotor Diameter

Length of air gap, $\delta = 0.2 + 2\sqrt{DL}$ mm, where $D$ and $L$ are in metres.

Rotor Diameter, $D_r =$ Stator inner diameter – 2(air gap length) $= D - 2\delta$

### 6.15.2 Number of Rotor Slots

The selection of number of rotor slots can be done by taking into consideration the following effects: Let $S_1$ is stator slots and $S_2$ is rotor slots.

(*i*) $S_1 \neq S_2$ to avoid cogging,

(*ii*) $S_1 - S_2 \neq \pm p, \pm 2p, \pm 5p$ due to fundamental belt locking,

(*iii*) $S_1 - S_2 \neq \pm 1, \pm 2$ or $\pm p \pm 1, \pm p \pm 2$ etc. to avoid too much noise and vibration,

(*iv*) $S_1 - S_2 \neq \pm 3p$ to avoid magnetic locking.

Generally, the rotor slots are selected by

$$q_1 - q_2 = \pm 1, \pm \frac{1}{3}, \pm \frac{2}{3}$$

where, $q_1$ = stator slots/pole/phase

$q_2$ = rotor slots/pole/phase

$q_1$ has already been estimated. So suitable number of rotor slots/pole/phase can be taken.

Number of rotor slots, $S_2 = 3 q_2 p$

### 6.15.3 Total Number of Rotor Conductors

***(i) For wound rotor induction motors***

$$Z_2 = S_2 \cdot N_{cr}$$

where, $N_{cr}$ is the rotor conductors per slot

$$\text{Conductors per phase, } Z_{2ph} = \frac{S_2 \cdot N_{cr}}{3}$$

$$\text{Rotor turns per phase, } N_{2ph} = \frac{Z_{2ph}}{2} = \frac{S_2 \cdot N_{cr}}{6}$$

$$\frac{\text{Stator voltage per phase}}{\text{Rotor voltage/phase at standstill}} = \frac{N_{ph}}{N_{2ph}}$$

or
$$\frac{V_{ph}}{V_{2ph}} = \frac{N_{ph}}{N_{2ph}} = \frac{N_{ph} \times 6}{S_2 \cdot N_{cr}} \qquad \text{...(6.11)}$$

$V_{2ph}$ can be selected but it should normally not exceed 500 V for delta connected and 290 V for star connected machines. With the help of equation (6.11), $N_{cr}$, $N_{2ph}$ and $Z_{2ph}$ can be calculated.

***(ii) Cage rotor induction motor:*** Total rotor conductors (or rotor bars) = Number of rotor slots

*i.e.,* $$Z_b = S_2$$

One rotor bar is provided in each slot of the rotor. So the total number of rotor bars will be equal to the number of rotor slots. They are then short-circuited at both the ends with end rings to make the cage.

Rotor pole pitch and slot pitch

$$\text{Rotor pole pitch, } \tau_{pr} = \frac{\pi D_r}{P}$$

$$\text{Rotor slot pitch, } \tau_{sr} = \frac{\pi D_r}{S_2}$$

### 6.15.4 Rotor Current

The entire stator m.m.f. does not get transferred to the rotor side. Around 85% of the stator m.m.f. gets transferred to the rotor side.

***(i) Wound rotor motor:*** Total ampere-conductor on rotor side $= 6N_{2ph} \cdot I_{2ph}$

85% of stator m.m.f. $= 0.85 \times 6 N_{ph} I_{ph}$

So, $6N_{2ph} \cdot I_{2ph} = 0.85 \times 6N_{ph} I_{ph}$

or $$I_{2ph} = \frac{0.85 \times N_{ph} I_{ph}}{N_{2ph}}$$

Rotor phase current, $I_{2ph}$, can be estimated by above equation.

***(ii) Cage rotor:*** Let current in the rotor bar is $I_b$

$$Z_b \cdot I_b = 0.85 \times 6N_{ph} I_{ph}$$

or $$I_b = \frac{0.85 \times 6N_{ph} I_{ph}}{Z_b}$$

***(iii) Current in the end ring:***

$$I_{er} = \frac{Z_b I_b}{\pi P},$$

where, $Z_b$ is number of rotor bars and $I_b$ is curent per bar.

### 6.15.5 Sectional Area of Rotor Conductor

***(i) Wound rotor machine:*** Sectional area of rotor conductor, $F_{c_2} = \frac{I_{2ph}}{\sigma}$

where, σ is the current density having values between 3 and 4 ampere/mm$^2$.

***(ii) Cage rotor machine:*** Sectional area of rotor bar, $F_{b_2} = \frac{I_b}{\sigma}$

$\sigma$ = 5 to 7 A/mm$^2$ for copper and

= 3 to 5 A/mm$^2$ for aluminium.

The current density for cage rotor bars is taken more than the wound rotor conductor since the ventilation in case of cage rotor is better. Depending upon the sectional area of the rotor conductor, the

**Fig. 6.12**(*a*) Single and double cages.

size (diameter for round wires and thickness and width for strip windings.) of the conductor can be estimated from any standard table (Table A.1).

For double cage 6.12(*a*) induction motors, the sectional area of inner cage is to be designed for normal operation and outer cage is designed depending upon starting torque required which is generally about 50% of maximum torque.

***(iii) Sectional area of each end ring:*** Sectional area of each end ring $= F_e = \dfrac{I_{er}}{\sigma_{er}}$

where, $I_{er}$ is the current in end ring and $\sigma_e$ is the current density in end ring.

Current density in end ring is taken slightly higher than that in rotor bars since ventilation is better for end rings.

### 6.15.6 Rotor Slot Design

***(i) Wound rotor motor:*** Three-phase windings are used. Closed slots are not used since leakage flux will be very high. Semi-closed slots are used similar as stator slots given in Fig. 6.12(*b*).

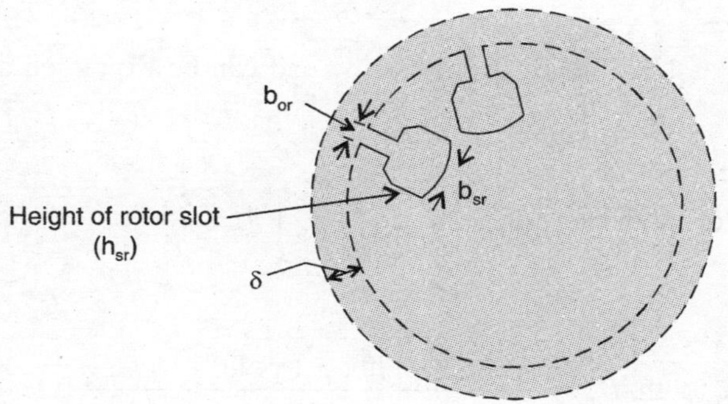


**Fig. 6.12**(*b*) Rotor slot.

Sectional area of copper per slot $= F_{c_2}$

Let $h_{sr}$ = height of rotor slot

$b_{sr}$ = width of rotor slot

Space required by each slot depends upon rotor winding space factor $(k_{fr})$.

Winding space factor depends upon the type of winding used.

$$\text{Area of slot} = h_{sr} \times b_{sr} = \frac{F_{c_2}}{k_{fr}} \qquad ...(6.12)$$

where, $k_{fr}$ = 0.3 to 0.4 for round wires

= 0.45 to 0.6 for strip conductors

and $h_{sr} = (2 \text{ to } 4) b_{sr}$ ...(6.13)

From equations (6.12) and (6.13), the height $(h_{sr})$ and width $(b_{sr})$ of rotor slot can be estimated.

***(ii) Cage rotor machine:*** Cross-sectional area of each rotor bar, $F_{b_2} = \frac{I_b}{\sigma}$

The size of the slot can be found out depending upon the shape of the slot to be selected. There is no insulation provided between bars and rotor core.

An air gap of 0.15 to 0.4 mm is maintained between rotor bars and core. Higher air gap is maintained for the machines having skewed slots. Generally, the slots are skewed by one slot pitch and the gap between rotor bar and core is kept around 0.4 mm.

### 6.15.7 Rotor Teeth Design

***Wound rotor:*** Rotor slot pitch at 1/3rd height from narrow end can be expressed as

$$\tau_{sr(1/3)} = \frac{\pi\left(D - 2\delta_0 - \frac{2}{3}h_{sr}\right)}{S_2}$$

Width of rotor teeth at 1/3rd height from narrow end can be expressed as

$$b_{sr(1/3)} = \tau_{sr(1/3)} - b_{sr}$$

$$\text{Sectional area of rotot tooth per pole} = \left[\tau_{sr(1/3)} - b_{sr}\right] \cdot L_i \times \frac{S_2}{P}$$

where, $L_i = k_i\, l$

$$\text{Flux density in rotor tooth, } B_{tr} = \frac{\text{Flux per pole}}{\text{Sectional area of rotor tooth/pole}}$$

$B_{tr}$ should not exceed 1.6 tesla.

### 6.15.8 Rotor Yoke Design

The height of rotor yoke (Fig. 6.13) is estimated on the fact that the half of the flux per pole is associated with rotor yoke

$$h_{yr} = \frac{\phi/2}{B_{yr} \times k_i\, l}$$

where, $B_{yr}$ is the flux density in the rotor yoke having values between 1.3 and 1.5 T. So, height of rotor yoke can be calculated.

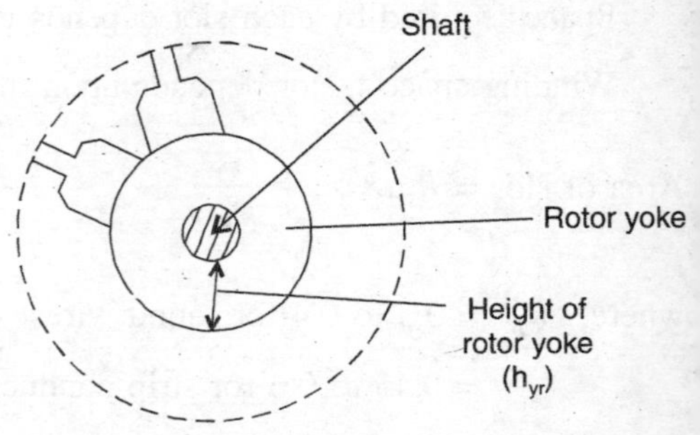


**Fig. 6.13** Rotor yoke.

### 6.15.9 Rotor Copper Loss

***(i) Wound rotor machine:*** Length of mean rotor turn,

$$L_{mt_2} = 2\left[L + 1.5\tau_{pr} + 0.12\right]$$

$\tau_{pr}$ = rotor coil span (Fig. 6.14)

Rotor resistance per phase

$$R_{2ph} = 0.021 \times \frac{N_{2ph} \times L_{mt_2}}{\text{Sectional area of rotor conductor}}$$

$$= 0.021 \times \frac{N_{2ph} \times L_{mt_2}}{F_{c_2}}$$

where, $L_{mt_2}$ is in m and $F_{c_2}$ is in mm$^2$.

Rotor Cu loss for 3-$\phi$ m/c = $3 \cdot I_{2ph}^2 \cdot R_{2ph}$

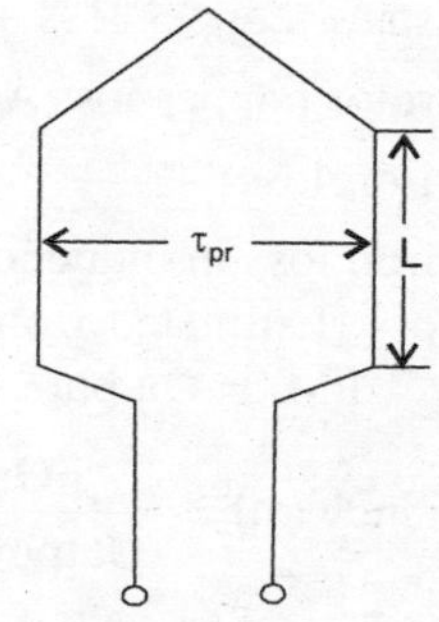


**Fig. 6.14** Rotor coil span.

***(ii) Cage rotor machine:*** Rotor bar resistance, $R_b = 0.021 \times \frac{L}{F_{b_2}}$

where, $L$ is in m and $F_{b_2}$ is in mm$^2$.

Cu loss in rotor bar $= I_b^2 \times R_b \times Z_b$ watt

End ring current, $I_{er} = I_b \cdot \frac{Z_b}{\pi P}$

Mean length of end ring $= \pi D_r$

Sectional area of end ring $= \frac{1}{2} \cdot \frac{F_{b_2} \times Z_b}{P}$ = Half of the sectional area of bars per pole.

Resistance of end ring, $R_e = 0.021 \times \frac{\pi D_r}{\text{Sectional area of end ring}}$

where, $D_r$ is in m and sectional area of end ring is in mm$^2$.

Cu loss in end rings $= 2 \cdot I_{er}^2 \cdot R_e$

Total rotor Cu loss = Cu loss in rotor bars + Cu loss in end rings.

Full load slip of motor,

$$S = \frac{\text{Rotor Cu loss}}{\text{Rotor intake}} = \frac{\text{Rotor Cu Loss}}{\text{F and W losses + Rotor Cu loss + Rotor output}}.$$

Usual value of slip = 2 to 5%. For larger motor there should be lower value of slip. If slip is not in the specified range, modify the parameters to modify the Cu losses. If slip is high, it can be reduced by decreasing rotor current and increasing sectional area of conductors and end rings.

## 6.16 EFFICIENCY

The different losses in induction motor (I.M.) have been estimated. Total loss in I.M. includes

Total stator copper loss +

Total rotor copper loss +

Total iron loss

Total iron loss in induction motor = (1.75 to 2.2) × iron loss in stator yoke and teeth + friction and windage loss (Estimated from Table 6.1).

Input of I.M. = Output + Total loss in I.M.

$$\text{Efficiency } (\eta) = \frac{\text{Output}}{\text{Output + Loss}}.$$

If $\eta$ is low efficiency can be improved by reducing stator and rotor Cu loss. Increase sectional area of conductors and end rings or decrease number of turns. Reduced iron loss by decreasing flux density will also improve $\eta$.

## 6.17 CONCEPT OF FLATTENED FLUX DENSITY $(B_{30^\circ}$ or $B_{60^\circ})$

In an induction motor, the flux is distributed approximately sinusoidally and m.m.f. also varies in similar way. If the permeability of the iron could be assumed constant, the value of magnetizing current would be obtained accurately by considering the mean m.m.f. When there is no saturation in iron parts of the magnetic ckt., the flux density distribution in space is sinusoidal. But due to saturation in iron parts, the m.m.f. produces a flat topped flux wave, which is shown in Fig. 6.15.

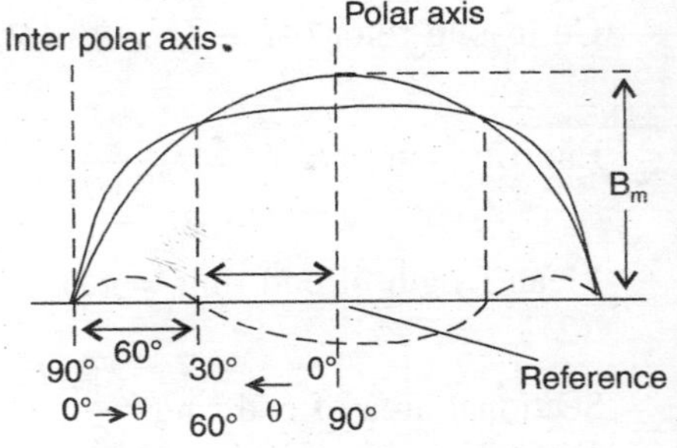


**6.15** Flattened flux density distribution.

The flat topped flux wave can be represented by

$$B_\theta = B_{m_1} \cos\theta + B_{m_3} \cos 3\theta + B_{m_5} \cos 5\theta + \ldots\ldots.$$

where, $B_\theta$ is flux density at an angel $\theta$ from reference axis. $B_{m_1}$, $B_{m_3}$, $B_{m_5}$ etc. are the maximum value of fundamental, 3rd harmonic, 5th harmonic flux densities respectively. Generally, the harmonics above the 3rd are small in magnitude and hence the flux density wave shape can be considered to consist of fundamental sine wave with a superimposed third harmonic.

When flux density is estimated at 30°, there is no effect of 3rd harmonic flux density, even if it is present because (Taking polar axis as reference).

$$B_{30^\circ} = B_{m_1} \cos 30^\circ + B_{m_3} \cos 90^\circ + \ldots\ldots$$

$$= B_{m_1} \cos 30^\circ + 0$$

$$= B_{m_1} \cos 30^\circ$$

or

$$B_{30^\circ} = B_{m_1} \cos 30^\circ$$

$$= \frac{\sqrt{3}}{2} B_{m_1}$$

$$= \frac{\sqrt{3}}{2} \cdot \frac{\pi}{2} B_{av}\text{, where, } B_{av} \text{ is the average flux density.}$$

$$= \frac{\sqrt{3}\pi}{4} B_{av}$$

$$= 1.36 \; B_{av}$$

When interpolar axis is taken as the reference axis, then

$$B_\theta = B_{m_1} \sin\theta + B_{m_3} \sin 3\theta + B_{m_5} \sin 5\theta$$

then $$B_{60°} = B_{m_1} \sin 60° + B_{m_3} \sin 3 \times 60°$$

$$= B_{m_1} \sin 60° + 0 = \frac{\sqrt{3}}{2} \cdot B_{m_1} = \frac{\sqrt{3}}{2} \cdot \frac{\pi}{2} B_{av}$$

$$= 1.36 \; B_{av}$$

## 6.18 NO LOAD CURRENT

The no load current ($I_0$) of an induction motor consists of two current components,

(1) Magnetizing current $(I_\mu)$, and

(2) Estimation of loss component of current ($I_c$).

The magnetic circuit *i.e.*, flux path links the following parts of the induction motors (*i*) Air gap, (*ii*) Stator teeth, (*iii*) Stator yoke, (*iv*) Rotor teeth, (*v*) Rotor yoke. Magnetizing current required at no load is equivalent to the m.m.f. required by the above portions of an I.M. The loss component of current includes the iron loss in I.M. and friction and windage loss at no load. The magnetizing current component lags behind the voltage by 90° and the loss component current is in phase with the voltage.

Chapter 6

### 6.18.1 Magnetizing Current $(I_\mu)$

To calculate $I_\mu$, the m.m.f. required by different portion of I.M. must be estimated.

***(a) M.M.F. for air gap:*** m.m.f. for air gap, $AT_g = 7.96 \times 10^5 \times B_{30°} \times \delta'$

where, $\delta'$ is the effective length of air gap and is given by

$$\delta' = k_c \cdot \delta$$

$k_c$ = gap contraction coefficient

$$= k_{c_1} \cdot k_{c_2}$$

$k_{c_1}$ = gap contraction coefficient due to stator slotting

$$= \frac{\tau_s}{\tau_s - k_{01} \cdot b_0}$$

$k_{01}$ = air gap coefficient for stator

$k_{c_2}$ = gap contraction coefficient due to rotor

$$= \frac{\tau_{sr}}{\tau_{sr} - k_{02} \cdot b_{0r}},$$

$b_{0r}$ = opening of rotor slot

$\tau_{sr}$ = rotor slot pitch

$k_{02}$ = air gap coefficient for rotor

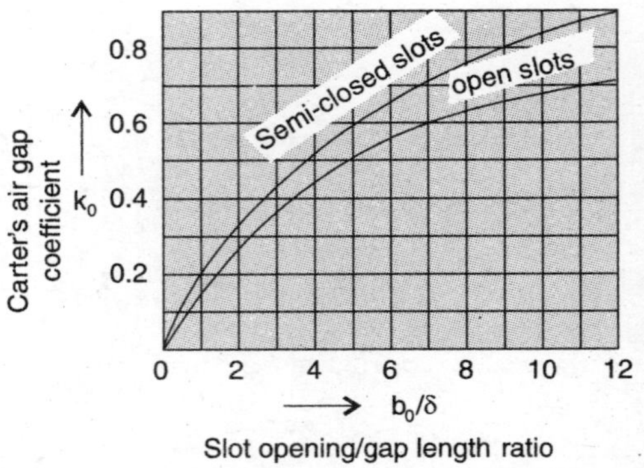

**6.16** Carter's air gap coefficient.

Estimate $k_{01}$ and $k_{02}$ from Fig. 6.16.

***(b) M.M.F. for stator teeth:*** Flux density at 1/3rd height of tooth from narrow end.

$$B_{t(1/3)} = \frac{\text{Flux per pole}}{\text{Sectional area of teeth per pole}}$$

$$= \frac{\phi}{\text{No. of teeth/pole} \times b_{t(1/3)} \times k_i\, l}$$

where, $$b_{t(1/3)} = \frac{\pi\left(D + \frac{2}{3} h_t\right)}{S_1} - b_s$$

$$B_{t(60°)} = 1.36\ B_{t(1/3)}$$

M.M.F. per metre of stator teeth $(at_t)$ corresponding to $B_{t(1/3)}$ is found from *B* vs *A-T* curve (Fig. 8.13)

So, M.M.F. required for stator teeth per pole, $AT_t = at_t \times h_t$

***(c) MMF for stator yoke:*** The mean length of flux path through stator yoke is taken 1/3rd pole

$$l_y = \frac{\pi\left(D_i + 2h_s + h_y\right)}{3P}$$

pitch at the mean yoke diameter m.m.f. per metre length of stator yoke $(at_y)$ corresponding to stator yoke flux density, $B_y$ can be found by Fig. 8.13.

∴ Total m.m.f. for stator yoke, $AT_y = at_y \times l_y$

***(d) MMF for rotor teeth:*** Flux density in rotor teeth at 1/3rd height from narrow end,

$$B_{tr(1/3)} = \frac{\text{Flux per pole}}{\left(\frac{S_2}{P}\right) \times b_{sr(1/3)} k_i l}$$

Now, $B_{tr(60°)} = B_{tr(1/3)} \times 1.36$

M.M.F. per metre of rotor teeth $(at_{tr})$ corresponding to $B_{tr(1/3)}$ is found from $B$ vs. $A$-$T$ curve (Fig. 8.13).

$\therefore$ M.M.F. required for rotor teeth, $AT_{tr} = at_{tr} \times h_{sr}$

***(e) M.M.F. for rotor core:*** The mean length of flux path in rotor core is taken 1/3rd pole pitch at the mean yoke diameter,

$$l_{yr} = \frac{\pi\left(D - 2\delta_0 - 2h_{sr} - h_{yr}\right)}{P}$$

m.m.f. per metre length of rotor yoke ($at_{yr}$) corresponding to rotor yoke flux density ($B_{yr}$) is estimated. Total m.m.f. for rotor core,

$$AT_{yr} = at_{yr} \cdot l_{yr}$$

Total m.m.f. per pole (at $B_{60°}$ or $B_{30°}$ )

$$AT_{60° \text{ or } 30°} = AT_g + AT_t + AT_y + AT_{tr} + AT_{yr}$$

Magnetizing current per phase,

$$I_\mu = \frac{0.427 \times P \times AT_{30°}}{k_w \cdot N_{ph}}$$

where, $k_w$ is winding factor.

### 6.18.2 Estimation of Loss Component of the Current ($I_c$)

$I_c$ = Iron loss + Friction and windage loss.

Total iron loss = 1.75 to 2.2 times the iron loss in stator teeth and yoke.

Friction and windage loss can be estimated by the Table 6.1.

**Table 6.1**

| *F and W losses as % of rated output* | *Output (kW)* |
|---|---|
| 6 | 0.5 |
| 5 | 1.0 |
| 3.5 | 3.70 |
| 2.7 | 7.50 |
| 1.5 | 37.00 |
| 1.2 | 75.00 |
| 1.0 | 150.00 |

Chapter 6

Total no load loss, $W_0$ = Total iron loss + F & W loss

So, $I_c = \dfrac{W_0}{3V_{ph}} = \dfrac{\text{Total no load loss}}{3 \times \text{Phase voltage}}$

Estimation of no load current as shown in Fig. 6.17 is done by

$$I_0 = \sqrt{I_\mu^2 + I_c^2}$$

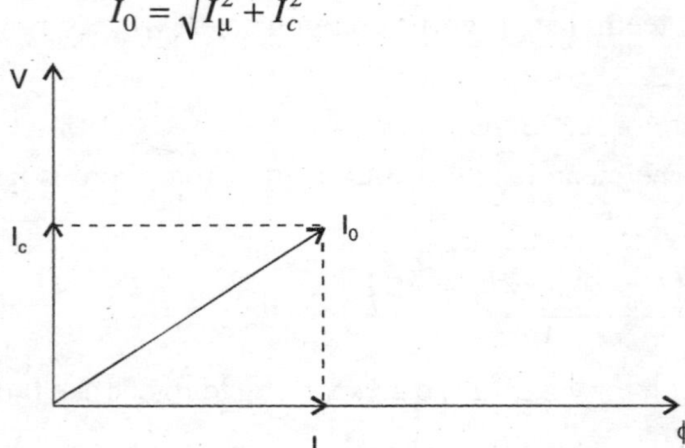


**Fig. 6.17** No load current.

**Table 6.2**

| *Output (in h.p.)* | *No load current as % of full load current* |
|---|---|
| 1–20 | 40–50 |
| 20–50 | 35–40 |
| 50–100 | 30–35 |
| above 100 | 25–30 |

No load current as percentage of full load current $= \dfrac{I_0}{I_{ph}} \times 100$

The approximate value of no load current experienced as percentage of full load current is shown in Table 6.2.

No load p.f., $\cos \phi_0 = \dfrac{I_c}{I_0}$

or $\phi_0 = \cos^{-1}\left(\dfrac{I_c}{I_0}\right)$

Also, $\tan \phi_0 = \dfrac{I_\mu}{I_c}$

or $\phi_0 = \tan^{-1}\left(\dfrac{I_\mu}{I_c}\right)$

## 6.19 ESTIMATION OF PERFORMANCE OF INDUCTION MOTOR

Short-circuit current or blocked rotor current: To find short-circuit current ($I_{sc}$), the value of leakage reactance and the resistance of windings should be calculated.

### 6.19.1 Estimation of Leakage Reactance

Leakage reactance consists of

(*a*) Stator slot leakage reactance.
(*b*) Rotor slot leakage reactance.
(*c*) Ovarhang or end turn leakage reactance.
(*d*) Differential, belt or harmonic leakage reactance.

***(a) Stator slot leakage reactance:*** Specific stator slot permeance for double-layer winding Fig. 6.18(*b*).

$$\lambda_{ss} = \mu_0 \left[ \frac{2h_1}{3b_s} + \frac{h_2}{4b_s} + \frac{h_3}{b_s} + \frac{2h_4}{b_0 + b_s} + \frac{h_5}{b_0} \right]$$

Specific stator slot permeance for single-layer winding Fig. 6.18(*a*).

$$\lambda_{ss} = \mu_0 \left[ \frac{h_1}{3b_s} + \frac{h_2}{b_s} + \frac{2h_3}{b_0 + b_s} + \frac{h_4}{b_0} \right]$$

Stator slot leakage reactance, $X_1 = 8\pi f \times \frac{N_{ph}^2}{P \cdot q_1} \cdot L \cdot \lambda_{ss}$

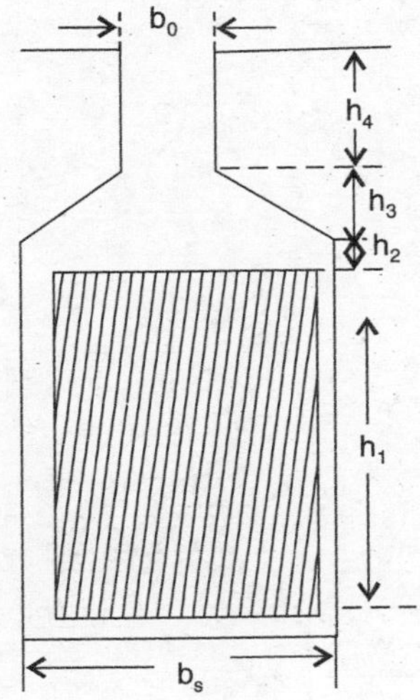


(*a*) Single layer winding

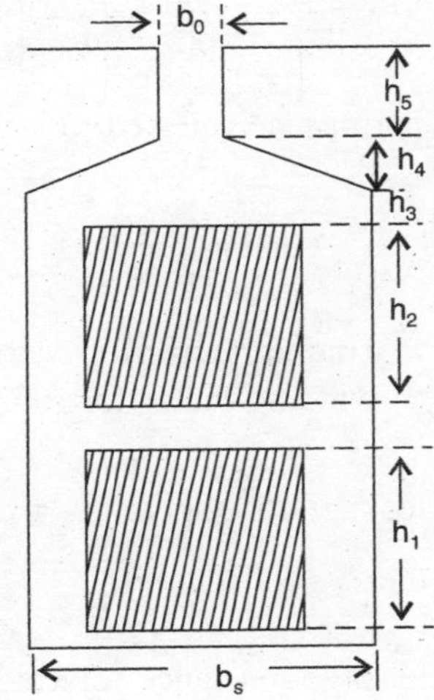


(*b*) Double layer winding

**Fig. 6.18** Different portions of stator slots.

***(b) Rotor Slot Leakage Reactance:*** For wound rotor I.M.: refer Fig. 6.19(*a*) specific slot permeance.

Chapter 6

$$\lambda_{rs} = \mu_0 \left[ \frac{h_{1r}}{3b_{sr}} + \frac{h_{2r}}{b_{sr}} + \frac{2h_{3r}}{b_{0r} + b_{sr}} + \frac{h_{4r}}{b_{0r}} \right]$$

For cage rotor I.M.: refer Fig. 6.19(*b*) specific slot permeance.

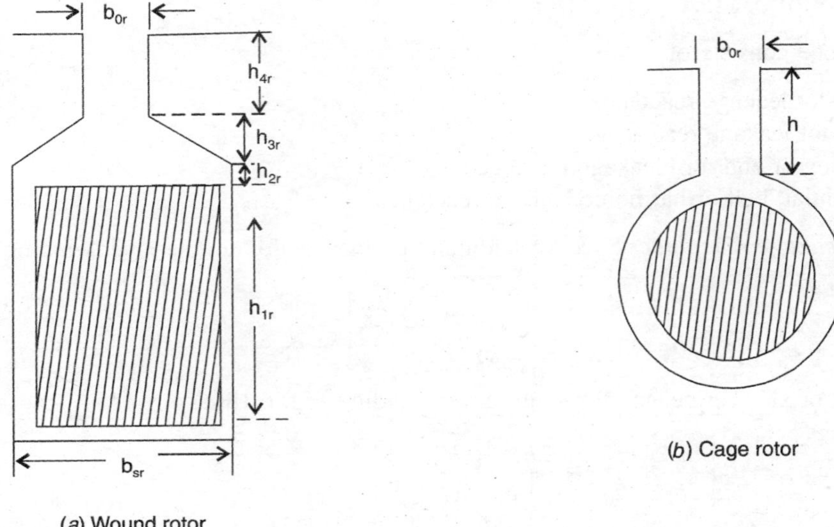


**Fig 6.19** Rotor slot.

$$\lambda_{rs} = \mu_0 \left[ 0.623 + \frac{h}{b_{0r}} \right]$$

Rotor slot permeance referred to stator,

$$\lambda'_{rs} = \left[ \frac{k_w}{k_{wr}} \right]^2 \frac{S_1}{S_2} \lambda_{rs}$$

where, $k_w$ = winding factor for stator and
$k_{wr}$ = winding factor for rotor.

Rotor slot leakage reactance, $X_2 = 8\pi f \times \dfrac{N_{2ph}^2}{P \cdot q_1} \cdot L \cdot \lambda_{rs}$

Rotor slot leakage reactance referred to stator, $X'_2 = \left[ \dfrac{k_w \cdot N_{ph}}{k_{wr} \cdot N_{2ph}} \right]^2 \cdot X_2$

***(c) Overhand or End Turn Leakage Reactance:*** Total Overhand Permeance

$$\lambda_0 = \mu_0 k_s \cdot \frac{\tau_P^2}{\pi l_0 \tau_s}$$

or $$l_0\lambda_0 = \mu_0 k_s \cdot \frac{\tau_P^2}{\pi\tau_s}$$

where, $k_s$ = slot leakage factor which depends upon a coil span to pole pitch ratio. It can be taken from Fig. 6.20(*a*) given as factor *k*..

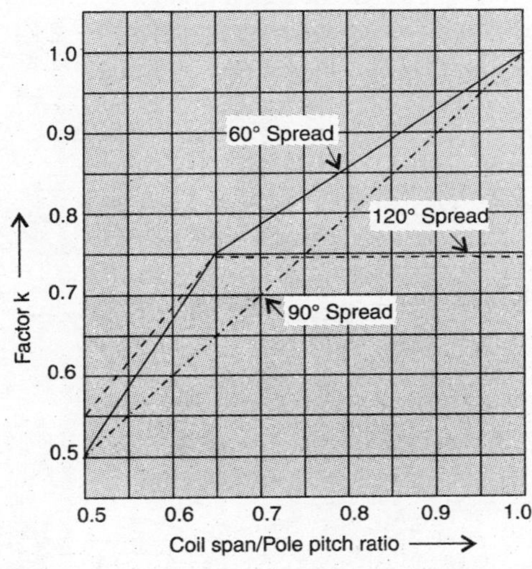


**Fig. 6.20**(*a*) Slot leakage factor ($k_s$).

$l_0$ is the length of winding conductor in overhang as shown in Fig. 6.20(*b*)

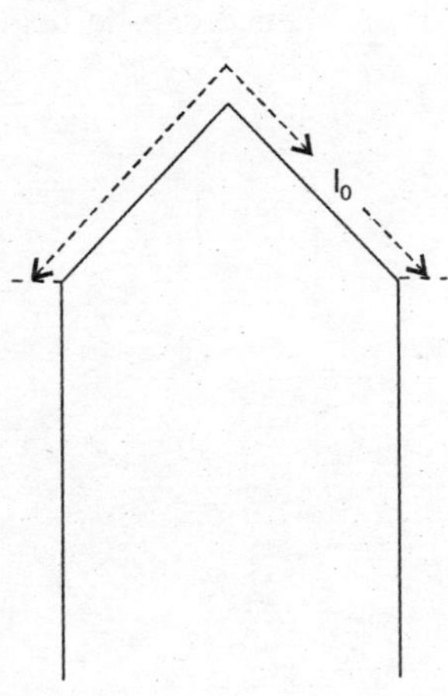


**Fig. 6.20**(*b*) Length of overhang conductor.

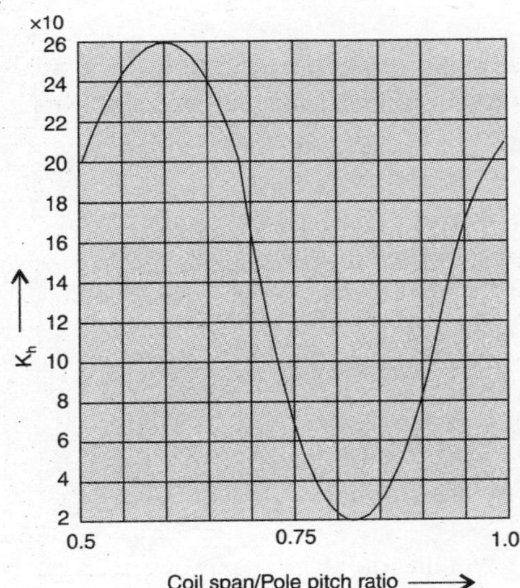


**Fig. 6.20**(*c*) Factor ($k_h$).

Chapter 6

Overhang leakage reactance,

$$X_0 = 8\pi f\left(\frac{N_{ph}^2}{P \cdot q_1}\right) \cdot l_0 \lambda_0$$

***(d) Zig-zag Leakage Reactance:*** Zig-zag leakage flux passes from one tooth to another tooth in zig-zag manner through the air gap. Its magnitude depends upon the length of air gap and relative position of tips of teeth.

Zig-zag leakage reactance,

$$X_3 = \frac{5}{6} X_\mu \left[\frac{1}{S_{1p}^2} + \frac{1}{S_{2p}^2}\right]$$

where, $S_{1p}$ = stator slot per pole = 3 . $q_1$
$S_{2p}$ = rotor slot per pole = 3 . $q_2$

$$X_\mu = \text{magnetizing reactance } = \frac{V_{ph}}{I_\mu}$$

***(e) Differential or belt or harmonic leakage:***

$$X_B = X_\mu [K_{hs} + K_{hr}]$$

where, $K_{hs}$ = constant for stator

$K_{hr}$ = constant for rotor

The values of $K_{hs}$ and $K_{hr}$ which depends upon coil span/pole pitch ratio, can be selected from Fig. 6.20(*c*).

Harmonic leakage reactance is very small in squirrel cage I.M. and hence can be neglected.

Total equivalent leakage reactance per phase referred to stator,

$$X_t = X_1 + X_2' + X_0 + X_3 + X_B$$

### 6.19.2 Calculation of Resistance

***(i) The resistance per phase***

$$R_s = 0.021 \times \frac{N_{ph} L_{mt}}{F_c}$$

where, $L_{mt}$ is in m and $F_c$ is in mm$^2$.

***(ii) Rotor Resistance***

(*a*) Wound rotor, $R_r = 0.021 \times \dfrac{N_{2ph} L_{mt_2}}{F_{c_2}}$

where, $L_{mt_2}$ is in m and $F_{c_2}$ is in mm$^2$.

The rotor resistance per phase referred to stator,

$$R_r' = \left(\frac{N_{ph}}{N_{2ph}}\right)^2 R_r$$

(*b*) Cage rotor: Rotor resistance referred to stator

$$R_r' = \frac{\text{Rotor Cu loss per phase}}{I_{ph}^2} \times \frac{1}{(\text{Full load } p.f)^2}$$

So, total equivalent resistance referred to stator per phase,

$$R_t = R_s + R_r'$$

Total equivalent impedance per phase referred to stator, $Z_s = \sqrt{R_t^2 + X_t^2}$

### 6.19.3 Short-Circuit Current and Power Factor

Short-circuit current and its power factor (p.f.) can be calculated for applied voltage and $Z_s$.

$$I_{sc} = \frac{V_{ph}}{Z_s'}$$

Short-circuit power factor $(\cos\phi_{sc}) = \dfrac{R_t}{Z_s}$

If p.f. is low, it can be modified by decreasing ampere-turn per pole or by decreasing air gap length or by decreasing air gap flux density.

### 6.19.4 Dispersion Coefficient

Dispersion coefficient $= \dfrac{I_\mu}{I_{sc}'}$

where, $I_{sc}'$ is ideal short-circuit current which can be estimated as

$$= \frac{V_{ph}}{X_t} \quad (R_t \text{ neglected})$$

So, Dispersion coefficient $= \dfrac{I_\mu X_t}{V_{ph}}$

## 6.20 CONSTRUCTION OF CIRCLE DIAGRAM FROM DESIGN DATA

The circle diagram for I.M. can be drawn with the help of no load current ($I_0$), no load p.f. ($\cos\phi_0$), short-circuit current ($I_{sc}$), short-circuit p.f. ($\cos\phi_{sc}$) etc., which have already been explained and estimation process given earlier. The construction steps of circle diagram is given below (refer Fig. 6.21).

(*i*) Represent *OY* as the applied voltage per phase.

(*ii*) Draw *OA* equal to no load current ($I_0$) at an angle ($\phi_0$) with *OY*. (Current scale should be approximately selected so that to accommodate the short-circuit current properly).

(*iii*) Draw *OB* equal to short-circuit current ($I_{sc}$) at an angle ($\phi_{sc}$) with *OY*.

(*iv*) Join *A* and *B*. *AB* represents motor output to power scale.

(*v*) Draw a line *AC* parallel to *OX*. Draw a perpendicular bisector of *AB* cutting the line *AC* at $O'$. Now draw a semi-circle with $O'$ as centre and $O'A$ as radius.

(*vi*) Draw a perpendicular *BE* on *AC*. Divide *BE* in the ratio so that

$$\frac{BF}{FE} = \frac{\text{Rotor resistance referred to stator}}{\text{Stator resistance}} = \frac{R_r'}{R_s}$$

(*vii*) Join *AF* which represents torque line. Line *AB* is known as output line.

**Steps to determine different parameters of designed motor:**

Let 1 cm = $k$ ampere

So 1 cm = $k \cdot V_{ph}$ watt per phase

Since circle diagram is drawn as per phase current and voltage basis so to get total power for a 3-phase machine, it should be multiplied by 3.

– Extend line *EB* and cut off *BG* = rated output per phase.

– Draw *GH* parallel to *AB*.

– Draw *HI* perpendicular to *OX*.

– Draw $O'N$ perpendicular on *AB*.

– Draw $O'S$ perpendicular on *AF*.

– *NP* and *SQ* perpendicular to *AD*.

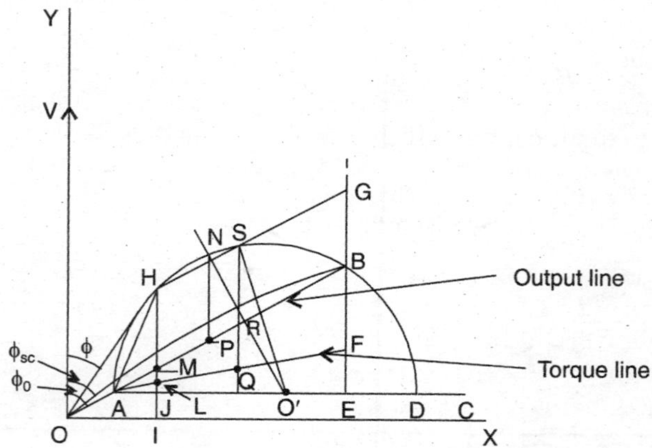


**Fig. 6.21** Circle diagram.

Full load stator current per phase = *OH*

Full load stator power factor, $\cos\phi = \dfrac{HI}{OH}$

Constant loss $= 3 \times IJ$

Full load rotor Cu loss $= 3 \times ML$

$$\text{Slip, } S = \frac{\text{Rotor Cu loss}}{\text{Rotor input}} = \frac{ML}{HL}$$

$$\text{Efficiency, } \eta = \frac{\text{Rotor output}}{\text{Sator input}} = \frac{HM}{HI}$$

Torque $= 3 \times HL$

Maximum output $= 3 \times NP$

Maximum torque $= 3 \times SQ$ watt

Starting torque $= 3 \times BF$

## 6.21 STATOR TEMPERATURE RISE

$$\text{Stator temperature, } Q°\text{C} = \frac{\text{Total loss to be dissipated from stator surface (in watt)}}{\text{Total loss dissipated from stator surface (in watt/°C)}}$$

**(*i*) Loss dissipated from outer cylindrical surface of stator and end surfaces:**

Outer cylindrical surface of stator yoke $= \pi D_0 L$

Area of end surfaces $= 2 \times \frac{\pi}{4}\left(D_0^2 - D^2\right)$

Cooling coefficient for outer surfaces can be calculated from Table 5.1 with peripheral velocity $\left(V_c = \frac{\pi D_0 N}{60} \text{ m/sec.}\right)$, where $N$ is in r.p.m.

Cooling coefficient for end surfaces can also be taken from Table 5.1 with $V_c = 0$.

$$\text{Loss dissipated from outer surface in watt/ °C} = \frac{\text{Outer cylindrical surface of stator}}{\text{Cooling coefficient for outer surface}}$$

$$\text{Loss dissipated from end surface in watt/ °C} = \frac{\text{Area of end surfaces of stator}}{\text{Cooling coefficient for end surfaces}}$$

Total loss dissipated per °C from outer and end surfaces can be estimated.

**(*ii*) Loss dissipated from inner cylindrical surface of stator:**

Inner cylindrical surface $= \pi D L$

Inner surface cooling coefficient from Table 5.1 with $V_c = V_p$

$$\text{Loss dissipated from inner surface in watt/ °C} = \frac{\text{Stator inner cylindrical surface}}{\text{Cooling coefficient for inner surface}}$$

**(*iii*) Loss dissipated by ducts:**

$$\text{Area of ducts} = \frac{\pi}{4}\left(D_0^2 - D^2\right)n_v$$

$$\text{Cooling coefficient of ducts} = \frac{0.1}{V_c}$$

where, velocity for ducts, $V_c = 0.1\ V_p$

$$\text{Loss dissipated from ducts in watt/°C} = \frac{\text{Area of ducts}}{\text{Cooling coefficient for ducts}}$$

Total loss dissipated for stator in watt/°C = Loss dissipated from (outer surfaces + end surfaces + inner surfaces + ducts)

**(*iv*) Estimate Cu loss in stator slot portion:** Iron loss in stator has already been estimated.

Total loss to be dissipated for stator surface

= Cu loss in slot portion of conductor + Iron loss in stator

$$\text{Stator temp. rise} = \frac{\text{Total loss to be dissipated for stator surface in watt}}{\text{Total loss dissipated from stator surface in watt/° C}}$$

## SOLVED PROBLEMS ON THREE-PHASE INDUCTION MOTORS AND OVERALL DESIGN OF INDUCTION MOTOR

**Q. 1** Calculate the main dimensions, turns per phase, total number of slots, sectional area of conductor and area of slot to accommodate the conductors and insulation in the slot for a 200kW, 3-phase, 50 Hz, 440 V, 1500 r.p.m. star connected slipring induction motor. Average flux density $\overline{B}$ is 0.52 Wb/m$^2$, $\overline{ac}$ = 32,000 A/m, efficiency is 0.88, winding space factor is 0.46, winding factor is 0.96, power factor 0.86 and current density in winding conductor is 4 A/mm$^2$. The design should be cheaper and hence $\frac{L}{\tau_p}$ ratio is taken 1.5, slot pitch is 20 mm.

**Solution:**

$$N_s = 1500 \text{ r.p.m.}$$

$$\therefore \quad \text{Number of poles } (P) = \frac{120 \times 50}{1500} = 4$$

We have output equation, $Q = CD^2LN$

where, $C = 18.3 \times 10^{-5} \cdot k_w \cdot \overline{B} \cdot \overline{ac} \cdot \eta \cdot \cos\phi$

$$= 18.3 \times 10^{-5} \times 0.96 \times 0.52 \times 32{,}000 \times 0.88 \times 0.86$$

$$= 2.21$$

$\therefore \quad 200 = 2.21 \times D^2 L \times 1500$

or $\quad D^2 L = \dfrac{200}{2.21 \times 1500} = 0.06$ ...(*i*)

$\dfrac{L}{\tau_P} = 1.5$ (given)

or $\quad L = 1.5 \times \dfrac{\pi D}{P} = 1.17\, D$ ...(*ii*)

From equations (*i*) and (*ii*)

$$D^3 = \frac{0.06}{1.17}$$

$\therefore \quad D = 0.37$ m

and $\quad L = 0.43$ m

Flux per pole, $\phi = 0.52 \times \tau_P \cdot L = 0.52 \times \dfrac{\pi \times 0.37}{4} \times 0.43$

$= 0.06$ Wb

Stator voltage per phase, $V_{ph} = \dfrac{440}{\sqrt{3}} = 254$ V

$\therefore$ Stator turns per phase,

$$N_{ph} = \frac{V_{ph}}{4.44 \times f \cdot k_w \cdot \phi} = \frac{254}{4.44 \times 50 \times 0.96 \times 0.06}$$

$$= \frac{254}{12.7} = 20$$

$\therefore$ Total number of stator conductors $= 3 \times 2 \times N_{ph}$

$= 120$

Slot pitch = 20 mm (given)

$\therefore$ Number of stator slots $= \dfrac{\pi \times 0.37}{20 \times 10^{-3}}$

$= 58$

$\therefore$ Conductors per slot $= \dfrac{120}{58}$

$= 2.06 \approx 2$

Current per phase, $I_{ph} = \dfrac{200 \times 10^3}{\sqrt{3} \times 440 \times 0.88 \times 0.86}$

$$= \frac{200 \times 10^3}{576} = 347 \text{ A}$$

Cross-sectional area of winding conductor, $F_c = \dfrac{I_{ph}}{\sigma} = \dfrac{347}{4}$

$= 86 \text{ mm}^2$

Since there are two conductors per slot.

∴ Area of conductor in one slot $= 2 \times 86 = 172 \text{ mm}^2$

∴ Total area of slot to accommodate winding and insulation $= \dfrac{\text{Conductor area}}{\text{Winding space factor}}$

$$= \frac{172}{0.46} = 373.9 \text{ mm}^2.$$

**Q. 2** If the induction motor given in question number 1 is delta connected, calculate the number of turns per phase, total number of slots, sectional area of conductors and area of slot to accommodate the conductors and insulation in the slots.

**Solution:**

Diameter, $D = 0.37$ m

Overall length, $L = 0.43$ m

Flux per pole = 0.06 Wb

Voltage per phase for delta connection = 440 V

∴ Stator turns per phase,

$$N_{ph} = \frac{V_{ph}}{4.44 f \phi k_w} = \frac{440}{4.44 \times 50 \times 0.96 \times 0.06} = \frac{440}{12.7}$$

$= 34$

∴ Total number of stator conductors $= 3 \times 2 \times N_{ph} = 6 \times 34 = 204$

Slot pitch = 20 mm (given)

No. of stator slots $= \dfrac{\pi \times 0.37}{20 \times 10^{-3}} = 58$

∴ Conductors per slot $= \dfrac{204}{58} = 3.51 \approx 4$

Current per phase, $I_{ph} = \dfrac{200 \times 10^3}{3 \times 440 \times 0.88 \times 0.86} = \dfrac{200 \times 10^3}{997.6} = 200.0$ A

Sectional area of conductor, $F_c = \dfrac{I_{ph}}{\sigma} = \dfrac{200}{4} = 50 \text{ mm}^2$

Area of conductor in one slot $= 4 \times 50 = 200 \text{ mm}^2$

Total area of slot to accommodate winding and insulation,

$$= \frac{200}{0.46} = 434 \text{ mm}^2.$$

**Q. 3** Calculate the following design informations for a 30kW, 440 V, 3-phase, 6 pole, 50 Hz delta connected squirrel cage induction motor:

(*i*) Main dimensions of stator frame.

(*ii*) Number of turns per phase in stator winding.

(*iii*) Number of stator slots.

Assume

Specific magnetic loading = 0.48 tesla
Specific electric loading = 26,000 ampere-conductor/m
Full load efficiency, $\eta$ = 0.88
Full load power factor = 0.86
Winding factor = 0.955

**(DDU, Gorakhpur Univ., 1996)**

**Solution:**

(*i*) Synchronous speed, $N = \dfrac{120 \times 50}{6} = 1000 \text{ r.p.m.}$

Output equation,

$$Q = CD^2 LN \text{ kilowatt}$$

or $$30 = 18.3 \times 10^{-5} \times 0.955 \times 0.48 \times 26{,}000 \times 0.88 \times 0.86 \times D^2 L \times 1000$$

$$= 1650$$

$$D^2 L = 0.018 \qquad \text{...(i)}$$

Let $$\frac{L}{\tau_P} = 1.2$$

or $$L = 1.2 \times \frac{\pi \times D}{P} = 0.628\, D \qquad \text{...(ii)}$$

From equations (*i*) and (*ii*)

$$D^3 = \frac{0.018}{0.628}$$

$\therefore$ $$D = 0.30 \text{ m}$$

and $$L = 0.19 \text{ m}$$

Chapter 6

(*ii*) Flux per pole $= 0.48 \times 0.19 \times \dfrac{3.14 \times 0.3}{6}$

$= 0.014$ Wb

Machine is delta connected. So, $V_{ph} = 440$ V

$$\text{Stator turns per phase, } N_{ph} = \frac{V_{ph}}{4.44 \times f \cdot \phi \, k_w}$$

$$\therefore N_{ph} = \frac{440}{4.44 \times 50 \times 0.014 \times 0.955} = \frac{440}{2.96} = 148$$

(*iii*) Let us assume the slot pitch = 18mm

$$\therefore \quad \text{Number of slots in stator} = \frac{3.14 \times 0.30}{18 \times 10^{-3}}$$

$$= 0.052 \times 10^3 = 52$$

$\therefore$ Number of stator slots = 52.

**Q. 4** Determine the main dimensions, turn per phase, number of slots, conductor section and slot area of a 3-phase, 5 H.P., 400 volts, 50 Hz, 1500 r.p.m. (synchronous), squirrel cage induction motor. The machine is to be started by a star-delta starter. The efficiency is 0.8 and power factor is 0.8 lagg at full load. **(DDU, GKP. Univ., 2001)**

**Solution:**

Let $k_w = 0.955$, $\phi = 0.50$ Wb

$\overline{ac} = 27{,}000$, Speed = 1500 r.p.m.

No. of poles (P) $= \dfrac{120 \times 150}{1500} = 4$

From output equation, $Q = CD^2LN$

$$5 \times 0.746 = 18.3 \times 10^{-5} \times 0.955 \times 0.5 \times 27{,}000 \times 0.8 \times 0.8 \times D^2 L \times 1500 = 2264 \times D^2L$$

$$\text{or} \qquad D^2 L = \frac{3.73}{2264} = 0.0016 \qquad \text{...(i)}$$

$$\text{Let} \qquad \frac{L}{\tau_P} = 1.0$$

$$\therefore L = \tau_P = \frac{3.14 \times D}{4} = 0.785\,D \qquad \text{...(ii)}$$

From equations (*i*) and (*ii*), we get

$$D^3 = \frac{0.0016}{0.785} = 2.03 \times 10^{-3}$$

or, $D = 0.13$ m

and $L = 0.1$ m

$$\text{Flux per pole} = 0.5 \times 0.10 \times \frac{3.14 \times 0.13}{4}$$

$$= 0.005 \text{ Wb}$$

Since star-delta starter is used, so the m/c winding is in delta.

$\therefore$ $V_{ph} = 400$

$$\therefore \quad \text{Turns per phase, } N_{ph} = \frac{400}{4.44 \times 50 \times 0.005 \times 0.955}$$

Total number of conductors $= 3 \times 2 \times 377 = 2262$

Let slot pitch is 15 mm.

$$\text{Number of slots in stator} = \frac{3.14 \times 0.13}{15 \times 10^{-3}}$$

$$= 27$$

Conductors per slot = 83

$$\text{Current per phase, } I_{ph} = \frac{3.73 \times 10^3}{3 \times 400 \times 0.8 \times 0.8}$$

$$= 4.8$$

Cross-sectional area of winding conductor,

$$= \frac{I_{ph}}{\sigma} = \frac{4.8}{3.5} = 1.3 \text{ mm}^2$$ (Let current density $\sigma = 3.5$ A/mm²)

Area of conductor in slot $= 83 \times 1.3 = 107.9$ mm²

Total area of slot (let winding space factor is 0.50)

$$= \frac{107.9}{0.5} = 215.8 \text{ mm}^2$$

**Q. 5** Estimate the main dimensions, number of stator slots and number of stator conductors per slot for a 140 H.P., 3300 volt, 50 c/sec., 12 poles, star-connected slipring induction motor. Assume

Average gap density = 0.4 Wb/m²

Ampere-conductor per metre = 25,000

Efficiency, $\eta$ = 90%

Power factor = 0.9

Winding factor = 0.96

**(DDU, GKP. Univ, 1995)**

**Solution:**

$$\text{Speed, } N = \frac{120 \times 50}{12} = 500 \text{ r.p.m.}$$

Output equation, $Q = CD^2LN$ kilowatt

$$140 \times 0.746 = 18.3 \times 10^{-5} \times 0.96 \times 0.4 \times 25{,}000 \times 0.9 \times 0.9 \times D^2 L \times 500$$

or $D^2 L = \dfrac{140 \times 0.746}{711}$

$$D^2L = 0.14 \qquad \text{...}(i)$$

Let $\dfrac{L}{\tau_P} = 1.1$

or $L = 1.1 \times \dfrac{\pi D}{P} = 1.1 \times \dfrac{3.14 \times D}{12}$

or $L = 0.287D \qquad \text{...}(ii)$

From equations (*i*) and (*ii*), we get

$$D^3 = \frac{0.14}{0.287} = 0.48$$

$\therefore \quad D = 0.78$ m

and $\quad L = 0.22$ m

$$\text{Flux per pole} = 0.4 \times \frac{3.14 \times 0.78}{12} \times 0.22$$

$$= 0.017 \text{ Wb}$$

$$\text{Stator voltage per phase, } V_{ph} = \frac{3300}{\sqrt{3}} = 1905 \text{ V}$$

$$\text{Stator turns per phase, } N_{ph} = \frac{1905}{4.44 \times 50 \times 0.96 \times 0.017}$$

$$= \frac{1905}{3.62} = 526$$

Total number of stator conductors $= 6 \times 526 = 3156$

Let slot pitch is 15 mm.

$$\text{Number of stator slots} = \frac{3.14 \times 0.78}{15 \times 10^{-3}} = 163$$

$$\text{Conductors per slot} = \frac{3156}{163} = 19$$

## OVERALL DESIGN OF INDUCTION MACHINE

Design a 3-$\phi$ I.M with following details:

| Rating | Voltage | Connection/type | Pole | Eff. | P.f. |
|---|---|---|---|---|---|
| 28 kW | 410V | Slipring Y–Y | 8 | 78% | 0.81 |

**Solution:**

Estimation of main dimension:

$$\frac{L}{\frac{\pi D}{P}} = 1.6$$

$$\Rightarrow L = \frac{1.6 \times \pi \times D}{8} = 0.628\,D$$

$$C = 18.3 \times 10^{-5} \times k_w \times B \times \overline{ac} \times \eta \times \cos\phi = 18.3 \times 10^{-5} \times 0.95 \times 0.45 \times 25{,}000 \times 0.78 \times 0.81$$

$$= 1.235 \qquad \left[\because\ \overline{B} = 0.45 = \text{average flux density in air gap}\right]$$

$$D^2 L = \frac{Q}{CN} = \frac{28}{750 \times 1.235} \qquad \left[\because\ N = \frac{120f}{P} = 750\right]$$

$$= 0.03022$$

$$\Rightarrow D^3 = \frac{0.03022}{0.628} = 0.0481$$

$$\boxed{D = 0.363\text{ m}}$$

$$\boxed{L = 0.228\text{ m}}$$

Air gap length, $\delta_0 = \left(0.2 + 2\sqrt{D \cdot L}\right) = 0.77\,\text{mm}$

No. of ventilating ducts $= n_v$

$$\therefore\ n_v = \left(\frac{L}{11} - 1\right) = \left(\frac{22.8}{11} - 1\right) = 1.07 \approx 1$$

Width of ventilating duct, $b_v$ = 1 cm

Chapter 6

$$b'_v = bv \cdot \frac{5}{5 + \left(\dfrac{b_v}{\delta_0}\right)} = 1 \cdot \frac{5}{5 + \dfrac{1}{0.077}}$$

$$= 0.277 \text{ cm}$$

$$= 2.77 \text{ mm}$$

Effective length,

$$l_e = L - n_v \times b'_v$$

$$= 22.8 - 0.277$$

$$= 22.5 \text{ cm}$$

$$= 0.225 \text{ m}$$

Gross iron length,

$$l = L - n_v b_v$$

$$= 22.8 - 1$$

$$= 21.8 \text{ cm}$$

$$= 0.218 \text{ m}$$

Actual iron length,

$$l_i = k_i \cdot l$$

$$= 0.91 \times 0.218 = 0.198 \text{ m}$$

### *Stator Design*

Estimation of $N_{ph}$, $Z_{ph}$ $Z_T$ and $N_c$: (stator side)

$$V_{ph} = \frac{410}{\sqrt{3}} = 236 \text{ V}$$

Turn per phase, $N_{ph_1} = \dfrac{V_{ph}}{4.44 \times k_w \cdot f \cdot \phi_1}$

$$= \frac{236}{4.44 \times 0.95 \times 50 \times 0.014}$$

$$= 79.9$$

$$= 80$$

$\because$ flux per pole,

$$\phi_1 = \overline{B} \tau_P l_e$$

$$= 0.45 \times 0.142 \times 0.225$$

$$= 0.014 \text{ Wb}$$

and $\tau_P = \dfrac{\pi D}{P} = 0.142 \text{ m}$

Number of conductors per phase,

$$Z_{ph} = 2 \times 80 = 160$$

Total no. of conductors,

$$Z_T = 160 \times 3 = 480$$

Selecting stator slot per phase per pole = $q_1$

$\therefore$ $q_1 = 2$

$\therefore$ Total no. of slots, $S_1 = 3 \times 2 \times 8$

$= 48$

Slot pitch, $\tau_{sg_1} = \dfrac{\pi D}{48} = 0.023$ m

$= 23.7$ mm

Conductor per slot, $Z_c = \dfrac{480}{48} = 10$.

Since integer value already come so, all the values above are corrected values.

### *Conductor Size*

Stator current per phase,

$$I_{ph} = \frac{k_w \times 10^3}{3V_{ph} \cdot \eta \cdot 6\cos\phi}$$

$$= \frac{28 \times 10^3}{3 \times 236 \times 0.78 \times 0.81} = 62.5 \text{ ampere}$$

Cross-sectional area of conductor,

$$F_c = \frac{I_{ph}}{\sigma}, \quad \text{But } \sigma = 3.5 \text{ Am/mm}^2$$

$$= \frac{62.5}{3.5} = 17.8 \text{ mm}^2$$

$$\therefore \quad \frac{\pi d^2}{4} = 17.8$$

$$\Rightarrow \quad d = \sqrt{\frac{17.8 \times 4}{\pi}}$$

$= 4.76$ mm

So, diameter of conductor, $d = 4.76$ mm.

### *Stator Slot Design*

Total no. of conductors in one slot is 10.

Let, $n_v$ = no. of vertical conductors = 5

$n_h$ = no. of horizontal conductors = 2

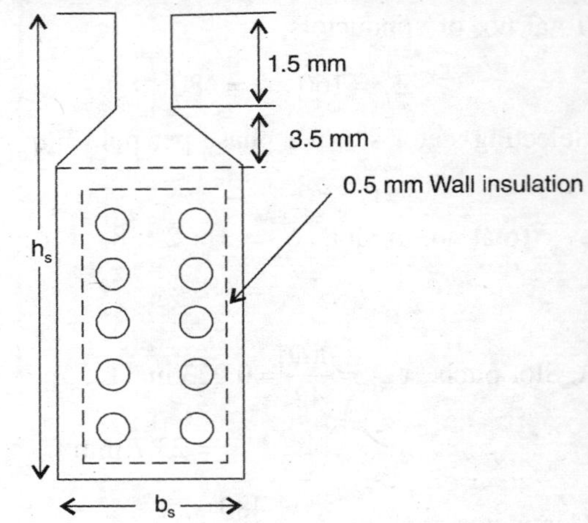


Height of slot, $h_s = n_v \times \text{diameter} + 0.5 + 0.5 + 3.5 + 1.5$

$$= (5 \times 4.76) + 0.5 + 0.5 + 3.5 + 1.5$$

$$= 29.8 \text{ mm}$$

Width of slot, $b_s = (n_h \times \text{diameter} + 0.5 + 0.5)$

$$= (2 \times 4.76) + 1 = 10.52 \text{ mm.}$$

***Teeth Design***

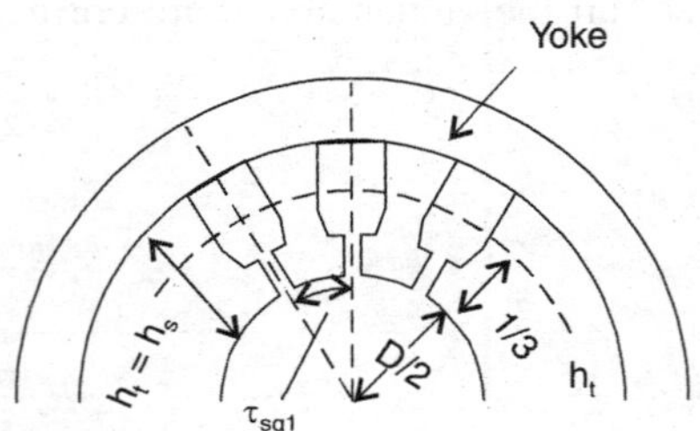


Diameter at 1/3rd height from narrow end of slot is,

$$= D + 2/3\ h_t$$

Width of stator teeth at 1/3rd height from the narrow end,

$$b_{ts1/3} = \frac{\pi\left(D + \frac{2}{3} h_t\right)}{S_1} - b_s$$

$$\therefore \quad b_{ts1/3} = \frac{\pi\left(363 + \frac{2}{3} \times 29.8\right)}{48} - 10.52$$

$$= 14.5 \text{ mm.}$$

***Height of Yoke, $h_y$***

Flux through yoke,

$$\frac{\phi}{2} = B_y \cdot h_y \cdot k_i l$$

$$\Rightarrow \quad h_y = \frac{\phi}{2 \cdot B_y \cdot k_i l} \qquad \text{Let, } B_y = 1.4 \text{ Wb/m}^2$$

$$= \frac{0.14}{2 \times 1.4 \times 0.198}$$

$= 0.025$ m

$= 25$ mm

$\therefore$ Outer diameter, $D_0 = D + 2h_t + 2h_y$

$= 363 + (2 \times 29.8) + (2 \times 25)$

$= 472.6$ mm

$= 0.472$ m.

### *Stator Winding Resistance*

$$R_{dc} = 0.021 \cdot \frac{L_{mt} \cdot N_{ph}}{F_c}$$

But, $L_{mt} = 2\left[L + 1.15\,\tau_p + 0.8\right]$ metre

$$= 2\left[0.228 + (1.15 \times 0.142) + 0.8\right]$$

$= 2.38$ m

So, $$R_{dc} = \frac{0.021 \times 2.38 \times 80}{17.8} = 0.224\,\Omega$$

$\therefore$ $R_{ac} = 0.224 \times 1.15$

$= 0.258\ \Omega$.

### *Stator Cu Loss*

$$= 3 \times I_{ph}^2 \cdot R_{ac}$$

$$= 3 \times (62.5)^2 \times 0.258$$

$= 3030$ watt

$= 3$ kW.

### *Stator Iron Loss*

Iron loss in stator yoke,

$B_y = 1.4$ Wb/m²

$\therefore$ $P_{iy} = 9$ watt/kg [From Fig. 5.30 for 0.35 mm thick lamination]

Chapter 6

***Volume of Yoke***

$$\frac{\pi}{4}\left[D_0^2-\left(D_0-2h_y\right)^2\right]\times l_i$$

$$=\frac{\pi}{4}\left[(0.472)^2-\{0.472-(2\times 0.025)\}^2\right]\times 0.198$$

$= 0.0069\ \text{m}^3$.

***Iron Loss in Yoke/Core***

$= 7600 \times 0.0069 \times 9$

$= 475$ watt.

***Iron Loss in Teeth***

Assume, $B_t = 1.6$ T

$\therefore\ P_t = 11$ watt/kg [From Fig. 5.30 for 0.35 mm lamination]

Volume of total teeth $= S_1\left[h_t \times b_{t_{1/3}} \times l_i\right]$

$= 48[0.0298 \times 0.0145 \times 0.198]$

$= 0.0041\ \text{m}^3$.

***Iron Loss in Teeth***

$= 7600 \times 0.0041 \times 11$

$= 343$ watt.

***Rotor Design***

Rotor Slot/pole/phase, $q_2 = (2 + 1)$

$\therefore\ q_2 = 3$

Total number of rotor slots, $S_2 = 3 \times P \times 3$

$= 3 \times 8 \times 3$

$= 72$

Total number of conductors,

$$Z_2 = S_2 \times Z_{c_2}$$

where, $Z_{c_2}$ = number of conductor per slot.

$$\Rightarrow \quad \frac{\text{Stator voltage/phase}}{\text{Rotor voltage/phase at standstill}}=\frac{N_{1ph}}{N_{2ph}}$$

$$\Rightarrow \qquad \frac{236}{200} = \frac{80}{N_{2ph}}$$

$$\Rightarrow \qquad N_{2ph} = \frac{80 \times 200}{236}$$ [Taking rotor phase voltage = 200]

$$= 67.7$$

Now, $N_{2ph} = \dfrac{S_2 \cdot N_{c_2}}{6}$

$$\therefore \qquad N_{c_2} = \frac{6 \times N_{2ph}}{S_2} = \frac{6 \times 67.7}{72}$$

$$= 5.6$$

$$\approx 6$$

Number of conductors per slot,

$$\left(N_{c_2}\right)_{\text{corr}} = 6$$

$$\left[Z_{T_{2\text{corr}}}\right] = 6 \times 72 = 432$$

$$Z_{ph_{2(\text{corr})}} = \frac{432}{3} = 144$$

$$\left(N_{ph}\right)_{\text{corr}} = \left(\frac{144}{2}\right) = 72$$

### *Rotor Current*

(85% of stator m.m.f. goes to rotor side)

$$6I_{2ph} \cdot N_{2ph} = 6I_{1ph} \cdot N_{1ph} \times 0.85$$

$$\therefore \qquad I_{2ph} = \frac{I_{1ph} \cdot N_{1ph} \times 0.85}{N_{2ph}}$$

$$= \frac{62.5 \times 80 \times 0.85}{72}$$

= 59 ampere.

$\therefore$ Area of conductor,

$$F_{c_2} = \frac{59}{3.5} \text{ mm}^2$$

$$= 16.8 \text{ mm}^2$$

Chapter 6

Diameter of rotor conductor = $\dfrac{\pi d^2}{4} = 16.8$

$$\Rightarrow d = \sqrt{\frac{16.8 \times 4}{\pi}} = 4.6 \text{ mm}.$$

### Rotor Slot Design

$n_v = 3$

$n_h = 2$

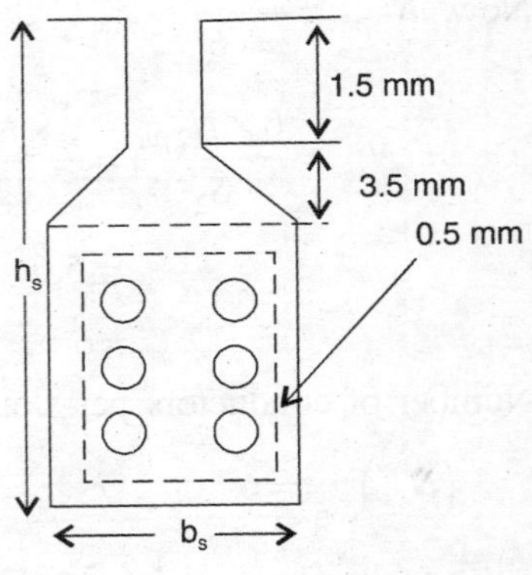


Height of slot, $h_s = (3 \times 4.6) + 0.5 + 0.5 + 3.5$

$= 18.3$ mm

Width of slot, $b_s = (2 \times 4.6) + 1 = 10.2$ mm

Rotor slot pitch, $\tau_{sg_2} = \dfrac{\pi(D - 2\delta)}{S_2} = \dfrac{\pi(363 - 2 \times 0.77)}{72}$

$= 15.7$ mm.

### Rotor Teeth Design

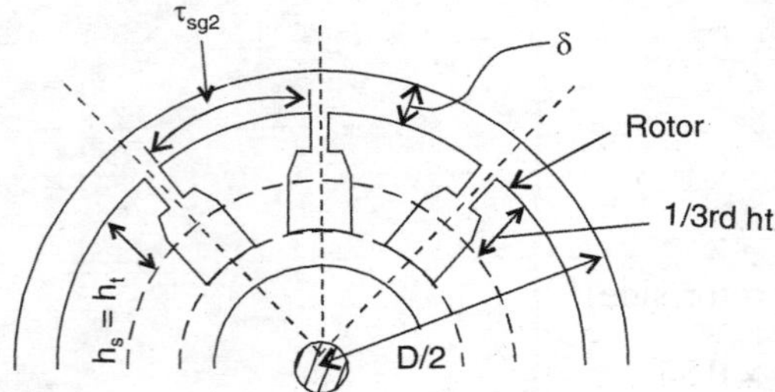


Width of rotor teeth at 1/3rd height from the narrow end

$$b_{tr1/3} = \frac{\pi\left(D - 2\delta - \dfrac{2h_t}{3}\right)}{S_2} - b_s$$

$$= \frac{\pi\left(363 - (2 \times 0.77) - \dfrac{2}{3} \times 18.3\right)}{72} - 10.2$$

$= 5$ mm

Area of teeth per pole,

$$= \frac{S_2}{P}\left[b_{tr1/3} \times l_i\right]$$

$$= \frac{72}{8}[0.005 \times 0.198]\ \text{m}^2$$

$$= 0.0089\ \text{m}^2$$

Average flux = 0.014 Wb

∴ Flux density per pole in rotor teeth,

$$= \frac{0.014}{0.0089} = 1.57\ \text{Wb/m}^2$$

### *Iron loss in Rotor Teeth*

There will be no significant iron loss in rotor during running condition because frequency is very less.

### *Cu Loss in Rotor*

$$R_{rdc} = 0.021 \times \frac{L_{mt2} \times N_{ph2}}{F_{c_2}}$$

$$L_{mt_2} = 2\left[L + 1.5\tau_p + 0.12\right] = 2\left[0.228 + 1.5 \times 0.142 + 0.12\right]$$

$$= 1.12\ \text{m}$$

$$R_{rdc} = \frac{0.021 \times 1.12 \times 144}{16.8}$$

$$= 0.2\ \Omega$$

$R'_{rdc}$ = referred to stator

$$= 0.2 \times \left(\frac{80}{72}\right)^2 = 0.24$$

$$R_{ac} = (1.15 \times 0.24)$$

$$= 0.28$$

Cu loss in rotor,

$$= 3I_{2ph}^2 R_{ac}$$

$$= 3 \times (59)^2 \times (0.28)$$

$$= 2925\ \text{watt}$$

Chapter 6

Total, Cu loss = Cu loss in stator + Cu loss in rotor

= 5955 watt

Total iron loss in stator (yoke + teeth),

= 818 watt

Actual iron loss is two times the above.

$\therefore$ Iron loss $= (2 \times 818) = 1636$ watt

Total loss = 7591 watt

$$\therefore \quad \eta = \frac{28}{28 + 7.69} \times 100$$

= 78.6%

### *Slip of Motor*

$$= \frac{\text{Rotor Cu loss}}{\text{Rotor intake } (\text{O}/P + \text{rotor Cu loss})}$$

$$= \frac{2.925}{28 + 2.925}$$

= 0.094

## Estimation of no Load Current

### *Magnetising current*

The m.m.f. required for various parts per pole pair are calculated below,

**(i) Air gap:** The flux tube crossing the air gap at 60° from the interpolar axis will always give good approximation.

We know,

$$B_{g60^\circ} = 1.36\,\overline{B}_{av}$$

$$= 1.36 \times 0.45$$

$$= 0.612$$

m.m.f. for air gap, $AT_g$ = 800,000 × $B_{g60^\circ}$ $k_c \cdot \delta$

where, $k_c = k_{c_1} \cdot k_{c_2}$

$$k_{c_1} = \frac{\tau_{sg_1}}{\tau_{sg_1} - \gamma_1 \delta}$$

$k_{c_1}$ = air gap coefficient for stator,

and $k_{c_2}$ = air gap coefficient for rotor.

$$= \frac{23.7}{23.7-(8.53\times 0.77)}$$

$$= 1.38$$

$$k_{c_2} = \frac{\tau_{sg2}}{\tau_{sg_2} - \gamma_2 \delta}$$

$$= \frac{15.7}{15.7-(8.53\times 0.77)}$$

$$= 1.71$$

But, $\gamma_1 = \dfrac{5\left(\dfrac{b_{01}}{\delta}\right)^2}{5+\left(\dfrac{b_{01}}{\delta}\right)} = \dfrac{5\left(\dfrac{3}{0.77}\right)^2}{5+\dfrac{3}{0.77}}$

$$= 8.53$$

Since, $b_{01} = b_{02}$ taken.

So, $\gamma_1 = \gamma_2 = 8.53$

$\therefore \quad k_c = k_{c_1} \cdot k_{c_2} = 2.35$

$\therefore \quad \delta' = \delta \times k_c = 1.8 \text{ mm} = 0.0018 \text{ m}$

$\therefore \quad AT_g = 800{,}000 \times 0.612 \times 0.0018$

$= 881.280$ AT.

**(ii) Stator teeth:** Flux density at 1/3rd height of tooth from narrow end,

$$B_{ts1/3} = \frac{\bar{\phi}}{\frac{S_1}{P}\left[b_{ts1/3} \times l_i\right]}$$

$$= \frac{0.014}{\frac{48}{8}[0.0145 \times 0.198]}$$

$$= 0.8 \text{ T}$$

$$B_{ts60^\circ} = 1.36 \times 0.8$$

$$= 1.1 \text{ T}$$

$\therefore$ MMF per metre, $at_{ts}$ = 200 AT.

MMF for stator teeth,

$$AT_{ts} = at_{ts} \times h_t$$

$$= 200 \times 0.0298$$

$$= 5.96 \text{ AT.}$$

**(iii) Stator core/yoke:** Sectional area of stator core,

$$h_y \times l_i$$

Chapter 6

Flux density in stator core,

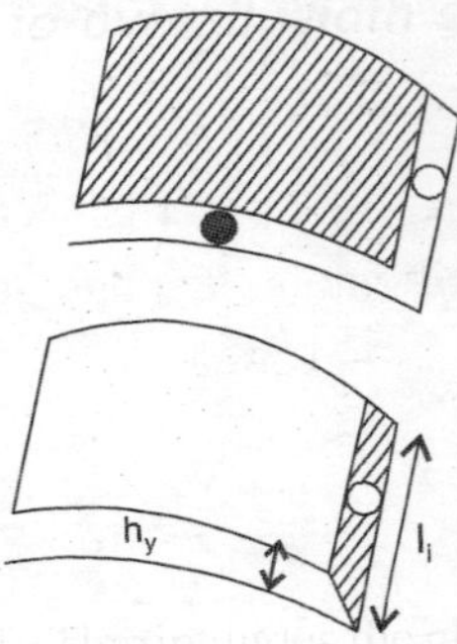

Half flux will pass through the core

$$B_{sy} = \frac{\phi/2}{h_y \times l_i}$$

$$= \frac{0.014}{2 \times 0.025 \times 0.198}$$

$$= 1.4 \text{ T}$$

$$\therefore \quad at_{ys} = 2000 \text{ AT/m}$$

Length of path through stator core or yoke,

$$L_y = \frac{\pi(D_0 - h_y)}{P} = \frac{\pi(0.472 - 0.025)}{8} = 0.175 \text{ m}$$

Ampere-turn for yoke,

$$AT_{ys} = 2000 \times 0.175$$

$$= 350 \text{ AT.}$$

**(iv) Rotor teeth:** Area of rotor teeth per pole,

$$= \frac{S_2}{P}\left[b_{tr1/3} \times l_i\right]$$

$$= \frac{7^2}{8}[0.005 \times 0.198] \text{ metre}$$

$$= 0.0089 \text{ m}$$

Flux density in rotor teeth at 1/3 height,

$$B_{tr1/3} = \frac{0.014}{0.0089} = 1.57 \text{ Wb/m}$$

$$B_{tr(60°)} = 1.36 \times 1.57 = 2.13 \text{ T}$$

$$\therefore \quad at_{tr} = 700 \times 100 = 70{,}000 \text{ AT}$$

$$\therefore \quad \text{Total, } AT_{tr} = 70{,}000 \times 0.018$$

$$= 1260 \text{ AT.}$$

**(v) Rotor core:** Depth of rotor core is taken equal to that of stator *i.e.*, $h_{sy} = h_{ry}$. The flux density and area of rotor core is same as that of stator core.

$$h_{sy} = h_{ry}$$

$$\therefore \quad B_{yr} = B_{sy} = 1.4$$

$$\therefore \quad at_{yr} = 2000 \text{ AT/m}$$

Length of flux path in rotor core per pole,

$$L_{ry} = \frac{\pi\left(D - 2\delta - 2h_{tr} - h_{ry}\right)}{P}$$

$$= \frac{\pi(0.363 - (2 \times 0.013) - 0.025)}{8}$$

$$= 0.1185 \text{ m} = 118.5 \text{ mm}$$

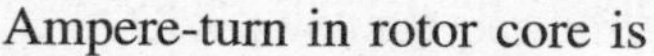

Ampere-turn in rotor core is

$$AT_{yr} = 2000 \times 0.118$$

$$= 236 \text{ AT}$$

Total magnetising m.m.f. per pole for $B_{60°}$,

$$AT_{60°} = AT_g + AT_{ts} + AT_{ys} + AT_{tr} + AT_{yr}$$

$$= 2733 \text{ AT}$$

Magnetising current per phase.

$$I_\mu = \frac{0.427 \times P \times AT_{60°}}{k_w \times N_{ph}}, \text{ where } N_{ph1} \text{ is the turns per phase in stator.}$$

$$= \frac{0.427 \times 8 \times 2733.00}{0.95 \times 80}$$

$$= \frac{9335.92}{76.00} = 122.84 \text{ A.}$$

### *Loss Component of Current*

Friction and windage loss = 1% of output

= 280 watt

Total no load loss = 7591 + 280

= 7871 watt

$$I_2 = \frac{\text{No load loss}}{3V_{ph1}} = \frac{7871}{3 \times 236} = 11 \text{ ampere.}$$

∴ No load current,

$$I_0 = \sqrt{I_m^2 + I_L^2}$$

$$= \sqrt{(34)^2 + (11)^2} = 139.4 \text{ A}$$

$$= 35.12$$

$$\cos\phi_0 = \frac{I_2}{I_0} = \frac{11}{35.12} = 0.31$$

## QUESTIONS WITH SHORT ANSWERS

**Q. 1 A 3-phase induction motor with large value of specific magnetic loading has a higher overload capacity. Explain.**

**Ans.** The value of air gap flux density determines the overload capacity. A high value of magnetic loading means that the flux per pole is large. Thus for the same voltage, the winding requires less turns per phase and when the number of turns is less, the leakage reactance becomes small. The circle diagram of the machine has a large diameter with small leakage reactance which means that the maximum output of the machine is large or in other words the machine has a large overload capacity.

**Q. 2 A 3-phase I.M. with high value of specific electric loading has a lower overload capacity. Explain.**

**Ans.** A large value of ampere-conductor would result in large number of turns per phase. This would mean that the leakage reactance of the machine becomes high and hence the diameter of circle diagram is reduced resulting in reduced value of overload capacity of the machine.

**Q. 3 A 3-phase I.M. with large number of slots has a higher overload capacity. Explain.**

**Ans.** If there are large number of slots, the width of insulation becomes more and this means that the leakage flux has a long path through air which results in the reduction of leakage flux. Therefore with large number of slots, the leakage flux and hence the leakage reactance is reduced. With small values of leakage reactance, the diameter of the circle diagram is large and hence the overload capacity increases.

**Q. 4 A 3-phase I.M. with large air gap has a high value of overload capacity. Explain.**

**Ans.** The length of air gap affects mainly the value of zig-zag leakage reactance which forms a large part of total leakage reactance in the case of induction motors. If the length of air gap is large, the zig-zag leakage flux is reduced resulting in a reduced value for leakage reactance. With decrease in the leakage reactance, the diameter of the circle diagram increases and hence the overload capacity increases.

**Q. 5 Explain types of a.c. armature windings.**

**Ans.** The following types of a.c. windings are used for armatures:

(*a*) *Concentric winding:* These types of windings are single layer windings which use concentric type of coils. The coil span in individual coil is different. These windings are so designed that the effective coil span of the winding is equal to that of winding as full pitch winding with some of the coils having a span greater than a pole pitch, some with less than a pole pitch. These are two types called as hemitropic and whole coil windings.

(*b*) *Mash winding:* This winding is very commonly used in small induction motors having circular conductors. This is single layer winding where all the coils have same span. Each coil wound as a former, making one coil side shorter than the other. The long and short sides occupy alternate slots. This type of winding is also known as basket winding.

(*c*) *Double layer windings:* Double layer windings are universally used for armature of synchronous generators and most induction motors of large and medium sizes. They can be either lap or wave type. These windings can be either integral type or fractional type.

**Q. 6 What are the factors to determine the rotor slots in induction motor?**

**Ans.** The factors determining the choice of rotor slots in induction motor are:

(*a*) The number of rotor slots should never be equal to stator slots but must either be smaller or larger. Satisfactory results are obtained when the number of rotor slots is 15 to 30 per cent larger or smaller than the number of stator slots.

(*b*) The difference between stator slots and rotor slots should not be equal to $p$, $2p$ or $5p$ to avoid belt locking.

(*c*) The difference between the number of stator and rotor slots should not be equal to $3p$ for 3-phase machines in order to avoid magnetic locking.

(*d*) The difference between number of stator slots and rotor slots should not be equal to 1, 2, $(p \pm 1)$ or $(p \pm 2)$ to avoid noise and vibrations.

**Q. 7 Why machines having long core length are provided with radial ventilating ducts while machines having deep cores are provided with axial cooling?**

**Ans.** Radial ventilating ducts provided with long core machines take away heat at a faster rate. In case of deep cores axial ventilating ducts dissipates heat. If such arrangement is not done, the central portion of long core and inner portion of deep core may attain a temperature higher than the permissible value and machine can be damaged.

**Q. 8 Why closed type slots are often used for small induction motors?**

**Ans.** Use of closed slots makes the reluctance more uniform as compared with open and semi-closed slots and gives reduced exciting current and better power factor.

**Q. 9 Why sheet steel with higher silicon content is used for transformer core whereas lower silicon content sheet for rotating machines?**

**Ans.** In rotating electrical machines, it is desirable to work the iron parts at higher values of flux density in order to achieve a higher output to weight ratio. The magnetic material should have a high saturation flux density and hence the presence of silicon is a disadvantage. Therefore, in rotating electrical machines sheet steel of low silicon content is used. Also increase of silicon content makes the steel more brittle.

For transformers sheet steel possessing higher silicon content is used because it has no rotating part and also the magnetizing current is not much important. It has high resistivity and hence eddy current losses are small.

**Q. 10 Why the length of air gap in induction motor is kept minimum possible whereas in a d.c. machine it is larger?**

**Ans.** Length of air gap in induction motor is kept small so that its no load current is small and hence motor has better power factor.

In d.c. machine, larger air gap length is provided so that it has large field m.m.f. and hence the distorting effect of armature reaction can be reduced.

## UNSOLVED QUESTIONS

1. Deduce for a 3-phase induction motor an expression showing the relationship between output, its main dimensions, speed, the specific electric and magnetic loading, efficiency and power factor. What are the various considerations in the selection of specific electric and magnetic loading for the design of a 3-phase induction motor?
2. What is gap contraction factor? How does it affect the calculation of ampere-turn of air gap of induction motor?
3. Explain various leakage reactances in a 3-phase induction motor.
4. Discuss causes of noise production in induction motors and their remedy.
5. Explain what considerations control the fixation of the ampere-conductor per metre of periphery for the stator of the induction motor.
6. When the performance was calculated at the design stage of a poly-phase induction motor, it was found to give a poor power factor. Discuss the modification which can probably be made to the design for improving its performance.
7. What changes would you suggest in the design of a three-phase squirrel cage induction motor to achieve increased starting torque?
8. Explain ventilation of electrical machines.
9. Determine the approximate diameter and length of the stator core, the number of stator slots for a 15 kW, 440 V, 3-phase, 4-pole, 1425 r.p.m induction motor with data given below:

   Specific magnetic loading = 0.48 tesla.

   Specific electric loading = 25,000 ampere-conductor per metre.

   Full load efficiency = 88 per cent.

   Full load power factor = 0.88.
10. A 3-phase, 400V, 50 Hz, 50 kW squirrel cage induction motor running at 1440 r.p.m. has full load efficiency of 86% and power factor of 0.85 lagging. Estimate the main dimensions and design the machine as far as you can. Flux density is 0.47 Wb/m$^2$, $\overline{ac}$ is 24,000 A/m, winding factor 0.955, slot pitch of 19 mm. Current density is 3 A/mm$^2$.
11. Calculate the following design informations for a 30 kW, 440 V, 3-phase, 6-pole, 50 Hz delta connected, squirrel cage induction motor:

    (*i*) Main dimensions of stator frame.

    (*ii*) Number of turns per phase in stator winding.

    (*iii*) Number of stator slots.

    Assume:

    Slot pitch = 28 mm, $L/\tau_p$ ratio = 1
    Specific magnetic loading = 0.48 tesla.
    Specific electric loading = 26,000 ampere-conductor/metre
    Full load efficiency = 0.88
    Full load power factor = 0.86
    Winding factor = 0.955

**12.** Estimate the main dimensions, number of stator slots and number of stator conductors per slot for a 100 H.P., 3300 volt, 50 C/sec., 12 poles star-connected slipring induction motor.

Assume:

Slot pitch = 16.0 mm
Average gap density = 0.4 Wb/m$^2$
Ampere-conductor per metre = 25,000

Efficiency, $\eta$ = 90%, $\dfrac{L}{\tau_p} = 1.2$

Power factor = 0.9
Winding factor = 0.96
Current density = 4 A/mm$^2$

**13.** Determine the main dimensions, turns per phase, number of slots, conductor section and slot area of : a 3-phase, 15 H.P., 400 volts, 50 Hz, 1500 synchronous r.p.m. squirrel cage induction motor. The machine is to be started by a star delta starter. The efficiency is 0.8 and power factor is 0.8 lagg at full load, slot pitch is 16 mm, $L/\tau_p$ ratio is 1.5 and current density of 3.5 A/mm$^2$, flux density 0.45 Wb and $\overline{ac}$ = 26,000 A/m.

**14.** Explain the factors which affect the fixation of the magnetic loading and ampere-conductor per metre of periphery for the stator of the induction motor.

For an induction motor of 270 kW, 3000 volts, 3-phase, 10-pole, 50 Hz.
Calculate:

(*i*) Suitable main dimensions;
(*ii*) Number of turns and slots.

Following data are given:
Flux density = 0.51 Wb/m$^2$.

Specific electric loading = 29,000 ampere-conductor per metre.

Efficiency = 0.92
Power factor = 0.88
Winding factor = 0.955

**15.** Determine the main dimensions, turns per phase, number of slots, conductor section and slot area of 200 H.P., 3-phase, 50 Hz, 400 volts, 1480 r.p.m slipring induction motor. Assume $B_{av}$ = 0.5 Wb/m$^2$, $\overline{ac}$ = 30,000 A/m, efficiency = 0.9 and power factor = 0.9, current density = 3.5 ampere per mm$^2$.

**16.** Calculate:

(*i*) Suitable main dimensions.

(*ii*) Number of stator turns and slots for a 3-phase, 3000 volts, 10 pole, 50 Hz induction motor developing 270 kW.

Flux density = 0.51 tesla
Specific electric loading = 29,000 ampere-conductor per metre
Winding factor = 0.955
Efficiency = 0.92
Power factor = 0.88

# 7

# Construction, Important Features and Design of D.C. Machines

## 7.1 INTRODUCTION

A d.c. machine is electromechanical energy conversion device. When mechanical energy is being converted into electrical energy by the machine, it is called a generator and when an electrical energy is being converted into mechanical energy then it is called as motor. During conversion of energy of one form into the other, a part of energy is lost as copper loss, iron loss, contact loss and friction and windage loss and converted into heat which increases the temperature of machine. In general, the generators and motors are very similar to each other in construction except due to some modifications for specific requirement in specific operation. The working principle of d.c. machines are based on Faraday's laws of electromagnetic induction and Lenz's law. According to this, the voltage induced in a coil is proportional to the rate of change of flux linking the conductor of coil and can be represented by

$$e = -N\frac{d\phi}{dt}$$

where, $N$ is number of turns in the coil.

$$\left(\frac{d\phi}{dt}\right) = \text{rate of change of flux linking the coil.}$$

Earlier the generation of electrical energy were d.c. generation and so the distribution systems was d.c. distribution. Wide use of d.c. motors were there during that period. But as the a.c. generation, transmission and distribution emerged, a.c. systems became popular due to its advantages over d.c. system and d.c. machines were replaced by a.c machines. There was almost total change in generation, distribution and operation pattern. Direct Current (d.c.) motors were also replaced by induction and synchronous motors. But due to some special characteristics and advantages d.c. motors are being used in various operations like traction, overhead cranes, rolling mills, process industry, battery drivers vehicles, machine tools with precise control and widespread range, cranes etc. Direct current (d.c.) generators are being used in aircraft, ships etc. to supply power.

## 7.2 TYPES OF D.C. MACHINES

Basically there are two types of d.c. machines:

(*i*) Homopolar machine.
(*ii*) Heteropolar machine.

### 7.2.1 Homopolar Machine

In such machine there is only one active pole. The conductors move across the field of constant value. Such machine is rarely used.

### 7.2.2 Heteropolar Machine

Such machine has a field system with a number of poles alternately arranged with North (N) and South (S) poles. The e.m.f. induced in each conductor alternates with a frequency of

$$f = \frac{pN}{120}$$

where, $p$ is the number of poles and $N$ is the speed in r.p.m.

Further the machines can be classified into two types according to excitation system:

(*i*) Permanent magnet type.
(*ii*) Electromagnet type.

Generally, the electromagnet type d.c. machines with heteropoles are being used.

## 7.3 CONSTRUCTION OF D.C. MACHINES

The different parts of a d.c. machine are:

(*i*) Field system or excitation system or stator.
(*ii*) Armature
(*iii*) Commutator
(*iv*) Brushes
(*v*) Bearings

### 7.3.1 Field System

It develops the required magnetic field in the machine for magnetization of different parts of the machine in a magnetic circuit. The field winding is placed on salient poles which are stationary and hence it is called as stator. Figures 7.1 and 7.2 show the magnetic circuit, and field system of large d.c. motor (Fig. 7.3).

The field system or stator of d.c. machine have frame or yoke, poles, field winding, interpoles and interpole winding.

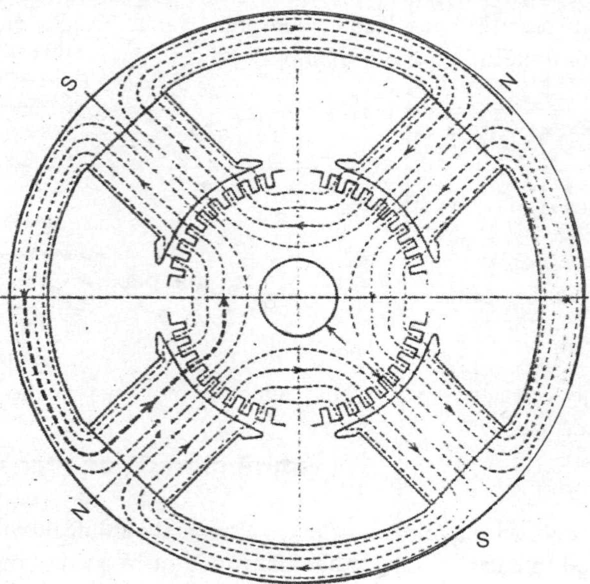


**Fig. 7.1** Magnetic circuit of 4-pole machine.

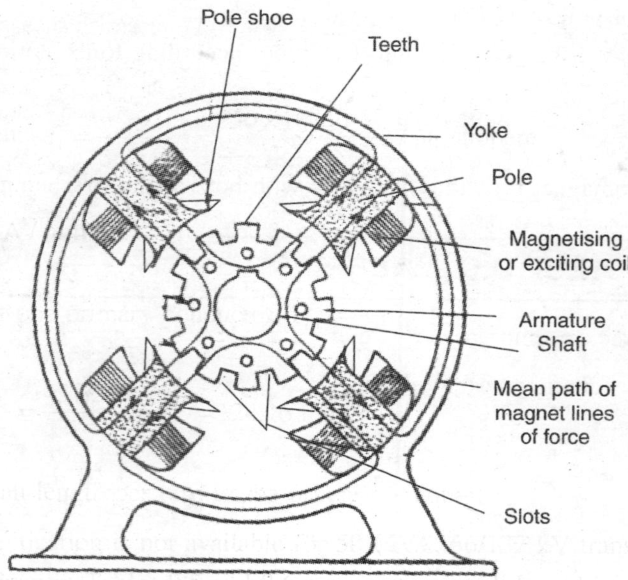


**Fig. 7.2** Magnetic circuit and field system of d.c. machine.

***(a) Yoke:*** It provides the magnetic path as well as the mechanical support to the whole machines. Earlier cast iron yokes were being used but now cast steel yokes are preferred due to larger permeability of cast steel. Due to this, the sectional area of cast yokes is considerably reduced to half of the sectional area of cast iron yokes. These days forged steel yokes are made. The details of the frame of a d.c. machine is shown in Fig. 7.4.

**Fig. 7.3** Field system of large d.c. motor showing interpoles and compensating winding.

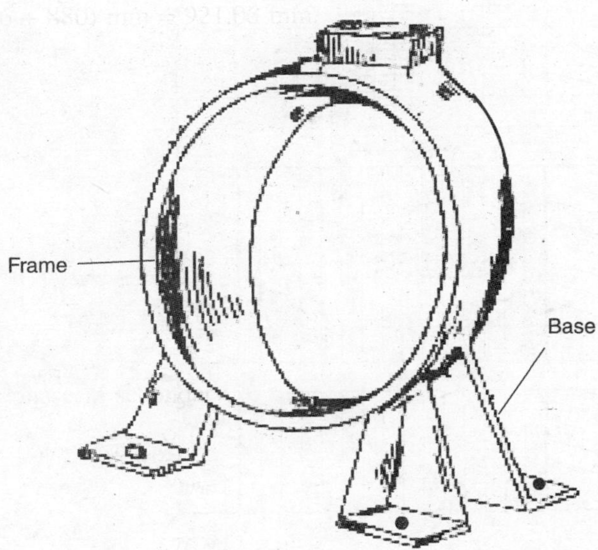


**Fig. 7.4** Frame of a d.c. machine.



***(b) Poles:*** The magnetic field pole consists of pole core or pole body and pole shoe. The poles shape is shown in Fig. 7.5.

The pole shoe of projecting poles or salient pole covers larger cross-section to achieve smooth spreading of flux in the air gap across long portion of air gap and hence reducing the reluctance of magnetic circuit. It also provides support to the field winding.

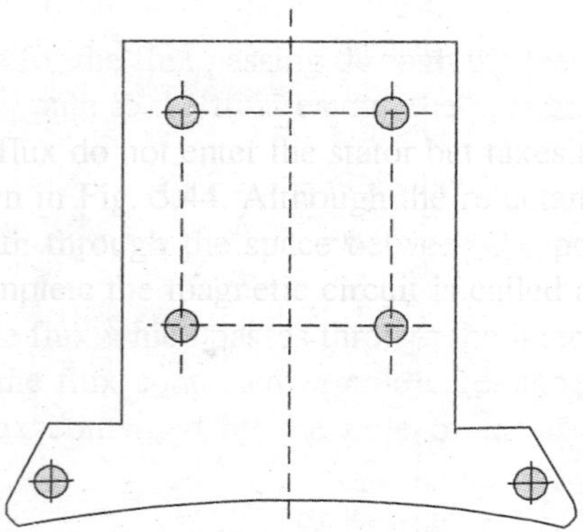

**Fig. 7.5** Pole of d.c. machine.

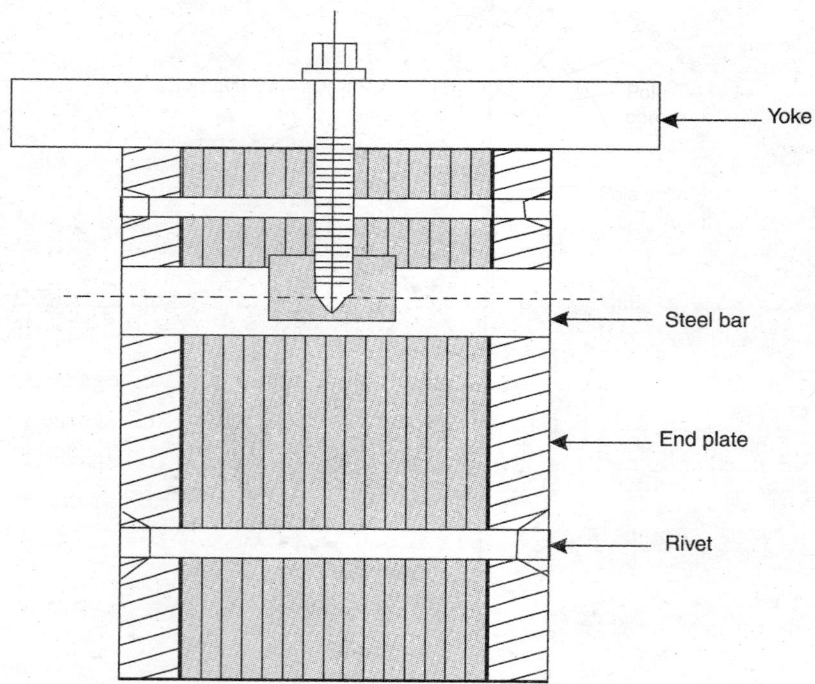


**Fig. 7.6** Fixing of pole core and yoke.

The pole core is made of solid cast iron or cast steel. But cast steel pole core with lamination are normally being used. The pole core and pole shoe are bolted together with hydraulic pressure. Further they are strengthened by screws bolted through the yoke in the pole body, steel bar, end plate and rivet as shown in Fig. 7.6.

***(c) Field Winding:*** Field coils of field winding are made of copper. The shape of winding conductors may be a round wire or a strip as required. The field winding consists of both shunt and series coils. The series field coils have larger conductor cross-section and are placed below the shunt field coils.

***(d) Interpoles:*** The interpoles are used in d.c. machines and are made from laminated steel or low carbon steel. Solid low carbon steel interpoles are used in small machines. Laminated interpoles are used in machines with commutation problem to obtain sparkles commutation. Interpoles may be parallel sided or tapered. The interpoles are made tapered and with larger cross-section to avoid saturation at the root as shown in Fig. 7.7.

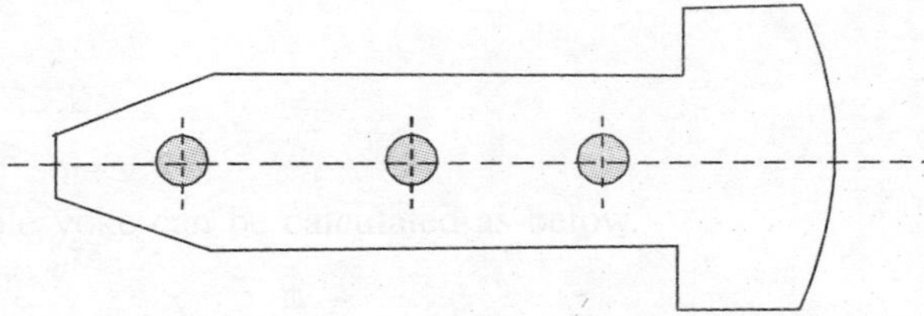

**Fig. 7.7** Commutating pole.

***(e) Interpole Winding:*** Generally, the round wires windings are used for interpole of small machines and copper strips with one or two layers for large d.c. machines. Nowadays epoxide resin through bonded coils with mica are used for insulation.

### 7.3.2 Armature

***(a) Armature Winding:*** It is a coil of large number of turns of insulated copper wire or copper strip wound on a laminated armature core. The coils are placed in the armature slots and fixed by slot liner of insulating material like latheroide paper. The slot liner is folded over the armature conductor. At the opening of the slot, the winding is fitted with wood or fibre wedges. Normally, two larger winding with diamond shape are used. One side of a coil is placed in a bottom of a slot and the other side of the same coil is placed at the top of a slot nearly one pole pitch away. There is small air gap between field poles and armature. The view of armature is shown in Fig. 7.8.

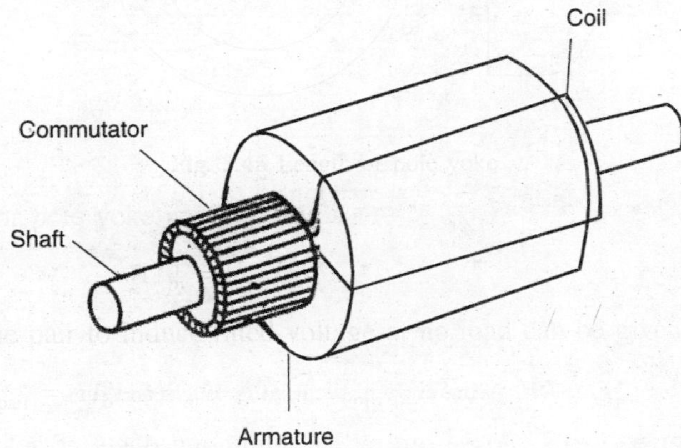


**Fig. 7.8** Armature view.

***(b) Armature:*** Teeth and slots are provided on the armature core as shown in Fig. 7.9. Armature winding is accommodated in the armature slots and e.m.f. is induced in the armature coils. It also provides a low reluctance path for magnetic flux through teeth and core. Armature is rotating at certain r.p.m. and e.m.f. induced in the conductor is alternating at some frequency. Hence the armature teeth and core cannot be made with solid material to avoid excessive eddy current loss. Hence, armature is laminated with 0.35 to 0.5 mm thickness of lamination with thin coating of varnish. Normally, sheet steels are used for armature core but for enclosed machines silicon steel is used to increase resistance and minimise the iron losses and to keep the temperature rise within specified limit.

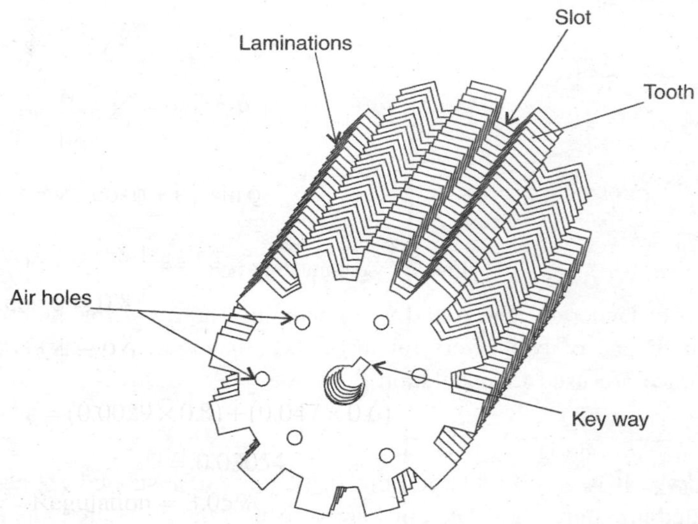


**Fig. 7.9** Armature teeth and slots.

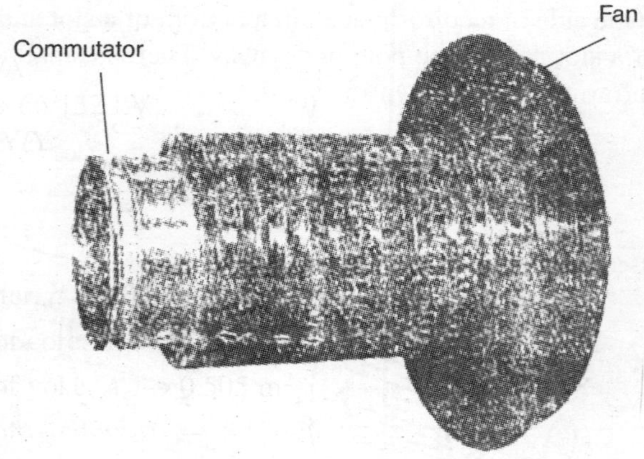


**Fig. 7.10** Armature with ventilating ducts and fans.

In the machines longer than 12 to 14 cm radial ventilating ducts of nearly 1 cm width for each length of 8 cm machine length are provided along its axial length for better cooling of machines. In machines, heat developed in the machine portions is dissipated partly by radiation and partly by conduction but the major portion of heat can be removed by means of regular supply of cool air circulating through the machine. In larger machines, the ventilation is achieved by ventilating fans as shown in Fig. 7.10. The fan is made of a diameter as large as is convenient for the end bracket, and is fitted on the next end from commutator.

### 7.3.3 Commutator

The function of the commutator is to collect and convert the alternating current, produced in the armature conductors, into direct current for the external circuit. It consists of (Fig. 7.11) wedge-shaped copper segments insulated from one another by thin layers of 0.8 mm mica and assembled side-by-side in cylindrical shape equal to commutator diameter required. Each segment has an extension, called as riser, and the leads of armature are soldered to risers. The number of segments in commutator is equal to the number of armature coils. The risers have air space between each other so that air is passed across the commutator to cool it. Mica layers are suitable for small machines only. Micanite is also used for small machines due to cost factor.

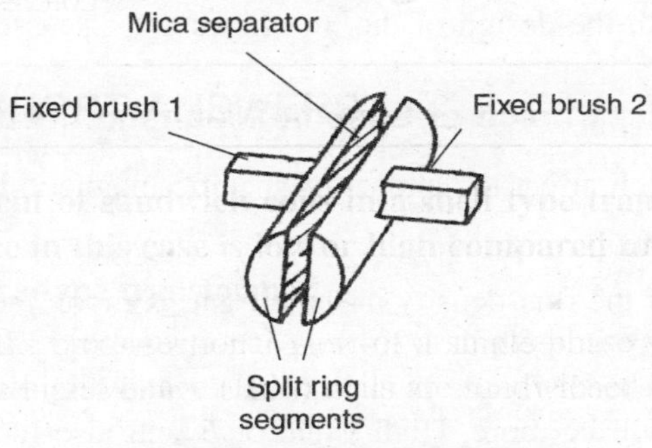


**Fig. 7.11** View of commutator segments.

### 7.3.4 Brushes

Brushes ride on the commutators and collect current from commutator and pass it to the external circuit. They are generally made of high grade carbon and are held in contact with the commutator by pressure springs. The two ends of armature coil are connected to the two semi-circular segments which also rotate as the coil rotates. The two segments are in contact with two fixed carbon brushes. The splitted-segments and brushes are so arranged that after each half rotation of the coil, the two segments change their contacting fixed brush pairs and hence current in external circuit is unidirectional. The commutator, brushes and brush holder are shown in Fig. 7.12.

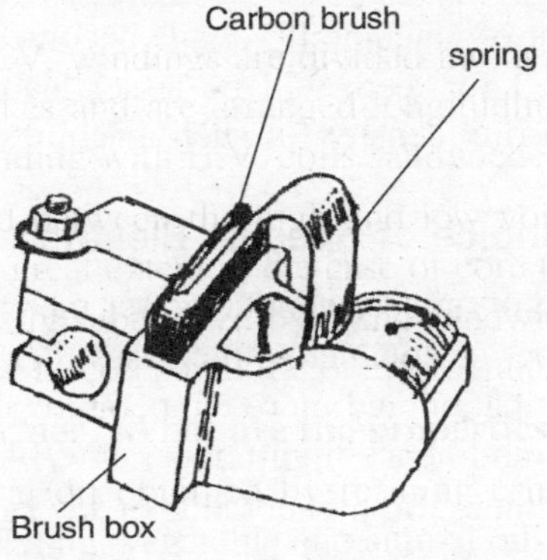


**Fig. 7.12** Commutator segments, brush and brush holder.

### 7.3.5 Bearings

The armature shaft is generally supported at the ends on ball or roller bearings. In small machines, ball bearings are used at both ends whereas in large machines roller bearing are used at the driving end and ball bearings are used at non-driving end or at commutator end. Pedestal bearings are also used for large machines. Bearings are used with hard oils or lubricants which provides silent running of machine.

## 7.4 IMPORTANT FEATURES OF D.C. MACHINES

Some features associated with the design of d.c. machines are discussed in this section.

### 7.4.1 Factors Affecting the Choice of Specific Magnetic Loading ($\overline{B}$)

There are several factors which affect the choice of specific magnetic loading or average flux density in the air gap ($\overline{B}$), given below:

***(a) Flux Density in Teeth:*** If the flux density in the air gap is more, the flux density in teeth will also become high. The flux density in teeth at 1/3rd height from narrow end should not be more than 2.2 Wb/m$^2$. High flux density in teeth means high iron losses and less efficiency. Also due to increased losses the temperature rise will be more. High value of $B$ will also require high m.m.f. for teeth and hence high field m.m.f. and high field copper loss. Specific magnetic loading should be selected so that flux density in teeth is below 2.2 Wb/m$^2$.

***(b) Frequency:*** The frequency of flux reversal in armature is $f = \dfrac{pN}{120}$. If frequency is high, iron losses in armature teeth and core will be high. Further if the flux density is more, losses will be more. Hence high value of flux density should not be taken in machines having high frequency of armature flux reversal.

***(c) Size of Machine:*** High value of flux density should not be used in machines with reduced size. Since the flux density will be more. Hence high value of $\overline{B}$ can be used with increased size of machine. The range of average flux density is given while designing steps are discussed.

***(d) Voltage:*** For specified diameter of machine if voltage becomes higher, space required by insulation will be more and hence space left for teeth will be lesser leading to narrow teeth and higher flux density in teeth. So, low value of flux density for such machines should be taken.

### 7.4.2 Factors Affecting the Choice of Specific Electric Loading ($\overline{ac}$)

***(a) Temperature Rise:*** High value of ampere-conductor per meter means larger conductors or reduced diameter of machine. When number of conductors increases then it will lead to high losses, increased insulation required, poor heat dissipation and more temperature rise. When diameter is reduced it will also lead to poor heat dissipation and hence high temperature rise. Semi-closed machines have better ventilation than totally closed machines. So higher value of $\overline{ac}$ can be taken for semi-closed machines.

***(b) Speed:*** If the speed of the machine is higher, the ventilation of machine is better so heat dissipated per unit area of machine will be more. So high value of $\overline{ac}$ for machines with high speed can be taken.

***(c) Voltage:*** For a given diameter of machine the high voltage means less space for copper winding and teeth due to increased insulation thickness. The size of teeth cannot be reduced due to limitation of flux density in teeth. Therefore small value of ampere-conductor per metre should be used.

***(d) Size of Machine:*** Larger the size of machine (*i.e.*, diameter of machine) large space is available for winding conductors. Hence larger value of $\overline{ac}$ can be taken for machines with large diameter.

***(e) Armature Reaction:*** High value of $\overline{ac}$ will cause to increase the armature ampere-turn ($AT_a$). Due to increased armature m.m.f., the armature reaction during load condition will increase and hence field form will be distorted. High field m.m.f. is required to compensate this. So cost of copper will increase in the machine.

***(f) Commutation:*** High value of $\overline{ac}$ will have either a large number of armature conductors or small diameter. High ampere-conductor will increase the reactance voltage. If small diameter is taken, it is not possible to use wide slots otherwise space left for teeth will be reduced and thereby increasing the flux density in teeth. Hence deeper slots will have to be used. Deeper slots also increase the reactance voltage. High reactance voltage results in bad commutation. So use of high value of $\overline{ac}$ will affect the commutation badly.

### 7.4.3 Selection of Air Gap Length (δ)

***(a) Armature Reaction:*** When the air gap length of machine is more, the field m.m.f. required will be more. The increased value of field m.m.f. will reduce the distorting effect of armature m.m.f. Hence with larger air gap the distortion of field form is reduced. But the increase in the field m.m.f. will result in increase in size and cost of machine.

***(b) Circulating Currents:*** If the air gap in the machine with multipolar lap winding is small, any irregularity may cause large circulating current. So air gap should be large in such machines.

***(c) Pole Face Losses:*** In rotating machines, the armature are slotted and on rotation there is rapid change of gap reluctance. This change of reluctance gives rise to flux pulsation which produces additional losses called pulsation losses in teeth and pole faces. This effect is considerably increased if the length of air gap is small as compared with slot openings. So the pulsation loss in the pole faces will decrease if the length of air gap is increased.

***(d) Noise:*** Quiet operation of machine is achieved with large air gap.

***(e) Cooling:*** Large air gap length in machine will provide better ventilation.

***(f) Mechanical Reasons:*** Machines without commutating poles have large value of air gap length to minimize distortion of field form. Due to use of commutating poles the length of air gap can be reduced to the minimum value from mechanical point of view. With smaller air gap length the possibility of unbalanced magnetic pull and rotor may collide with stator. So, the air gap length should be suitably taken to avoid this.

### 7.4.4 Effect of Armature Reaction

Armature reaction affects on following:

***(a) EMF:*** When machine is running under saturation condition then flux per pole decreases due to armature reaction. If the machine is loaded, the e.m.f. generated in the machine decreases due to decrease in flux per pole. This reduction in value of $\overline{ac}$ flux depends upon the degree of saturation. The value of flux will reduce much if the machine is heavily loaded.

***(b) Iron Losses:*** The iron losses in teeth and pole shoe depends upon the flux density in them. Due to the field distorting effect of armature reaction, the flux density at load condition increases too much with respect to flux density at no load. If the flux density increases at load, the losses will also be more on load as compared with no load. The iron losses at load are nearly 1.5 times the iron losses at no load.

***(c) Sparking and Ring Fire:*** As armature reaction increases, the maximum value of air gap flux density increases at load with respect to no load. The maximum value of voltage between adjacent segments of commutator will increase with increase in air gap flux density. If the voltage between adjacent commutator segments increases beyond certain limit (30V), there is possibility of spark between adjacent segments. This spark may spread over the whole commutator and a ring of fire may be formed around the commutator.

***(d) Commutation:*** The reversal of current from one direction to a direction opposite to it is called commutation. For good commutation, the coils undergoing commutation should not have any generated e.m.f. in them. The coils undergo commutation at the brushes placed at the geometrical neutral axis. When the coil is undergoing commutation then the flux density at the neutral axis (interpolar axis) should be zero so that there is no e.m.f. generation in the coils. This will be a good commutation condition. But due to armature reaction the flux density at neutral axis is not zero. The flux density at neutral axis will generate an e.m.f. in the coils undergoing commutation. The e.m.f. induced in the coil undergoing commutation at neutral axis tries to maintain the current in original direction and hence delays the commutation. So, armature reaction delays the commutation.

### 7.4.5 Armature Winding Terms

The different terms related to armature windings are conductor, turn, coil, coil side, overhang coil span etc. These terms are briefly described below:

***(a) Conductor:*** The length of winding wire across the axial length of machine is called conductor as shown in Fig. 7.13(*a*).

***(b) Turn:*** A turn consists of two conductors accommodated in the armature slots approximately a pole pitch apart as shown in Fig. 7.13(*b*). If two conductors of a turn are placed in the slots at a pole pitch apart, the resultant emf induced in the two conductors will be the sum of e.m.fs. induced in each conductor.

***(c) Coil:*** A coil consists of one or more turns. A coil with single turn is called single turn coil as in Fig. 7.13(*a*) and coil with more turns is called as multi-turn coil as in Fig. 7.13(*b*).

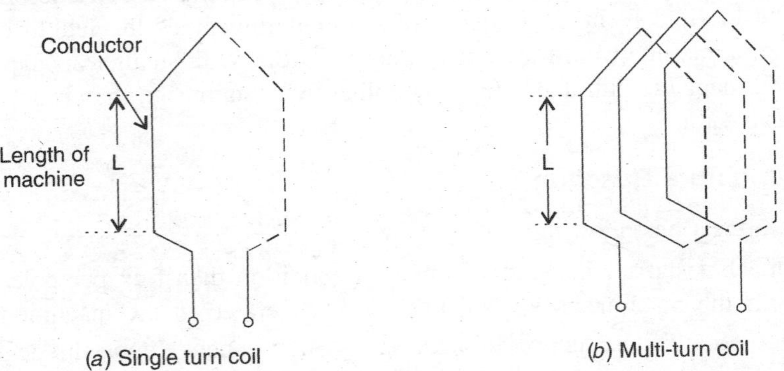


**Fig. 7.13** Single and multi-turn coil.

***(d) Coil Side:*** The coil have two sides. The two sides of a coil are placed in the different armature slots. One side of a coil is placed in the upper portion of one slot and the other side is placed in the lower portion of the other slot at a pole pitch apart in a double layer winding normally used in d.c. machines. The coil side placed on upper portion of slot is known as upper coil side and the coil side placed on lower portion of slot is known lower coil side.

***(e) Overhang:***The winding length required to connect the two conductors at the end of machine is known as overhang of the winding.

***(f) Coil Span:*** The peripheral distance between two slots in which the two sides of a coil are placed is called coil span. It is generally expressed in terms of number of slots or conductors.

***(g) Full Pitch Coil:*** When the two coil sides of a coil are placed in the slots over the periphery of armature exactly a pole pitch apart then the coil is called full pitch coil. In full pitch coil, the coil span is equal to the pole pitch of the machine.

***(h) Short Pitched or Chorded Coil:*** When the coil span is less than the pole pitch of the machine, the coil is called a chorded coil or short pitched coil.

### 7.4.6 Single Layer and Double Layer Winding

***(a) Single Layer Winding:*** When the whole slot of the machine is accommodated by one side coil of only one coil, the winding is known as single layer winding. Single layer winding is normally not used due to commutation problems.

***(b) Double Layer Winding:*** When upper portion of a slot (top layer and nearer to air gap) is occupied by the upper coil side of a coil and the lower portion (bottom layer) of the same slot is occupied by the lower coil side of another coil, winding is known as double layer winding. The two layers of the windings in the same slot are insulated by an insulation strip called separator. Double layer windings are most commonly used in d.c. machines. It gives satisfactory arrangement of end connections. For double layer winding, the number of armature slots is equal to the number of coils in the armature winding. But in larger machines, the number of coils may become larger and number of slots may be limited due to design considerations. In such machines, the slot may accommodate two or more than two coil sides. The numbering of conductors should be done in such a way that odd numbered conductors should occupy top layers of the slot portion and even numbers should occupy the bottom layer of slot portion. Figure 7.14 shows the slots with 2, 4 and 6 conductors.

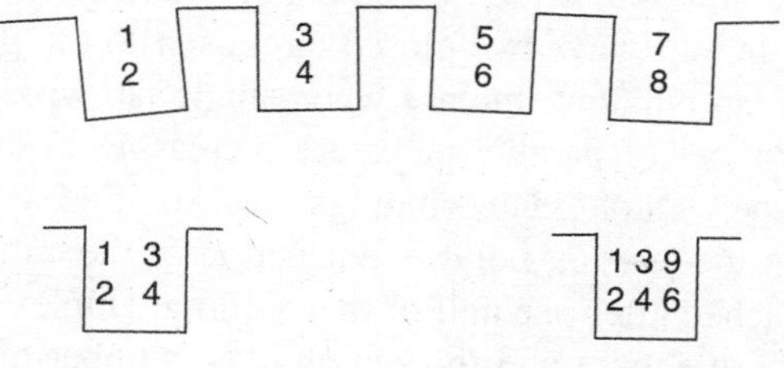


Armature slots and conductors

**Fig. 7.14** Winding arrangement.

### 7.4.7 Types of Armature Winding

All the coils in one parallel path are connected in series and so the e.m.fs. induced in these coils are

Chapter 7

added to get the resultant e.m.f., which is equal to voltage rating of machine. Generally, two types of windings are used in d.c. machines and they are lap winding and wave winding.

***(a) Lap Winding:*** When end of a coil one is connected to the starting of coil two like coil one is lapping over the coil two. The other coils of the winding are connected in similar way to form a closed winding. Since one coil is lapping over the next coil, the winding is known as lap winding. Figure 7.15 (*a*) shows a portion of lap winding with single turn coil. Coils *a* and *b* consist of coil sides $a_1$, $a_2$ and $b_1$, $b_2$ respectively. The end of coil side $a_2$ is connected to $b_1$ which is under the same pole as the starting of coil *a*. Other coils of the winding is connected in similar way and the end of the last coil is connected to the starting of the first coil and closing the winding. The lap winding arrangement can be done by two ways *i.e.*,

– Simplex lap winding, and
– Duplex lap winding.

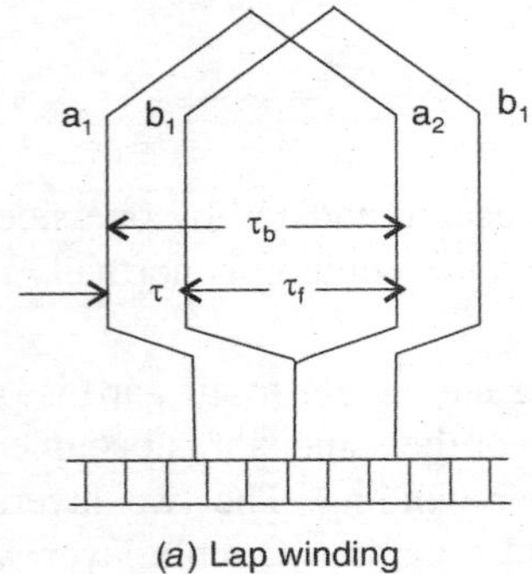

(*a*) Lap winding

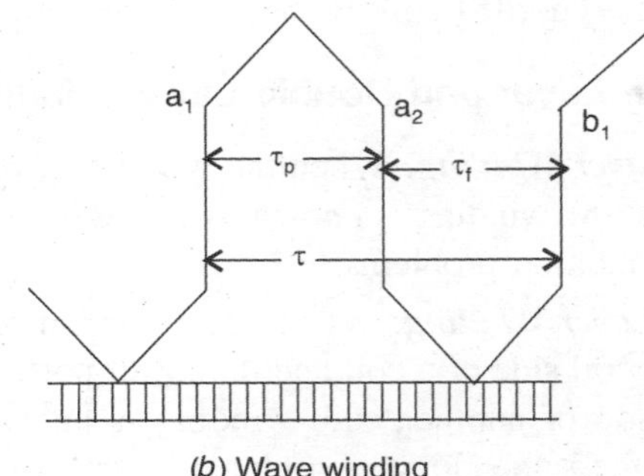

(*b*) Wave winding

**Fig. 7.15** Lap and wave winding.

***Simplex Lap Winding:*** The armature lap winding, which forms only one closed circuit is called simplex lap winding. The commutator pitch of simplex lap winding is unity. The number of parallel circuits in simplex lap winding is equal to the number of poles in the machine. Generally, simplex lap windings are being used in d.c. machines.

***Multiplex Lap Winding:*** Armature lap winding with several closed circuits is called multiplex lap winding. It is used in the machines in which commutator pitch is ±2 and is known as duplex lap winding. Thus a duplex lap winding have two closed circuits with the parallel circuits equal to twice of parallel circuits in simplex lap winding. In case when single lap winding is not suitable double lap winding is used. Since the number of parallel paths are increasing in duplex lap winding, it may be used for heavy current machines. Duplex lap windings are classified as singly re-entrant and doubly re-entrant. In singly re-entrant the starting point is reached after the entire winding and in doubly re-entrant the starting point is reached after one half of the winding. Duplex lap winding with odd number of commutator bars are singly re-entrant and those with even number of commutator bars are doubly re-entrant.

***(b) Wave Winding:*** Wave windings can be classified into simplex wave winding and multiplex wave winding. Figure 7.15 (*b*) shows a portion of wave winding with single turn coil with coils *a* and *b*. The end $a_2$ of a coil which lies in south pole is connected to $b_1$, the starting of coil *b*, which lies in North pole. Other coils are connected in similar fashion.

***Simplex Wave Winding:*** Armature winding with wave winding and having only one closed circuit is known as simplex wave winding. The commutator pitch for simplex wave winding depends upon the number of poles and commutator segments. The number of parallel circuits (parallel paths) in simplex wave winding is always two irrespective of number of poles.

***Multiplex Wave Winding:*** Armature wave winding with several closed circuits is termed as multiplex wave winding. In a multiplex wave winding after making one round of the commutator, one arrives at the commutator segment which is away from the initial segment by 2, 3, ..., *m* segments, thus obtaining 2, 3, ..., *m* windings, each simplex wave winding.

Duplex winding equivalent to two simplex wave winding is used in d.c. machines where simplex winding is not suitable. The number of parallel circuits (parallel path) in duplex wave winding is always 4 irrespective of number of poles (double of parallel path in simplex wave winding). Duplex wave winding is singly re-entrant when commutator bars are odd and it is doubly re-entrant if the commutator bars are even.

### 7.4.8 Winding Pitches

Different winding pitches are explained below:

***Back Pitch ($\tau_b$):*** The distance between the two coil sides of a coil at the back of the commutator is called the back pitch, so the distance between $a_1$ and $a_2$ (Fig. 7.15) will be the back pitch ($\tau_b$).

***Front Pitch ($\tau_f$):*** Front pitch is the distance between the end of a coil (conductor) and the starting of the next coil (conductor) connected to the same commutator segment as shown in Fig. 7.15. The distance between $a_2$ of coil *a* and $b_1$ of coil *b* is the front pitch ($\tau_f$).

***Resultant Pitch ($\tau$):*** The distance between the starting coil (conductors) of the two consecutive coils is termed as resultant pitch. In Fig. 7.15 the resultant pitch is shown by $\tau$ between $a_1$ and $b_1$.

## 7.5 OUTPUT EQUATION

### Derivation of Output Equation

The output equation of d.c. machines is derived which relates power developed by armature and main dimensions of the machines. In d.c. machines the fundamental equation for power developed by armature is being converted into output equation of machine to relate its main dimensions.

Let power developed by the armature = $P_a$ kilowatt

$$P_a = \text{Generated e.m.f.} \times \text{Armature current} \times 10^{-3}$$

$$P_a = E \times I_a \times 10^{-3} \text{ kilowatt} \qquad ...(7.1)$$

where, $E$ is in volts and $I_a$ in ampere.

Generated e.m.f. in a d.c. machine,

$$E = \frac{p\phi NZ}{60 \times a}$$

Putting this value of $E$ in equation (7.1), we get

$$P_a = \frac{p\phi NZ}{60 \times a} \times I_a \times 10^{-3} \qquad ...(7.2)$$

where,

$p$ = number of poles
$\phi$ = flux per pole in air gap in weber
$N$ = speed in r.p.m.
$Z$ = total number of armature conductors
$a$ = number of parallel paths

Current in each conductor, $I_z = \dfrac{I_a}{a}$ ...(7.2 (*a*))

If number of poles is $p$ then total magnetic loading of the machine will be $p\phi$.

So, $$p\phi = \overline{B} \cdot \pi DL$$ ...(7.2 (*b*))

or, $$\overline{B} = \frac{p\phi}{\pi DL}$$

where, $\overline{B}$ = average value of flux density in air gap, also known as specific magnetic loading
$D$ = armature diameter, in metre
$L$ = armature core length in metre

If specific electric loading *i.e.*, ampere-conductor per metre length of periphery along air gap of the machine is $\overline{ac}$, then

Total ampere-conductor = $\overline{ac} \cdot \pi D$
Also total ampere-conductor = $I_z \cdot Z$

So, $I_z \cdot Z = \overline{ac} \cdot \pi D$ ...(7.2 (*c*))

So, $$\overline{ac} = \frac{I_z \cdot Z}{\pi D}$$

Now, putting the value of $(p\phi)$, $(I_z \cdot Z)$ from equations (7.2*b*) and 7.2 (*c*), in equation (7.2), we get,

$$P_a = \left(\pi DL\overline{B}\right)\left(\pi D\overline{ac}\right) \times \frac{N}{60} \times 10^{-3}$$

or $$P_a = \left(\pi^2 \overline{B}\,\overline{ac} \times 10^{-3}\right) D^2 L \cdot \frac{N}{60}$$

or $$P_a = \left(\pi^2 \overline{B}\,\overline{ac} \times 10^{-3}\right) D^2 L \cdot n$$

where, $n$ is the speed in r.p.s.

$$\therefore\ P_a = CD^2 Ln$$ ...(7.3)

where, $C = \pi^2 \overline{B}\,\overline{ac} \times 10^{-3}$ and called as output coefficient and equation (7.3) called the output equation of d.c. machine.

## 7.6 DETERMINATION OF $P_a$ FOR DIFFERENT MACHINES

Power developed by armature is different for generator and motor.

***(i) For Generator:*** Power developed by the armature of generator.

$$P_a = \text{Input power} - \text{Friction, windage and iron loss}$$

$$= \frac{\text{Output}}{\text{Efficiency}} - \text{F \& W and iron losses}$$

$$= \frac{P}{\eta} - \text{F \& W and iron losses}$$

For large generators friction, windage and iron losses can be neglected hence for larger generators,

$$P_a = \frac{P}{\eta} \qquad \text{...(7.4)}$$

For smaller generators rotational losses (F & W and iron loss) cannot be neglected and these losses are taken as 1/3rd of total losses.

$$\text{F \& W and iron loss} = \frac{1}{3}P\left(\frac{1-\eta}{\eta}\right)$$

Since total loss = Input – Output

$$\text{So, total loss } = \frac{P}{\eta} - P = P\left(\frac{1-\eta}{\eta}\right)$$

So, for small generators,

$$P_a = \frac{P}{\eta} - \text{F \& W and iron loss}$$

$$\text{or} \quad P_a = \frac{P}{\eta} - \frac{1}{3}P\left(\frac{1-\eta}{\eta}\right)$$

$$\text{or} \quad P_a = P\left(\frac{2+\eta}{3\eta}\right) \qquad \text{...(7.5)}$$

***(ii) For Motors:*** In d.c. motor friction and windage and iron losses are supplied by the armature. So for a d.c. motor,

$$P_a = P + \text{F \& W and iron losses}$$

$$= P + \frac{1}{3}P\left(\frac{1-\eta}{\eta}\right)$$

For large motors F & W and iron losses can be neglected, so

For large d.c. motors, $P_a = P$ ...(7.6)

and for small motors, $P_a = P + \frac{1}{3}P\left(\frac{1-\eta}{\eta}\right)$

Chapter 7

or $P_a = \left(\frac{1+2\eta}{3\eta}\right)P.$ ...(7.7)

So, equations (7.4) , (7.5), (7.6) and (7.7) give the value of $P_a$ for different types of d.c. machines.

## 7.7 SELECTION OF $\overline{B}$ AND $\overline{ac}$

### 7.7.1 Selection of Specific Magnetic Loading ($\overline{B}$)

The value of maximum magnetic flux density in the air gap ($B_m$) usually varies from 0.55 to 1.10 Wb/m$^2$ and corresponding value of average flux density ($\overline{B}$) in the air gap is from 0.4 to 0.8 Wb/m$^2$.

But, $$B_m = \frac{\overline{B}}{\psi}$$

where, $\psi = \frac{\text{Pole arc}}{\text{Pole pitch}} = \frac{b_p}{\tau_p}$.

**Table 7.1:** Different value of $B_m$

| *Output in kW* | $B_m$ *(Wb/m$^2$)* | *Output in kW* | $B_m$ *(Wb/m$^2$)* |
|---|---|---|---|
| 5 | 0.57 | 1000 | 0.96 |
| 10 | 0.65 | 2000 | 0.97 |
| 20 | 0.70 | 3000 | 0.99 |
| 50 | 0.77 | 4000 | 1.00 |
| 100 | 0.82 | 5000 | 1.05 |
| 200 | 0.87 | | |
| 500 | 0.90 | 10,000 | 1.10 |
| 750 | 0.95 | | |

Smaller machines have low value of flux density. As rating of machine increases, the flux density increases. Table 7.1 shows different values of maximum flux density for different machine rating or output.

### 7.7.2 Selection of Specific Electric Loading ($\overline{ac}$)

The value of specific electric loading *i.e.*, ampere-conductor per metre peripheral length along the air gap of machine varies between 15,000 to 51,000 ampere-conductor/m. Low rating machines have lower value of $\overline{ac}$ and higher rating machines have high value of $\overline{ac}$. Table 7.2 gives the value of $\overline{ac}$ for different output of the machine.

**Table 7.2:** Different values of $\overline{ac}$

| *Rating in kW* | $\overline{ac}$ | *Rating in kW* | $\overline{ac}$ |
|---|---|---|---|
| 5 | 15,000 | 750 | 38,000 |
| 10 | 17,500 | 1000 | 40,000 |
| 20 | 19,000 | 2000 | 43,000 |
| 50 | 25,000 | 3000 | 46,500 |
| 100 | 27,500 | 4000 | 48,000 |
| 200 | 31,000 | 5000 | 49,000 |
| 500 | 35,500 | 10000 | 51,000 |

## 7.8 FACTORS EFFECTING THE SELECTION OF NUMBER OF POLES

In a.c. machines, number of poles ($p$) are given by $\frac{120f}{N}$ whereas in d.c. machines any number of poles can arbitrarily be taken. However economical justification suggest for small range of number of poles for d.c. machine. Factors which effects the selection for suitable number of poles are given below.

### 7.8.1 Frequency

The frequency ($f$) of flux reversal in the armature $\left(\frac{Np}{120}\right)$ is proportional to the number of poles in the machine for any given speed $N$. For a given speed if the number of poles becomes larger then the frequency of flux reversal will also be more and may result in large core losses. The frequency of flux reversal in the armature core generally lies between 25 and 50 Hz. Lower values (25 to 35 Hz) frequency should be used for large machines and high frequency in the range of 50-75 Hz may be used for smaller machines.

### 7.8.2 Weight of Iron Parts

The weight of iron in different portion of the magnetic circuit of the machine is being affected by the number of poles. The effect of number of poles on weight of iron of different parts like yoke and armature core is discussed below.

***(a) Yoke:*** As it is clear from Fig. 7.16($a$) that in a machine with two poles if flux per pole is = $\phi$, then flux in yoke = $\phi/2$. Since the total flux around the air gap remains constant so if the number of poles of this machine is increased to 4.

Then flux per pole = $\frac{\phi}{2}$ and hence,

flux in yoke = $\frac{\phi}{4}$.

Chapter 7

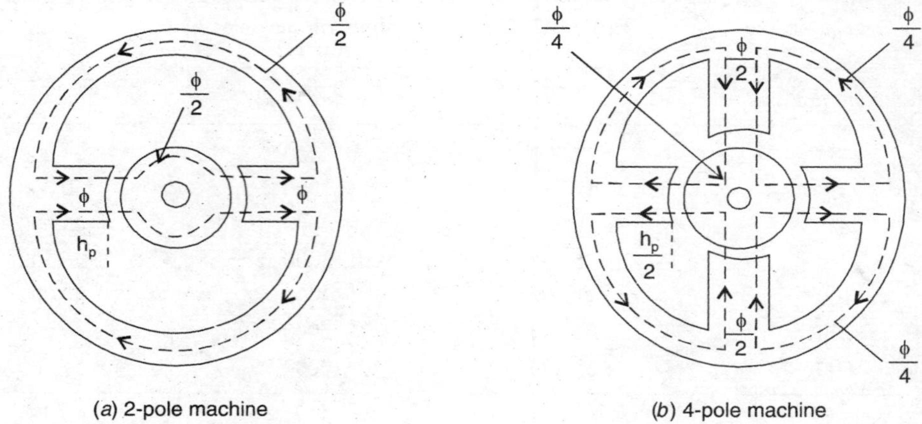


**Fig. 7.16** Reduction in machine dimension with increase in number of poles.

So it is observed that when number of poles are doubled then the flux in the yoke is halved. For same flux density in yoke, the weight of iron in yoke can be reduced in proportion to increase in no. of poles.

(*b*) *Armature Core:* From Fig. 7.16 it is clear that flux in armature core in 2-pole machine is ϕ/2 and in 4-pole machine, flux in armature core is ϕ/4. So by increasing the number of poles the size of the armature core is reduced but at the same time increases the frequency of flux reversal in the armature core causing increased core losses. Thus the loss should also be taken into consideration while selecting larger no. of poles.

**(i) Eddy Current Loss**

For 2-pole machine:

$$P_e \propto B^2 f_1^2$$

$$P_e \propto \frac{\phi^2}{4A_1^2} f_1^2 \qquad \ldots(7.8) \qquad \left[\because B = \frac{\phi/2}{A_1}, \text{ for 2-pole machine}\right]$$

where, $A_1$ is the armature core cross-sectional area and $f_1$ is the frequency of flux reversal.

For 4-pole machine:

$$P_e' \propto B^2 f_2^2$$

$$P_e' \propto \frac{\phi^2}{4A_2^2} f_1^2 \qquad \ldots(7.9) \qquad \left[\begin{array}{l}\because B = \dfrac{\phi/4}{A_2} \quad \text{and} \\ f_2 = 2f_1, \text{ for 4-pole machine}\end{array}\right]$$

where, $B$ = flux density in core

$A_2$ = cross-section of armature core

$f_2$ = frequency of flux reversal

From eqns. (7.8) and (7.9), it is evident that to get same eddy current loss there is no change in cross-sectional area of armature core due to increase in number of poles.

**(ii) Hysteresis's Loss**

For 2-pole machine:

$$P_h \propto B^{1.6} f_1$$

or, $$P_h \propto \left(\frac{\phi/2}{A_1}\right)^{1.6} f_1 \qquad \left[\because B = \frac{\phi/2}{A_1}, \text{ for 2-pole machine}\right]$$

$$\Rightarrow \quad P_h \propto \left(\frac{\phi}{A_1}\right)^{1.6} \left(\frac{1}{2}\right)^{1.6} f_1 \qquad ...(7.10)$$

For 4-pole machine:

$$P_h' \propto B^{1.6} f_2$$

or, $$P_h' \propto \left(\frac{\phi/4}{A_2}\right)^{1.6} f_2 \qquad \left[\because B = \frac{\phi/4}{A_2}, \text{ or and } f_2 = 2f_1, \text{ for 4-pole machine}\right]$$

$$P_h' \propto \left(\frac{\phi}{A_2}\right)^{1.6} \left(\frac{1}{4}\right)^{1.6} \cdot 2f_1 \qquad ...(7.11)$$

From equations (7.10) and (7.11), it is clear that the hysteresis's loss for 4-pole machine will be less than that for 2-pole machine. So for same hysteresis loss, the size of the armature core can be reduced by increasing the number of poles.

**(*c*) *Magnets:*** Ampere-turn developed by field coils are normally fixed and it is independent of number of poles. As the number of poles will increase, the ampere-turn developed by each coil will be reduced in same proportion. Hence the height of the pole for same depth of the winding will be reduced and the iron required will also be reduced. Thus the overall diameter of the machine will decrease when the number of poles will increase.

### 7.8.3 Weight of Copper

**(*a*) *Armature Copper:*** Due to increase in the number of poles, the pole pitch $\left(\frac{\pi D}{p}\right)$ will be reduced for a given diameter of the machine. Due to decrease in pole pitch, the length of overhang portion decreases and copper required in overhang portion reduces and hence overall copper required will be decreased.

**(*b*) *Field Copper:*** It has already been discussed that due to increase in number of poles, the field ampere-turn per pole reduces and the area of cross-section per pole required will also reduce in the same ratio. The copper length and the weight of copper will thus reduce when the number of poles will increase.

### 7.8.4 Commutator Length

**(*a*) *For 2-pole Machine:*** There are two brush arms in 2-pole machine.

So, current per path in armature $= \frac{I}{2}$ and

Current per brush arm $= I$

***(b) For 4-pole Machine:*** There are four brush arms in 4-pole machine.

So, current per path in armature $= \frac{I}{4}$ and

Current per brush arm $= \frac{I}{2}$

So, the current in each brush arm in 4-pole machine reduces and hence the area in each arm will also be reduced. Normally, the thickness of brushes are not appreciably reduced when the number of poles are increased. Therefore the length of brushes for each brush arm is reduced with increased number of poles. This results, in reduction in length of commutator and also reduction in overall length of machine.

### 7.8.5 Labour Charges

#### *Armature Coils*

$$\text{e.m.f.}, \; E = \frac{p\phi NZ}{60 \cdot a}$$

Since $p\phi$ is assumed to be constant.

$$\text{So,} \; E \propto \frac{NZ}{60 \cdot a}$$

In lap winding, $a = p$

$$\text{So} \; E \propto \frac{NZ}{60 \cdot p}$$

$$\text{or} \; Z \propto \frac{60 \cdot p \cdot E}{N} \qquad ...(7.12)$$

From equation (7.12), it is clear that number of conductor ($Z$) increases, proportional to the number of poles. In multipolar machine using lap winding, single turns are normally used. So the number of armature coil will increase with increase in armature conductors. Number of commutator segments are same as the number of armature coils hence the number of commutator segments will also increase. In wave winding, the number of conductors remains the same with the change in number of poles and hence the number of coils remains same.

### 7.8.6 Flash Over

The number of brush arm is equal to the number of poles. If the number of poles are increased then the number of brush arms will increase. So for the same diameter of the commutator, the distance between two adjacent brush arms will be reduced. This will increase the possibility of flash over between brushes.

### 7.8.7 Important Features for Selection of Number of Poles

(*i*) The frequency of flux reversal in the armature core should normally lies between 25 to 50 Hz. From 25 to 35 Hz frequency should preferably be selected for large machines. Higher frequency range can be taken for small machines.

(*ii*) The value of current per brush arm should be limited to 400 amperes and current per parallel path in the armature should be 200 amperes.

(*iii*) The value of armature mmf per pole should normally be selected according to Table 7.3. Armature m.m.f. per pole.

$$AT_a = \left(\frac{\overline{ac}}{2}\right) \times \text{pole pitch} = \frac{\overline{ac}}{2} \cdot \frac{\pi D}{p} = \frac{AC}{2p}$$

where, AC is the total ampere-conductor.

**Table 7.3:** Armature m.m.f. per pole

| *Output (kW)* | *Armature mmf per pole (amp.)* |
|---|---|
| up to 100 | below 5000 |
| 100–500 | 5000–7000 |
| 500–1500 | 7000–10,000 |
| above 1500 | up to 12,000 |

**Table 7.4:** Number of poles

| *Output (kW)* | *No. of poles* | *Speed in r.p.m.* |
|---|---|---|
| less than 2 | 2 | more than 1250 |
| 2 to 75 | 4 | 900–1750 |
| 75–200 | 6 | up to 1200 |
| 200–650 | 6–8 | up to 1200 |
| 650–1500 | 8–12 | up to 900 |
| 1500–2500 | 12–14 | up to 500 |
| 2500–5000 | 14–24 | up to 375 |

## 7.9 SELECTION OF CORE LENGTH AND DIAMETER



### 7.9.1 Factors for Selection of Core Length

#### *(i) Commutation*

Conductor length is longer for longer machine length which results in larger e.m.f. induced per conductor and increased voltage across the adjacent commutator segments. Increase in voltage across the commutator segment should not exceed 20 V in open circuited condition.

### *(ii) Cost*

The overhang portion of the winding will be less in longer machine so less amount of copper will be required. This is due to the fact that for the same rating of the machine if length of machine is more, diameter will be less and if diameter is less, pole pitch will be less. Hence longer machines will be less costly.

### *(iii) Ventilation*

Ventilation of central portion of the longer machine is difficult. The temperature in this portion of machine may increase more than the desirable value.

### 7.9.2 Selection of Armature Diameter

Following factors are considered while selecting the armature diameter:

***(i) Peripheral speed:*** Peripheral speed, $V_p = \dfrac{\pi DN}{60}$

where, $D$ is the armature diameter and $N$ is in r.p.m.

The range of peripheral speed is 15–30 m/sec without special construction of the machine. If special construction is done then the peripheral speed can be taken upto 50 m/sec. The centrifugal force increases as the peripheral velocity increases so if the peripheral speed has to be increased, the special construction to maintain the end connections should be employed.

***(ii) Pole pitch:*** The pole pitch obtained after selecting a suitable diameter, may be used as a check for the number of poles. Table 7.5 is given for this.

**Table 7.5:** Pole pitch

| *No. of Poles* | *Pole pitch (cm)* |
|---|---|
| 2 | up to 24 |
| 4 | 24–40 |
| 6 | 35–45 |
| above 6 | 45–55 |

## 7.10 ESTIMATION OF MAIN DIMENSION (*D* & *L*) – (SEPARATION OF *D* AND *L*)

Output equation can be estimated as

$$P_a = CD^2LN \qquad \text{...(7.13)}$$

where, $C$ is known as output coefficient and its value is given by

$$= \pi^2 \overline{B}\,\overline{ac} \times 10^{-3}$$

Select suitable value of $\overline{ac}$ and estimate

$$D^2L = \frac{P_a}{CN} \qquad \text{...(7.14)}$$

Usually, square pole face construction is used since length of mean turn of winding is reduced in such construction and saving in cost of conductors used is achieved for the field coils of the machine.

$$\frac{\text{Pole arc}}{\text{Pole pitch}} = \frac{b_p}{\tau_p} = \psi, \text{ where, } b_p \text{ is pole arc length.}$$

Usually, the value of $\psi$ varies from 0.65 to 0.72.

For square pole face construction, the pole arc length is equal to the length of machine.

*i.e.,* $b_p = L$

So, for square pole face construction,

$$\frac{b_p}{\tau_p} = \frac{L}{\tau_p} = 0.65 \text{ to } 0.72 \qquad \text{...(7.15)}$$

$$= \psi$$

In long pole construction, the length of the machine is taken up to twice of the width of the pole body ($b$) as shown in Fig. 7.20.

So, $L = b \text{ to } 2b$

$$= (0.45 \text{ to } 1.1) \times \tau_p$$

or $\dfrac{L}{\tau_p} = 0.45 \text{ to } 1.1$

The value of $\dfrac{L}{\tau_p}$ is selected between 0.70 and 0.95.

So, usually $\dfrac{L}{\tau_p} = 0.70 \text{ to } 0.95$ ...(7.16)

But, $\tau_p = \dfrac{\pi D}{p}$

- For square pole face construction, main dimensions $D$ and $L$ can be calculated from equations (7.14) and (7.15).
- For long pole construction, the main dimensions can be calculated by equations (7.14) and (7.16).

Checks: The feasibility of diameter ($D$) and length ($L$) estimated can be checked by

(*i*) Peripheral speed, $V_p = \dfrac{\pi DN}{60}$ to be within specified limit of 15 to 30 m/sec and up to 50 m/sec with special construction.

(*ii*) It can be seen for number of poles as

$$\tau_p = \frac{\pi D}{p}$$

(*iii*) Armature mmf per pole can be taken as

$$AT_a = \frac{\overline{ac} \cdot \tau_p}{2}$$

The method will be more clear while solving the numerical problems for machine design.

## 7.11 VENTILATING DUCTS

The cooling of armature of the machine is essential. The ventilating ducts are provided for cooling of armature. If the length of the armature core exceeds about 12 cm, then the ventilating ducts are provided in the core. Radial ducts are provided along the axial length of the machine. Ventilating ducts of about 0.9 to 1 cm width at every 7–8 cm of axial length of the armature core are being provided in the machine.

Total armature core length estimated = $L$

Gross core iron length, which includes the length of iron and also insulation length between laminations as shown in Fig. (7.17).

Gross iron length $l = l_1 + l_2 + l_3 + l_4 + l_5 + \cdots$

Gross iron length = $l = L - n_v\, b_v$

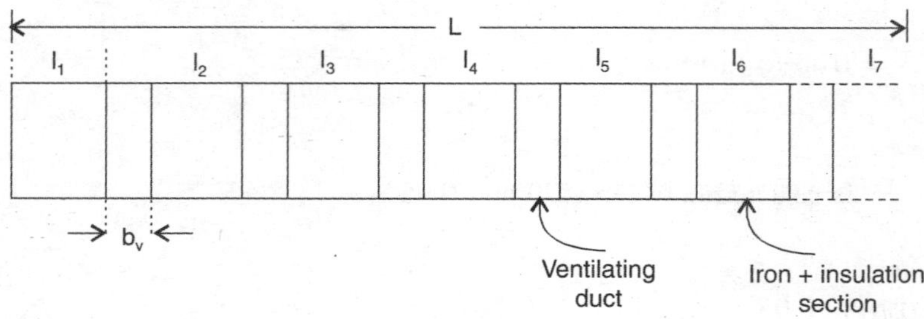


**Fig. 7.17** Gross iron length and ventilating ducts.

where, $n_v$ = number of ventilating ducts, and

$b_v$ = width of ventilating ducts.

Actual iron length in the core,

$$L_i = k_i\, l$$

where, $k_i$ is stacking factor, having values between 0.88 to 0.92.

## 7.12 CALCULATION OF AIR GAP LENGTH

MMF required for air gap for salient pole machine can be expressed as

$$AT_\delta = 800{,}000 \times B_m\, k_c\, \delta \qquad ...(7.17)$$

$k_c$ = gap contraction coefficient having value 1.15.

$\delta$ = air gap length at the centre of pole,

$B_m$ = maximum flux density in air gap $= \dfrac{\overline{B}_{corr}}{\psi}$.

Armature m.m.f. per pole, $AT_a = \dfrac{\overline{ac}}{2} \cdot \tau_p = \dfrac{\overline{ac} \cdot \pi D}{2 \times p}$

The value of air gap m.m.f. is usually between 0.5 and 0.7 of armature m.m.f.

$$AT_\delta = (0.5 \text{ to } 0.7) \times AT_a$$

$$= (0.5 \text{ to } 0.7) \times \frac{\overline{ac} \cdot \pi D}{2 \cdot p} \qquad \text{...(7.18)}$$

From equations (7.17) and (7.18), we get,

Length of air gap, $\delta = \dfrac{(0.5 \text{ to } 0.7)\,\overline{ac} \cdot \pi D}{2p \times 800{,}000 \times B_m\, k_c}$

The air gap length at the pole tip is generally 1.5 to 2 times $\delta$.
*i.e.*, maximum air gap length = (1.5 to 2) $\delta$

## 7.13 DETAILS OF ARMATURE DESIGN

### 7.13.1 Armature Winding

Multiplex windings and simplex windings are done in the d.c. machines. In case of multiplex windings equalizer connections are being used which increases the cost of the machine. So simplex windings are generally used in armature. There are two types of simplex windings *i.e.*, simplex lap winding and simplex wave winding. Some salient points regarding the simplex lap and simplex wave windings are discussed below.

| *Sl. No.* | *Simplex lap winding* | *Simplex wave winding* |
|---|---|---|
| 1. | Number of parallel path = $p$ | Number of parallel path = 2 |
| 2. | Current in each path = $\frac{1}{p}$ (full load current) | Current in each path = $\frac{1}{2}$ (full load current) |
| 3. | E.m.f. in each parallel path = $E$ | E.m.f. in two paths = $E$ |
| 4. | Total number of conductors large | Total number of conductors less |
| 5. | Current per path less | Current per path high |
| 6. | Less conductor cross-sectional area | High cross-sectional area |
| 7. | Equalizer connections needed (More expensive machine) | Equalizer connections not needed |
| 8. | Generally not used for small machine | Generally used for small machine |

With above comparison, it is clear that the wave winding is superior than lap winding but in some cases and above certain limit lap winding is better as compared to wave winding.

- Wave windings are used in the machines which have rated current less than 400 A and above 400 ampere rating lap windings are used.
- Since equalizer connections in lap winding are needed, so windings can be short pitched. This reduces overhang portion of the winding. This reduces weight of machine. This is preferred in machines like fraction motors where weight should be reduced by any means.
- In lap winding generally even coils per slot are used. Eddy current loss may be reduced by using more than one arrangement of the conductors in the slot.

### 7.13.2 Number of Armature Conductors

The total number of conductors on armature can be obtained by the equation as

$$Z = \frac{E \times 60a}{p\phi N} \qquad \text{...(7.19)}$$

Number of parallel path can be known by the type of winding.
Induced e.m.f. for generator, $E = V$ (terminal voltage) + voltage drop
Induced e.m.f. for motor, $E = V$ (supply voltage) – voltage drop

The voltage drop in the armature winding, interpole winding, series field winding and contact drop at brushes may be considered about 2 to 3% of the terminal voltage for large machines and 5 to 8% for medium and small machines. By knowing the voltage drop and voltage rating of the machine, the number of conductor, $Z$, can be calculated with equation (7.19). Field current may be related between 0.4 and 3 per cent of line current.

$$\text{Number of coils for single turn winding} = \frac{Z}{2}$$

$$\text{Number of coils} = \frac{Z}{2T_c} \quad \text{(for multi-turn coils)}$$

where, $T_c$ is the number of turns per coil.

### 7.13.3 Number of Commutator Segments

After estimating the number of coils, the commutator pitch should be calculated to see whether the commutator pitch is within the specified limit.

Number of coils = Number of commutator segment. Peripheral length of commutator segment at outer surface should be 3 to 4 mm.

Mica insulation between adjacent segments = 0.8–1.0 mm. So, pitch of commutator segment = 4 to 5 mm.

Commutator diameter, $D_c = (0.62 \text{ to } 0.75) \times D$
Commutator peripheral length = $\pi D_c$
Let $C$ is the number of commutator segments.

$$\text{Commutator pitch} = \frac{\pi D_c}{C}$$

This should be within the specified limit *i.e.*, between 4 and 5 mm.

### 7.13.4 Factors Effecting the Choice of Number of Slots

**(i) *Slot pitch:*** Let $\tau_s$ is slot pitch and $S$ is the no. of total slots

So, $$\tau_s = \frac{\pi D}{S}$$

Slot pitch varies from 1.5 to 4 cm but usually it is taken between 2.5 and 3.5 cm.

**(ii) *Slot Loading or total ampere-conductor per slot:*** Total ampere-conductor per slot should not exceed 1500 A.

So, $$I_z \cdot N_c \ngtr 1500\,\text{A}$$

where, $N_c$ is the number of conductor per slot and $I_z = \dfrac{I_a}{a}$ is conductor current.

**(iii) *Commutation:*** The number of slots per pole should not be less than 9 (in small machines, it may taken upto 8 since the internal resistance is high in that case) to prevent sparking. For better commutation the number of slots per pole should be 9 to 16.

*i.e.*, $$\frac{S}{p} = 9 \text{ to } 16$$

Number of slots per pole for different machine ratings are given in Table 7.6.

**Table 7.6:** Armature slots per pole

| *Rating (kW)* | *Slots per pole* |
|---|---|
| less than 5 | 8 |
| 5-50 | 10 |
| above 50 | 12 or more |

**(iv) *Flux pulsations:*** The number of slots per pole pair should be an odd integer in order to minimize pulsation losses.

(*a*) Number of slots per pole $= \dfrac{S}{p} = \text{integer} + \dfrac{1}{2}$

(*b*) Number of slots under pole shoe (arc) = integer + $\dfrac{1}{2}$

or Number of slots under pole shoe (arc) = integer

(*v*) Number of slots suitable for winding

Following points should be taken into consideration:

(*a*) D.C. windings are mostly double layer type.

(*b*) Number of slots should be so taken that the number of conductors per slot is an even integer.

(*c*) Number of conductors per slot should be divisible by the number of coil sides per slot to get the number of turns per coil as an integer.

(*d*) For lap winding number of slots mmf be multiple of number of pair of poles.

(*e*) In case of wave winding, number of slots should not be multiple of pair of poles.

Chapter 7

### 7.13.5 Estimation of Corrected Values

From corrected number of conductors, calculate corrected value of flux per pole,

$$\phi_{corr} = \phi \times \frac{\text{Original conductors}}{\text{Corrected conductors}}$$

and $$\overline{B}_{corr} = \frac{\phi_{corr}}{\tau_p \cdot L} \text{ and } \overline{ac}_{corr} = \frac{I \cdot Z_{corr}}{\pi \cdot D}, \text{ then}$$

check for slot loading = $I \cdot N_c$

### 7.13.6 Cross-sectional Area of Armature Winding Conductor

Armature current, $I_a = \dfrac{P_a \times 10^3}{E}$

We have $I_a = I_l + I_f$; for generator

$= I_l - I_f$; in case of motor

where, $I_a$ is the line current and $I_f$ is shunt field current.

We get conductor current as $I_Z = \dfrac{I_a}{a}$

So, conductor cross-sectional area, $F_c = \dfrac{I_Z}{\sigma}$ sq. millimetre

where, $\sigma$ is the current density in armature conductor in A/mm$^2$

### 7.13.7 Selection of Current Density

If current density is higher, then the cross-sectional area is reduced resulting in less cost of copper winding. But due to higher current density and reduced cross-sectional area, the resistance increases and hence copper loss increases. This results in increasing the temperature rise and decrease in efficiency. So the current density should be so optimally selected that it is suitable for efficiency and temperature rise conditions. Higher current density could be taken when the ventilation and cooling of the machines is better since the temperature rise can be controlled by better heat dissipation.

The current density can be taken as below:

$\sigma$ = 4.5 A/mm$^2$, for large strip-wound armature with very good normal ventilation.
$\sigma$ = 5 A/mm$^2$, for small wire wound armature with very good normal ventilation.
$\sigma$ = 6-7 A/mm$^2$, for high speed fan ventilated machine.

d.c.c. or enamel covering insulation are used for round wire conductors having cross-sectional area less than 10 mm$^2$.

For the cross-sectional area of more than 10 mm$^2$, square or rectangular conductors are used insulated with d.c.c. or braided cotton or enamel covering.

### 7.13.8 Limitation on Conductor Size

The depth of single conductor should be limited according to following due to eddy current loss.

(*i*) up to 19 mm when frequency is less than 25 Hz.
(*ii*) up to 15 mm when frequency is more than 25 Hz.
(*iii*) sectional area is limited to 50 mm$^2$ for a frequency of 50 Hz.

### 7.13.9 Insulation of Armature Winding

Nowadays class *E*, *B* and *F* insulations are being preferred in inspite of class *A* to get higher output from given size of the machine. The medium and larger machines are generally designed for class *F* insulation. The details of slot dimension and insulations are shown in Fig. 7.18.

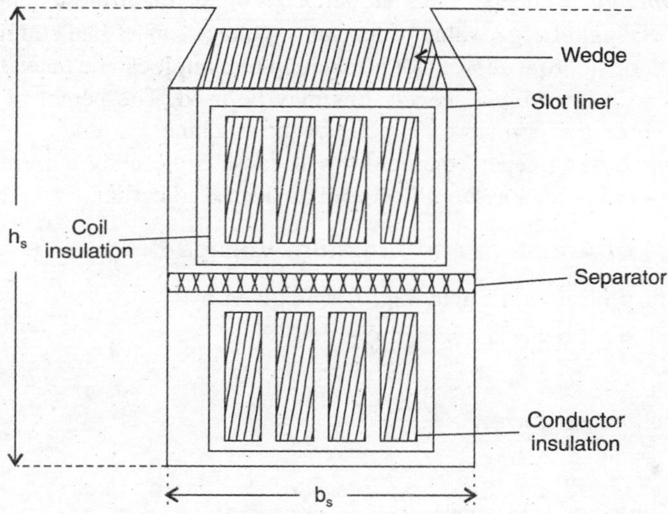


**Fig. 7.18** Slot with insulation.

Earlier the individual conductors were being covered by cotton tape of 0.2 mm thick layer but now enamel covering of about 0.1 mm to 0.15 mm is used. The slot is insulated by slot liner of latheroid paper of 0.2 mm thickness. The medium and large machines have rectangular conductors and double layer winding. Each coil of the machine is insulated by wrapping around the conductors 3 turns of mica wrap of 0.2 to 0.25 mm thickness and one turn of manila paper of 0.15 mm thickness. The wedge on the top of the slot is made of hardwood or bakelite having thickness of 4 to 5 mm to cover the opening of the slot. A separator is used between two layers of the winding. Separator is made of micanite or latheroid of 1.0 mm thick strip.



### 7.13.10 Dimensions of the Slot

(*i*) ***Flux density:*** The size of the slots should be such that the flux density in armature teeth does not increase too much. The flux density at 1/3rd height from top of teeth should not be more than 2.0 Wb/m$^2$. Open slots causes more flux pulsation in the air gap so slots should not be much wider.

**Table 7.7:** Approximate slot height for armature

| *Armature dia. (cm)* | *Slot height (cm)* |
|---|---|
| 15 | 2.20 |
| 20 | 2.70 |
| 25 | 3.20 |
| 30 | 3.70 |
| 40 | 4.25 |
| 50 | 4.50 |

***(ii)* *Reactance Voltage:*** In deeper slots the ampere-conductor of the machine increases and reactance voltage is high. Larger value of reactance voltage causes bad commutation. Generally, the ratio of height of slot to width of slot is taken up to 3 but it can be taken up to 4 in several cases. In machines with interpoles, deeper slots may be used. The height of slot can be seen from Table 7.7 according to armature diameter of the machine.

***(iii)* *Eddy Current Loss:*** Deeper conductors will cause more eddy current loss in conductor. So suitable depth of conductor should be used as discussed earlier.

### 7.13.11 Resistance of Armature and Armature Voltage Drop

The length of mean turn of armature winding, $L_{mt}$, is given by

$$L_{mt} = 2L + 2.3\tau_p + 5h_s \quad \text{...(7.20)}$$

$$\text{Resistance of each conductor} = \frac{1}{2} \times \frac{\rho L_{mt}}{F_c}$$

$$\text{Emperical expression for resistance} = \frac{\rho l}{A}$$

$$\text{Resistance of each parallel path} = \frac{Z}{a} \times \frac{\rho L_{mt}}{2F_c}$$

Since '*a*' paths are connected in parallel.

∴ Resistance of armature, $R_a$ can be expressed as

$$R_a = \frac{1}{a}\left(\frac{Z}{a} \times \frac{\rho L_{mt}}{2F_c}\right)$$

$$= \frac{Z}{2} \times \frac{\rho L_{mt}}{a^2 \cdot F_c}$$

If $L_{mt}$ is in metre and $F_c$ in mm², then resistivity of copper, ρ is 0.021Ω at around 75°C.

**Armature voltage drop:**

Armature voltage drop = $I_a \cdot R_a$ volt

### 7.13.12 Copper Weight and Copper Loss in Armature Winding

Total length of conductor = Length of mean turn × Number of turn

$$= L_{mt} \times \frac{Z}{2}$$

Total volume of copper in armature winding = Cross-sectional area of conductor × Total length

$$= F_c \times L_{mt} \times \frac{Z}{2}$$

$$\text{Weight of copper} = 8900 \times F_c \times L_{mt} \times \frac{Z}{2} \text{ kilogram}$$

$$\text{Armature winding copper loss} = I_a^2 \cdot R_a \text{ watt}$$

### 7.13.13 Height of Armature Core

Half of the useful flux per pole ($\phi$) passes through the armature core. So, flux through the armature core, $\phi_c = \dfrac{\phi}{2}$

Let $h_c$ = height of armature core below the slot, and

$B_c$ = flux density in armature core having values between 1 to 1.5 Wb/m$^2$.

$$\text{So, } \frac{\phi}{2} = \text{Flux density} \times \text{Area}$$

$$= B_c \times h_c \times L_i$$

$$\text{or, } h_c = \frac{\phi}{2B_c L_i}$$

Select suitable value of $B_c$, generally around 1.25 Wb/m$^2$, and calculate $h_c$.

Inner diameter of armature core = $D_i = D - 2(h_s + h_c)$

***Flux Density in the Tooth:*** Flux density in the tooth $= \dfrac{\phi}{\text{Area of teeth per pole}}$

$$\phi = \bar{B}\,\tau_p\,L$$

Number of teeth = Number of slots

So, number of teeth per pole = $\dfrac{S}{p}$

Number of teeth per pole/pole arc,

$$= \frac{b_p}{\tau_p} \times \frac{S}{p}$$

$$= \psi \times \frac{S}{p}$$

Let $b_{t(1/3)}$ is the width of tooth at 1/3rd height from narrow end.

Slot pitch = Peripheral distance between centres of two adjacent slots or peripheral length per slot.

$$\text{Slot pitch at 1/3rd height of tooth} = \frac{\pi D_{(1/3)}}{S}$$

where, $D_{(1/3)}$ = diameter of armature at 1/3rd height from narrow end.

Since slots are considered to be parallel sided given in Fig. 7.19.

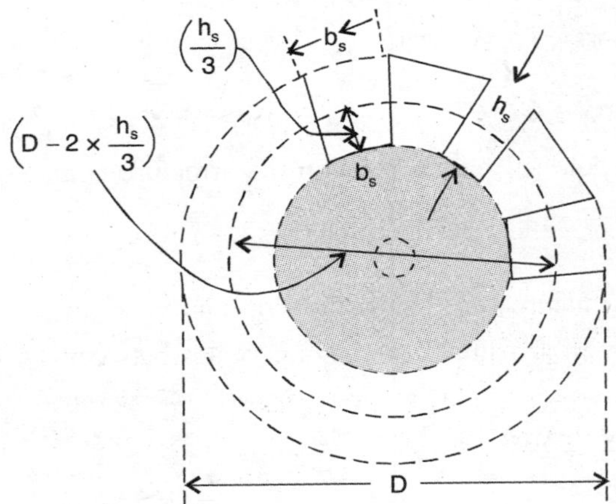


**Fig. 7.19** Detail of tooth and slot.

So,

$$b_{t(1/3)} = \frac{\pi D_{(1/3)}}{S} - b_s$$

where, $b_s$ = width of slot and $D_{(1/3)} = \left[D - \left(2 \times \left(\frac{2}{3}\right) h_s\right)\right]$

The maximum value of flux density at 1/3rd height from narrow end should not exceed 2.2 Wb/m$^2$. This can be estimated by the flux density in teeth at 1/3rd height.

$$B_{t1/3} = \frac{p\phi_{corr}}{\psi \cdot S \cdot l_i \times b_{t1/3}}$$

If flux density is more than specified limit then increase the width of tooth to reduce the flux density in teeth.

## 7.14 ARMATURE TEMPERATURE RISE

(*i*) Outside cylindrical surface = $\pi DL$ square metre

Cooling coefficient = (refer Table 5.1)

$$\text{Loss dissipation} = \frac{\pi DL}{\text{Cooling coefficient}} \text{ watt/°C}$$

(*ii*) Inside cylindrical surface = $\pi D_i L$ square metre

Estimate peripheral velocity for $D_i$ and then estimate cooling coefficient for that peripheral velocity (refer Table 5.1)

$$\text{Loss dissipation} = \frac{\pi D_i L}{\text{Cooling coefficient}} \text{ watt/°C}$$

(*iii*) Cooling surface of one duct and two end surfaces $= 3 \times \frac{\pi}{4}\left(D^2 - D_i^2\right)$ square metre

Velocity of air in ducts = 0.1 × Peripheral velocity ($V_p$)

Calculate cooling coefficient for this velocity of air in ducts then

$$\text{Loss dissipation} = \frac{\text{Cooling surface}}{\text{Cooling coefficient}} \text{ watt/°C}$$

Calculate total loss dissipation = Sum of all loss dissipation

Total loss to be dissipated (in watt) = Copper loss in active portion of winding + Iron loss

$$\text{Temperature rise of armature} = \frac{\text{Total loss to be dissipated (in watt)}}{\text{Total loss dissipation (in watt per °C)}}$$

## 7.15 DESIGN OF FIELD SYSTEM

The design of poles and fields winding includes the calculation of area of cross-section of poles, height of pole, yoke and design of series and shunt field winding.

### 7.15.1 Sectional Area of Pole

Flux in the pole is sum of useful flux per pole and leakage flux.

Total flux in pole,

$$\phi_p = \text{Useful flux} + \text{Leakage flux}$$

$$= \phi + \phi_{el}$$

To determine the total flux per pole the determination of leakage flux is necessary. Leakage flux can be determined depending upon the dimensions of the pole. Total flux per pole can be estimated as

Total flux per pole, $\phi_p$ = Leakage coefficient × $\phi$

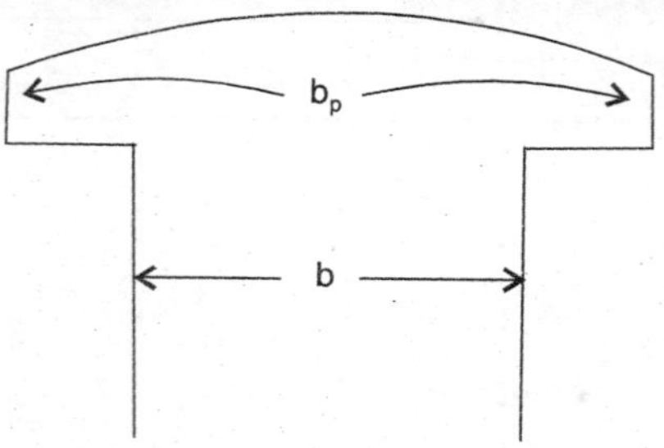


**Fig. 7.20** Pole shoe and core.

The estimation of leakage flux is explained earlier. A suitable value of leakage coefficient can be assumed as given in Table 7.8.

**Table 7.8:** Leakage coefficient for estimation of flux per pole

| *Output (kW)* | *Leakage coefficient* |
|---|---|
| 50 | 1.12 – 1.25 |
| 100 | 1.12 – 1.22 |
| 200 | 1.10 – 1.20 |
| 500 | 1.08 – 1.18 |
| 1000 | 1.06 – 1.16 |
| 1500 | 1.05 – 1.14 |
| 2000 | 1.04 – 1.12 |

Let $B_p$ is flux density in pole body having values between 1.0 and 1.7 Wb/m$^2$.

Flux in pole body,

$$\phi_p = B_p \cdot b \cdot k_i \, l_p$$

– Select leakage coefficient and estimate $\phi_p$.

– Select $B_p$ and take $l_p = L - (1 \text{ to } 1.5 \text{ cm})$.

The axial length of pole is taken less than the machine length to permit end play and to avoid magnetic centering of armature length of pole.

So, $$b = \frac{\phi_p}{B_p \cdot k_i \cdot l_p}$$

where, $k_i$ is the stacking factor, having values between 0.90 to 0.94.

For machines with laminated pole cores, the axial length of poles is equal to the length of armature core *i.e.*, $l_p = L$.

### 7.15.2 Design of Height of Pole

The height of the pole is determined on the basis of the m.m.f. to be provided by the field winding on the pole at full load. So first full load field m.m.f. should be estimated. To estimate full load field m.m.f., we should know the open circuit characteristics (O.C.C.) or magnetization curve of the machine. To know the O.C.C., m.m.f. required by the different portion of the machine (pole height teeth and yoke etc.) in the magnetic circuit. Thus it is seen that full load field m.m.f. and height of pole are related with each other and any one can be estimated only when one of them is known.

To proceed the design of d.c. machine, we assume the value of full load field m.m.f. To nullify the effects of armature reaction, the field should be so designed that m.m.f. developed by the field coils is sufficient enough as compared with m.m.f. developed by the armature at full load.

Generally,

$$\frac{\text{Field m.m.f. at full load}}{\text{Armature m.m.f. at full load}} = \frac{AT_{fl}}{AT_a} = 1.1 \text{ to } 1.25$$

or, $$\frac{AT_{fl}}{AT_a} = 1.1 \text{ to } 1.25, \text{ for non-inter polar machine}$$

$$= 0.65 \text{ to } 1.05, \text{ for inter-polar machine.}$$

## Design of Field Winding

Let $h_f$ = height of field coil in metre

$d_f$ = depth of field winding in metre

$S$ = cooling surface of field coil in metre square

$L_{mtf}$ is the length of mean turn of field coil as given while cooling of field winding is considered, the top and bottom surfaces are neglected. So cooling surface of field winding, $S = 2\,L_{mtf} \cdot h_f$

Permissible copper loss in each field coil,

$$= 2\,L_{mtf} \cdot h_f \cdot q_f$$

where, $q_f$ is permissible loss per unit winding surface for normal temperature rise in watt/m$^2$.

Sectional area of field winding coil = $h_f \cdot d_f$

Actual area of copper in each coil = $k_f \cdot h_f \cdot d_f$

where, $k_f$ is the winding space factor, having values from 0.4 to 0.65 for round coils and from 0.75 to 0.80 for rectangular conductors.

But area of copper in each field coil can also be given by $T_f\,F_f$

where, $T_f$ = number of turns in each field coil,

$F_f$ = sectional area of field conductor in metre square.

So, $$T_f F_f = k_f \cdot h_f \cdot d_f \qquad \text{...(7.21)}$$

Copper loss in each field coil, $Q_f = I_f^2\,R_f$

$$= I_f^2 \times \frac{T_f \cdot \rho \cdot L_{mtf}}{F_f}$$

$$= \left(F_f \cdot \sigma_f\right)^2 \frac{T_f \cdot \rho \cdot L_{mtf}}{F_f}$$

$$= \sigma_f^2 \cdot T_f \cdot \rho \cdot L_{mtf} \cdot F_f \qquad \text{...(7.22)}$$

Volume of copper in each field coil = Sectional area × Length of mean turn × Number of turns

$$= F_f \cdot L_{mtf} \cdot T_f$$

Copper loss in each field coil, $Q_f = \sigma_f^2 \rho \times$ Volume of copper

*i.e.*, if the volume of copper is fixed, then the copper loss is proportional to square of current density.

From equations (7.21) and (7.22)

$$Q_f = \sigma_f^2 \rho L_{mtf} k_f h_f \cdot d_f \qquad \text{...(7.23)}$$

The copper loss should not exceed the permissible limit of copper loss given by 2 $L_{mtf} \cdot h_f \cdot q_f$ and equating this to equation (7.23)

$$2 L_{mtf} \cdot h_f \cdot q_f = \sigma_f^2 \rho L_{mtf} \cdot k_f \cdot h_f \cdot d_f$$

or $$\sigma_f = \sqrt{\frac{2 q_f}{\rho \cdot k_f \cdot d_f}}$$ where, $\sigma_f$ is the current density. ...(7.24)

$$= 3.0 \text{ A/mm}^2$$

m.m.f. per metre height of field winding $= \dfrac{AT_{fl}}{h_f}$

$$= \frac{I_f T_f}{h_f}$$

$$= \frac{\sigma_f F_f \cdot T_f}{h_f}$$

Putting the value of $F_f T_f$ from equation (7.21), we get

$$= \frac{\sigma_f \cdot k_f \cdot h_f \cdot d_f}{h_f}$$

$$= \sigma_f \cdot k_f \cdot d_f$$

$$= k_f \cdot d_f \times \sqrt{\frac{2 q_f}{\rho \cdot k_f \cdot d_f}} \qquad \left[ \because \sigma_f = \sqrt{\frac{2 q_f}{\rho \cdot k_f \cdot d_f}} \text{ from (7.24)} \right]$$

$$= \sqrt{\frac{2 q_f \cdot k_f \cdot d_f}{\rho}}$$

where, $\rho = 2 \times 10^{-8}\,\Omega$-m, $q_f$ in W/m$^2$, $d_f$ in metre.

$$\text{m.m.f. per metre height of field} = 10^4 \times \sqrt{q_f \cdot k_f \cdot d_f} \qquad \text{...(7.25)}$$

The value of $q_f$ is generally taken as 700 W/m$^2$. The depth of the winding can be selected according to Table 7.9.

By selecting suitable value of $k_f$, $q_f$ and $d_f$, the height of field winding can be estimated.

**Table 7.9:** Depth of field winding ($d_f$)

| *Armature diameter (cm)* | *Winding depth (cm)* |
|---|---|
| 20 | 3.0 |
| 35 | 3.5 |
| 50 | 4.0 |
| 65 | 4.5 |
| 100 | 5.0 |
| above 100 | 5.5 |

Height of field winding,

$$h_f = \frac{AT_{fl} \times 10^{-4}}{\sqrt{q_f\, k_f d_f}} \qquad \text{...(7.26)}$$

Total height of pole,

$h_p = h_f$ + Height of pole shoe + Height of insulation and height wasted due to curvature of yoke.

But, height of pole shoe = (0.10–0.20)$h_p$ and

height of insulation and space wasted due to curvature = (0.10–0.15)$\tau_p$.

### 7.15.3 Yoke Design

The flux in the yoke is the sum of leakage flux and half of the useful flux per pole. The leakage coefficient for yoke is slightly higher than that for pole body. The height of yoke are determined on the basis of flux contained by the flux in yoke ($\phi_y$).

$$\phi_y = \frac{\phi}{2} + \text{Leakage flux}$$

or $\quad \phi_y = \text{Leakage coefficient} \times \dfrac{\phi}{2}$

Let $B_y$ = Flux density in yoke

= 1.0 to 1.2 Wb/m$^2$, for cast steel yokes and

= 1.4 to 1.55 Wb/m$^2$, for laminated yokes

Chapter 7

Let, $h_y$ = Height of pole yoke, $L_{yi}$ = Axial iron length of yoke

$\therefore \quad \phi_y = h_y \cdot L_{yi} \cdot B_y,$ and $L_{yi} = k_i \, l$

$$h_y = \frac{\phi_y}{B_y \cdot L_{yi}}$$

$L_{yi}$ = $L$ for cast steel yoke,
$L_{yi}$ = Actual iron length in yoke for machine with laminated yoke.

Outer diameter of yoke, $D_y = D + 2(\sigma + h_p + h_y)$.

## 7.16 DESIGN OF SERIES FIELD WINDING

The series field ampere-turn per pole

$$AT_{flse} = (0.15 \text{ to } 0.20) \times AT_a$$

Since series field current, $I_{se} = I_a$
So, series field turns,

$$T_{se} = \frac{AT_{flse}}{I_{se}}$$

$\therefore$ Sectional area of series field winding,

$$F_{se} = \frac{I_{se}}{\sigma_{se}}$$

where, $\sigma_{se}$ is the current density in series field winding. Current density in case of series field winding can be taken slightly more than that in shunt field winding. Cooling is better in this case since it is placed over the shunt field winding, $\sigma_{se}$ = up to 4 $\text{A/mm}^2$.

## 7.17 MMF OF THE MAGNETIC CIRCUIT AND OCC

The different parts of the machine in the magnetic circuit are armature core, armature teeth, pole, yoke and air gap. MMF per pole pair for any portion can be estimated by knowing magnetic flux density in that portion and then corresponding ampere-turn per metre length of that portion from *B-H* curve. The ampere-turn per metre length of that portion multiplied by the length of that portion will give the ampere-turn required by that portion. The different portion of the magnetic circuit are shown in Fig. 7.2.

***(i) MMF for Air Gap:***

$$AT_\delta = 800{,}000 \; B_m \, k_c \cdot \delta$$

where, $k_c$ is the gap contraction factor and $\delta$ is the air gap length.

Gap contraction factor is estimated for slots as well as for ducts. The detail of this is given in overall design problem at the end of this chapter.

***(ii) MMF for Pole Body:***

$B_p$ = Flux density in pole

$at_p$ *i.e.*, m.m.f. per metre length of pole body can be estimated corresponding to $B_p$ from *B-H* curve.

MMF for pole body, $AT_P = at_p \times h_p$

***(iii) MMF for Yoke:***

Estimate, $at_y$ = MMF per metre length of yoke corresponding to $B_y$ from *B-H* curve.

$B_y$ = Flux density in yoke.

MMF required for yoke, $AT_y = at_y - l_y$

where, $l_y$ is the length of flux path in yoke and is given by

$$= \frac{\pi(D + 2\delta + 2h_p + h_y)}{p}$$

***(iv) MMF for teeth:***

$at_t$ = MMF per metre length of teeth corresponding to flux density, $B_{t(1/3)}$, in tooth at 1/3rd height from narrow end of tooth.

MMF for teeth, $AT_t = at_t \times h_t$

where, height of tooth $h_t$ = height of slot, $h_s$.

***(v) MMF for armature core:***

$at_c$ = MMF per metre length of core corresponding to flux density in core ($B_c$) for *B-H* curve.

MMF for armature core, $AT_c = at_c \times l_c$

where, $l_c$ is the length of flux path in armature core and is given by

$$= \frac{\pi(D - 2h_s - h_c)}{p}$$

Total m.m.f. per pole at no load,

$$AT_0 = AT_\delta + AT_p + AT_y + AT_t + AT_c \quad \text{...(7.27)}$$

## Open Circuit Characteristics (O.C.C.)

The e.m.f. induced is proportional to flux if the speed of the machine is constant. To obtain O.C.C. or magnetization curve, m.m.f. per pole is required to establish different values of useful flux. Generally, the variation of flux is done between 70 to 120 per cent of the flux at rated voltage. Below 70 per cent of the useful flux at rated voltage, the reluctance of iron portion is negligible with respect to the reluctance of air gap hence O.C.C. curve below 70 per cent becomes straight line. MMF per pole are generally calculated from 70 to 120 per cent of rated voltage and magnetization curve is plotted from the result.

## 7.18 DESIGN OF SHUNT FIELD WINDING

Varnish covered wire conductors coils of few hundred turns are used for shunt field winding in small machines. The insulated conductors are housed in an epoxy resin moulding and fixed with pole. Rectangular conductors are used for large machines. Sectionalized coils are used in the machines with fan ventilation. In sectionalized coils, the coils are wound in sections inspite of having conductors on one piece. Different sections are separated by a distance and so the ventilation spaces are available which results in improved ventilation.

The entire space along the height of the pole is taken by the winding in shunt machines. In compound machines, 80 per cent of the winding space is taken by the shunt field winding and 20 per cent by series field winding.

Following steps are adopted to design a shunt field winding:

(*i*) The voltage across the shunt field winding is taken equal to 80 to 85% of terminal voltage. This is done to allow for voltage regulation. In generators, 15 to 20 per cent of rated voltage is considered for the field rheostat. In field control motors this margin depends upon the range of speed control.

$\therefore$ Voltage across shunt field winding = (0.80 to 0.85) volt

$$\text{Voltage across each shunt field coil, } E_f = \frac{(0.80 \text{ to } 0.85)}{p} \text{ volt} \qquad ...(7.28)$$

Since the number of shunt field coils are equal to the number of poles and they all are connected in series.

(*ii*) Select suitable value of winding depth, $d_f$

(*iii*) Estimate length of mean flux,

$$L_{mt} = 2\left(L_p + b + d_f\right)$$

$$= 2\left(L_p + b + d_f\right) \qquad ...(7.29)$$

(*iv*) Resistance of each field coil,

$$R_f = \frac{T_f \cdot \rho \cdot L_{mt}}{F_f} = \frac{E_f}{I_f}$$

Sectional area of shunt field conductor,

$$F_f = \frac{I_f T_f \cdot \rho L_{mt}}{E_f}$$

$$= \frac{AT_{fl} \cdot \rho \cdot L_{mt}}{E_f} \qquad ...(7.30)$$

From equation (7.30), sectional area of conductor is estimated. With this sectional area suitable cross-section of conductor can be chosen from Table A.1. Rectangular conductors must be taken for longer sectional area.

(*v*) Winding height

$h_f$ = Height of pole – Height of shoe – Height of insulation –Clearance.

$= h_p - (0.10 \text{ to } 0.20) \times h_p - (0.10 \text{ to } 0.15) \times \tau_P$

(*vi*) Number of turns, $T_f = \dfrac{k_f \cdot h_f \cdot d_f}{F_f}$

where, $k_f$ is the winding space factor depending upon type of winding whether round wire or rectan-gular.

(*vii*) Resistance of each field coil,

$$R_f = \frac{T_f \, \rho \, L_{mt}}{F_f}$$

Shunt field current, $I_f = \dfrac{E_f}{R_f}$

Check for current density.

(*viii*) Calculate m.m.f.,

$$AT_{fl} = I_f \cdot T_f$$

If it is less than the required value of m.m.f. then increase the winding depth and if it is more the winding depth should be decreased.

(*ix*) Ohmic loss in each field coil = $I_f^2 \, R_f$

or $$Q_f = I_f^2 \, R_f$$

(*x*) Cooling surface of each coil,

$$= 2 \, L_{mt} \left( h_f + d_f \right)$$

Cooling coefficient, $C = \dfrac{0.14 \text{ to } 0.16}{1 + 0.1 \, V_p}$

refer to Table 5.1

(*xi*) Temperature rise of the coil $= \dfrac{Q_f \cdot C}{\text{Cooling surface of each coil}}$

$$= \frac{Q_f \cdot C}{2 \, L_{mt} \left( h_f + d_f \right)}$$

If temperature rise is more than the specified limit, then increase the winding depth.

## 7.19 DESIGN OF COMMUTATOR

The commutator is constructed with hard drawn copper segments separated by thin sheets of mica or micanite separator. The details of commutator are given as under.

(*i*) Generally commutator pitch should be more than 4 mm.
(*ii*) Thickness of surface 3.2 mm and 0.8 mm mica separator.
(*iii*) Commutator diameter = 0.6 to 0.8 times armature diameter.
(*iv*) Peripheral speed around 15 m/sec.
(*v*) Number of commutator segments = Number of coils.
(*vi*) The length depends upon space required by the brushes, the number of brushes per arm and surface area sufficient to dissipate the heat produced by the losses in the commutator. The minimum number of segments is that which gives a voltage of 15 V between segments at no load and can be given by $\frac{E.p}{15}$.

***Design of Brushes:*** Different materials like carbon, carbon graphite, metal graphite are being used for brushes. The current density in the brushes ($\sigma_b$) depends upon the material of the brush.

$$\sigma_b = 0.08 \text{ to } 0.20 \text{ A/mm}^2$$

$$\text{Current carried by each brush spindle} = \frac{2I_a}{p}$$

Each spindle (brush arm) may have many brushes. Generally, the area of each individual brush should be so chosen so that it does not carry a current of more than 70 A. It is better to use a large number of brushes of smaller width. The commutator and brushes are shown in Figs. 7.11 and 7.12.

Total brush contact area per arm,

$$A_b = \frac{2I_a}{p \cdot \sigma_b} \qquad ...(7.31)$$

Number of brushes per arm,

$$n_b = \frac{A_b}{w_b \cdot t_b} \qquad ...(7.32)$$

where, $w_b$ is the width of one brush and
$t_b$ is the thickness of one brush.

Contact area of each brush, $a_b = w_b\, t_b$

$w_b$ and $t_b$ should be selected from Table 7.10

**Table 7.10:** Different sizes of brushes

| *Width, $w_b$ (mm)* | *Length (mm)* | *Thickness, $t_b$ (n* |
|---|---|---|
| 16 | 30 | 6, 8, 16 |
| 24 | 35 | 9, 15, 17 |
| 30 | 42 | 12. 16, 18, 23, 26 |

Length of commutator excluding risers,

$$L_c = n_b(w_b + C_b) + C_1 + C_2 \quad ...(7.33)$$

where, $C_b$ = clearance between the brushes, usually 4 to 5 mm,

$C_1$ = clearance allowed for staggering the brushes,

= usually 10 mm for small machines and 30 mm for large machines.

and $C_2$ = clearance for allowing the end play

= 10 to 25 mm.

Overall length of commutator $= L_c + 20$ mm for risers.

***Commutator Losses:*** There are brush contact losses and the brush friction losses at the commutator. Brush contact loss, $P_{bc}$ = Voltage drop per brush set × $I_a$ (arm. current) voltage drop per brush arm set *i.e.*, one +ve and one – ve brush arm is 2 V for carbon/graphite brushes. So, $P_{bc}$ = 2 × Armature current = 2 × $I_a$. The brush contact drop for different material is given in Table 7.11.

**Table 7.11:** Brush contact drop for different material

| *Different materials* | *Brush contact drop (V)* | *Current density (A/mm²)* | *Pressure (kN/m²)* | *Commutator speed (m/s)* | *Coeff. of friction* |
|---|---|---|---|---|---|
| Natural graphite | 0.7–1.2 | 0.1 | 14 | 50–60 | 0.1–0.2 |
| Hard carbon | 0.7–1.8 | 0.065–0.085 | 14–20 | 20–30 | 0.15–0.25 |
| Electro graphite | 0.7–1.8 | 0.085–0.11 | 18–21 | 30–60 | 0.1–0.2 |
| Metal graphite | 0.4–0.7 | 0.1–0.2 | 18–21 | 20-30 | 0.1–0.2 |

Brush friction loss, $P_{bf}$ can be calculated by

$$P_{bf} = \mu P_b \, p A_b \, V_c$$

where, μ is the coefficient of friction having values from 0.10 to 0.25 and $P_b$ is brush contact pressure on commutator, measured in N/m².

Total contact area of all brushes = $p.\ A_b$

$V_c$ = peripheral speed of commutator in, m/sec.

The coefficient of friction basically depends upon the peripheral speed of commutator, $V_c$. This decreases at high speeds.

***Temperature rise of commutator:*** Commutator temperature rise can be calculated as below:

Total commutator loss = $P_{bc} + P_{bf}$

Commutator cooling surface $= \pi D_c \cdot L_c$

where, $D_c$ = commutator diameter = (0.6–0.8)$D$

$L_c$ = length of commutator

$$\text{Commutator temperature rise} = \frac{\text{Total loss dissipated from commutator} \times \text{Cooling coefficient}}{\text{Commutator cooling surface}}$$

$$= \frac{(P_{bc} + P_{bt}) \times \text{Cooling coefficient}}{\pi D_c L_c}$$

Cooling coefficient can be calculated by Table 5.1.

## 7.20 LENGTH AND DESIGN OF INTERPOLES

### 7.20.1 Length of Inter Poles

Inter poles are constructed with cost steel or punched sheet steel. There is no pole shoe in the inter poles *i.e.*, the length and width of pole body is equal to the length and width of pole shoe. The role of commutator is to make the current unidirectional obtained from armature winding where current changes its direction as conductor pass through poles of opposite polarity. The current is collected by the brushes which may cover two to three segments of commutator. During commutation, segments are short- circuited by the brush and during a short period between closing and opening of the short-circuit, the current in the coil change from a steady value of one polarity (+I) to the steady value of other polarity (–I). This change of current is opposed by the self induced e.m.f. and is called as reactance voltage.

To reverse the current in the coil during commutation, reversing e.m.f. should be developed in the coil sufficient to overcome the self induced e.m.f. The reversing e.m.f. may be produced by giving brushes lead or interpoles or commutating poles between the two adjacent poles. In generator, the polarity of the inter pole is same as the main pole just ahead and in motor the polarity of the interpole is same as that of main pole just behind.

MMF required at interpoles should be sufficient to neutralize the effect of armature reaction and also to overcome the reactance voltage due to commutation. The length of interpole is so selected that length of mean turn is shorter. Less copper will be required and hence it will be economical, less losses and leakage flux. The flux density in the interpoles should be below the saturation level. This is required to achieve the proper compensation of reactance voltage at all loads. For larger laminated machines with variable loads, the length of interpole is taken equal to the length of main pole. But we can take length of interpole $L_{ip}$ = (0.50 to 0.70) · $L$

### 7.20.2 Design of Interpoles

Armature m.m.f. per pole, $AT_a = \frac{I_z \cdot z}{2p}$

MMF required for a flux density ($B_{mi}$) at interpolar air gap,

$$= 800{,}000\, B_{mi}\, k_{ci}\, \delta_i$$

where, $k_{ci}$ = gap contraction coefficient at inter pole

$\delta_i$ = air gap length at interpole

= (1.5 to 1.75) $\delta$

MMF required to overcome armature reaction,

$$= \frac{I_z \cdot Z}{2p} \text{ for machines with no compensating winding}$$

$$= (1-\psi)\frac{I_z \cdot Z}{2p} \quad \text{for machines with compensating winding}$$

where, $\psi = \dfrac{b_p}{\tau_p} = \dfrac{\text{Pole arc}}{\text{Pole pitch}}$

$\therefore$ MMF required for interpole.

$$AT_i = 800{,}000 B_{mi} k_{ci} \cdot \delta_i + \frac{I_z \cdot Z}{2p} \quad \text{for machines with no compensating winding.}$$

$$= 800{,}000\, B_{mi} k_{ci} \delta_i + (1-\psi)\frac{I_z \cdot Z}{2p} \quad \text{for machines with compensating winding.}$$

The width of the interpole, $b_i$ is equal to distance through which the armature moves during commutation of the complete slot group.

$$b_i = \left[\text{Brush thickness} + (N'-1) \times \text{Commutator segment pitch}\right] \times \frac{D}{D_c}$$

where, $N'$ is the number of coil sides per layer in the slot.

The reactance voltage can be obtained by calculating the permeance coefficient for slot, tooth and overhang leakage.

Permeance coefficient for slot can be given by the expression,

$$\lambda_s = \left[\frac{h_1}{3b_s} + \frac{h_2}{b_s} + \frac{2h_3}{b_s + b_0} + \frac{h_4}{b_0}\right] \quad \text{(refer Fig. 7.21)}$$

Permeance coefficient for tooth top,

$$\lambda_t = \frac{b_i}{6 \cdot \delta_i}$$

Permeance coefficient for overhang portion,

$$\lambda_0 = \frac{L_0}{L}\left(0.23 \log_{10} \frac{L_0}{b_e} + 0.07\right)$$

where, $L_0$ is the length of overhang and $b_e$ is the periphery of one complete layer of the winding.

Total permeance coefficient,

$$\lambda = \lambda_s + \lambda_f + \lambda_0$$

The effective value of leakage flux established for a slot containing $Z_s$ conductors each carrying a current of $I_z$ is

$$= \text{m.m.f. per slot} \times \mu \times \lambda \times \text{Length}$$

$$= I_z \cdot Z_s \cdot \mu \cdot \lambda \cdot L$$

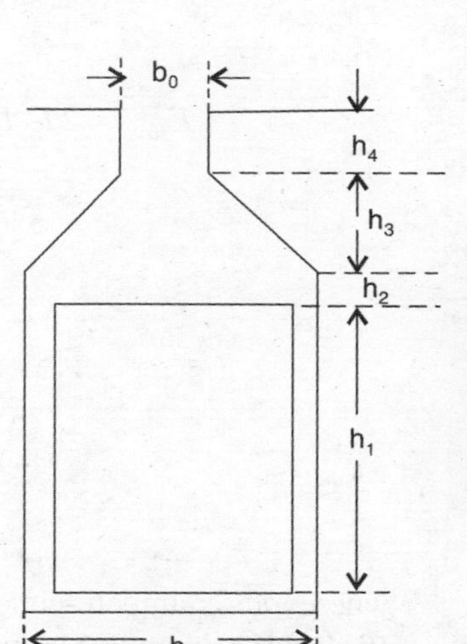

**Fig. 7.21** Parallel sided semi-closed slot.

Chapter 7

Since there are two sides per coil, the total flux linked with a coil $= 2\mu\lambda L I_z \cdot Z_s$
Flux density in the interpolar air gap,

$$B_{mi} = \frac{2 \cdot I_z \cdot Z_s \cdot \mu\lambda L}{L_{ip} \cdot b_i}$$

Number of interpolar turns,

$$T_i = \frac{AT_i}{I_a}$$

Current density in interpolar winding may be taken from 2.5 to 4 A/mm$^2$.
Pitchel Mayer's equation can also be used to find the reactance voltage in a coil.
The average reactance voltage in a coil,

$$E_{rave} = L \cdot \frac{di}{dt}$$

$$= 2T_c^2 \cdot L \cdot \frac{2I_z}{t_c} \cdot \mu\lambda$$

where, $T_c$ is the number turns in a coil and $t_c$ is time of commutation.
Let us suppose that brush thickness is equal to the commutator segment.
So,

$$t_c = \frac{\pi D_c}{n_c \times \pi D_c \cdot n}$$

where, $n_c$ is the number of commutation segments,

$$n = \text{r.p.s}, \; 2n_c \cdot T_c = Z$$

So,

$$E_{rave} = 2T_c^2 L \cdot \mu\lambda \times 2I_z \cdot n_c \cdot n$$

$$= 2T_c L \cdot \mu\lambda \cdot I_z \cdot (2n_c \cdot T_c) \cdot n$$

$$= 2T_c L \cdot \mu\lambda \cdot I_z \cdot Z \cdot n$$

$$= 2T_c L \cdot \mu \cdot \lambda \cdot \left(\frac{I_z \cdot Z}{\pi D}\right) \cdot \pi D n$$

$$= 2T_c L \cdot \mu \cdot \lambda \cdot \overline{ac} \cdot V_p$$

where $\overline{ac}$ is ampere-conductor per metre length of periphery and $V_p$ is the peripheral velocity.
Resistance of interpole winding at 75°C,

$$= \frac{\text{No. of poles} \times \text{Turns on each pole} \times L_{mt} \times \rho}{\text{Sectional area of conductor}}$$

## 7.21 COMPENSATING WINDING

Armature reaction causes distortion of the field form. It is desirable to prevent the distortion of the field form and for this the effect of armature reaction should be neutralized over the pole pitch. Compensating winding of concentric coil type are used for this purpose. The coils span along the interpolar space and hence their magnetization axis coincides with neutral axis. The coil sides are placed in axial slots in the pole shoes and carry currents in opposite to that in armature. The total ampere-conductor per pole for the compensating winding should be equal to the ampere-conductor of the armature of the position under pole shoe.

MMF for compensating winding,

$$AT_c = \frac{I_z \cdot Z}{2p} \times \frac{\text{Pole arc}}{\text{Pole pitch}} = \frac{I_z \cdot Z}{2p} \cdot \psi$$

## 7.22 LOSSES AND EFFICIENCY

Losses in d.c. machine should be estimated to know the efficiency and temperature rise in the machine. There are losses given below:

(*i*) Ohmic losses ($I^2R$ loss).
(*ii*) Rotational losses.

Rotational losses consist of iron losses, friction and windage losses.

***I*²*R* Losses:** This loss is occurring in

(*i*) Armature winding.
(*ii*) Interpolar winding.
(*iii*) Series field winding.
(*iv*) Compensating winding
(*v*) Brush contacts

and in short field circuit, the losses occurs in the winding and in the field regulator.

***Rotational Losses – Iron Loss:*** The iron loss per kg weight of iron is given by the following expression:

$0.06 f B_m^2 + 0.008 f^2 B_m^2 t^2$ ; for teeth of machine.

and $0.06 f B_m^2 + 0.005 f^2 B_m^2 t^2$ ; for core of the m/c.

where, $B_m$ is taken in Wb/m$^2$ and $f$ is in Hz and thickness of lamination, $t$ is in mm.

The pulsation loss in the pole faces should be taken as 25 to 50 per cent of the sum of iron losses in core and teeth.

***Friction and Windage Loss:*** The friction losses occur in bearings and commutator. The value of bearing friction losses depends upon the pressure in bearing, the peripheral speed of the shaft at bearing and the coefficient of friction between the bearing and the shaft. The windage losses produced by rotation depends upon peripheral speed of the rotor, the rotor diameter, the core length and on the construction of machines. Friction and windage losses are shown in Table 7.12.

**Table 7.12:** Bearing friction and windage losses

| *Peripheral speed (m/sec)* | *Losses as % of output* |
|---|---|
| 10 | 0.2 |
| 20 | 0.4 |
| 25 | 0.5 |
| 30 | 0.6 |
| 35 | 0.8 |
| 40 | 0.9 |
| 50 | 1.2 |

***Stray Load Losses:*** Some losses are occurring in machine which cannot be calculated but it appears when the machine is loaded. This type of loss is called as stray load losses. This loss is due to large iron loss, eddy current and when a coil is under commutation. The stray load losses may be assumed as

= 0.5 per cent of rated output for machines with compensating winding.

= 1.0 per cent of rated output for machines without compensating winding.

## Efficiency

$$\text{Efficiency, } \eta = \frac{\text{Output}}{\text{Input}} = \frac{\text{Output}}{\text{Output} + \text{Losses}}$$

$$= 1 - \frac{\text{Losses}}{\text{Input}}$$

***Condition for Maximum Efficiency:*** For the machines having constant speed and constant voltage, the losses can be divided into following groups:

(*i*) Constant losses ($P_1$): These losses do not change with change in load current. In shunt machine, such losses are bearing friction and windage, brush friction, shunt excitation losses and no load iron losses.

(*ii*) Losses proportional to armature current ($P_2$): Such losses varies linearly with load current like brush contact loss.

(*iii*) Losses proportional to square of current ($P_3$): In shunt machines such losses are $I^2R$ losses and stray load losses.

Sum of all the losses,

$$P_t = P_1 + P_2 + P_3$$

Writing in terms of proportionality constants, we get

$$P_t = k_1 + k_2 I + k_3 I^2$$

where, $I$ is the current drawn by the armature of motor.

The motor input power = $VI$

So, $\eta = 1 - \dfrac{k_1 + k_2 I + k_3 I^2}{VI}$

$$= 1 - \left(\frac{k_1'}{I} + k_2' + k_3' \cdot I\right)$$

where, $k_1' = \dfrac{k_1}{V}$, $k_2' = \dfrac{k_2}{V}$ and $k_3' = \dfrac{k_3}{V}$.

Now, $\eta$ will be maximum when $\dfrac{d\eta}{dI} = 0$

By differentiating the above equation for $\eta$, we get,

$$k_1' = k_3' I^2$$

or $\dfrac{k_1}{V} = \dfrac{k_3' I^2}{V}$

or $k_1 = k_3 I^2$

So, the condition for maximum efficiency in the machine is that the constant losses should be equal to the losses proportional to square of current.

## SOLVED PROBLEMS

Solved problems on main dimensions, no. of poles, armature turns, air gap length, field winding etc., of d.c. machine and overall design of a d.c. generator are being presented in this section.

**Q. 1** Calculate the main dimensions of a 12 kW, 220 V, 4-pole, 1200 r.p.m. shunt generator. The specific electric loading is 18,000, average flux density is 0.46 Wb/m$^2$, full load efficiency is 0.86 and pole arc to pole pitch ratio is 0.68. The length of pole arc is equal to the length of the armature core. Friction and windage losses are neglected.

### Solution:

Since F&W losses are neglected. So, armature power,

$$P_a = \frac{P}{\eta} = \frac{12}{0.86} = 13.95 \text{ kW}$$

Speed, $n = \dfrac{1200}{60} = 20$ r.p.s.

Output coefficient, $C = \pi^2 \overline{B}\,\overline{ac} \times 10^{-3}$

$$= (3.14)^2 \times 0.46 \times 18{,}000 \times 10^{-3}$$

$$= 81.6$$

Chapter 7

Output equation, $P_a = CD^2Ln$

or $$D^2L = \frac{P_a}{Cn} = \frac{13.95}{81.6 \times 20}$$

$$= 0.0085 = 8.5 \times 10^{-3}$$

Also, $$b_p = L \text{ (given)}$$

So, $$\frac{L}{\tau_p} = 0.6$$

or $$L = 0.68 \times \frac{\pi D}{p}$$

For $p = 4$,

$$L = 0.53\ D$$

$\therefore$ $$0.53D^3 = 8.5 \times 10^{-3}$$

$$D^3 = \frac{8.5 \times 10^{-3}}{0.53} = 16 \times 10^{-3}$$

or $D = 0.25$ m

and $L = 0.13$ m **Ans.**

**Q. 2** Estimate the diameter and the length of armature core of a 12 kW, 220 V, 4-pole, 1200 r.p.m. shunt generator. The specific electric loading is 18,000, average flux density is 0.46 Wb/m$^2$, full load efficiency is 0.86 and long pole construction is adopted. The length to pole pitch ratio is 0.92. Friction and windage losses are neglected.

**Solution:**

Armature power, $$P_a = \frac{P}{\eta} = \frac{12}{0.86} = 13.95$$

Speed, $$n = \frac{1200}{60} = 20 \text{ r.p.s}$$

Output coefficient,

$$C = \pi^2 \overline{B}\,\overline{ac} \times 10^{-3}$$

$$= \pi^2 \times 0.46 \times 18{,}000 \times 10^{-3}$$

$$= 81.6$$

From output equation, $P_a = CD^2Ln$

$$D^2L = \frac{P_a}{Cn} = \frac{13.95}{81.6 \times 20}$$

$$= 0.0085$$

For long pole construction,

$$\frac{L}{\tau_p} = 0.92 \text{ (given)}$$

or $$L = 0.92 \times \frac{\pi D}{p}$$

$$= 0.72D$$

∴ $$0.72 \times D^3 = 8.5 \times 10^{-3}$$

or $D = 0.22$ m

and $L = 0.16$ m **Ans.**

**Q. 3** Calculate diameter and length of armature core of 70 kW, 240 V, 900 r.p.m., 4-pole d.c. shunt generator. The average air gap flux density is 0.70 Wb/m² and $\overline{ac}$ is 34,000. Core length to pole pitch ratio is 0.80. Full load armature voltage drop is 9.6 V and field current is 3.0 ampere.

**Solution:**

Output coefficient, $C = \pi^2 \overline{B}\,\overline{ac} \times 10^{-3}$

or, $$C = 9.86 \times 0.70 \times 34{,}000 \times 10^{-3}$$

$$= 234.66$$

Speed, $$n = \frac{900}{60} = 15 \text{ r.p.s.}$$

$$E = 240 + 9.6 = 249.6 \text{ V}$$

Full load current $$= \frac{70 \times 1000}{240}$$

$$= 291.6 \text{ A}$$

Armature current, $I_a$ $= 291.6 + 3.0$

$$= 294.6 \text{ A}$$

Power developed by armature, $P_a = E \times I_a \times 10^{-3}$

$$= 294.6 \times 294.6 \times 10^{-3} \qquad \left[\because \; E = 249.6 \text{ V and } I_a = 293.6 \text{ A}\right]$$

$$= 73.28 \text{ kW}$$

From output equation, we have

$$D^2 L = \frac{P_a}{Cn} = \frac{73.28}{234.66 \times 15}$$

$$= 0.02 \text{ m}^3$$

Also, $\dfrac{L}{\tau_p} = 0.80$ (given)

or $L = 0.80 \times \dfrac{\pi D}{p}$

$= 0.63D$

$\therefore$ $0.63\,D^3 = 0.02$

$D^3 = \dfrac{0.02}{0.63} = 0.0317$

or $D = 0.316$ m

and $L = 0.199$ m **Ans.**

**Q. 4** Calculate the main dimensions of 100 kW, 300 V, 900 r.p.m., 6-poles d.c. shunt generator. The maximum air gap flux density is 0.90 Wb/m² and specific electric loading, $\overline{ac}$ is 36,000. Square pole body construction is preferred and pole arc to pole pitch ratio is 0.69. Full load armature voltage drop is 9.6 voltage and field current is 3.3 ampere.

**Solution:**

Given, $B_m = 0.90$ Wb/m²

$$\frac{\overline{B}}{B_m} = \psi \Rightarrow \overline{B} = \psi B_m, \text{ and } \psi = \frac{b_p}{\tau_p}$$

So, output coefficient, $C = \pi^2 \overline{B}\, \overline{ac} \times 10^{-3}$

$$= \pi^2 \psi B_m \overline{ac} \times 10^{-3}$$

$$= \pi^2 \times 0.69 \times 0.90 \times 36{,}000 \times 10^{-3}$$

$$= 220.42$$

Speed, $n = \dfrac{900}{60} = 15$ r.p.s.

$E = 300 + 9.6 = 309.6$

Full load current $= \dfrac{100 \times 1000}{300} = 333.33$

Armature current, $I_a = 333.33 + 3.3$

$= 336.63$ A

Power developed by armature,

$P_a = E \times I_a \times 10^{-3}$

$$= 309.6 \times 336.63 \times 10^{-3}$$

$$= 104.22$$

From output equation, $D^2 L = \dfrac{P_a}{Cn}$

$$= 0.031 \text{ m}^3$$

Also, $\dfrac{b_p}{\tau_p} = 0.69$ (given)

For square pole body construction,

$$\frac{b_p}{\tau_p} = \frac{L}{\tau_p}$$

So, $\dfrac{L}{\tau_p} = 0.69$

or $\quad L = 0.69 \times \dfrac{\pi D}{p}$

$$= 0.36D$$

$\therefore \quad 0.36\, D^3 = 0.031$

or $\quad D^3 = \dfrac{0.031}{0.36}$

$$= 0.086$$

$$\left.\begin{array}{ll} \text{or} & D = 0.44 \text{ m} \\ \text{and} & L = 0.16 \text{ m} \end{array}\right\} \textbf{Ans.}$$

**Q. 5** Estimate the diameter, length of armature core, number of poles and length of air gap of a 400 kW, 400 V, 1200 r.p.m. d.c. generator. The maximum air gap flux density is 0.92 Wb/m$^2$ and ampere-conductor per metre length of periphery is 40,000. The ratio of pole arc to pole pitch is 0.78 and efficiency is 0.90. Peripheral speed should be less than 36 m/sec and frequency of flux reversal in armature should also be less than 48 Hz. Also, m.m.f. required for air gap is 0.65 times the armature m.m.f. The air gap construction coefficient is taken as 1.15, long pole construction with $L/\tau_p = 0.90$.

Chapter 7

### Solution:

Output coefficient,

$$C = \pi^2 \overline{B}\, \overline{ac} \times 10^{-3}$$

$$= (3.14)^2 \psi B_m \cdot \overline{ac} \times 10^{-3}$$

$$= (3.14)^2 \times 0.78 \times 0.92 \times 40{,}000 \times 10^{-3}$$
$$= 283.0$$

Power developed by the armature,

$$P_a = 444.44 \text{ kW}$$

Speed, $n = \dfrac{1200}{60} = 20$ r.p.s.

From output equation, $D^2 L = \dfrac{P_a}{Cn}$

$$= \frac{444.44}{283.0 \times 20}$$
$$= 0.078$$

Frequency for 4-pole, $f = \dfrac{p \cdot n}{2} = \dfrac{4 \times 20}{2} = 40$ Hz

Frequency for 6-pole, $f = \dfrac{6 \times 20}{2} = 60$ Hz

So, 6-pole construction will exceed the frequency of 48 Hz.

So, 4-pole construction can be selected.

Also, $\dfrac{L}{\tau_p} = 0.90$ (given)

or $L = 0.90 \times \dfrac{\pi D}{p}$

when $p = 4$, then $L = 0.70\ D$

$\therefore$ $0.70 D^3 = 0.078$

or $D = 0.48$ m

and $L = 0.34$ m

Peripheral velocity,

$$V_p = \pi D n$$
$$= 30.14 \text{ m/sec.}$$

Armature m.m.f., $AT_a = \dfrac{\overline{ac} \cdot \tau_p}{2} = 7536$ A $\left[\because\ \tau_p = \dfrac{\pi D}{p} = 0.37 \text{ m}\right]$

Length of air gap,

$$B_m = 0.90 \text{ Wb/m}^2 \text{ (given)}$$

m.m.f. required for air gap,

$$AT_\delta = 800{,}000 \times 0.90 \times 1.15 \times \delta$$

$$= 828{,}000 \times \delta$$

Also given that

$$AT_\delta = 0.65 \times AT_a$$

$$= 0.65 \times 7536$$

$$= 4898.4$$

So, length of air gap,

$$\delta = \frac{4898.4}{828{,}000} = 5.9 \times 10^{-3}\ \text{m}$$

So, the no. of poles, $p = 4$

Since frequency and peripheral speed are within the specified limit,

Diameter of core = 0.48 m

Length of core = 0.34 m

Length of air gap = $5.9 \times 10^{-3}$ m

**Q. 6** Calculate mean e.m.f. per conductor, total flux and the number of series conductors of a 800 kW, 440 V, 600 r.p.m. generator with armature diameter 0.59 m and length 0.76 m. The pole arc to pole pitch ratio is 0.72 and maximum flux density in air gap is 0.96 Wb/m$^2$. The armature drop is 8.5 volts.

## Solution:

Speed, $n = \frac{600}{60} = 10$ r.p.s.

Average flux density,

$$\overline{B} = \psi B_m$$

and $\psi = \frac{b_p}{\tau_p} = 0.72$

So, $\overline{B} = 0.72 \times 0.96$

$$= 0.69$$

Mean e.m.f. per conductor $= \overline{B} \cdot \pi D n$

$$= 0.69 \times 3.14 \times 0.59 \times 10$$

$$= 12.7\ \text{V}$$

Total flux $= \overline{B} \cdot \pi DL$

$$= 0.69 \times 3.14 \times 0.59 \times 0.76$$

$$= 0.97 \text{ Wb.}$$

Induced e.m.f. on full load = 440 + 8.5

$$= 448.5 \text{ V}$$

$$\text{Number of series conductors} = \frac{\text{Induced e.m.f. on full load}}{\text{Mean e.m.f. per conductor}}$$

$$= \frac{448.5}{12.7}$$

$$= 35$$

**Q. 7** Estimate specific electric loading and specific magnetic loading of a 200 kW, 410 V, 480 r.p.m., 4-pole generator having diameter of 0.80 m and length of 0.20 m. It has wave winding with 450 conductors. Voltage drop in armature is assumed to be negligible.

**Solution:**

Armature current,

$$I_a = \frac{200 \times 1000}{410} = 487.8 \text{ A}$$

For wave wound armature number of parallel path = 2

$$\therefore \qquad \text{Current in conductor, } I_z = \frac{487.8}{2} = 243.9 \text{ A}$$

Total number of conductors = 450

Specific electric loading,

$$\overline{ac} = \frac{I_z \cdot Z}{\pi \cdot D}$$

$$= \frac{243.9 \times 450}{3.14 \times 0.80}$$

$$= 43692.2 \text{ A/m}$$

We have total flux, $p\phi = \overline{B} \cdot \pi DL$

or $$\overline{B} = \frac{p\phi}{\pi DL}$$

But from equation, $E = \dfrac{p\phi nZ}{a}$

$$\Rightarrow \qquad p\phi = \frac{Ea}{nZ}$$

$$\therefore \qquad \overline{B} = \frac{Ea}{\pi DL \cdot nZ}$$

$$= \frac{410 \times 2}{450 \times 8 \times 3.14 \times 0.80 \times 0.20}$$

$$= 0.45 \text{ Wb/m}^2.$$

**Q. 8** Calculate the total m.m.f. required per pole to produce a flux of 0.12 Wb in a machine with following details:

The length of machine is 0.31 m and pole arc to pole pitch ratio is 0.64. There are 12 slots per pole. The width and height of tooth are 12 mm and 36 mm respectively. Length of flux path between two adjacent poles in core is 0.20 m, sectional area of core is 0.038 m$^2$, area of pole face is 0.10 m$^2$, length of air gap 4.80 mm. Sectional area of pole body is 0.076 m$^2$, height of pole 0.22 m, sectional area of yoke 0.046 m$^2$. Mean length of flux path in yoke is 0.43 m between two adjacent poles. The relative permeability of teeth is 88 and for rest of the magnetic circuit is 1150. Stacking factor is considered as 0.92 and leakage flux is neglected.

**Solution:**

$$\text{MMF for air gap, } AT_\delta = \phi_m \frac{\delta}{\mu_0 A_\delta}$$

$$= \frac{0.12 \times 0.0048}{4\pi \times 10^{-7} \times 0.10} = \frac{5.76 \times 10^{-4}}{4 \times 3.14 \times 10^{-7} \times 0.10}$$

$$= \frac{5.76 \times 10^{-4}}{1.256 \times 10^{-7}} = 4585.9 \text{ A}$$

$$\text{MMF per pole, } AT_p = \frac{\phi_p \times h_p}{\mu_r \mu_0 \times \text{Sectional area of pole body}}$$

$$= \frac{0.12 \times 0.22}{1150 \times 4 \times 3.14 \times 10^{-7} \times 0.076}$$

$$= \frac{0.0264}{1097.7 \times 10^{-7}}$$

$$= 240.5 \text{ A}$$

$$\text{Flux in core } = \frac{0.12}{2} = 0.06 \text{ Wb}$$

Length of flux path in core = 0.20 m

Sectional area of core = 0.038 m (given)

$$\therefore \text{ MMF for core, } AT_c = \frac{0.06 \times 0.20}{1150 \times 4 \times 3.14 \times 10^{-7} \times 0.038}$$

$$= \frac{0.012}{548.8 \times 10^{-7}} = 218.6 \text{ A}$$

$$\text{Flux in yoke} = \frac{0.12}{2} = 0.06$$

Length of flux path in yoke = 0.43 m

Sectional area of yoke = 0.046 $m^2$ (given)

$$\therefore \text{ MMF for yoke, } AT_y = \frac{0.06 \times 0.43}{1150 \times 4 \times 3.14 \times 10^{-7} \times 0.046}$$

$$= \frac{0.0258}{14{,}444 \times 10^{-7} \times 0.046}$$

$$= \frac{0.0258}{664.4 \times 10^{-7}}$$

$$= 388.3 \text{ A}$$

$$\text{Teeth per pole arc} = \psi \times \frac{S}{p}$$

$$= 0.64 \times 12$$

$$= 7.68$$

$$\approx 8$$

$$\text{Flux in one tooth} = \frac{0.12}{8} = 0.015 \text{ Wb}$$

$\therefore$ MMF for teeth,

$$AT_t = \frac{0.015 \times \text{Height of tooth}}{88 \times 4\pi \times 10^{-7} \times \text{Stacking factor} \times \text{Width of teeth} \times \text{Length of machine}}$$

$$= \frac{0.015 \times 36 \times 10^{-3}}{88 \times 4 \times 3.14 \times 10^{-7} \times 0.92 \times 0.31 \times 0.012}$$

$$= \frac{0.54 \times 10^{-3}}{1.89 \times 10^{-7}}$$

$$= 1428 \text{ A}$$

Total m.m.f. per pole,

$$AT_t = AT_\delta + AT_p + AT_c + AT_y + AT_t$$

$$= 4585.9 + 240.5 + 218.6 + 388.3 + 1428$$

$$= 6861.3 \text{ A}$$

**Q. 9** If the m.m.f. per pole estimated in Q.8 is assumed to be referred to shunt field coil of a 350 V, 4-pole generator with other details as below. Calculate, diameter of conductor, height of coil, number of turns and field current.

m.m.f. per pole = 6861 A, depth of winding = 42 mm, length of inner turn is 0.98 m, length of outer turn = 1.26 m, loss dissipated from outer surface excluding ends is 1250 watt/m$^2$, space factor = 0.60, resistivity of conductor is 0.021 Ω/m/mm$^2$. Consider a voltage drop of 60 V across the field regulator.

**Solution:**

Voltage across shunt field winding = 350 – 60 = 290 V.

Voltage across each field coil, $E_f = \dfrac{290}{4} = 72.5$

Length of mean turn, $L_{mt} = \dfrac{0.98 + 1.26}{2} = 1.12$ m

Area of field conductor, $F_f = \dfrac{AT_{fl} \cdot \rho \cdot L_{mt}}{E_f}$

$$= \frac{6861 \times 0.021 \times 1.12}{72.5}$$

$$= \frac{161.3}{72.5} = 2.22 \text{ mm}^2$$

Sectional area of field conductor, $F_f = 2.22$

So, taking round wire and let diameter of conductor is $d$. So, $F_f = \dfrac{\pi}{4} d^2$

So, $$d = \sqrt{\frac{4 \times 2.22}{3.14}} = 1.68 \text{ mm}$$

So, diameter of conductor = 1.68 mm

Number of turns, $T_f = \dfrac{k_f \cdot h_f \cdot d_f}{F_f}$

Chapter 7

or $$T_f = \frac{0.60 \times h_f \times 0.042}{2.22 \times 10^{-6}} = 0.011 \times 10^6 h_f$$

$$= 1.1 \times 10^4 h_f \qquad ...(i)$$

Area of outer surface excluding ends = 1.26 $h_f$

∴ Permissible loss

$$Q_f = 1250 \times 1.26\ h_f$$

$$= 1575\ h_f$$

Field current,

$$I_f = \frac{Q_f}{E_f}$$

$$= \frac{1575\ h_f}{72.5}$$

$$= 21.7\ h_f$$

We know that

$$AT_{fl} = I_f \cdot T_f$$

$$= 21.7 h_f T_f$$

∴ $$21.7 h_f \cdot T_f = 6861$$

Number of turns in each field coil, $T_f = \dfrac{316.1}{h_f}$ ...(*ii*)

From equations (*i*) and (*ii*), we get

$$T_f^2 = 1.26 \times 10^4 \times 316.1$$

$$= 398.38 \times 10^4$$

or $$T_f = 19.95 \times 10^2$$

$$= 1995$$

Height of field coil, $h_f = \dfrac{1995}{1.1 \times 10^4} = 0.18$ m

Resistance of each field coil,

$$R_f = \frac{T_f \cdot \rho L_{mt}}{F_f}$$

$$= \frac{1995 \times 0.021 \times 1.12}{2.22}$$

$$= \frac{46.92}{2.22} = 21.13\ \Omega$$

So, field current, $I_f = \frac{E_f}{R_f} = \frac{72.5}{21.13}$

$= 3.43$ A

**Q. 10** If a generator of 350 V, 4-pole as given in Q.9 develops m.m.f. per pole by field coils equal to 6881 A, with depth of winding 42 mm, length of inner turn 0.98 m, length of outer turn in 1.26 m, height of winding 0.18 m has to dissipate 200 W in an environment where the temperature difference between machine and ambient air is 40°C. The heat dissipation from outer surface excluding top and bottom surfaces of coil is 25 watt/m$^2$ . The resistivity is 0.021 Ω/m/mm$^2$. Calculate current density and space factor.

**Solution:**

Length of mean turn, $L_{mt} = \frac{0.98 + 1.26}{2} = 1.12\ \text{m}$

Temperature difference = 40°C

Loss dissipated per unit surface = $25 \times 40 = 1000\ \text{W/m}^2$

Permissible loss = Heat dissipated

$Q_f = 200$ watt

We can have

$$Q_f = I_f^2 R_f = I_f^2 \cdot \frac{\rho \cdot T_f \cdot L_{mt}}{F_f} \qquad \left\{ \text{Resistance} = \frac{\rho \cdot \text{length}}{\text{Area}} \right\}$$

$$= (I_f \cdot T_f) \cdot \frac{I_f}{F_f} \cdot \rho \cdot L_{mt}$$

$$= AT_{fl} \cdot \frac{I_f}{F_f} \cdot \rho \cdot L_{mt}$$

$$= 6861 \times 0.021 \times 1.12 \cdot \frac{I_f}{F_f}$$

Since $Q_f = 200$

So,

$$\frac{I_f}{F_f} = \frac{200}{161.37}$$

Chapter 7

So, current density, $\sigma = \dfrac{I_f}{F_f} = 1.24 \text{ A/mm}^2$ **Ans.**

$$\text{Total area of conductor} = T_f \cdot F_f \times 10^{-6}$$

$$= T_f \cdot \frac{I_f}{\sigma} = \frac{AT_{fl}}{\sigma}$$

$$= \frac{6861}{1.24} \times 10^{-6} = 533 \times 10^{-6} \text{ m}^2$$

Total area of winding = $h_f \cdot d_f$

= 0.18 × 0.042

= 0.007 m²

$$\text{Space factor} = \frac{\text{Total area of conductor}}{\text{Total area of winding}}$$

$$= \frac{0.0055}{0.007} = 0.78$$

## OVERALL DESIGN OF A D.C. GENERATOR

**Q. 11** Design as far as possible a d.c. generator with following specifications:

| *Rated power output (kW)* | *No. of poles* | *Voltage (V)* | *RPM* |
|---|---|---|---|
| 80 | 4 | 250 | 900 |

Design for main dimensions, armature, field systems etc. Assume suitable data required in design efficiency of the generator should be 0.90 and armature temperature rise should be within specified limit.

### Solution:

Main dimensions:

Rated power output = 80 kW

Speed, $n = \dfrac{900}{60} = 15$ r.p.s.

Taking an efficiency of 0.90.

Power developed by the armature,

$$P_a = P\left(\frac{2+\eta}{3\eta}\right) = 85.92 \text{ kW}$$

Specific magnetic winding, $\overline{B} = 0.5\ \text{Wb/m}^2$

and specific electric loading $\overline{ac} = 34{,}000\ \text{A/m}$

So, output coefficient, $C = \pi^2 \times 0.51 \times 34{,}000 \times 10^{-3}$

$= 170.96$

Output equation can be written as

$$P_a = CD^2 Ln$$

or $$D^2 L = \frac{P_a}{Cn} = \frac{85.92}{170.96 \times 15} = \frac{85.92}{2564.4}$$

$$D^2 L = 0.03 \quad \text{...}(i)$$

Square pole construction is preferred for pole body.

So, for square pole body, $\dfrac{L}{\tau_p} = \dfrac{b_p}{\tau_p}$.

Let us assume that pole arc to pole pitch ratio,

$$\frac{b_p}{\tau_p} = 0.66$$

$$\therefore \quad \frac{L}{\tau_p} = 0.66$$

or $$L = 0.66 \times \frac{\pi \times D}{4}$$

$$= 0.52\ D$$

Now, putting this value in equation ($i$), we get

$$0.52D^3 = 0.03$$

or $D^3 = 0.05$

or $D = 0.38$ m

and length $L = 0.19$ m

Hence,

Diameter of armature = 0.38 m

Length of core = 0.19 m

Peripheral speed, $V_p = \pi Dn = 3.14 \times 0.38 \times 15$

$= 17.8$ m/sec

Pole pitch, $\tau_p = \dfrac{\pi D}{p} = 0.29$ m

For pole arc, given that $\frac{b_p}{\tau_p} = 0.66$

So, pole arc, $b_p = 0.66 \times 0.29 = 0.19$ m

The core length is 0.19 m and hence one ventilating duct have to be provided. Let the width of ventilating duct is 1 cm.

So, gross iron length,

$$l = L - n_v \cdot b_v$$

where, number of ventilating duct, $n_v$ = 1 and width of ventilating duct, $b_v$ = 1 cm = 0.01 m

$$\therefore \quad l = 0.19 - 0.01 \times 1$$

$$= 0.18 \text{ m}$$

Net iron length

$$l_i = k_i \, l$$

Taking stacking factor, $k_i$ = 0.90

$$l_i = 0.90 \times 0.18$$

$$= 0.16 \text{ m}$$

The frequency of flux reversal, $f = \frac{pn}{2} = \frac{4 \times 15}{2} = 30$ Hz

The thickness of lamination is preferred for 0.35 mm.

Line current, $I_l = \frac{P \times 1000}{\eta \cdot V} = \frac{80 \times 1000}{0.90 \times 250} = \frac{80{,}000}{225.0}$

$$= 355.5 \text{ A}$$

Taking field current of 1.4% of line current,

$$I_f = 4.9 \text{ A}$$

Considering internal voltage drop of 5 per cent

Internal voltage drop, = 17.7 volts

Armature current, $I_a = I_l + I_f$

$$= 255.5 + 4.9$$

$$I_a = 260.4$$

Generated e.m.f.,

$$E = \text{Terminal voltage (V)} + \text{Voltage drop}$$

$$= 250 + 17.7$$

$$= 267.7 \text{ V}$$

***Type of Armature Winding:*** Since armature current is less than 400 A, a simplex wave winding can be selected for this machine.

So, the number of parallel path = 2

and current per parallel path $= \dfrac{360.4}{2} = 180.2$ A

So, $I_z$ = 180.2 A

Useful flux per pole, $\phi = \overline{B} \cdot \tau_p \cdot L$

$$= 0.51 \times 0.29 \times 0.19$$

$$= 0.028 \text{ Wb.}$$

Number of armature conductors,

$$Z = \frac{2 \cdot E}{\phi \cdot p \cdot n} = \frac{267.7 \times 2}{0.028 \times 4 \times 15} = \frac{267.7 \times 2}{1.68}$$

So, $Z = 318.6$

$$\approx 318$$

Number of slots is selected as below :

The slot pitch should be between 2.5 and 3.5 cm. So, the number of slots should be between

$$\frac{\pi D}{\text{Slot pitch}} = \frac{3.14 \times 380}{35} \text{ to } \frac{3.14 \times 380}{25}$$

$$= 34 \text{ to } 47$$

- Number of slots per pole, $\dfrac{S}{p} = 9$ to 16
- So, number of slots should be within 36 to 64.
- In case of wave winding, the number of slots should not be multiple of 2. So, number of slots should not be an integer.
- With $\psi = 0.66$, the number of slots per pole shoe are nearby an integer for 39 to 43.
- The number of 280 coil sides per slot is 2, 4, 6. Hence 4 slots with coil sides per slot $n = 4$ with two turn coils can be taken.

Total number of slots, $S = 41$

Total number of coils $= \dfrac{1}{2} \times 4 \times 41 = 82$

Number of armature conductor actually used, Z = 2 × 2 × 82

So, corrected number of conductors = 328 = $Z_{corr}$

Number of conductors per slot, $N_c = \dfrac{328}{41} = 8$

Slot pitch $= \dfrac{\pi D}{S} = \dfrac{3.14 \times 0.38}{41} = 0.029$ m

Chapter 7

Corrected value of flux per pole,

$$\phi_{corr} = 0.028 \times \frac{318}{328} = 0.027 \text{ Wb}$$

The corrected value of $\overline{B}$ and $\overline{ac}$ are

$$\overline{B}_{corr} = \frac{\phi_{corr}}{\tau_p \cdot L} = \frac{0.027}{0.29 \times 0.19} = 0.55$$

and

$$\overline{ac}_{corr} = \frac{I_z \cdot Z_{corr}}{\pi \cdot D} = \frac{180.2 \times 328}{3.14 \times 0.38} = \frac{59{,}105}{1.19} = 49{,}668.$$

Slot loading $= N_c \cdot I_z = 8 \times 180.2 = 1441.6$ which is less than the specified limit of 1500 A.

Commutator diameter, $D_c = 0.63 \times 0.38 = 0.23$ m

Commutator peripheral length, $\pi \cdot D_c = 0.72$ m

Let the number of commutator segments, $C = 150$

So, commutator pitch $= \frac{\pi D_c}{C} = \frac{0.72}{150} = 4.8 \times 10^{-3}$ m

This is in specified limit.

The current density in armature winding = 5.5 A/mm$^2$

$$\text{Armature conductor cross-section, } F_c = \frac{\text{Current per parallel path}}{\sigma} = \frac{I_z}{5.5}$$

$$= \frac{180.2}{5.5} = 32.76 \text{ mm}^2$$

Refering to Table A.1 a rectangular copper conductor of size $3.0 \times 11 \text{ mm}^2$ can be used. The area given in table is 32.5 mm$^2$.

### *Slot Size*

Width of Slot: Bare conductor = $3.0 \times 4 = 12$ mm

Conductor insulation = 1.4

Slot insulation = $2 \times 1.0 = 2.0$ mm

Total width = 15.4

Width of slot, $b_s$ = 15.4 mm

### *Height of Slot*

Bare conductor = 2 × 11 = 22.0 mm

Conductor insulation = 4 × 0.35 = 1.40 mm

Slot insulation = 2 × 1.0 = 2.0 mm

Separator of 1 mm = 1.0 mm

Stack of 0.6 mm = 0.6 mm

Total = 27 mm

Wedges at slot opening = 4.0 mm

So, total height for slot = 27 + 4 = 31.0 mm

So, height of slot, $h_s$ = 31.0 mm

$$\text{Slot pitch } = \frac{\pi D}{S} = \frac{3.14 \times 380}{41} = 29.1 \text{ mm}$$

Flux density in teeth at 1/3rd height from narrow end:

Slot pitch at 1/3rd height from narrow end

$$= \frac{\pi\left[D - \left(2 \times \frac{2}{3} h_s\right)\right]}{41}$$

$$= \frac{3.14\left[380 - \left(2 \times \frac{2}{3} \times 31\right)\right]}{41}$$

= 25.9 mm

Width of tooth at 1/3rd height from narrow end,

$$b_{t(1/3)} = 25.9 - 15.4 = \frac{\pi D_{1/3}}{S} - b_s$$

= 10.5 mm

$$\text{Flux density at 1/3rd height, } B_{t(1/3)} = \frac{p\phi_{corr}}{\psi S L_i \times 0.0105}$$

$$= \frac{4 \times 0.027}{0.66 \times 41 \times 0.16 \times 0.0105}$$

$$= \frac{4 \times 0.027}{0.0455}$$

= 2.3 Wb/m$^2$

This is slightly higher than the specified limit. The tooth width should be increased to decrease the flux density in the tooth. But the design in this particular problem is continued with this flux density.

Armature m.m.f. per pole, $AT_a = \frac{\overline{ac} \cdot \pi D}{2 \cdot p} = \frac{49{,}668 \times 3.14 \times 0.38}{2 \times 4}$

$= 7407$ A

Ampere-turn required for air gap is taken 0.55 times of $AT_a$,

$$AT_\delta = 0.55 AT_a = 0.55 \times 7407$$

$$= 4073 \text{ A}$$

Maximum flux density in the air gap,

$$B_m = \frac{B_{corr}}{\psi} = \frac{0.55}{0.66} = 0.83 \text{ Wb/m}^2$$

Let gap contraction coefficient, $k_c = 1.15$

We have, $AT_\delta = 800{,}000 \times B_m k_c \cdot \delta$

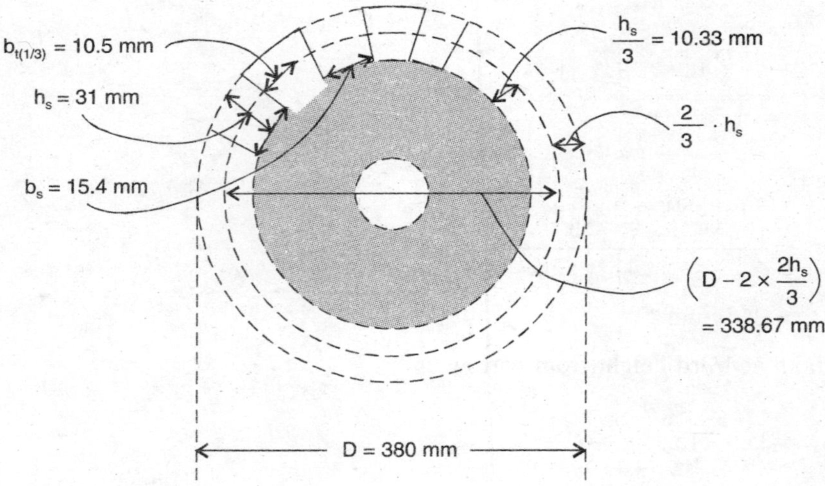


So,

$$\delta = \frac{AT_\delta}{800{,}000 \times B_m \cdot k_c}$$

$$= \frac{4073}{800{,}000 \times 0.83 \times 1.15}$$

$$= \frac{4073}{763{,}600} = 0.0053 \text{ m}$$

or air gap length, $\delta = 5.3$ mm .

Armature core :

Flux in armature core, $\phi_c = \dfrac{\phi_{corr}}{2} = \dfrac{0.027}{2} = 0.0135$

$$= 13.5 \times 10^{-3} \text{ Wb/m}^2$$

Let flux density in armature core, $B_c$ = 1.25 Wb/m$^2$

$$\therefore \qquad \frac{\phi_{corr}}{2} = B_c \cdot h_c \cdot L_i$$

So, height of armature core, $h_c = \dfrac{\phi_{corr}}{2 \cdot B_c \cdot L_i}$

$$= \frac{0.027}{2 \times 1.25 \times 0.16} = \frac{0.027}{0.4}$$

$$= 0.067 \text{ m}$$

So, height of armature core, $h_c$ = 67 mm

Internal diameter of armature core,

$$D_i = D - 2h_s - 2h_c$$

$$= 380 - 2 \times 31 - 2 \times 67$$

$$= 380 - 62 - 134$$

$$= 184 \text{ mm}$$

Resistance and voltage drop of armature:

$$L_{mt} = 2L + 2.3\tau_p + 5 \times h_s$$

Length of mean turn, $L_{mt} = 2 \times 0.19 + 2.3 \times 0.29 + 5 \times 0.031$

$$= 0.38 + 0.66 + 0.15$$

$$= 1.19 \text{ m}$$

Resistance of each conductor at 75°C, $R = \dfrac{\rho \cdot L_{mt}}{2 \cdot F_c} = \dfrac{0.021 \times 1.19}{2 \times 32.76}$

$$= \frac{0.021 \times 1.19}{65.52} = 4.8 \times 10^{-4}$$

For wave winding, resistance of armature,

$$R_a = \frac{1}{2}\left(\frac{Z}{2} \cdot \frac{\rho L_{mt}}{2 \cdot F_c}\right)$$

Chapter 7

$$= \frac{1}{2} \times \frac{328}{2} \times 4.8 \times 10^{-4}$$

$$= 3.9 \times 10^{-2}\ \Omega$$

Armature voltage drop $= I_a \cdot R_a = 360.4 \times 3.9 \times 10^{-2}$

$= 14.05$

$= 14.0$ V

### *Copper Loss and Copper Weight in Armature*

Weight of copper in armature $= 8900 \times F_c \cdot L_{mt} \cdot \frac{Z}{2}$

$$= 8900 \times 32.76 \times 10^{-6} \times 1.5 \times \frac{328}{2}$$

$= 71.7$ kg

Armature copper loss $= I_a^2 \cdot R_a$

$$= (360.4)^2 \times 3.9 \times 10^{-2}$$

$= 5.00 = 5.0$ kW

Field System:

Pole Section:

Taking leakage coefficient for pole = 1.16

and flux density in pole body, $B_p$ = 1.45 Wb/m$^2$

Flux in pole body,

$$\phi_p = 1.16 \times 0.027$$

$= 0.031$ Wb

Axial length of pole, $L_P = L = 0.19$

$$\phi_p = B_p \cdot b \cdot k_i L$$

So, width of pole body,

$$b = \frac{\phi_p}{B_p \cdot k_i \times L} = \frac{0.031}{1.45 \times 0.90 \times 0.19}$$

$$= \frac{0.031}{0.24} = 0.129\,\text{m}$$

$= 129$ mm

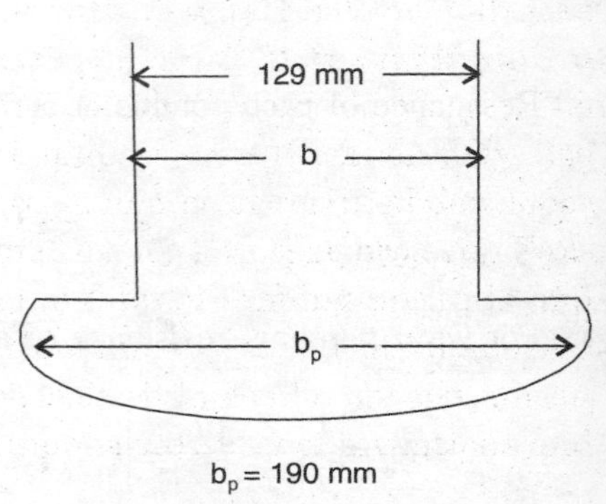

### Field Winding

Selecting, $AT_{fl}$ at full load $= 0.9 \times AT_a$

$\therefore$ Field m.m.f. at full load, $AT_{fl} = 0.9 \times 7407$

$$= 6666 \text{ A}$$

Selecting depth of field winding, $d_f = 40$ mm

Space factor, $k_f = 0.75$,

Loss dissipated ($q_f$) excluding top and bottom is taken as 700 W/m$^2$.

So, m.m.f. per metre height of field winding,

$$= 10^4 \sqrt{q_f \cdot k_f \cdot d_f}$$

$$= 10^4 \sqrt{700 \times 0.75 \times 0.04}$$

$$= 4.58 \times 10^4$$

Height of field winding,

$$h_f = \frac{AT_{fl} \times 10^{-4}}{\sqrt{q_f \cdot k_f \cdot d_f}}$$

$$= \frac{6666 \times 10^{-4}}{4.58} = 0.14 \text{ m}$$

The height of pole shoe is taken 0.020 m and space for insulation is 0.01 m.

Total height of pole,

$h_p = 0.14 + 0.020 + 0.010 = 0.17$ m

### Yoke

Flux in yoke,

$$\phi_y = \frac{\text{Leakage coefficient} \times \phi_{corr}}{2}$$

$$= \frac{1.15 \times 0.027}{2} = 0.015$$

and $\phi_y = h_y \cdot L_{yi} \cdot B_y$

Taking $B_y = 1.5$ Wb/m$^2$ and $L_{yi} = l_i = 0.16$

Height of yoke,

$$h_y = \frac{\phi_y}{L_{yi} \cdot B_y}$$

Chapter 7

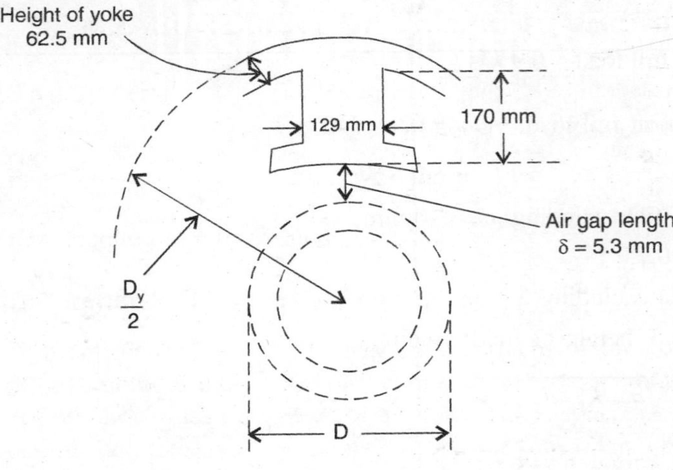


$$= \frac{0.015}{0.16 \times 1.5} = 0.0625$$

$$= 62.5 \text{ mm}$$

$\therefore \quad h_y = 62.5$ mm

Outer diameter of yoke, $D_y = D + 2\left(\delta + h_p + h_y\right)$

$\therefore \quad D_y = 0.38 + 2(0.17 + 0.005 + 0.062)$

$= 0.854$ m

### MMF Required for Open Circuit Condition

*(i) Ampere-turn required for air gap*

Maximum air gap flux density, $B_m = 0.83$ Wb/m²

Slot opening to air gap length ratio $= \frac{15.4}{5.3} = 2.9$

Carter's air gap coefficient for this ratio is 0.37 (from Fig. 4.3)

Gap contraction factor for slots $= \frac{\text{Slot pitch}}{\text{Slot pitch} - 0.37 \times b_s}$

$$= \frac{0.029}{0.029 - 0.37 \times 0.015} = \frac{0.029}{0.129 - 0.0055}$$

$$= \frac{0.029}{0.023} = 1.26$$

Duct width to air gap length ratio $= \frac{10}{5.3} = 1.88$

Carter's air gap coefficient for this ratio is 0.3 (from Fig. 4.3)

Gap contraction factor for ducts,

$$= \frac{L}{L - 0.3 \times h_v \times b_v} = \frac{190}{190 - 0.3 \times 1 \times 0.01}$$

$$= 1.0$$

Overall gap contraction factor, $k_g = 1.2 \times 1.0 = 1.2$

Ampere-turn for air gap, $AT_\delta = 800{,}000 \times 0.83 \times 1.2 \times 5.3 \times 10^{-3}$

$$= 4223.0 \text{ A}$$

### (ii) MMF for Pole Body

$B_p$ is 1.45 Wb/m²,

∴ $at_p$ corresponding to $B_p$ from *B-H* curve is 900 A/m.

Total m.m.f. for pole body,

$$AT_p = at_p \times h_p = 900 \times 0.17$$

∴ $AT_p = 153$ A

### (iii) MMF for Pole Yoke

$$AT_y = at_y \cdot l_y$$

$$l_y = \frac{\pi D_y}{p} = \frac{3.14 \times 0.854}{4} = 0.67 \text{ m}$$

$$B_y = 1.5 \text{ Wb/m}^2$$

So, $at_y = 1000$ A/m

$$AT_y = 1000 \times 0.67 = 670 \text{ A}$$

### (iv) MMF for Teeth

$$B_{t(1/3)} = 2.3 \text{ Wb/m}^2$$

$$at_t = 17{,}000 \text{ A/m}$$

MMF required for teeth,

$$AT_t = 527 \text{ A}$$

### (v) MMF for Core

Flux density in core,

$B_c = 1.25$ Wb/m².

Corresponding to this flux density,

$at_c = 350$ A/m

Length of flux path in core, $L_c = \dfrac{\pi(D - 2h_s - h_c)}{4}$

$$L_c = \frac{\pi(380 - 2 \times 31 - 67)}{4}$$

$= 0.197$ m

Total m.m.f. for teeth, $AT_c = at_c \times L_c$

$= 350 \times 0.197$

$= 68.9$ A

Total m.m.f. required,

$$AT_T = AT_\delta + AT_p + AT_y + AT_t + AT_c$$

$= 5641$ A

Assuming that the full load field m.m.f. is 1.15 times the no load field m.m.f.

∴ Field m.m.f. required at full load, $AT_{fl} = 1.15 \times 5641$

∴ $AT_{fl} = 6487$ A.

### *Field Winding*

The voltage across the shunt field winding is 250 volt. 20% of this voltage is kept as reserve for speed regulators.

Voltage across each shunt field coil is

$$E_f = \frac{0.80\,V}{p} = \frac{0.80 \times 250}{4}$$

$= 50$ V

m.m.f. of each coil at full load $= 6487$ A

*i.e.,* $AT_{fl} = 6487$ A

Height of field coil, $h_f = 0.14$ m $= 140$ mm

Depth of winding, $d_f = 40$ mm

Axial length of pole, $L_p = L = 190$ mm

Width of pole shoe, $b_p = 190$ mm

Width of pole body, $b = 129$ mm

The length of mean turn,

$$L_{mt} = 2\left[L_p + b + d_f\right]$$

$= 2\ [190 + 129 + 40]$

$= 718$ mm $= 0.718$ m

Sectional area of shunt field conductor,

$$F_f = \frac{AT_{fl} \cdot \rho \cdot L_{mt}}{E_f}$$

*i.e.*, $$F_f = \frac{6487 \times 0.021 \times 0.718}{50}$$

where, $\rho$ = 0.021 Ω/m/mm² at 75°C.

So, $F_f$ = 1.96 mm²

Area of conductor = 2.02 mm²

From standard size Table A.1

Space factor, $k_f = 0.75$

Number of turns, $$T_f = \frac{k_f \cdot h_f \cdot d_f}{F_f}$$

$$= \frac{0.75 \times 0.14 \times 0.04}{2.02 \times 10^{-6}}$$

$$= 2.0 \times 10^3$$
$$= 2000$$

Resistance of each coil at 75°C, $$R_f = \frac{T_f \cdot \rho \cdot L_{mt}}{F_f}$$

$$= \frac{2000 \times 0.021 \times 0.718}{2.02}$$

$$= 14.9\ \Omega$$

Field current, $$I_f = \frac{E_f}{I_f} = 3.35\text{ A}$$

∴ Field m.m.f. provided, $AT_{fl} = 3.35 \times 2000$

$= 6700$ A

Loss in each field coil, $Q_f = I_f^2 R_f = (3.35)^2 \times 14.9$

= 162.7 watt

Cooling surface of each coil,

$$= 2\,L_{mt}\left(h_f + d_f\right)$$

$$= 2 \times 0.718(0.14 + 0.04)$$
$$= 0.25\text{ m}^2$$

$$\text{Cooling coefficient} = \frac{0.16}{1+0.1V_p}$$

$$= \frac{0.16}{1+0.1\times 17.8} = \frac{0.16}{2.78} = 0.05$$

$$\text{Temperature rise} = \frac{\text{Loss in each field coil}\times\text{Cooling coefficient}}{\text{Cooling surface}}$$

$$= \frac{162.7\times 0.05}{0.25}$$

$$= 32.5°\text{C}$$

This is within limit.

### *Commutator*

Commutator diameter calculated earlier,

$D_c$ = 0.23 m

Peripheral speed, $V_c$ = 10.8 m/sec

Commutator pitch = 4.8 mm

Number of commutator segments = 150

Armature current, $I_a$ = 355.5 + 3.35

= 358.8 A

$$\text{Current per brush arm} = \frac{2\times 358.8}{4} = 179.4 \text{ A}$$

Select current density in brushes = 0.1 A/mm$^2$.

Current in each brush should be less than 70 A.

Using 4 brushes per arm,

$$\text{So, current in each brush} = \frac{179.4}{4} = 44.8 \text{ A}$$

$$\text{Area of each brush, } a_b = \frac{44.8}{0.1} = 448$$

$$= 448 \text{ mm}^2$$

Brush should cover atleast 2.5 segments.

So, thickness of each brush $= 4.8\times 2.5 = 12$ mm

$$\text{Width of each brush} = \frac{448}{12} = 37 \text{ mm}$$

Area of each brush used = 448 mm$^2$

Area of brushes in each brush arm,

$$A_b = 4 \times a_b = 4 \times 448 = 1792 \text{ mm}^2$$

Providing 5 mm clearance between brushes and 20 mm for staggering and end play each.

Length of commutator, $L_c = n_b(w_b + C_b) + C_1 + C_2$

$$= 4(37 + 4.8) + 10 + 10$$

$$= 187.2 \text{ mm}$$

$$L_c = 0.187 \text{ m}$$

Taking 20 mm for risers, overall length of commutator can be calculated as below.

Overall length of commutator = 0.187 + 0.02

= 0.207 m

### Losses

The brush contact drop is taken as 1 volt per brush

$\therefore$ brush contact loss $= 2 \times 1 \times 358.8$

= 717 watts

Selecting

Brush pressure = 20 kN/m$^2$

Coefficient of friction = 0.16

From Table 7.11

Brush friction loss = $\mu P_b \, p A_b V_c$

$$= 0.16 \times 20 \times 10^3 \times 4 \times 0.0017 \times 10.8$$

= 247.4 watts

Total loss at commutator = 717 + 247 = 964 watts

Commutator cooling surface $= \pi \cdot D_c \cdot L_c$

$$= 3.14 \times 0.23 \times 0.187$$

$$= 0.137 \text{ m}^2$$

Cooling coefficient for commutator,

$$= \frac{0.025}{1 + 0.1 V_c} = \frac{0.025}{1 + 0.1 \times 10.8}$$

$$= \frac{0.025}{1 + 1.08} = \frac{0.025}{2.08} = 0.0125$$

Temperature rise in commutator,

$$= \frac{\text{Total losses} \times \text{Cooling coefficient}}{\text{Cooling surface}}$$

$$= \frac{964 \times 0.0125}{0.137} = 87°\text{C}$$

***Interpole***

Length of air gap under interpole, $\delta_i = 1.3\ \delta$

$$= 1.3 \times 5.3 = 6.89 \text{ mm}$$

$$\delta_i = 6.89 \text{ mm}$$

Width of interpole $= 1.5 \times$ Slot pitch

$$= 1.5 \times 29.1 = 43.6 \text{ mm}$$

With these details, the specific permeance are estimated for slot, tooth top and overhang. The total permeance is obtained as

$$\lambda = 3.28 \times 10^{-6}$$

Time of commutation = $2.23 \times 10^{-3}$ sec.

Average reactance voltage in coil,

$$E_{r\,ave} = 2T_c^2 \cdot L \cdot \frac{2I_z}{t_c} \cdot \mu\lambda$$

$$= 1.05 \text{ V}$$

Maximum reactance voltage is taken as 1.35 V

$\therefore$ Maximum flux density under interpole, $B_{mi} = \dfrac{1.35}{0.19 \times 15} = 0.47 \text{ Wb/m}^2$

$\therefore$ $B_{mi} = 0.47 \text{ Wb/m}^2$

The gap contraction factor for interpoles is less than the gap contraction factor for poles, since the air gap length is more under inter pole. So, gap contraction factor for interpoles will be less than 1.2.

Contraction factor, $k_{ci} = 1.15$

MMF required under interpole,

$$AT_i = 800{,}000 \times B_{mi} \cdot k_{ci} \cdot \delta_i + \frac{I_z \cdot Z}{2p}$$

$$= 800{,}000 \times 0.47 \times 1.15 \times 6.89 \times 10^{-3} + \frac{I_z Z}{2p}$$

$$= 2979.0 + \frac{180 \times 328}{2 \times 4}$$

$$= 2979.0 + 7380$$

$$= 10359$$

Number of turns on each interpole,

$$= \frac{10359}{360} = 28$$

When current density is 2.24 A/mm$^2$, then the sectional area of interpole winding can be estimated as

$$\text{Sectional area} = \frac{360}{2.24} = 160 \text{ mm}^2.$$

## Efficiency

### (i) Copper Losses

Armature copper loss = 5000 watt (calculated earlier)

$$\text{Copper loss in shunt field } = 4 \times I_f^2 \times R_f$$

$$= 4 \times (3.35)^2 \times 14.5 = 652 \text{ watt}$$

Interpole copper loss = 835 watt

Brush contact loss estimated earlier = 717 watt

Total copper loss + Brush contact loss = 5000 + 652 + 835 + 717 = 7204 watt

### (ii) Iron Losses

$$\text{Average width of teeth} = \frac{\pi(D - h_s)}{S} - b_s$$

$$= \frac{3.14(380 - 31)}{41} - 15.4$$

$$= 26.7 - 15.4 = 11.3 \text{ mm}$$

Weight of armature teeth,

$$= 41 \times 0.19 \times 0.011 \times 0.031 \times 7600$$

$$= 20.1 \text{ kg}$$

Specific iron loss in teeth,

$$= 0.06 + B_m^2 + 0.008\, f^2 B_m^2\, t^2$$

$$= 0.06 \times 30 \times (2.3)^2 + 0.008 \times 30^2 \times (2.3)^2 \times (0.35)^2$$

$$= 9.5 + 4.6 = 14.1 \text{ watt/kg}$$

So, iron loss in teeth $= 20.1 \times 14.1 = 283.4$ watt

$$\text{Weight of armature core } = \pi(0.38 - (2 \times 0.031) - 0.067) \times 0.067 \times 0.19 \times 7600$$

$$= 76 \text{ kg}$$

$$\text{Specific iron loss in core } = 0.06\, f\, B_m^2 + 0.005\, f^2 B_m^2\, f^2$$

Chapter 7

$$= 0.06 \times 30 \times (1.25)^2 + 0.005(30)^2 \times (1.25)^2 \times (0.35)^2$$

$$= 2.8 + 0.86 = 3.6 \text{ watt/kg}$$

$\therefore$ Iron loss in armature core = 76 × 3.6

= 273.6 watt

Total iron loss in armature = 283 + 273

= 556 watt

### *(iii) Friction and Windage Loss*

Brush friction loss = 247 watt (calculated earlier)

The peripheral speed of armature is 17.8 m/sec.

From Table 7.12 given for bearing friction and windage losses, taking friction losses in bearing as 0.4% of output of machine.

$$\text{Bearing friction and windage loss} = \frac{0.4}{100} \times 80 \times 1000$$

= 320 watt

Total friction and windage loss = 247 + 320

= 567 watt

Total losses = (Total copper loss + Brush contact loss) + Total iron loss + Total F & W loss

= 7204 + 556 + 567 = 8327.00 watt

Input = 80,000 + 8327 = 88,327 watt

$$\text{Efficiency, } \eta = \frac{\text{Output}}{\text{Input}} \times 100 = \frac{80{,}000}{88{,}327} \times 100$$

= 90.6%

## Check

### *Temperature Rise of Armature*

(*a*) Outside cylindrical surface of armature $= 3.14 \times 0.38 \times 0.19$

$= 0.22 \text{ m}^2$

$$\text{Cooling coefficient} = \frac{0.03}{1 + 0.1 \times V_p} = \frac{0.03}{(1 + 0.1 \times 17.8)} \quad \text{(from Table 6.1)}$$

$$= \frac{0.03}{2.78} = 0.01$$

$$\text{Loss dissipated from outer surface of armature} = \frac{0.22}{0.01} = 22 \text{ watt/°C}$$

(*b*) Inner surface of armature $= \pi D_i L$

$$= 3.14 \times 0.18 \times 0.19$$
$$= 0.107 \text{ m}^2$$

Peripheral speed of inner portion $= \pi D_i n$

$$= 3.14 \times 0.18 \times 15$$
$$= 8.4 \text{ m/sec.}$$

$$\text{Cooling coefficient} = \frac{0.03}{1 + 0.1 \times 8.4} = \frac{0.03}{1 + 0.84} = \frac{0.03}{1.84}$$
$$= 0.016$$

$$\text{Loss dissipated from inner surface} = \frac{0.107}{0.016} = 6.6 \text{ W/°C}$$

(*c*) Cooling surface of one duct and two end surfaces,

$$= 3 \times \frac{\pi}{4}\left(D^2 - D_i^2\right)$$

$$= 3 \times \frac{3.14}{4}\left[(0.38)^2 - (0.184)^2\right]$$

$$= 3 \times \frac{3.14}{4}[0.144 - 0.0338]$$

$$= 3 \times \frac{3.14}{4} \times 0.110 = 0.26 \text{ m}^2$$

Velocity of air in ducts $= 0.1 \times 17.8$

$$= 1.78 \text{ m/sec}$$

$$\text{Cooling coefficient} = \frac{0.10}{1.78} = 0.056$$

Loss dissipated from duct and ends,

$$= \frac{0.26}{0.056} = 4.6 \text{ watt/°C}$$

Total loss dissipated from armature,

$$= 22 + 6.6 + 4.6$$
$$= 33.2 \text{ watt/°C}$$

Total loss to be dissipated:
The copper loss in active portion of the armature winding is considered for temperature rise.
Total loss to be dissipated = Copper loss + Iron loss

$$= 5000 \times \frac{2 \times 0.19}{\text{Length of mean turn of arms winding}} + \text{Iron loss}$$

$$= 5000 \times \frac{2 \times 0.19}{1.19} + 556$$

$$= 1596 + 556 = 2152.0$$

$$\text{Temperature rise of armature} = \frac{\text{Loss to be dissipated}}{\text{Loss dissipated}}$$

$$= \frac{2152.0}{33.2} = 64.8°\text{C}$$

## DESIGN SHEET

### *Loadings:*

| | | | |
|---|---|---|---|
| 1. | Specific magnetic loading | $\overline{B}$ | 0.51 Wb/m$^2$ |
| 2. | Specific electric loading | $\overline{ac}$ | 34,000 A/m |
| 3. | Output coefficient | $C$ | 170.96 |

### *Ratings:*

| | | | |
|---|---|---|---|
| 1. | Output power | | 80 kW |
| 2. | Voltage | | 250 V |
| 3. | Speed | | 900 rpm |
| 4. | Armature current | $I_a$ | 360.4 A |
| 5. | Current per parallel path | $I_z$ | 180.2 |

### *Main dimensions:*

| | | | |
|---|---|---|---|
| 1. | Diameter of armature | $D$ | 0.38 m |
| 2. | Length of armature | $L$ | 0.19 m |
| 3. | Peripheral speed | $V_p$ | 17.8 m/sec |
| 4. | Pole pitch | $\tau_p$ | 0.29 m |
| 5. | Pole arc | $b_P$ | 0.19 m |
| 6. | Gross iron length | $l$ | 0.18 m |
| 7. | Actual iron length | $l_i$ | 0.16 m |
| 8. | Frequency of flux reversal | $f$ | 30 Hz |
| 9. | Line current | $I_l$ | 355.5 A |
| 10. | Number of ventilating duct | | 1 |

### *Armature winding, slot, teeth etc:*

| | | | |
|---|---|---|---|
| 1. | Type of winding | | Simplex wave |
| 2. | Armature conductors | $Z$ | 328 |

| | | | |
|---|---|---|---|
| 3. | Number of slots | $S$ | 41 |
| 4. | Number of coils | | 82 |
| 5. | Slot pitch | | 0.029 m |
| 6. | Width of slot | $b_s$ | 15.4 mm |
| 7. | Height of slot | $h_s$ | 31.0 mm |
| 8. | Current per slot | | 1441 A |
| 9. | Air gap length | $\delta$ | 5.3 mm |
| 10. | Armature m.m.f. per pole, $AT_a$ | $AT_a$ | 7407 A |
| 11. | Internal diameter of armature core | $D_i$ | 0.184 m |
| 12. | Length of mean turn of armature | $L_{mt}$ | 1.19 m |
| 13. | Armature copper loss | | 5.0 kW |
| 14. | Armature temperature rise | | 64.8°C |

***Field system:***

| | | | |
|---|---|---|---|
| 1. | Width of pole body | $b$ | 0.129 m |
| 2. | Height of field winding | $h_f$ | 0.14 m |
| 3. | Height of pole | $h_p$ | 0.17 m |
| 4. | Depth of field winding | $d_f$ | 40 mm |
| 5. | Height of yoke | $h_y$ | 62.5 mm |
| 6. | Outer diameter of yoke | $D_y$ | 0.854 m |
| 7. | Total m.m.f. required at no load | $AT_T$ | 5641 A |
| 8. | Full load field m.m.f. | $AT_{fl}$ | 6487 A |
| 9. | Length of mean turn of field coil | | 0.718 m |
| 10. | Sectional area of field conductor | | 1.96 mm$^2$ |
| 11. | No. of field turns | | 2000 |
| 12. | Temperature rise in field winding | | 32.5°C |

***Commutator:***

| | | | |
|---|---|---|---|
| 1. | Diameter of commutator | $D_c$ | 0.23 m |
| 2. | Commutator peripheral speed | $V_c$ | 10.8 m/sec. |
| 3. | Commutator pitch | | 4.8 mm |
| 4. | No. of commutator segments | | 150 |
| 5. | Current per brush arm | | 179.4 A |
| 6. | Area of brushes in each brush arm | $A_b$ | 1792 mm$^2$ |
| 7. | Length of commutator | $L_c$ | 0.187 m |
| 8. | Total loss at commutator | | 964 watts |
| 9. | Temperature rise in commutator | | 87°C |

***Inter pole:***

| | | | |
|---|---|---|---|
| 1. | Length of air gap under interpole | | 6.89 mm |
| 2. | Width of interpole | | 43.6 mm |
| 3. | No. of turns on each interpole | | 28 |

***Efficiency:***

| | | |
|---|---|---|
| 1. | Total copper loss | 6552 watt |
| 2. | Brush contact loss | 652 watt |
| 3. | Total iron loss | 556 watt |
| 4. | Total friction and windage loss | 567 watt |
| 5. | Total losses | 8327 watt |
| 6. | Input | 80,000 watt |
| 7. | Efficiency | 90.6% |

## PROBLEMS WITH SHORT ANSWERS

**Q. 1 Write short note on reduction of effects of armature reaction in d.c. machines.**

**Ans.** The methods adopted to reduce the effects of armature reaction are:

(*i*) *Increasing the length of air gap at the pole tips.* Air gap in the polar region offers reluctance to path of armature flux. Therefore, if the length of air gap at the polar region increased, effect of armature reaction will be reduced. The length of air gap at the pole tips is made 1.5 to 2 times the length of air gap at the centre.

(*ii*) *Increase in reluctance of pole tips.* By increasing the reluctance of pole tips, the distorting effect of the armature reaction is reduced.

(*iii*) *By using compensating winding.* Such windings are used to neutralize the effect of armature reaction. The conductors of this winding carry currents in opposite direction to that of the adjacent armature conductors in order to nullify the effect of armature m.m.f.

To make it effective at all load, the compensating winding is connected in series with armature winding.

(*iv*) *Use of interpoles.* Interpoles provide an m.m.f. at the brush which is equal and opposite to that of armature m.m.f. This helps in commutation to take place at magnetic neutral axis. These poles are placed at the geometric neutral axis.

**Q. 2 Why interpoles are desirable in d.c. machines?**

**Ans.** Interpoles are desirable in d.c. machines since these causes the following effects:

(*i*) Sparkless commutation can be obtained up to 35 per cent overload with fixed brush position. This ensures automatic neutralisation of reactance voltage which is also due to armature current.

(*ii*) Interpoles neutralize the cross magnetizing effect of armature reaction. Hence brushes are not to be shifted from their original position. The cross magnetization is nullified automatically for all loads because both are produced by the same armature current.

**Q. 3 Why the yoke of a d.c. machine is not laminated whereas the armature core is laminated? Explain.**

**Ans.** Flux in the case of d.c. machine is steady and unidirectional flux or flux of constant value. The armature is a rotating part so the flux in the slot and teeth region changes because of change of reluctance of the magnetic circuit. Whereas in stator there are no such changes, hence yoke of a d.c. machine is not laminated but armature core is laminated so that iron losses in the armature can be reduced.

**Q. 4 Why the brushes are made of carbon and not of copper for a d.c. machine?**

**Ans.** Brushes are made of carbon and not of copper, because the brushes are stationary part and commutator is rotating part, if brushes are made of copper, the wear and tear of commutator surface due to rubbing action will be large and commutator surface will be damaged. Hence, when brushes are made of carbon which is a softer material than copper, there will be no damage to the commutator surface. Brushes can easily be replaced when it is required to replace.

**Q. 5 Why mica is used as the insulation between commutator segments?**

**Ans.** Mica is used to insulate the commutator segments in d.c. machines because it has very high specific weight as well as it is mechanically strong. It is used in thickness of 0.8 mm. It is made by sticking splittings on sheets of paper by means of a drying varnish. These sheets are pressed between steam heated plates and then cooled.

**Q. 6 Why the brushes are staggered?**

**Ans.** In order to prevent formation of ridges on the commutator surface due to all brushes following the same track, the brushes are staggered. This means they are being set slightly different axial positions. This causes uniform wear of the commutator and absence of ridges also.

**Q. 7 Explain the reactance voltage in a d.c. machine.**

**Ans.** The current in the coil undergoing commutation changes from $+I$ to $-I$ and as the coil has a self inductance owing to slot leakage and overhang leakage flux, a voltage is induced in the coil due to change of flux linkage. When there are several coil sides per slot e.m.f. is also induced due to mutual inductance between coil sides in the same coil. This is known as reactance voltage.

**Q. 8 Why riser is provided on the commutator?**

**Ans.** The armature conductors are connected to the commutator with the help of risers. The outer end of riser is shaped in form of a clip into which the armature conductors can be soldered. As the diameter of commutator is about 0.7 times that of rotor so for convenience of armature conductors connection to the commutator risers are used. These are made up of copper strips.

**Q. 9 Write the factors governing the number of poles and also write the advantages of increase in number of poles.**

**Ans.** Factors governing the number of poles are

(*i*) Frequency of flux reversal
(*ii*) Weight of iron parts
(*iii*) Weight of copper
(*iv*) Overall size of machine
(*v*) Commutator length
(*vi*) Flash over between brushes
(*vii*) Labour charges

Advantages of increasing number of poles are reduction in

(*i*) Weight of iron parts and copper
(*ii*) Length of commutator
(*iii*) Overall size of the machine.

## UNSOLVED QUESTIONS

1. Explain the factors which influence the choice of average gap density and ampere-conductor per metre, in the design of d.c. machines.
2. Mention the important guidelines for selecting the number of armature slots.
3. Explain specific electric and magnetic loadings.
4. What is output coefficient? Explain. Find out the expression for the output coefficient for a d.c. machine.
5. What are the factors which influence the choice of number poles in a d.c. machine?
6. Write the various parts of a magnetic circuit of d.c. machine.
7. What are the advantages of selecting large number of poles in a d.c. machine?
8. Explain the different factors which are considered while selecting armature diameter.
9. Explain why the pole shoe is longer than the pole core?
10. Calculate the length and diameter of the armature of a 7.5 kW, 220 V, 1000 r.p.m., 4 pole d.c. shunt motor, with the following data:

    Full load efficiency = 0.83; maximum flux density in the air gap = 0.9 Wb/m$^2$; $\overline{ac}$ = 30,000 A/m, Field form factor = 0.7; maximum efficiency occurs at full load; the field current is 2.5% of the rated current; a square pole face is used. **(AMIE W 1990)**
11. Calculate the main dimensions of the armature core of a 6 pole, 32 h.p., 440 V, 900 r.p.m shunt motor with 84 per cent efficiency. Square pole construction is adopted and pole arc to pole pitch ratio is 0.66. Average flux density is 0.52 Wb/m$^2$ $\overline{ac}$ is 18,000.
12. Explain the power developed by the armature ($P_a$) in a d.c. machine and show that

    (*i*) $P_a = \dfrac{P}{\eta}$, for large generators

    (*ii*) $P_a = P\left(\dfrac{2+\eta}{3\eta}\right)$, for small generators

    (*iii*) $P_a = P$, for large motors

    (*iv*) $P_a = P\left(\dfrac{1+2\eta}{3\eta}\right)$, for small motors.

    where, $P$ is the power output and $\eta$ is the efficiency.
13. Estimate the main dimensions of a 170 kW, 4-pole 450 V, 600 r.p.m generator. Average flux density is 0.6 Wb/m$^2$ and $\overline{ac}$ is 22,000. Pole arc to pole pitch ratio is 0.68 and 86 per cent efficiency.
14. Calculate the diameter and length of armature core of a 120 kW, 220 V, 750 r.p.m., 4-pole generator with the following details:

    Efficiency = 88%

    Average flux density, $\overline{B}$ = 0.56 Wb/m$^2$

    Specific electric loading, $\overline{ac}$ = 26,000 A/m

    Frequency range = 25–40 Hz

    Pole arc to pole pitch ratio = 0.65

# 8

# Constructional Features and Design Procedure of Transformers

## 8.1 INTRODUCTION

The bulk of a.c. power are being generated by the large size generators situated at remote places. The site of the generating stations depends upon the availability of water head for hydropower station, coal or oil for thermal power station etc. The electricity is being generated at maximum of 15.75 kV and mostly used at 400 and 220 volt. The generators are situated far away from the load centres. The electrical power has to be transmitted over the long distance and it is evident that the transmission of electrical power at high voltage is economical. So, we have to step-up the voltage level at generating station then transmit it near to load centres and then step-down the voltage for the use of consumers at 400 or 220 volts.

Transformer, a static electromagnetic device, can change the voltage level *i.e.*, it can step-up the voltage or step-down the voltage. It transfers electrical energy from one circuit to the other circuit without being electrically connected and also without changing the supply frequency. The use of transformer has made the a.c. power preferred commercially. Transformers have very high efficiency.

## 8.2 TYPES OF TRANSFORMERS

Classification on the basis of core:

(*i*) Core type transformer.
(*ii*) Shell type transformer.

Classification on the basis of voltage level:

(*i*) Step-up transformer.
(*ii*) Step-down transformer.

Classification on the basis of service:

(*i*) Distribution transformer.
(*ii*) Power transformer.

Each transformer can be used as step-up transformer or step-down transformer. The larger the rating of the transformers the higher efficiency is obtained. It consists of windings wound on a laminated core. The core provides the path for magnetic flux. The windings are low-voltage winding and high voltage winding usually made of copper. Aluminium windings are also used in small transformers. Rectangular conductors are used for the winding of L.V. winding and H.V. winding of large transformers. Circular conductors are used for H.V. winding of small and medium transformers. Double cotton or single cotton with an under layer of enamel or synthetic enamel insulations is used for conductor insulation.

### 8.2.1 Core Type Transformers

In core type transformers windings encircle the core as shown in Fig. 8.1(*b*). Each core of the transformer is wound with both primary and secondary (or L.V. and H.V.) windings in order to reduce the magnetic leakage. Usually, the coils are cylindrical and are placed one over the other with proper insulation. The portion over which windings are placed are called as the core and the portion which connects the cores is called the yoke. L.V. winding is placed nearer to the core and then H.V. winding over the L.V. winding as shown in Fig. 8.2.

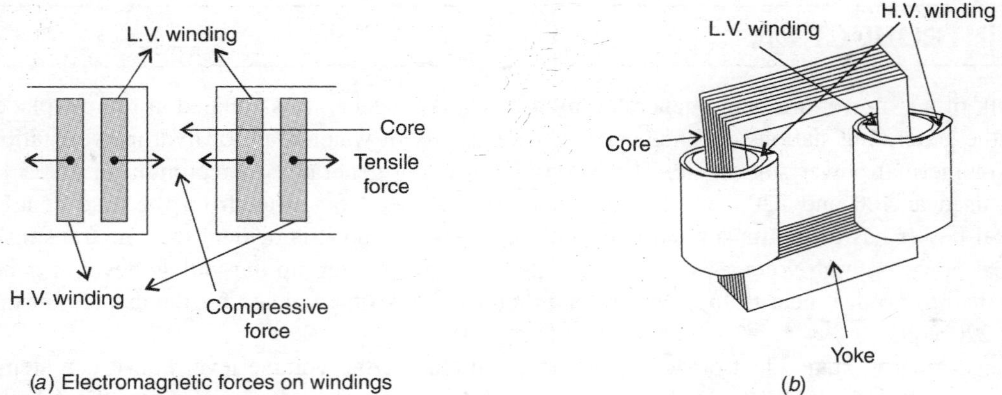


**Fig. 8.1** Single-phase core type transformer.

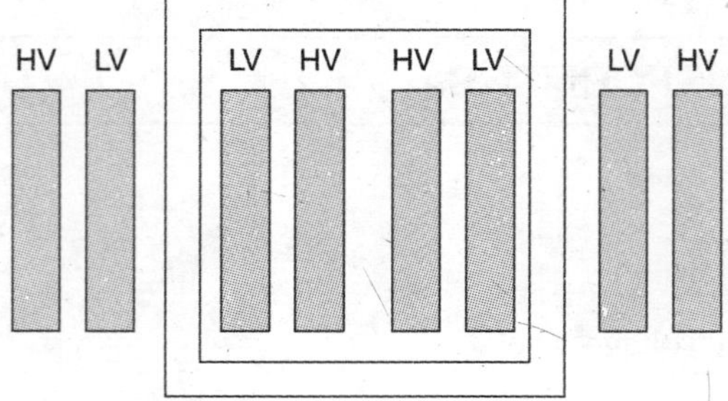


**Fig. 8.2** L.V. and H.V. winding of single-phase core type transformer.

Three-phase core type transformer consists of three cores and two yokes at the top and the bottom. L.V. and H.V. winding of one phase are wound on one core as shown in Fig. 8.3.

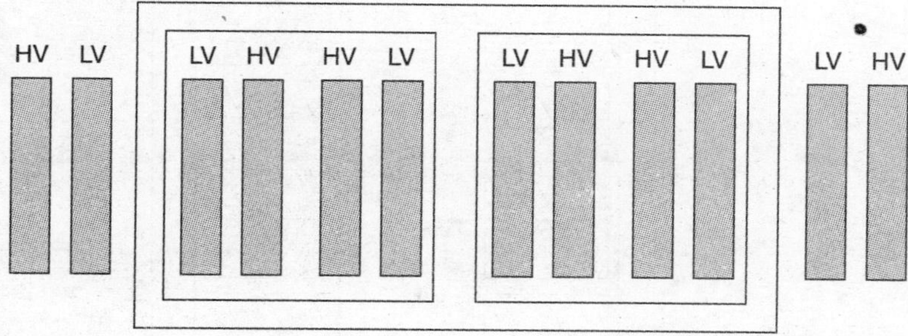


**Fig. 8.3** 3-phase core type transformer.

### 8.2.2 Shell Type Transformers

In shell type transformers the core encircles the windings which is clear from the Fig. 8.4. The windings are wound around the central core of the transformer and the flux path is being completed through side cores. High Voltage (H.V.) and Low Voltage (L.V.) windings are divided into a number of coils and are arranged alternately along the height of core.

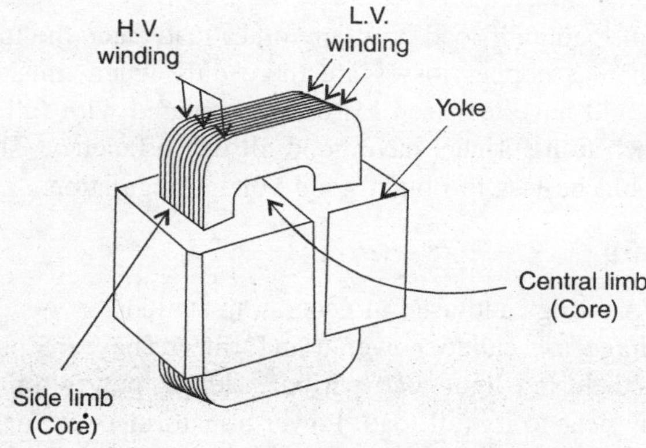


**Fig. 8.4** Single-phase shell type transformer.

Three-phase shell type transformer:
The arrangement of winding in a 3-phase shell type transformer is shown in Fig. 8.5.

## 8.3 DISTRIBUTION AND POWER TRANSFORMERS

### 8.3.1 Distribution Transformers

Transformers up to 500 kVA rating used to step-down the high voltages (transmission level) to a standard service voltages (distribution level) for the consumers. Distribution transformers are located near the consumers. The distribution transformers should be designed to have its maximum efficiency at half of the full level since they are kept in service for whole day (24 hours) irrespective of the load.

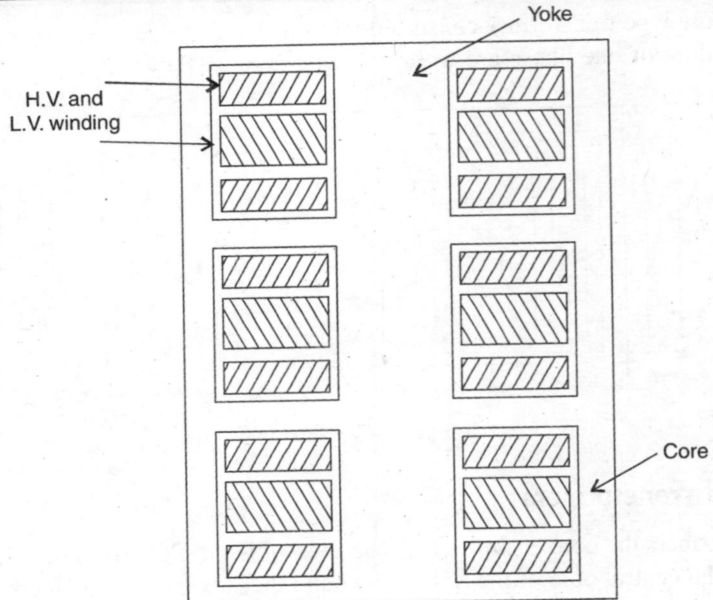


**Fig. 8.5** 3-phase shell type transformer.

It may have to operate at full load, half load or at any other load since the load is varying. Iron losses take place for 24 hours whereas copper losses are there only when transformer is loaded. So, the distribution transformers should have less iron losses as compared with full load copper loss. Due to lower value of iron loss, such transformers have good all day efficiency. The leakage reactance of a distribution transformer should be less to obtain good voltage regulation.

### 8.3.2 Power Transformers

Transformers above 500 kVA ratings and used in generating stations or in sub-stations for stepping-up and stepping-down the voltage are called power transformer. They are put in service during load periods and are disconnected during light load periods. Hence, power transformers are designed to have its maximum efficiency near to its full load. Power transformers are designed for higher leakage reactance than that of distribution transformer. Since the voltage regulation is not so important as the current limiting effect of higher leakage reactance is important in case of a power transformer.

## 8.4 CORE AND YOKE

Eddy current losses and hysteresis losses occur in core and yoke. Eddy current losses are proportional to the square of thickness of lamination. So, the lamination of 0.35 mm of silicon steel sheets are being used to minimize the eddy current loss. Addition of silicon increases the resistance and hence reducing the eddy current loss. Silicon content is up to 3–4%, above this the steel becomes brittle. The core should have minimum iron losses and minimum no load current, so proper core material which should have high permeability, high resistivity and low coercive force should be used. Earlier hot rolled steel were being used working up to a flux density of 1.4 Wb/m$^2$. But due to development of cold rolled grain oriented (CRGO) silicon steel which works up to 1.7 tesla. By using CRGO the specific losses are reduced to approximately 50% than that in hot rolled steels. This reduces the cost also. The laminations are insulated with each other by thin film varnish. Double sided heat resistant coating on sheet steals results higher space factor.

## 8.5 STEPPED CORE AND YOKE

Usually, circular coils are used in H.V. and L.V. winding of transformers. Circular coils have better mechanical strength than that of rectangular coils. It is desirable to use circular core but it is difficult to manufacture a circular core. Rectangular or square core can be used. Square cores are used for small transformers. In case a single rectangular or square core is being used for large transformer, the utilization of space is not proper. To utilize the space available and also to minimize the mean length of transformer turns, stepped core are being used as shown in Fig. 8.6. If the stepped core is used, the diameter of the circumscribing circle over the core will be less. Hence, there is saving in copper for

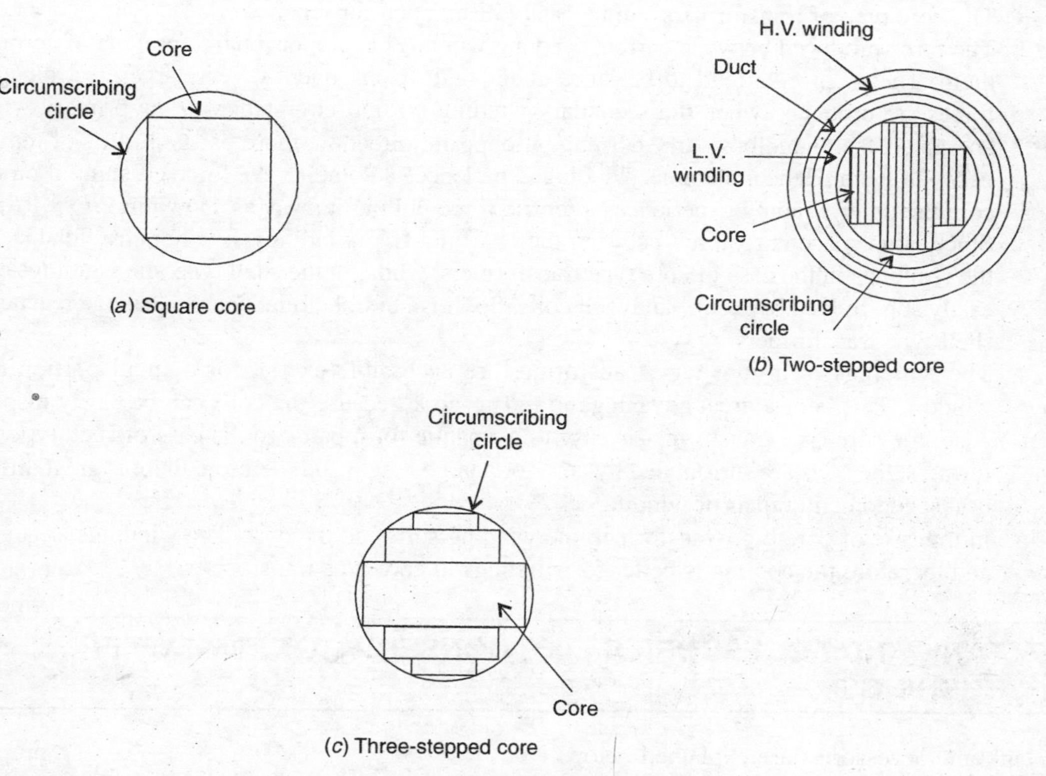


**Fig. 8.6** Square, two-stepped and three-stepped cores.

the windings. When number of steps increases charges for shearing and assembling of lamination increases considerably. Number of steps depends upon the kVA rating in the gross core sectional area. Table 8.1 gives the number of steps for various values of gross core-section.

Rectangular yoke is being used for small transformers and two or three-stepped yoke is usually used for medium and large transformers. The sectional area of yoke is taken 10 to 15% more than the

**Table 8.1:** Number of steps in core

| Gross core section ($m^2 \times 10^{-3}$) | <3 | 3–5 | 5–7 | 7–15 | 15–45 | 45–80 | 80–200 |
|---|---|---|---|---|---|---|---|
| Number of steps | 1 | 2 | 3 | 4 | 5 | 6 | 7 |

sectional area of core so that the iron losses are reduced in yoke. No load current is also reduced in yoke.

## 8.6 COMPARATIVE CHARACTERISTICS OF CORE AND SHELL TYPE TRANSFORMERS

The comparative analysis between core and shell type transformers are given below:

1. Core type transformers are simpler in design and have easier assembly and insulation of windings. The core type of transformers can be easily dismantled for repair work.
2. The force produced between current carrying windings is proportional to product of currents in them. These currents tend to be large under fault. Consequently very large electromagnetic forces are produced when the secondary winding is short-circuited with the primary winding energized. The windings carry currents in opposite direction, there exists a force of repulsion between them. Hence, the inner winding experiences a compressive force crushing it on to the core the outer winding experiences a tensile force pulling it away as shown in Fig. 8.1 (*a*).
3. Since large space is required between the H.V. and L.V. windings, it is not easy to sub-divide the windings in the case of core type transformers, while, in the shell type, the windings can be easily sub-divided by using sandwich coils. So, it is possible to reduce the leakage reactance of shell type transformers.
4. The windings of a core type transformer are accessible except for a small portion in the window. This is of a great advantage in repair work because the coils can be easily inspected. Also, the core type transformer is easy to dismantle for repairs. In the case of shell type transformers, the coils are surrounded by core for a large length and therefore there is great difficulty in inspection and repair of windings.
5. In the case of core type transformer, the winding surround the core. The windings are exposed and therefore the cooling is better for windings in core type transformer.

## 8.7 TANK AND ACCESSORIES LIKE CONSERVATOR, BREATHER, BUSHINGS

The tank and accessories are explained below:

The transformer winding is placed in a tank, filled with transformer oil, with good insulating properties. Function of oil is to transfer the heat by convection from the heated surface to the tank surface, so cooling various parts of the transformer.

Tanks for small transformers are fabricated from welded sheet steel, and for large transformers from plain boiler plates. The lids of these transformer tanks can be of cast iron. A water proof gasket is used at the joints.

Other different accessories with transformer tank are, thermometer pockets, drain cock, rollers or wheels for moving the transformer from one place to another, breather, bushings and buchholz relay. Conservator, breather and buchholz relay provide protection to the transformer.

For cooling, cooling tubes are welded with the tank, but for radiators, separate radiators are individually welded and then bolted-on to the transformer tank.

Transformers with voltage rating of 6 kV and the output rating of 25 kVA and more should be provided with oil conservator.

The conservator is of airtight cylindrical metal drum which is supported horizontally on a neighbouring wall. When the transformer is working, its temperature rises, rise being excessive in case of over loads. Due to these changes of temperature the oil in the tank of the transformer undergoes the process of expansion and contraction.

Conservators are provided over the transformer tanks to absorb this expansion and contraction of oil, without allowing the oil to come in contact with the air. Usually the conservator capacity should be 10 to 12 per cent of the oil volume in the tank.

A breather mounted on the transformer tank contains calcium chloride or silica gel, which extracts the moisture from the air. Due to changes in the oil volume the displacement of air due to these changes of oil volume takes place through the breather, which can extract the moisture from the air. The silica gel extracts the moisture from air.

The terminal connection of the windings are taken to the insulator bushings mounted on the transformer tank. The bushings consist of a current carrying part in the form of a conducting rod, and a porcelain cylinder installed in the hole of the cover and used to isolate the current-carrying part. Porcelain insulators are used up to a voltage rating of 33 kV. When the voltage of the H.V. winding of the transformer exceeds the above value, the condenser bushings or oil filled terminal bushings are normally used. For certain cases, a combination of the two types of bushings is also used.

## 8.8 OUTPUT EQUATION

The design of a transformer is initiated with the output equation for that transformer. The output equation for a transformer gives the relation for output of transformer, supply frequency, net cross-section of core, maximum flux density, current density in windings, window space factor and area of window. The fundamental equation for the transformer rating is used to derive the output equation for that transformer. Based on the fundamental equation for transformer rating and other parameters, different output equations are obtained for different types of transformers. The derivation of output equation for different types of transformers are explained below.

To find out the output equation of a transformer the output equation of single-phase core type transformer is first explained. Rating for a single phase transformer is given by

$$Q = V_1 I_1 \times 10^{-3} \text{ kilovolt ampere}$$

where, voltage $V_1 = 4.44 f\phi_1 N_1$

So,

$$Q = 4.44 f\phi_1 N_1 I_1 \times 10^{-3} \quad \text{...(8.1)}$$

where, $Q$ = rating in kVA

$I_1$ = primary winding current,

$I_2$ = secondary winding current

$N_1$ = number of primary winding turn

$N_2$ = number of secondary winding turn

$\phi_1$ = peak value of fundamental flux in core, measured in Wb.

$\phi_1 = A_i B_m$

$B_m$ = peak value of permissible flux density

$A_i$ = net cross-section of core

Net cross-section of the core means the cross-section of actual iron in the core of transformer excluding insulation provided in laminations of the core. When, the insulation is also taken into considerable and the cross-sectional area of core is estimated, it gives the gross cross-sectional area of the transformer core.

$$\text{Gross cross-section of core} = \frac{\text{Net cross-section of core}}{\text{Stacking factor}}$$

But, stacking factor = 0.9, for 0.5 mm and 0.35 mm thickness of lamination,

0.95, for 1.0 mm thickness of lamination.

So, $A_i$ = Stacking factor × Gross core-section

= Stacking factor × $A_g$

$$A_i = kd^2$$

where, $k$ is the factor which depends upon the number of steps in the core and $d$ is the inner diameter of the circumscribing circle of the winding on the core.

The values of $k$ for different steps in the core is given in Table 8.2.

**Table 8.2:** Values of $k$ for different core steps

| *No. of steps in core* | *Square* | *Two-stepped (Cruciform)* | *Three-stepped* | *Four-stepped* |
|---|---|---|---|---|
| Values of $k$ | 0.45 | 0.56 | 0.6 | 0.62 |

$$A_y = 1.15\, A_i$$

where, $A_y$ is cross-sectional area of yoke.

The yoke section is taken 15% more than the core section so that flux density in yoke will be less and hence iron loss and magnetizing current will be less. In the transformers in which C.R.G.O. material is used, the cross-sectional area of the yoke may be taken equal to the cross-sectional area of core *i.e.*,

$$A_y = A_i$$

The width of the largest stamping of the core inside the winding can be estimated depending upon the inner diameter of the circumscribing circle, $d$ of the winding on the core. The width of the core, $W_c$ will be different for different stepped core. The width for largest steps in the transformer can be taken from following relation:

$W_c = 0.71d$, for square core,

$= 0.85\, d$, for two stepped core (or cruciform section),

= 0.90 $d$, for three stepped core,

= 0.93 $d$, for four stepped core.

Estimation of window space factor:

Window space factor can be defined as the ratio of actual copper section in the window to the actual window area of the transformer and can be written as

$$K_w = \frac{\text{Actual copper sectional area in the window}}{\text{Actual window area of transformer}}$$

$$= \frac{\text{Winding conductor area in window}}{\text{Actual window area}}$$

### 8.8.1 Single-phase (1-ϕ) Core Type

In a single-phase core type transformer there is one L.V. winding and one H.V. winding accommodated in a single window. Half of L.V. winding and half of H.V. winding are placed on one core and the other half of L.V. and H.V. windings are placed on other core of the transformer as shown in Fig. 8.7. L.V. windings are placed nearer to the core.

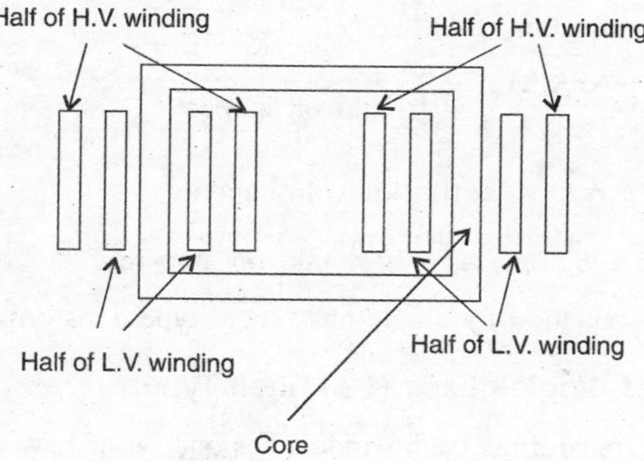


**Fig. 8.7** Single-phase core type transformer.

Let, cross-sectional area of primary winding conductor,

$$a_1 = \frac{I_1}{\sigma}$$

Cross-sectional area of secondary winding conductor,

$$a_2 = \frac{I_2}{\sigma}$$

where, σ is the average value of current density permissible in L.V. winding and H.V. winding depending upon cooling of transformer. Although the current density in the L.V. winding and H.V. winding are different. Current density in H.V. winding is slightly higher than the current density in L.V. winding. Since the H.V. winding is placed on outer portion and hence cooling is better. Due to better cooling the current density in H.V. winding may be more than the current density in L.V. winding.

Total conductor sectional area in each window, $A_c = a_1 N_1 + a_2 N_2$

For an ideal transformer, $I_1N_1 = I_2N_2$

So, window space factor,

$$K_w = \frac{a_1N_1 + a_2N_2}{A_w} \quad ...(8.2)$$

where, $A_w$ is the sectional area of the window.

From equation (8.2), we get,

$$a_1N_1 + a_2N_2 = K_wA_w$$

or
$$\frac{I_1}{\sigma} \cdot N_1 + \frac{I_2}{\sigma} \cdot N_2 = K_wA_w$$

or
$$2\frac{I_1N_1}{\sigma} = K_wA_w \quad \{\because I_1N_1 = I_2N_2\}$$

or
$$I_1N_1 = \frac{\sigma \cdot K_wA_w}{2} \quad ...(8.3)$$

Putting the value of $I_1\,N_1$ in equation (8.1), we get,

$$Q = 4.44\,f\,\phi_1 \frac{\sigma\,K_wA_w}{2} \times 10^{-3} \text{ kilovolt ampere}$$

$$= 2.22\,f\,\phi_1 \sigma\,K_wA_w \times 10^{-3} \text{ kilovolt ampere} \quad \{\text{since } \phi_1 = A_i \cdot B_m\}$$

$$Q = 2.22\,f\,A_i\,B_m\,\sigma\,K_wA_w \times 10^{-3} \text{ kilovolt ampere} \quad ...(8.4)$$

Equation (8.4) is the output equation for single-phase core type transformers.

### 8.8.2 Output Equation of Single-phase (1-$\phi$) Shell Type

In a single-phase shell type transformer each window has one set of L.V. winding and one set of H.V. winding as given in Fig. 8.8.

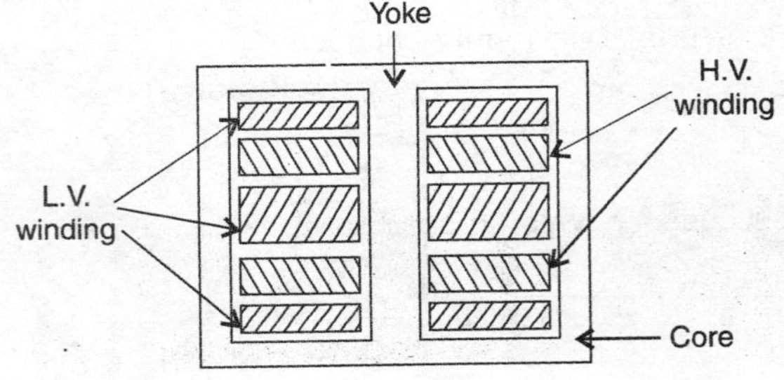


**Fig. 8.8** Single-phase shell type.

So, total conductor area in each window,

$$A_c = a_1N_1 + a_2N_2$$

Window space factor of single-phase shell type transformer,

$$K_w = \frac{a_1 N_1 + a_2 N_2}{A_w}$$

or $$a_1 N_1 + a_2 N_2 = K_w A_w$$

or $$\frac{I_1}{\sigma} \cdot N_1 + \frac{I_2}{\sigma} \cdot N_2 = K_w A_w$$

Since $I_1 N_1 = I_2 N_2$

$\therefore$ $$2 I_1 N_1 = \sigma K_w A_w$$

$$I_1 N_1 = \frac{\sigma K_w A_w}{2} \quad ...(8.5)$$

Now the rating of single-phase shell type transformer is given by

$$Q = V_1 I_1 \times 10^{-3} \text{ kilovolt ampere}$$

$$= 4.44 f \phi_1 N_1 I_1 \times 10^{-3} \text{ kilovolt ampere}$$

Now, putting the value of $I_1 N_1$ from equation (8.5), in above equation, we get

$$Q = 4.44 f \phi_1 \frac{\sigma K_w A_w}{2} \times 10^{-3} \text{ kilovolt ampere}$$

$$= 2.22 f \phi_1 \sigma K_w A_w \times 10^{-3} \text{ kilovolt ampere}$$

Since $\phi_1 = A_i \, B_m$

So, $$Q = 2.22 f A_i B_m \sigma K_w A_w \times 10^{-3} \text{ kilovolt ampere} \quad ...(8.6)$$

Equation (8.6) is the output equation for a single-phase shell type transformer.

### 8.8.3 Output Equation for a 3-phase (3-$\phi$)Core Type Transformer

In case of a 3-phase core type transformer each window contains two L.V. windings and two H.V. windings as shown in Fig. 8.9.

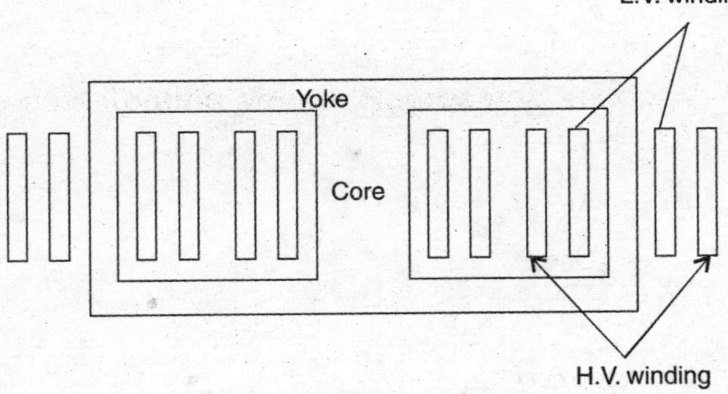


**Fig. 8.9** 3-phase core type transformer.

Chapter 8

Total conductor area in each window,

$$A_c = 2(a_1 N_1 + a_2 N_2)$$

So, window space factor,

$$K_w = \frac{2(a_1 N_1 + a_2 N_2)}{A_w}$$

$$= \frac{2}{A_w}\left(\frac{I_1}{\sigma} \cdot N_1 + \frac{I_2}{\sigma} \cdot N_2\right) \qquad [\because I_1 N_1 = I_2 N_2]$$

or
$$I_1 N_1 = \frac{\sigma K_w A_w}{4} \qquad \text{...(8.7)}$$

Now the rating of 3-phase core type transformer

$$Q = 3 V_1 I_1 \times 10^{-3} \text{ kilovolt ampere}$$

$$= 3 \times 4.44 f \phi_1 N_1 I_1 \times 10^{-3} \text{ kilovolt ampere} \qquad \text{...(8.8)}$$

Putting the value of $I_1 N_1$ from equation (8.7) in equation (8.8), we get,

$$Q = 3 \times 4.44 f \phi_1 \frac{\sigma K_w A_w}{4} \times 10^{-3} \text{ kilovolt ampere}$$

Also,
$$\phi_1 = A_i B_m$$

$$Q = 3.33 f A_i B_m \sigma K_w A_w \times 10^{-3} \text{ kilovolt ampere} \qquad \text{...(8.9)}$$

Equation (8.9) is the output equation for a 3-phase core type transformer.

### 8.8.4 Output Equation of a 3-phase (3-ϕ) Shell Type Transformer

In 3-phase shell type transformer each window contains one L.V. winding and one H.V. winding as shown in Fig. 8.10.

Total conductor area in the window, $A_c = a_1 N_1 + a_2 N_2$

Window space factor, $K_w = \dfrac{a_1 N_1 + a_2 N_2}{A_w}$

or
$$a_1 N_1 + a_2 N_2 = K_w A_w$$

or
$$\frac{I_1}{\sigma} \cdot N_1 + \frac{I_2}{\sigma} \cdot N_2 = K_w A_w$$

Since
$$I_1 N_1 = I_2 N_2$$

So,
$$I_1 N_1 = \frac{\sigma K_w A_w}{2}$$

Rating of 3-ϕ shell type transformer,

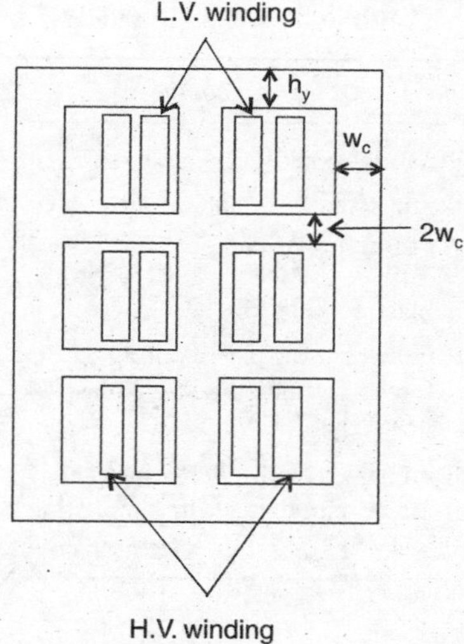


**Fig. 8.10** 3-phase shell type transformer.

$$Q = 3V_1 I_1 \times 10^{-3} \text{ kilovolt ampere}$$

$$= 3 \times 4.44 f \phi_1 N_1 I_1 \times 10^{-3} \text{ kilovolt ampere} \quad ...(8.10)$$

Putting the value of $I_1 N_1$ from equation (8.10) and $\phi_1 = A_i B_m$, we have

$$Q = 3 \times 4.44 f \phi_1 \frac{\sigma K_w A_w}{2} \times 10^{-3}$$

$$= 6.66 f A_i B_m \sigma K_w A_w \times 10^{-3} \text{ kilovolt ampere} \quad ...(8.11)$$

Equation (8.11) is the output equation for a 3-$\phi$ shell type transformer.

## 8.9 WINDOW SPACE FACTOR, CURRENT DENSITY AND FLUX DENSITY

(*i*) Selection of window space factor:

$$K_w = \frac{8}{30 + kV}, \text{ for small transformers up to 10 kVA,}$$

where, kV is the phase voltage of H.V. winding.

$$K_w = \frac{10}{30 + kV}, \text{ for transformers up to 250 kVA,}$$

and $$K_w = \frac{12}{30 + kV}, \text{ for transformers above 250 kVA.}$$

(*ii*) Selection of current density:

The current density may be approximately selected for a transformer to be designed.

**Table 8.3:** Current density

| | *Type of transformer cooling* | $\sigma$ *(A/mm²)* |
|---|---|---|
| Natural cooling | Air natural cooling<br>Oil immersed with natural air cooled<br>(100 kVA–1 MVA) | 1.2–2.3 |
| Forced cooling | Air blast cooling<br>Oil immersed air-blast cooling<br>Oil forced circulated and air blast cooling<br>(2–3 MVA) | 2.2–4.0 |
| Cooling with circulating water | Oil immersed with circulating water<br>Oil forced circulated with circulating water<br>(100 MVA) | 4.5–6 |

(*iii*) Choice of flux density:

$B_m$ = 0.9 to 1.1 T, for distribution transformer,

= 1.1 to 1.4 Wb/m², for power transformer,

= 1.2 to 1.4 T, for hot rolled Si steel,

= 1.4 to 1.7 T, for cold rolled grain oriented (CRGO) Si steel.

For CRGO transformer up to 132 kV = 1.55 Wb/m²

For CRGO transformer up to 275 kV = 1.6 T

For CRGO transformer up to 400 kV = 1.7 T

The distribution transformer has to be designed for high all day efficiency so, $B_m$ should be low to keep low iron losses. Tesla (T) and Weber/m² both are same.

## 8.10 EMF PER TURN ($E_t$)

KVA rating per phase is given by

$$Q = VI \times 10^{-3}$$

E.M.F. per turn, $E_t = 4.44 f \phi_1$ ...(8.12)

Or $E_t = 4.44 f B_m A_i$

$$Q = 4.44 f \phi_1 I N \times 10^{-3} \quad ...(8.13)$$

For a given transformer, the ratio of magnetic loading to current loading is constant. This characteristics is obtained because the ratio of cross-sectional area of core and cross-sectional area of winding is constant for a given transformer.

So, $\dfrac{A_i}{A_c} = \text{Constant}$

$$\frac{\phi_1 \times \sigma}{B_m \times IN} = \text{Constant}(\gamma)$$

Since $\sigma$ and $B_m$ are nearly constant.
Hence

$$\frac{\phi_1}{IN} = \text{Constant}$$

or $$\frac{\phi_1}{IN} = \gamma \qquad ...(8.14)$$

or $$IN = \frac{\phi_1}{\gamma} \qquad ...(8.15)$$

The value of $IN$ can be put in equation (8.13) and

$$Q = 4.44\, f \phi_1 \frac{\phi_1}{\gamma} \times 10^{-3} \text{ kilovolt ampere}$$

$$= 4.44\, f \frac{\phi_1^2}{\gamma} \times 10^{-3} \text{ kilovolt ampere}$$

or $$\phi_1^2 = \frac{Q.\gamma \times 10^3}{4.44\, f}$$

or $$\phi_1 = \sqrt{\frac{Q\, \gamma \times 10^3}{4.44\, f}} \qquad ...(8.16)$$

Putting the value of $\phi_1$ in equation (8.12), then e.m.f. per turn,

$$E_t = 4.44\, f \sqrt{\frac{Q\, \gamma \times 10^3}{4.44\, f}}$$

$$= \sqrt{4.44 \times f \times Q \times \gamma \times 10^3}$$

$$= K_t \sqrt{Q} = K_t \sqrt{kVA \text{ rating}} \qquad ...(8.17)$$

where, $K_t = \sqrt{4.44\, f \gamma \times 10^3}$ is constant for a transformer. $Q$ is the kVA per phase.

The value of $K_t$ depends upon

(*i*) Type of transformer (core or shell type). It is more for shell type than the core type for same rating since shell type transformer require more magnetic material.

(*ii*) Material employed in construction.

(*iii*) Choice of electric and magnetic cooling.

The value of $K_t$ for different types of transformer is given in Table 8.4.

**Table 8.4:** Value of $K_t$

| *Types of transformer* | $K_t$ |
|---|---|
| 3–ϕ shell type | 1.3 |
| 3–ϕ core type power | 0.6–0.7 |
| 3–ϕ core type distribution | 0.45 |
| 1–ϕ shell type | 1–1.2 |
| 1–ϕ core type | 0.75–0.85 |

## 8.11 DIFFERENT DIMENSIONS OF TRANSFORMER

Different portions of single-phase core type and three-phase core type transformer are shown in Figs. 8.11(*a*) and (*b*). Similarly the different dimensions of 1-ϕ and 3-ϕ shell type transformers are shown in Figs. 8.11(*c*) and (*d*).

$W_w$ = Width of window

$h_w$ = Height of the window

$D$ = Distance between centers of two adjacent limbs

$W$ = Overall width of transformer

$H$ = Overall height of transformer

$d$ = Inner diameter of the circle enclosing winding

$A_y$ = Cross-sectional area of yoke

$A_i$ = Cross-sectional area of core

$W_c$ = Width of longest stamping of core inside the circumscribing circle of the winding

$h_y$ = Height of yoke of transformer

$d_0$ = Outer diameter of H.V. winding (winding placed on outer portion)

(*i*) Different dimensions of a 1-ϕ core type transformer Fig. 8.11(*a*)

$h_w$ = Height of window

$W_w$ = Width of window

$W_c$ = Width of core

$h_y$ = Height of yoke

$D$ = Distance between centres of two adjacent limbs = $W_w + W_c$

$W$ = Overall width of transformer = $D + W_c$

$H$ = Overall height of transformer = $h_w + 2h_y$

$d$ = Inner diameter of the circle enclosing winding

(*ii*) Different dimensions of a 3-ϕ core type transformer Fig. 8.11 (*b*)

$h_w$ = Height of window

$W_w$ = Width of window

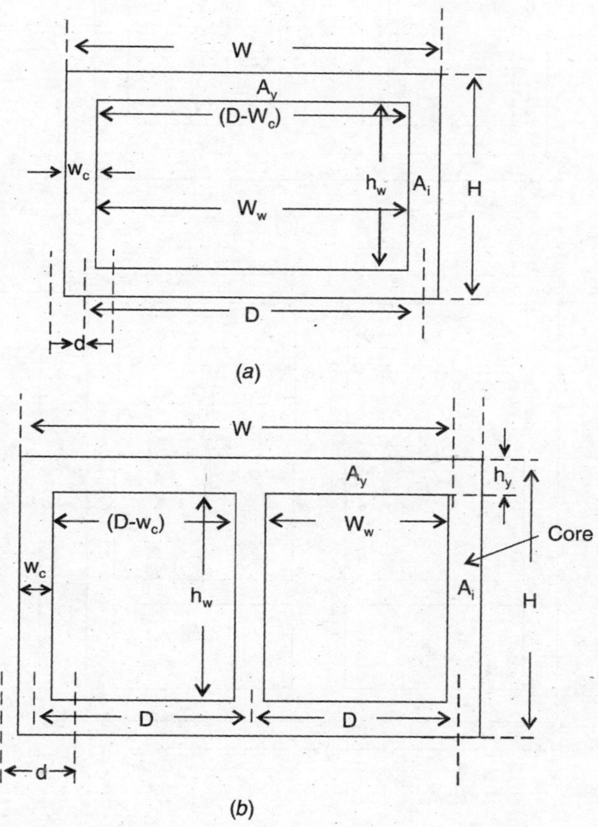


**Fig. 8.11** Different dimensions of 1-ϕ and 3-ϕ core type transformers.

$W_c$ = Width of core

$h_y$ = Height of yoke

$d$ = Inner diameter of the circle enclosing winding

Distance between two adjacent limbs, $D = W_w + W_c$

Overall width of transformer, $W = 2D + W_c$

Overall height of transformer, $H = h_w + 2h_y$

(*iii*) Different dimensions of 1-ϕ shell type transformer Fig. 8.11 (*c*)

$h_w$ = Height of window

$h_y$ = Height of yoke

$W_w$ = Width of window

$W_c$ = Width of core

Chapter 8

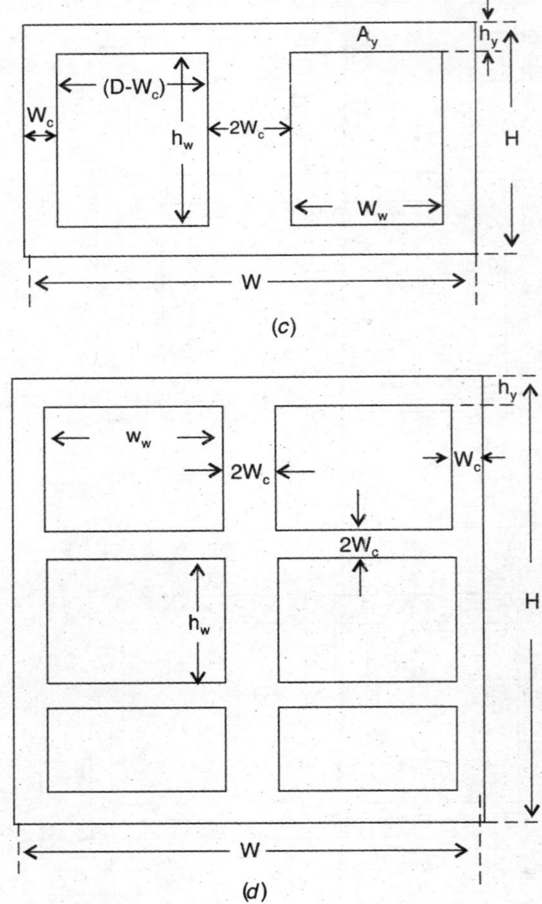


**Fig. 8.11** Different dimensions of 1-φ and 3-φ shell type transformers.

Overall width of transformer, $W = 2W_w + 4W_c$

Overall height of transformer, $H = h_w + 2h_y$

$d$ = Inner diameter of the circle enclosing winding

(*iv*) Different dimensions of 3-φ shell type transformer Fig. 8.11 (*d*)

$d$ = Inner diameter of circle enclosing winding

$W_c$ = Width of core

$h_y$ = Height of yoke

$W_w$ = Width of window

$h_w$ = Height of window

Overall width of transformer, $W = 2W_w + 4W_c$

Overall height of transformer, $H = 3h_w + 6W_c$

### *Main Dimensions of Transformer*

The main dimensions of a transformer are

(*i*) Inner diameter of circumscribing circle

(*ii*) Width of window

(*iii*) Height of window

(*iv*) Overall width of transformer

(*v*) Overall height of transformer

## 8.12 STEPS TO DESIGN A TRANSFORMER

Different dimensions of a transformer can be estimated with the following steps:

(*i*) Select $K_t$ for a particular transformer. Estimate $E_t$ (e.m.f. per turn) from equation (8.17), for given kVA rating.

(*ii*) Estimate iron cross-section, $A_i$ from equation (8.12) by selecting suitable value of $B_m$.

(*iii*) Calculate inner diameter of circle enclosing the winding, $d$ from $A_i = kd^2$. Estimate the width of largest step, $W_c$.

(*iv*) Estimate window area, $A_w$ from output equation by selecting suitable value of current density and suitable value of window space factor, $K_w$.

(*v*) Select suitable value of window space factor $K_w$.

(*vi*) Window area, $A_w = h_w \times W_w$

(*vii*) The ratio of height of windows to width of window is taken from 2 to 4.

$$\frac{h_w}{W_w} = 2 \text{ to } 4$$

(*viii*) Select suitable value of $h_w / W_w$ ratio and calculate height, $h_y$ and width, $W_w$ of core from points (*vi*) and (*vii*).

(*ix*) Overall width of transformer can be calculated.

(*x*) Width of yoke is calculated as width of yoke

$W_y = 0.9d$, for 3-stepped core.

$= 0.85d$, for cruciform section.

$= 0.71d$, for square section.

Generally, rectangular section is considered for making small and medium transformers:

(*xi*) Sectional area of yoke, $A_y$ is taken 15% more than $A_i$.

$$A_y = 1.15\, A_i$$

If C.R.G.O. is used, $A_y = A_i$

(*xii*) Height of the yoke, $h_y$ is calculated from $A_y = h_y \times W_y$

## 8.13 ESTIMATION OF NO LOAD CURRENT: ($I_0$)

No load current, $I_0 = \sqrt{I_\mu^2 + I_c^2}$

where, $I_c$ is the core loss component of no load current and $I_\mu$ is the magnetising component of no load current.

***(i) Core loss component ($I_c$):*** $B_m$ is the flux density in core of transformer.

In transformer if $A_y = 1.15\ A_i$

Flux density in yoke, $B_y = B_m \dfrac{A_i}{A_y}$ tesla

Volume of iron in core = Number of core in transformer × Height of window × Core section

$$V_c = \text{Number of core} \times h_w \times A_i$$

Weight of iron in core = 7600 × $V_c$

Let $p_{ic}$ is the specific iron loss per kg weight of iron (watt/kg) in core corresponding to the flux density, $B_m$ in the core and thickness of lamination of the core (Fig. 8.12).

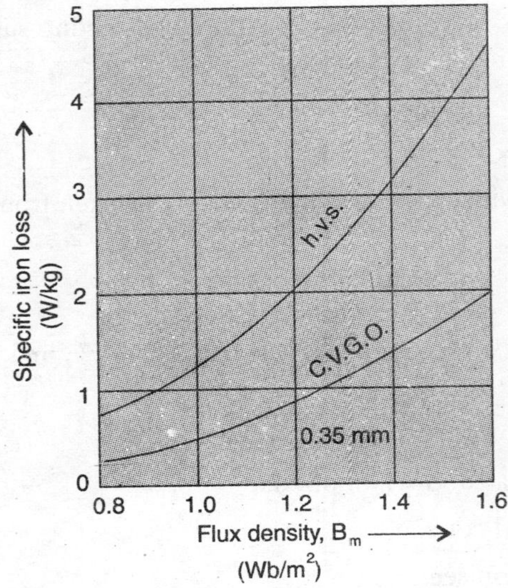


**Fig. 8.12** Specific iron loss curve for transformers.

So, iron loss in transformer core,

$$W_{ic} = p_{ic} \times 7600 \times V_c \text{ watt}$$

Volume of iron in yoke, $V_y$ = Number of yoke × $W$ × $A_y$

Weight of iron in yoke = 7600 × $V_y$

If $p_{iy}$ is the specific iron loss (watt/kg) in yoke corresponding to flux density in yoke, $B_y$ then

Iron loss in yoke, $W_{iy} = p_{iy} \times 7600 \times V_y$

Total iron loss in transformer, $W_i = W_{ic} + W_{iy}$ watt

$$= p_{ic} \times 7600 \times V_c + p_{iy} \times 7600 \times V_y$$

Core loss component,

$$I_c = \frac{W_i}{3 \times V_1} \quad \text{...(8.18)}$$

***(ii) Magnetizing component ($I_\mu$):*** Corresponding to *B*, $at_c$ can be obtained from *B-H* curve as given in Fig. 8.13.

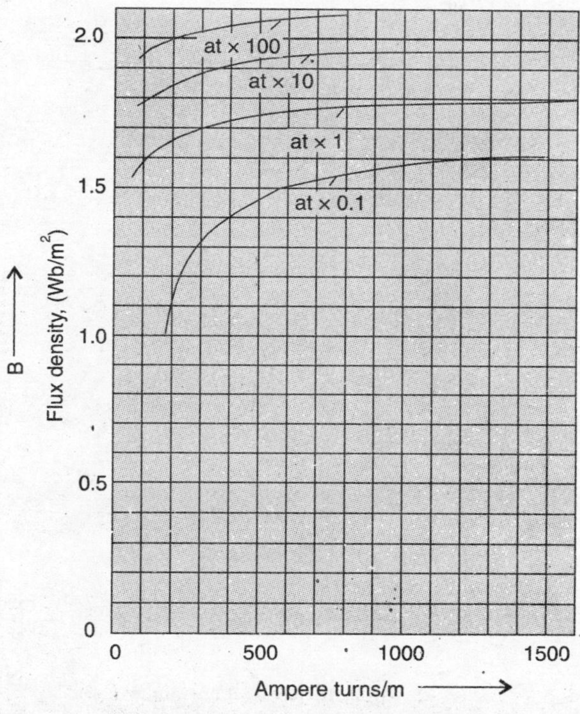


**Fig. 8.13** B-H curve for CRGO steel.

$at_c$ = Ampere-turn per meter length of core for producing flux density, $B_m$ in the core.

Ampere-turn or m.m.f. required for core,

$$AT_c = h_w \times at_c \times \text{Number of core in transformer}$$

Let $at_y$ is the ampere-turn per meter length of yoke for producing flux density, $B_y$ in yoke.

Ampere-turn required for yoke is

$$At_y = \text{Number of yoke} \times W \times at_y$$

Total m.m.f. required for core and yoke of transformer is

$$AT_{cy} = AT_c + AT_y$$

For joints in transformers, 5% additional ampere-turn for core and yoke can be taken

So, total m.m.f., $AT_T = 1.05\,[AT_{cy}]$

Magnetizing component current is calculated on the basis of maximum flux density, $B_m$. So, RMS value of magnetizing current.

(*i*) For a 3-$\phi$ transformer, $I_\mu = \dfrac{AT_T}{3N_1}$ ...(8.19)

where, $N_1$ is the number of turns in primary.

(*ii*) For a 1-$\phi$ transformer,

$$I_\mu = \frac{AT_T}{\sqrt{2}\,N_1} \qquad ...(8.20)$$

$$I_0 = \sqrt{I_c^2 + I_\mu^2} \qquad ...(8.21)$$

$$\cos\phi_0 = \frac{I_c}{I_0} \text{ (Fig. 8.14)} \qquad ...(8.22)$$

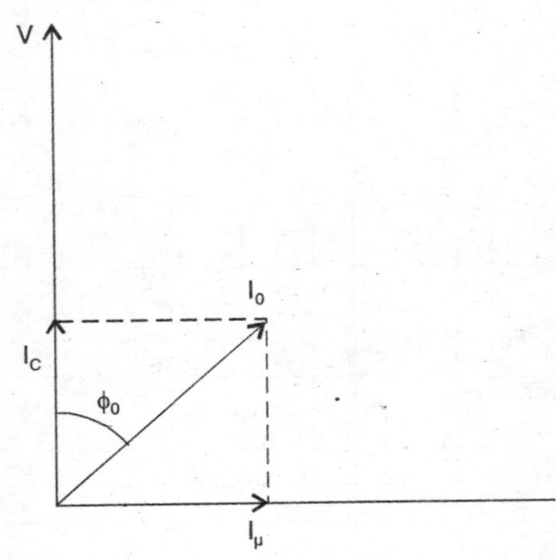


**Fig. 8.14** No load current.

If magnetizing VA is given, value of magnetizing current component can be calculated by

$$I_\mu = \frac{\text{Magnetizing VA / kg} \times \text{weight of core}}{\text{Number of phases } \times \text{Voltage / phase}} \qquad ...(8.23)$$

## 8.14 DETERMINATION OF NUMBER OF TURNS AND LENGTH OF MEAN TURNS OF WINDINGS

***(i) Inside diameter and outside diameter of L.V. winding***

Estimate L.V. winding phase voltage.

$$\text{Estimate L.V. winding turns per phase} = \frac{\text{L.V. winding phase voltage}}{\text{Voltage per turn}} \qquad ...(8.24)$$

Calculate L.V. winding phase current

$$\text{For 1-}\phi\text{ transformer, phase current} = \frac{\text{kVA or VA rating}}{\text{Phase voltage}}$$

$$\text{For 3-}\phi\text{ transformer, phase current} = \frac{\text{kVA or VA rating}}{\text{Phase voltage}}$$

Select (or given) a suitable current density of L.V. winding.
Cross-sectional area of L.V. winding conductor,

$$a_l = \frac{\text{L.V. winding phase current}}{\text{Current density}}$$

From Table A.1 size of bare conductor can be selected and then the correct area of bare conductor can be calculated. With this calculated area of bare conductor, corrected current density can be calculated. Now the conductor (winding) has to be placed on the core of the transformer. Number of turns and size of conductor are known. The arrangement of L.V. winding conductors and insulation along the height and width of the transformer window can be done. This is done in detail in overall design problem. Radial width and axial height of L.V. winding can be estimated as below:

Radial width of L.V. winding = Number of layers along width of window × Radial width of conductor + Insulation thickness between all the layers.

Axial height of L.V. winding = Number of layers along height of window × Axial height of conductor.

Inside diameter of L.V. winding = Diameter of circumscribing circle + 2 × Wrap thickness of insulation between L.V. winding and core (of suitable thickness depending upon L.V. winding voltage).

Outside diameter of L.V. winding = Inside diameter of L.V. winding + 2 × Radial width of L.V. winding.

***(ii) Inside diameter and outside diameter of H.V. winding***

Estimate H.V. winding phase voltage according to the type of connection whether star or delta.

$$\text{Number of H.V. winding turns/phase} = \frac{\text{H.V. winding phase voltage} \times \text{L.V. winding turns per phase}}{\text{L.V. winding phase voltage}} \qquad ...(8.25)$$

Tapping on H.V. winding are to be provided. The number of turns will increase according to percentage of tappings to be provided on H.V. winding. Estimation of number of coils, voltage per coil, turns per coil, layers per coils, turns per layer, H.V. winding phase current. Cross-sectional area of H.V. winding conductor. Size of bare conductor, size of conductors with insulation covering, corrected value of current density, insulations required, inside diameter of H.V. winding and outside diameter of H.V. winding are given in the overall design problem.

The thickness of major insulation of L.V. windings and H.V. windings can be approximately selected from Table 8.5 and Table 8.6. For the transformers having voltage more than 33 kV, the thickness of insulation can be suitably selected depending upon the voltage of H.V. and L.V. windings. The details of different insulations to be provided for a 33 kV transformer is shown in Fig. 8.22.

**Table 8.5:** Insulation of L.V. windings up to 33 kV. (All dimensions in mm)

| Rating kVA | Voltage kV | From winding to core | |
|---|---|---|---|
| | | $S_1$ | $K_1$ |
| 25–630 | up to 1 | 5 | – |
| 26–630 | 3.3 and 6.6 | 12 | 2.5 |
| 800 and above | up to 1, 3.3 and 6.6 | 15 | 5.0 |
| 25–630 | 11 | 18 | 3.0 |
| 800 and above | | 18 | 5.0 |
| 25–630 | 15 | 21 | 4.0 |
| 800 and above | | 23 | 5.0 |
| Any kVA | 33 | 27 | 5.0 |

**Table 8.6:** Insulation of H.V. windings up to 33 kV (All dimensions in mm)

| Rating kVA | Voltage kV | Between H.V. and L.V. | | From winding end to yoke | | Between phases | |
|---|---|---|---|---|---|---|---|
| | | $S_2$ | $K_2$ | $h$ | $w$ | $S_3$ | $K_3$ |
| 25–100 | | 8.5 | 2.5 | 20 | – | 10 | 2 |
| 125–630 | 3.3 and 6.6 | 12.0 | 2.5 | 20–30 | – | 10 | 2 |
| 25–630 | 11 | 12.0 | 3.0 | 30 | – | 14 | 2 |
| 800 and above | | 17.0 | 5.0 | 30–50 | – | 14 | 2 |
| 25–630 | 15 | 15.0 | 3.5 | 40 | – | 17 | 2 |
| 800 and above | | 17.0 | 5.0 | 40–50 | – | 17 | 2 |
| 10–800 | 33 | 27.0 | 5.0 | 60 | 2 | 30 | 3 |
| 1000 and above | | 27.0 | 5.0 | 75 | 2 | 30 | 3 |

## 8.15 RESISTANCE OF H.V. WINDING AND L.V. WINDING

Mean diameter of H.V. winding, $d_{mhv} = \dfrac{\text{Inner dia. of H.V. winding} + \text{Outer dia. of H.V. winding}}{2}$

Mean turn length of H.V. winding, $L_{mhv} = \pi \times d_{mhv}$

Resistance of H.V. winding, $R_{hv}$ (at 75°C) $= \dfrac{N_h \rho L_{mhv}}{a_h}$ ...(8.26)

where, $N_h$ is the number of H.V. winding turns and $a_h$ is the cross-sectional area of H.V. winding conductor.

Mean diameter of L.V. winding, $d_{mlv} = \dfrac{\text{Inner dia. of L. V. winding + Outer dia. of L. V. winding}}{2}$

Length of mean turn of L.V. winding, $L_{mlv} = \pi \times d_{mlv}$

Resistance of L.V. winding $R_{Lv} = \dfrac{N_l \rho L_{mlv}}{a_l}$ ...(8.27)

where, $N_L$ is the number of L.V. winding turns and $a_l$ is the cross-sectional area of L.V. winding conductor.

Resistance of L.V. winding referred to H.V. winding side, $R'_{lV} = R_{lV} \times \left(\dfrac{N_h}{N_l}\right)^2$

Total resistance referred to H.V. winding side, $R_{ThV} = R_{hV} + R'_{lV}$

$$R_{Thv} = R_{hv} \times \left(\frac{N_h}{N_l}\right)^2 \times R_{lv} \qquad \text{...(8.28)}$$

Per unit (P.U.) resistance of transformer $= \dfrac{I_h\ R_{Thv}}{V_{hv}}$ ...(8.29)

## 8.16 LEAKAGE REACTANCE

The magnetizing current in the transformer produces the flux. The flux set-up by the magnetizing current is basically divided into two components given as following:

(*i*) The portion of flux linking both the windings of the transformer is called as useful flux, $\phi$. Useful flux remains constant at all loads of transformer.

(*ii*) The flux linking with one winding only is called as leakage flux, $\phi_l$. The amount of leakage flux is proportional to the load on the transformer so the value of leakage flux linking one winding depends upon the total ampere-turns of that winding. When the transformer is operating at no load, the leakage flux will be quite less. As the load on the transformer is increased, leakage fluxes will increase, since both the transformer windings will carry larger amount of current. Leakage fluxes produce a self-induced back e.m.f. in their respective windings. The reactance of this is called the leakage reactance of the windings. The different values of leakage reactance are observed for L.V. winding and H.V. winding. The effect of leakage reactance is to cause a voltage drop in the respective windings.

The calculation of leakage reactance is the estimation of the distribution of leakage flux and the resulting flux leakages of the L.V. and H.V. windings. Generally two types of windings namely concentric winding and sandwich winding are used for the transformers. Normally concentric windings are used for core type of transformers and sandwich windings are used for shell type transformers. In the case of concentric winding of core type of transformer the leakage flux is mainly confined into the space between windings and runs parallel with the core for nearly the full length of the coil (Fig. 8.15). The leakage flux in case of sandwich winding of shell type transformer is confined between the windings and is in parallel along the width of the coils (Fig. 8.19). The perfect distribution of leakage flux is very difficult in both the above cases. The field is symmetrical, therefore the mathematical expressions for the leakage reactance can be developed on the basis of considerable simplifying assumptions.

Estimation of leakage reactance of core type transformers: The following assumptions are made to simplify the calculation of leakage flux and then the leakage reactance of the core type transformer.

(*i*) $\mu_{Fe} = \infty$

(*ii*) Leakage flux lines are parallel to core

(*iii*) $L_{mt_1} = L_{mt_2} = L_{mt}$ [Estimated with the help of mean of diameter of windings.

$L_{mt} = \pi \times$ mean of $L_{mt_1}$ and $L_{mt_2}$

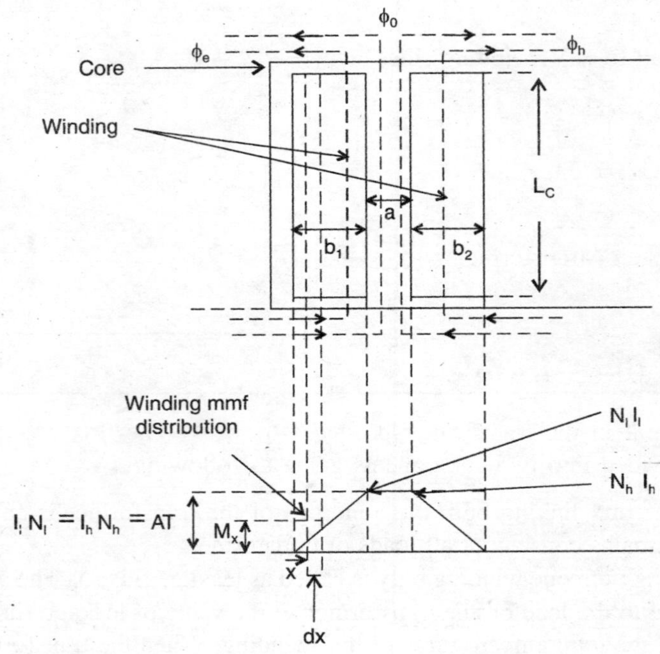


**Fig. 8.15** Distribution of leakage flux and m.m.f. in core type transformer.

(*iv*) $I_l N_l = I_h N_h = AT$ ( total m.m.f.). So, magnetizing m.m.f. and hence magnetizing current is zero.

(*v*) Half of the leakage flux in duct links with each winding.

(*vi*) The reluctance of flux path through yokes in negligible.

(*vii*) The windings are uniformly distributed and hence the winding m.m.f. varies linearly from one end to the other.

$\phi_l$ and $\phi_h$ are the leakage fluxes in the L.V. winding and the H.V. winding respectively and $\phi_0$ is flux through the space (duct).

$L_m$ = Mean circumference of duct

$L_c$ = Axial height of windings

$b_1, b_2$ = Radial width of L.V. winding and H.V. winding

$a$ = Width of radial duct

Consider an elementary path of leakage flux at a distance $x$ of thickness $dx$ (Fig. 8.16)

Area of elementary path $= L_{mt} \cdot dx$

Permeance for elementary path of thickness, $dx$

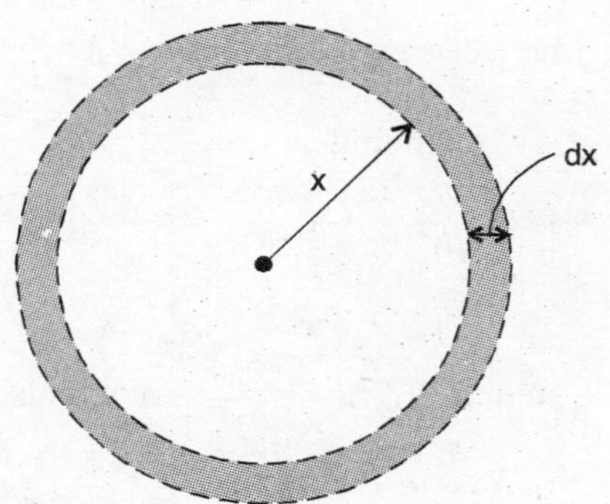


**Fig. 8.16** Elementary path of thickness dx.

$$dP_x = \frac{\mu_0 L_{mt} . dx}{L_c} \qquad \left\{ \text{From expression } \frac{\mu_0 A}{l} \right\}$$

m.m.f. at distance $x$, $M_x = \dfrac{x}{b_1} \cdot N_l I_l$

Leakage flux through the elementary path,

$$d\phi = M_x \cdot dP_x = \frac{x}{b_1} \cdot N_l I_l \mu_0 \frac{L_{mt} \cdot dx}{L_c}$$

Since this flux links with $\dfrac{x}{b_1} \cdot N_l$ turns of primary, so, flux linkages due to elementary path of flux,

$$= \frac{x}{b_1} \cdot N_l \cdot \frac{x}{b_1} \cdot N_l I_l \mu_0 \frac{L_{mt} \cdot dx}{L_c}$$

$$= \mu_0 N_l^2 \cdot \frac{L_{mt}}{L_c} \cdot I_l \left( \frac{x}{b_1} \right)^2 \cdot dx$$

Flux linkages due to L.V. winding (in space $b_1$)

$$\phi_1 = \mu_0 N_l^2 \cdot \frac{L_{mt}}{L_c} \cdot I_l \int_0^{b_1} \left( \frac{x}{b_1} \right)^2 \cdot dx$$

$$\phi_1 = \mu_0 N_l^2 \cdot \frac{L_{mt}}{L_c} \cdot I_l \cdot \frac{1}{b_1^2} \left[ \frac{x^3}{3} \right]_0^{b_1}$$

$$\phi_1 = \mu_0 N_l^2 \cdot \frac{L_{mt}}{L_c} \cdot I_l \cdot \frac{b_1}{3}$$

Permeance of the space (duct) between two windings $= \mu_0 \frac{L_{mt} \cdot a}{L_c}$

Flux linkages in space between two windings,

$$\phi_0 = N_l I_l \ \mu_0 \frac{L_{mt} \cdot a}{L_c}$$

Half of flux links with each winding. So, flux $\frac{(\phi_0)}{2}$ links with $N_l$ turns of L.V. winding.

So, flux linkages of L.V. winding due to flux $\phi_0$ is

$$\phi_1' = \frac{\phi_0}{2} N_l$$

So, total flux linkages for L.V. winding, $\phi_l = \phi_1 + \phi_1'$

$$\phi_l = \mu_0 N_l^2 \cdot \frac{L_{mt}}{L_c} \cdot I_l \frac{b_1}{3} + \frac{1}{2} \mu_0 N_l^2 \cdot \frac{L_{mt}}{L_c} \cdot a \ I_l$$

$$\phi_l = \mu_0 N_l^2 \cdot \frac{L_{mt}}{L_c} I_l \left( \frac{b_1}{3} + \frac{a}{2} \right) \qquad ...(8.30)$$

L.V. winding leakage inductance, $L_l = \frac{\phi_l}{I_l}$

$$= \mu_0 N_l^2 \cdot \frac{L_{mt}}{L_c} \left[ \frac{b_1}{3} + \frac{a}{2} \right] \qquad ...(8.31)$$

L.V. winding leakage reactance,

$$X_l = 2\pi f L_l$$

$$= 2\pi f \mu_0 N_l^2 \cdot \frac{L_{mt}}{L_c} \left[ \frac{b_1}{3} + \frac{a}{2} \right] \qquad ...(8.32)$$

Similarly, we can estimate the total flux linkage for H.V. winding.

Flux linkage due to H.V. winding, $\phi_2 = \mu_0 N_h^2 \cdot \frac{L_{mt}}{L_c} \cdot I_h \cdot \frac{b_2}{3}$

Flux linkages of H.V. winding due to flux $\phi_0$ is $\phi_2' = \frac{\phi_0}{2} \cdot N_h$

So, $\phi_h = \phi_2 + \phi_2'$

$$\phi_h = \mu_0 N_h^2 \cdot \frac{L_{mt}}{L_c} \cdot I_h \frac{b_2}{3} + \frac{1}{2}\mu_0 N_h^2 \cdot \frac{L_{mt}}{L_c} \cdot a I_h$$

$$= \mu_0 N_h^2 \cdot \frac{L_{mt}}{L_c} \cdot I_h \left[\frac{b_2}{3} + \frac{a}{2}\right] \qquad ...(8.33)$$

H.V. winding leakage inductance, $L_h = \dfrac{\phi_h}{I_h}$

$$= \mu_0 N_h^2 \cdot \frac{L_{mt}}{L_c}\left[\frac{b_2}{3} + \frac{a}{2}\right] \qquad ...(8.34)$$

H.V. winding leakage reactance,

$$X_h = 2\pi f L_h$$

$$= 2\pi f \mu_0 N_h^2 \frac{L_{mt}}{L_c}\left[\frac{b_2}{3} + \frac{a}{2}\right] \qquad ...(8.35)$$

The H.V. winding leakage reactance referred to L.V. winding side

$$X_h' = 2\pi f L_h \left(\frac{N_l}{N_h}\right)^2$$

$$= 2\pi f \mu_0 N_l^2 \frac{L_{mt}}{L_c}\left[\frac{b_2}{3} + \frac{a}{2}\right] \qquad ...(8.36)$$

Total leakage reactance of transformer referred to L.V. winding side (per phase)

$$X_{tl} = X_l + X_h'$$

$$= 2\pi f \mu_0 N_l^2 \frac{L_{mt}}{L_c}\left[\frac{b_1 + b_2}{3} + a\right] \qquad ...(8.37)$$

Per unit reactance $= \dfrac{I_l X_{tl}}{V_l}$

$$= 2\pi f \mu_0 \frac{I_l N_l^2}{V_l} \frac{L_{mt}}{L_c}\left[\frac{b_1 + b_2}{3} + a\right]$$

$$= 2\pi f \mu_0 \frac{AT}{E_t} \frac{L_{mt}}{L_c}\left[\frac{b_1 + b_2}{3} + a\right] \qquad ...(8.38)$$

where, $\dfrac{N_l}{V_l} = E_t$ and $N_l I_l = AT$.

If the leakage reactance is very high, it can be reduced by

(*i*) Increasing the window height [due to increasing window height, window width will decrease]. So, $L_c$ will increase and $L_{mt}$ will decrease.

(*ii*) Space between two windings (*a*) may be reduced.

(*iii*) $b_1$ and $b_2$ may be reduced.

In the transformers in which H.V. winding is partitioned into two parts as shown in Fig. 8.17.

The leakage reactance per phase of such transformers referred to L.V. winding side is given by,

$$X_l = 2\pi f \mu_0 N_l^2 \frac{L_{mt}}{L_c}\left[a + \frac{b_1 + b_2 + b_2'}{3} + \frac{a'}{4}\right]$$

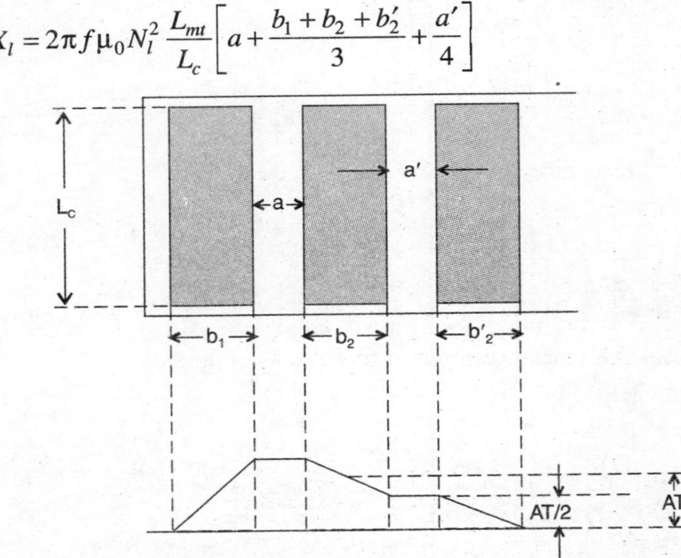


**Fig. 8.17** Distribution of m.m.f. in a transformer in which H.V. winding is divided in two parts.

In some transformer, (like distribution transformer) it is required to keep the leakage reactance so small that voltage is maintained within 5% of rated value. To achieve this the L.V. winding is divided in two parts and H.V. winding is sandwiched between these two parts of L.V. winding. The total leakage reactance of such transformer (per phase) referred to L.V. winding side as shown in Fig. 8.18 can be given by

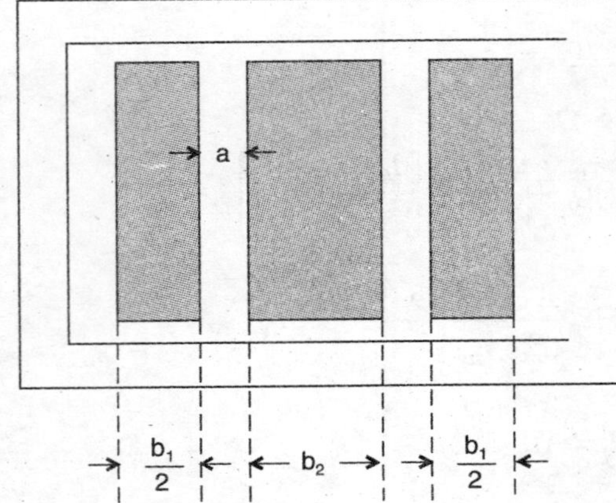


**Fig. 8.18** Transformer with H.V. winding sandwiched between two parts of L.V. winding.

$$X_l = \mu_0 \pi f N_l^2 \frac{L_{mt}}{L_c}\left[\frac{b_1 + b_2}{6} + a\right] \qquad ...(8.39)$$

***Leakage Reactance of Shell Type of Transformers:*** The arrangement of sandwich winding in a shell type transformer is shown in Fig. 8.19. If H.V. winding has $n$ coils, $L_c$ is the width of the coil and other details as shown in Fig. 8.19, the total reactance of transformer referred to L.V. winding side is

$$X_l = 2n \times 2\pi f \mu_0 \left(\frac{N_l}{2n}\right)^2 \frac{L_{mt}}{L_c}\left(\frac{b_1 + b_2}{6} + a\right)$$

$$= \mu_0 \pi f \frac{L_{mt}}{L_c} \cdot \frac{N_l^2}{n}\left(\frac{b_1 + b_2}{6} + a\right) \qquad ...(8.40)$$

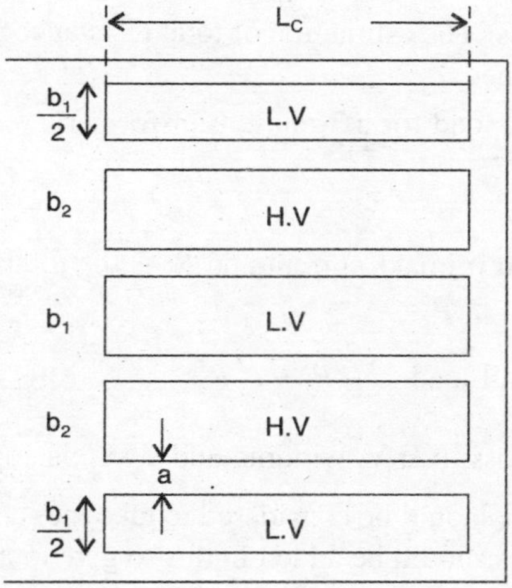

**Fig. 8.19** Shell type transformer with sandwich winding.

## 8.17 REGULATION

The regulation of a transformer is defined as the ratio of change in secondary terminal voltage between no load and full load conditions to the secondary no load voltage. The primary voltage is assumed to be constant. The regulation of a transformer can be expressed as below:

$$\text{Regulation} = \frac{\text{No load secondary voltage} - \text{Full load secondary voltage}}{\text{Secondary no load voltage}}$$

$$= \frac{E_s - V_s}{E_s} \qquad ...(8.41)$$

$$\% \text{ regulation} = \frac{E_s - V_s}{E_s} \times 100 \qquad ...(8.42)$$

Per unit regulation of a transformer for full load rated output, $Q$ and full load current, $I_p$ can be given

$$\text{Per unit regulation, } E = \frac{I_p R_p \cos\phi + I_p X_p \sin\phi}{V_p}$$

$$= \varepsilon_r \cos\phi + \varepsilon_x \sin\phi \qquad \text{...(8.43)}$$

where, $\varepsilon_r$ = P.U. resistance

$\varepsilon_x$ = P.U. reactance

## 8.18 ESTIMATION OF LOSSES AND EFFICIENCY OF A TRANSFORMER

**(i) $I^2R$ Loss or Ohmic Loss:** The estimation of total resistance referred to H.V. winding side, $R_{Thv}$, has already been explained.

$I^2R$ loss (at 75°C) at full load for a 3-phase transformer,

$$W_c = 3I_p^2 R_p$$

where, $I_p$ is the full load primary current and $R_p$ is the total resistance referred to primary side of transformer.

$$\text{or total copper loss at full load} = I_1^2 R_1 + I_2^2 R_2$$

where, $I_1^2 R_1$ is copper loss in H.V. winding and $I_2^2 R_2$ is copper loss in L.V. winding.

Some additional amount should be considered to take into account the stray load loss. About 15 to 20 per cent of $I^2R$ loss should be added in this to give total $I^2R$ loss including stray load loss.

$$\text{Total } I^2R \text{ loss, } W_c = (1.15 \text{ to } 1.20) \times 3I_p^2 R_p \qquad \text{...(8.44)}$$

**(ii) Core Loss:** The method to estimate total iron loss in the transformer, $W_i$ has been given earlier. The total copper losses of small transformer may vary from 0.8 to 1.5% of rated output. Whereas the iron losses of small transformer varies from 0.5 to 1% of the rated output.

For medium and large transformers the total copper losses should be in the range of 0.4 to 1% of the rated output and the total iron losses should vary within 0.2 to 0.6% of the rated output. Copper loss, $W_c$ estimated is the loss at full load. The copper loss at various other loads can be estimated as below

$$\text{Copper loss at half the full load} = \left(\frac{1}{2}\right)^2 W_c \qquad \text{...(8.45)}$$

$$\text{Copper loss at } \left(\frac{1}{4}\right)^{th} \text{ the full load} = \left(\frac{1}{4}\right)^2 W_c \qquad \text{...(8.46)}$$

**(iii) Efficiency:** Total loss at full load = Total $I^2R$ loss + Total core loss

$$\text{Efficiency of a transformer} = \frac{\text{Output}}{\text{Input}} = \frac{\text{Output}}{\text{Output + Losses}}$$

$$\%\ \text{efficiency} = \frac{\text{Output}}{\text{Output + Losses}} \times 100 \qquad \text{...(8.47)}$$

***Weight of transformer:*** Weight of copper $= 8900 \times [L_{mt_1} \times N_1 F_{c_1} \times 10^{-6} + L_{mt_2} \times N_2 F_{c_2} \times 10^{-6}]$ ...(8.48)

where, $L_{mt_1}$ = length of mean turn of L.V. winding

$N_1$ = number of turns in L.V. winding

$F_{c_1}$ = cross-sectional area of L.V. winding conductor

Similarly, $L_{mt_2}$, $N_2$ and $F_{c_2}$ are denoted for H.V. winding

Weight of iron in transformer = Weight of core + Weight of yoke (calculated earlier).

Weight of oil in transformer = 330 × Volume of oil in transformer.

Weight of transformer = Weight of copper in windings + Weight of iron + Weight of oil + Weight of different fittings and fixtures etc.

## 8.19 DESIGN OF TANK

When the transformers are put into operation, losses are occuring in the transformer cores and windings which are converted into thermal energy and cause heating in that part of transformer. The temperature of that part will rise. To keep the temperature rise within the safe limit the heat produced should be dissipated safely. The heat is transferred to a cooling medium like air or water depending upon the method of transformer cooling. In smaller transformers the losses are comparatively lesser and hence the problem of cooling in such cases are solved by natural air cooling. This can be used where the total loss is below 4 kW. As the transformer becomes larger, the natural air cooling method is not effective and hence oil immersed transformers with various arrangements for cooling are being used which has already been explained earlier.

In the oil immersed transformers, the heat developed in the cores and windings is passed to the oil and then through oil to the walls of the tank from which it is dissipated to the air or water. The advantages of oil cooling over air cooling include freedom from the possibility of dust clogging the cooling ducts, or of moisture affecting the insulation, and hence the transformers for higher voltages can be designed. The oil in the ducts adjacent to the cores and windings take up heat by conduction and rises to upper level and cool oil from bottom of the tank comes to take its place. Oil has large coefficient of volume expansion with increase of temperature and hence substantial circulation is easily achieved.

The best dissipator of external heat is a plain tank. But the size of transformer cannot be increased indefinitely and hence the improvement in heat dissipation (by increasing heat dissipating surface) can be obtained by providing cooling tubes. The specific heat dissipation due to convection of oil is

$$\lambda = 40.3\left(\frac{\theta}{H}\right)^{1/4} \text{ watt/m}^2\text{-°C}$$

where, $\theta$ is the temperature rise of the transformer above ambient temperature and $H$ is the height of the dissipating surface in metre.

Heat dissipation by convection by oil is fairly large as compared with heat dissipation due to convection by air. This justify the use of oil as cooling medium in the transformers. Experiments show that the working temperature of oil is 50 to 60°C. Tank walls and tubes dissipate heat by radiation and convection. The plain tank surface dissipates 6 watt by radiation and 6.5 watt by convection per sq. metre surface of tank per °C. Thus if the cooling surface of a transformer plan tank is $S_t$ sq. metre. (Heat dissipation from top and bottom surfaces are neglected).

Total heat dissipated = $12.5 \times S_t$ watt per 0°C

$$\text{Temperature rise without tubes, } \theta = \frac{\text{Total loss}}{\text{Heat dissipated per}^\circ\text{C}}$$

$$= \frac{W_c + W_i}{12.5 \cdot S_t}$$

Let area of tubes = $x \cdot S_t$ (As given in Fig. 8.20)

Heat dissipated by tubes by convection = $8.78 \times (8.78 \times x \times S_t) \times S_t$ watt/ °C. When the cooling tubes are provided the cooling surface increases. But the heat dissipation is more than what can be expected by merely increase in the surface area. This is achieved by improved oil circulation due to more effective hinds of pressure produced by columns of oil in tubes. Approximately 35% more should be considered for loss dissipated by tubes by convention due to this improvement in oil circulation.

Heat dissipation by tubes by convention = $1.35 \times 6.5 \times x \times S_t$

$= 8.78 \times x \times S_t$

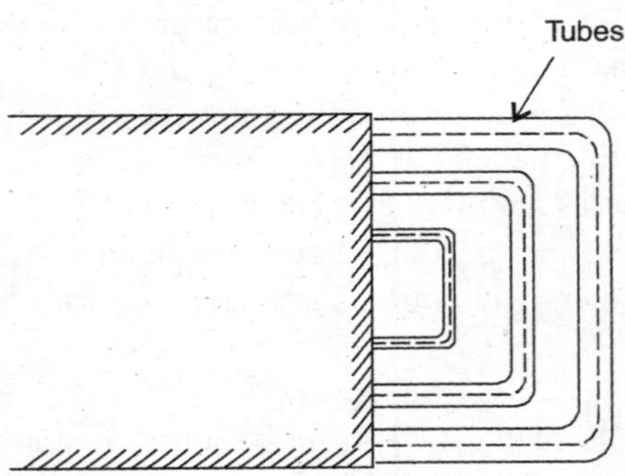


**Fig. 8.20** Transformer tubes.

Total heat dissipated by tank with tubes or radiators,

$$= 12.5 \times S_t \times \theta + (6.5 \times x \times S_t \times \theta) \times 1.35 \text{ watt}$$

$$= 12.5 \times S_t \times \theta + 8.78 \times x \times S_t \times \theta$$

Total heat to be dissipated = Total losses = $W_c + W_i$

$$\text{Temperature rise with tubes, } \theta = \frac{W_c + W_i}{12.5 S_t + 8.78 x\, S_t}$$

or $x = \frac{1}{8.78}\left[\frac{W_c + W_i}{S_t \theta} - 12.5\right]$

So, area of tubes as assumed earlier $= xS_t = \frac{1}{8.78}\left[\frac{W_c + W_i}{S_t\, \theta} - 12.5\right] S_t$

$$= \frac{1}{8.78}\left[\frac{W_c + W_i}{\theta} - 12.5\ S_t\right]$$

Let, length of tube $= l_t$

Diameter of tube $= d_t$

Then area of one tube $= \pi\, d_t\, l_t$

Total area of tubes is available and area of one tube is calculated. So, number of tubes can be calculated as

$$n_t = \frac{1}{8.78\ \pi\, d_t\, l_t}\left[\frac{W_c + W_i}{\theta} - 12.5\ S_t\right] \qquad ...(8.49)$$

The cooling tubes may be circular or elliptical. The normal spacing is 5 cm diameter tubes placed at 7.5 cm space.

### 8.19.1 Main Dimensions of Tank

The main dimensions of tank *i.e.*, width of the tank, length of the tank and height of the tank can be estimated as below.

***(i) Width of the Tank:***

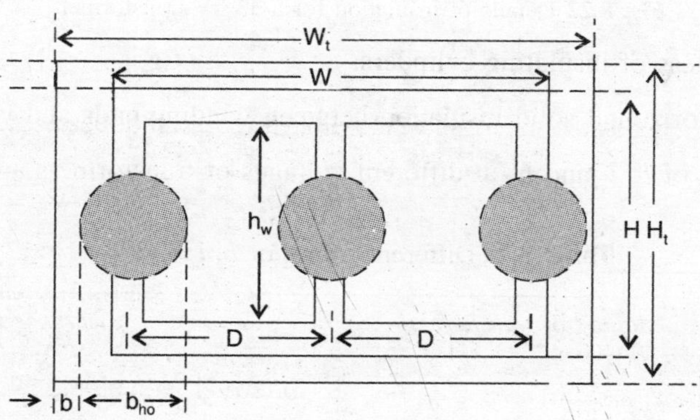


**Fig. 8.21** Main dimensions of tank.

As given in Fig. 8.21 the width of the tank

$$W_t = 2D + d_{ho} + 2b$$

where, $D$ = distance between centres of adjacent limbs

$d_{h0}$ = external diameter of H.V. winding

$b$ = space between H.V. winding and tank

Chapter 8

***(ii) Length (thickness of the side of transformer perpendicular to overall width side) of tank:***

$$L_t = d_{h0} + 2l_s$$

where, $l_s$ is the clearance on each side of transformer between H.V. winding and tank along the width.

***(iii) Height of the tank:***

Height of the tank = $H + h_d$

where, $H$ = height of transformer,

$h_d$ = Total space (clearance) between assembled transformer and the tank. This includes the clearance at the base, oil height at the assembled transformer, space for leads and tap changing gears etc.

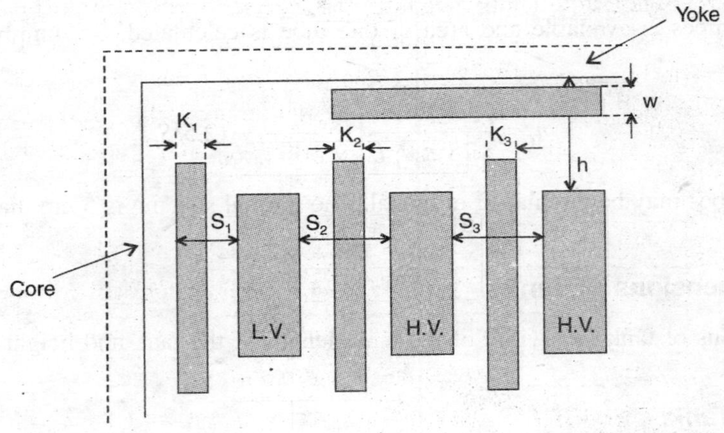


**Fig. 8.22** Details of insulation for a 33 kV transformer.

$K_1$, $K_2$, $K_3$ = thickness of insulating cylinders,

$w$ = thickness of horizontal solid insulation between winding ends and yoke

The different values of $b$, $l_s$ and $h_d$ at different voltages of transformers are given in Table 8.7.

**Table 8.7:** Different values of $b$, $l_s$ and $h_d$

| *H.V. side voltage (kV)* | *Rating of transformer (kVA)* | *Clearance (cm)* | | |
|---|---|---|---|---|
| | | $b$ | $l_s$ | $h_d$ |
| 11 kV or less | less than 1000 | 6 | 5.5 | 45 |
| | 1000–5000 | 8 | 9 | 48 |
| 11 kV–33 kV | less than 1000 | 10 | 12 | 55 |
| | 1000–5000 | 15 | 20 | 60 |

## SOLVED PROBLEMS ON TRANSFORMERS

**Q. 1** Estimate the main core dimensions, of a 25 kVA, 3-phase, 6600/440 volts, delta/star. 50 Hz, core type transformer with the following data:

Stepped core for which area factor = 0.56
Spaces factor for window = 0.25
Voltage per turn = 21.0 V
Current density = 2.36 A/mm$^2$
Flux density = 1.1 tesla **(DDU, GKP University, 1999)**

**Solution:**

Given voltage per turn, $E_t = 21 = 4.44\, f\, \phi_m$

$$\text{or } 21 = 4.44 \times f \times B_m A_i = 4.44 \times 50 \times 1.1 \times A_i$$

$$\text{or } A_i = \frac{21}{4.44 \times 50 \times 1.1} = 0.085 \quad \text{m}^2$$

Diameter of circumscribing circle,

$$d = \sqrt{\frac{A_i}{0.56}} = \sqrt{\frac{0.085}{0.56}} = 0.39 \text{ m}$$

Width of largest step in core may be equal to, $W_c = 0.85 \times d = 0.33$
Output equation can be expressed as

$$Q = 3.33\, f A_i B_m \sigma K_w \cdot A_w \times 10^{-3}$$

$$\text{or} \quad 25 = 3.33 \times 50 \times 0.085 \times 1.1 \times 2.36 \times 0.25 \times A_w \times 10^{-3}$$

$$\therefore \; A_w = 0.027 \text{ m}^2$$

$$\text{or } h_w \times W_w = 0.027 \quad \text{m}^2$$

$$\text{Let, } \frac{\text{Height of window}}{\text{Width of window}} = 2 = \frac{h_w}{W_w}$$

$\therefore$ Height of window = $2 \times W_w$ or $h_w = 2 \times W_w$
Putting the value of $h_w$ in above equation, we get

$$\text{or } W_w^2 = 0.0135$$

$$\text{or } W_w = 0.11 \text{ m}$$

$$\therefore \quad h_w = 0.22 \text{ m}$$

$\therefore$ Distance between two adjacent cores,

$$D = W_w + W_c = 0.11 + 0.33 = 0.44 \text{ m}$$

Overall width of transformer, $W = 2D + W_c$

$= 0.88 + 0.33$

$= 1.11$ m

Overall height of transformer $= h_w + 2h_y$

$= 0.22 + 2 \times 0.33$

$= 0.88$ m $\quad \left[\begin{array}{l} \textit{Taking } A_y = A_i \textit{ and } W_c = W_y \\ \therefore\ h_y = 0.33 \end{array}\right]$

**Q. 2** Estimate the main dimensions of a 500 kVA, 6600/400 V, 3-phase, 50Hz core type oil immersed self cooled distribution transformer. Assume suitable data if needed.

**(DDU GKP Univ. 1994)**

Let voltage per turn, $E_t = 20$ V

Area factor for stepped core = 0.56

Window space factor = 0.30

Current density = 3.0 A/mm$^2$

Flux density, $B_m = 1.2$ Wb/m$^2$

Width of largest step, $W_c = 0.85 \cdot d$

Ratio height of window to width of window = 2.1

Distance between two adjacent limbs = $1.85 \times d$

Use $A_y = A_i$

**Solution:**

Voltage per turn, $E_t = 4.44\, f\, \phi_m = 4.44 \times f\, B_m A_i$

or $\quad A_i = \dfrac{E_t}{4.44\, f\, B_m} = \dfrac{20}{4.44 \times 50 \times 1.2} = 0.075\,\text{m}^2$

Diameter of circumscribing circle, $d = 0.36$ $\quad \left\{\because\ d = \sqrt{\dfrac{A_i}{0.56}}\right.$

Width of largest step, $W_c = 0.85 \times d = 0.30$ m

Distance between two adjacent limbs, $D = 1.85 \times 0.30 = 0.56$ m

Width of window, $W_w = 0.26$ m $\quad [\because\ W_w = D - W_c]$

We have output equation,

$$Q = 3.33 f\, A_i B_m \sigma K_w A_w \times 10^{-3}$$

$$500 = 3.33 \times 50 \times 0.075 \times 3.0 \times 0.30 \times 1.2 \times A_w \times 10^{-3}$$

$$\therefore\quad A_w = 0.037\ \text{m}^2$$

Area of window,

$$A_w = h_w \times W_w$$

$$\therefore \quad h_w = \frac{A_w}{W_w} = 0.14 \text{ m}$$

Since $A_y = A_i$ and let $W_c = W_y$

$$\therefore \quad h_y = 0.30 \text{ m}$$

Overall width of transformer,

$$W = 2D + W_c = 0.56 \times 2 + 0.30$$

$$= 1.42 \text{ m}$$

Overall height of frame,

$$H = h_w + 2h_y$$

$$= 0.14 + 2 \times 0.30$$

$$= 0.74 \text{ m}$$

**Q. 3** A 300 kVA, 6600/440 volt, three-phase delta/star core type transformer has a maximum flux density of 1.35 Wb/m$^2$ and the total weight of core is 650 kg. The magnetizing VA/kg and the iron loss/kg corresponding to 1.35 Wb/m$^2$ are 30 and 2.5 watt respectively. Calculate the no load current if the m.m.f. required for joints is 2.5 per cent of that iron. Assume other data needed.

**(DDU, GKP, 1998)**

**Solution:**

Total iron loss, $W_i = (625 \times 2.5) = 1625$ watt

Core loss component, $I_c = \dfrac{W_i}{3V_1} = 0.082$

Total magnetizing VA will be 2.5% more so

Total = 1.025 × (magnetizing VA/kg × weight of core)

Magnetizing current component,

$$I_\mu = \frac{\text{Magnetizing VA/kg} \times \text{Weight of core}}{\text{No. of phases} \times \text{Voltage/phase}}$$

$$= \frac{1.025 \times 30 \times 650}{\text{Phases} \times \text{Voltage/phase}}$$

$$= 1.00$$

No load current,

$$I_0 = \sqrt{I_c^2 + I_\mu^2}$$

$$= \sqrt{(0.082)^2 + (1.0)^2}$$

$$I_0 = 1.0 \text{ ampere}$$

**Q. 4** Design a tank for a 1250 kVA, 6600/440 V, 50 Hz, 3-phase, delta/star, core type oil immersed natural cooled transformer having distance between centres of adjacent limbs 0.48 m, outer diameter of H.V. winding 0.46 m, height of frame 1.28 m, core loss 3.8 kW, $I^2R$ loss 11.0 kW. Specific heat dissipation from tank walls due to radiation and convection is 6 watt/m$^2$-°C and 6.5 watt/m$^2$-°C respectively. The temperature rise should not exceed 37°C. The clearance between windings and tank is 72 mm and 95 mm in each side along width and length respectively. The conduction is improved by 35% when tubes are provided.

**Solution:**

Width of tank $= 2 \times 0.48 + 0.46 + 2 \times 0.072$

$= 1.564$ m

Length of tank $= 0.46 + 2 \times 0.095 = 0.65$ m

Providing 0.55 mm for base and 0.325 mm for oil above the frame.

Height of oil level from bottom tank $= 1.28 + 0.55 + 0.325$

$= 1.36$ m

0.350 mm height is also given for 0 (zero) etc.

Height of tank $= 1.36 + 0.35$

$= 1.71$ m

Dissipating surface of plain tank,

$$S_t = 2(1.564 + 0.65) \times 1.71$$

$$S_t = 7.57 \text{ m}^2$$

Let the area of tubes be $x \cdot S_t$

Total loss dissipated by tank wall and tubes $= 12.5\, S_t + 8.78\, x \cdot S_t$

$= S_t\,(12.5 + 8.78\,x)$

$= 7.57 \times (12.5 + 8.78\,x)$

$\therefore$ Total dissipating area $= (1 + x)\, S_t$

$= 7.57\,(1 + x)$ sq. metre

$$\text{Specific loss dissipation} = \frac{12.5 + 8.78\,x}{1 + x}\left\{\frac{7.57(12.5 + 8.78\,x)}{7.57(1 + x)}\right\}$$

Total loss $= 3.8 + 11 = 14.8$ kW, temperature rise should be below 37°C.

$$\therefore \text{Specific loss dissipation} = \frac{14.8 \times 10^3}{7.57(1 + x) \times 37} = \frac{12.5 + 8.78\,x}{1 + x}$$

Hence, $x = 4.32$

Area of tubes = $xS_t$ = 32.7 sq. metre

Let height of tube is 1.45 m and diameter is 50 mm.

Distributing area of each tube = $\pi \times 1.45 \times 0.05$

$= 0.22$

Hence, number of tubes required $= \dfrac{32.7}{0.22} = 148$

## OVERALL DESIGN OF A TRANSFORMER

Design a 3-f power, core type transformer with following specifications:

| Rating | Voltage ratio | Connection | Tapping | Efficiency |
|---|---|---|---|---|
| 50 MVA | 66/132·kV | Y/Y | 5% | 99% |

**Solve:** Estimation of main dimension:

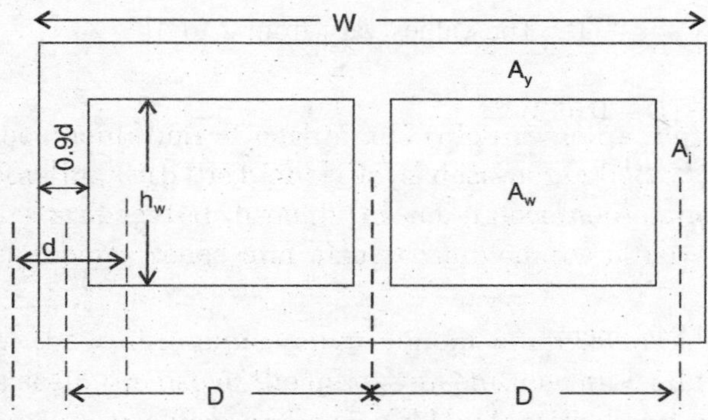


### *Output Equation*

$$Q = 3.33 A_i \cdot f \cdot B_m \cdot A_w \cdot K_w \cdot \sigma \times 10^{-3} \qquad ...(1)$$

where, $\sigma = 3.5\ \text{A/mm}^2$, for forced cooling,

$B_m = 1.55\,\text{T}$ for CRGO (power transformer upto 132 kV),

Voltage per turn, $E_t = K_t\sqrt{Q}$ where, $K_t = 0.67$, for 3-$\phi$ core power transformer

Sectional area of core, $A_i = \dfrac{E_t}{4.44 \times 50 \times 1.55}$

$= 0.4359$ sq. metre

Winding inner diameter, $d = \sqrt{\dfrac{A_i}{k}}$ where, $k = 0.56$, for 2 step transformer.

$= 0.88$ sq. metre

Chapter 8

Window space factor, $K_w = \frac{12}{30 + kV}$, where, $kV$ is the phase voltage of H.V. winding.

$$= 0.112.$$

Each window area,

$$A_w = \frac{Q}{3.33 \cdot A_i \cdot f \cdot B_m \cdot K_w \cdot \sigma \times 10^{-3}}$$

$$= 1.136 \text{ sq. metre}$$

$D$ = distance between two adjacent cores,

$W$ = overall width,

and $h_w$ = height of window

Now, $A_w = h_w (D - 0.9d)$

$\Rightarrow \quad 1.136 = h_w (D - 0.792)$ ...(*i*)

Again,

$$\frac{h_w}{D - 0.9d} = 3 \quad [\because \text{ The values vary from 2 to 4}]$$

$\Rightarrow \quad h_w = 3(D - 0.792)$ ...(*ii*)

From equations (*i*) and (*ii*),

$$1.136 = \frac{h_w^2}{3}$$

$$\therefore \quad h_w = 1.846 \text{ metre}$$

$$D = \frac{h_w}{3} + 0.792 = 1.407 \text{ metre}$$

$\therefore \quad W = 2D + 0.9d = 3.599$ metre

## Estimation of no Load Current

Sectional area of yoke, $A_y = 1.15 A_i$

$$= 1.15 \times 0.4395$$

$$= 0.505 \text{ sq. metre}$$

Flux density in yoke, $B_y = B_m \times \frac{A_i}{A_y} = 1.336$ T

Volume of iron in core = $3 \times h_w \times A_i = 2.41$ cu. metre

Wt. of iron in core = $7600 \times 2.41$ = 18,346 kg

$B_m = 1.55$ T, (From Fig. 8.12)

Specific iron loss in core *i.e.*, in watt/kg = $p_{ic}$

$\therefore\ p_{ic}$ = 1.98 watt/kg when $B_m$ = 1.55

Total iron loss in core, $W_{ic}$

$= p_{ic} \times$ Total weight of core

= 36,326 watt

Total weight of iron in yoke = $(2 \times W \times A_y) \times 7600$

= 27,625 kg

Specific iron loss, $p_{iy}$ in yoke for flux density

*i.e.*, $B_y$ = 1.336 T, $p_{iy}$ = 1.1 watt/kg

Total iron loss in yoke, $W_{iy}$ = (1.1 × total wt.)

= 30,388 watt

Total iron loss in transformer, $W_i = W_{ic} + W_{iy}$

= 66,714 watt

**Core loss component of current,** $I_c = \dfrac{\text{Total iron loss}}{3V_1}$

where, $V_1$ is the primary phase voltage in volts, having value of 0.583 A.

**For magnetizing current component ($I_\mu$):**

$B_m$ = 1.55, corresponding to $B_m$ find $at_c$ *i.e.*, ampere-turn required by core.

$$at_c = (900 \times .1) \text{ ampere-turn/metre} \quad \text{(From Fig. 8.13)}$$

= 90 ampere-turn/metre

**Total AT for core,**

$$AT_c = 3 \times h_w \times at_c = 3 \times 1.846 \times 90 = 498.42 \text{ AT}$$

Similarly, for, $B_y$ = 1.336 T,

$at_y$ = (320 × .1) = 32 at/mt.

Total ampere-turn required for yoke, is

$AT_y = 2 \cdot W \cdot at_y$ = 230 AT

Total *AT* required for core and yoke is,

$$AT_{cy} = AT_c + AT_y = 498 + 230 = 728\,AT$$

Some extra ampere-turn is taken (*i.e.*, 5% of $AT_{cy}$) due to the joints in the transformer.

$$\therefore \quad AT_T = 1.05 \times AT_{cy} = \mathbf{764.4\ AT}.$$

Magnetizing current component,

$$I_\mu = \frac{AT_T}{3N_1}$$

where, $N_1$ is the number of turns/phase in primary winding

$$\therefore \quad I_\mu = 3 \text{ ampere}$$

$$V_{1Ph} = \frac{66}{\sqrt{3}} = 38{,}105 \text{ volts}$$

$$\therefore \quad N_1 = \frac{V_{1Ph}}{E_t} = 254$$

No load current,

$$I_0 = \sqrt{I_c^2 + I_\mu^2} = 3.056 \text{ ampere}$$

$$\text{No load p.f.} = \frac{I_c}{I_0} = 0.19$$

## Estimation of Mean Length per Turn in Primary ($L_{mt_1}$)

L.V. winding (Primary Winding), 66/132 kV,

$$66 \text{ kV} = V_1, \; V_{1Ph} = \frac{66}{\sqrt{3}} = 38.1 \text{ kV} = 38{,}105 \text{ V.}$$

$$\text{Number of turns/phase in primary, } N_1 = \frac{V_{1ph}}{E_t} = 254$$

$$\text{Primary phase current, } I_1 = \frac{Q}{3 \times V_{1ph}} = \frac{50 \times 10^6}{3 \times 38{,}105} = 437.38 \text{ ampere}$$

$$\because \quad \sigma = 3.5 \text{ A/mm}^2.$$

$$\therefore \text{ Sectional area of one primary conductor, } F_{c1} = \frac{I_1}{\sigma}$$

$$\therefore \quad F_{c1} = \frac{I_1}{\sigma} = \frac{437.38}{3.5} \text{ mm}^2 = 124.96 \text{ mm}^2$$

Now, $L_{mt_1}$ is the mean length per turn of the primary winding.

Since insulation specification is not available for 50 MVA, 66/132 kV transformer.

So, by analogy of given in Tables 8.5 and 8.6 we approximated the values.

$t_1$ = width of winding, $d$ = inner diameter.

Total cross-sectional area of primary winding (L.V.), $F_{c_1'}$

$$\therefore \quad F_{c_1'} = N_1 \times 124.96 \times 2 \text{ mm}^2 \text{ (taking both sides on core)}$$

$$= 63479.68 \text{ mm}^2 = 0.063 \text{ sq. metre}$$

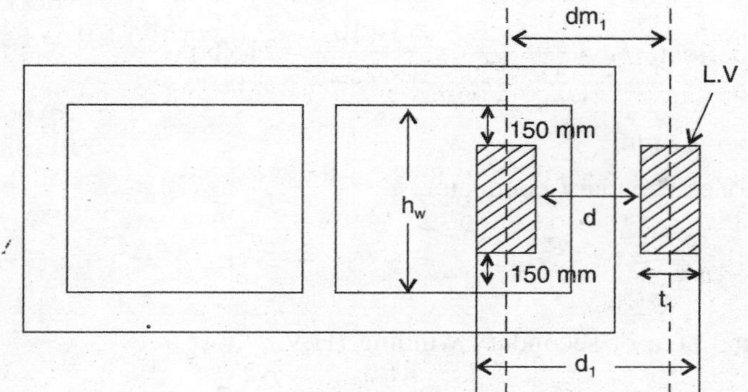


With the help of Figure (8.24), we get,

$$(h_w - 300) \times 2t_1 = F_{c_1'} \quad \Rightarrow 2t_1 = \frac{F_{c_1'}}{(h_w - 300)} = 41.06 \text{ mm}$$

$\therefore$ outer diameter of primary winding *i.e.*, (L.V. side).

$$\Rightarrow d_1 = 2t_1 + d$$

$$= (41.06 + 880) \text{ mm} = 921.06 \text{ mm.}$$

Mean diameter of primary winding, $dm_1 = \dfrac{d + d_1}{2}$

$\therefore L_{mt_1} = \pi dm_1 = 2839.17 \text{ mm} \cong 2.83 \text{ metre}$

## Estimation of Mean Length per Turn in Secondary (H.V. Side)

$$V_2 = 132 \text{ kV}, \quad V_{2Ph} = \frac{132}{\sqrt{3}} = 76.21 \text{ kV}$$

$N_2$ = number of turns/phase in secondary

$$\frac{V_{1Ph}}{V_{2Ph}} = \frac{N_1}{N_2} \quad \Rightarrow N_2 = N_1 \times \frac{V_{2Ph}}{V_{1Ph}}$$

$$= 254 \times \frac{76.21}{38.1}$$

$$= 508.$$

Since 5% tapping is provided and we know tapping is provided to H.V. side only. So, turns of high voltage side will increase by 5%.

$\therefore$ New, $N_2 = (1.05 \times 508) = 533.$

Chapter 8

Secondary phase current, $I_2 = \dfrac{Q}{3 \times V_{2Ph}} = \dfrac{50 \times 10^3}{3 \times 76.21} = 218.6$ A

$\sigma = 3.5 \text{ A / mm}^2$

Sectional area of one secondary conductor,

$$F_{c_2} = \frac{218.6}{3.5} = 62.45 \text{ mm}^2$$

Total cross-sectional area of secondary winding (H.V.), $F_{c_2'}$

$\therefore\ F_{c_2'} = 2 \times N_2 \times 62.45 = 66{,}579 \text{ mm}^2 = 0.0665 \text{ m}^2$ (For both sides on core)

$t_2$ = Width of H.V. winding (Secondary)

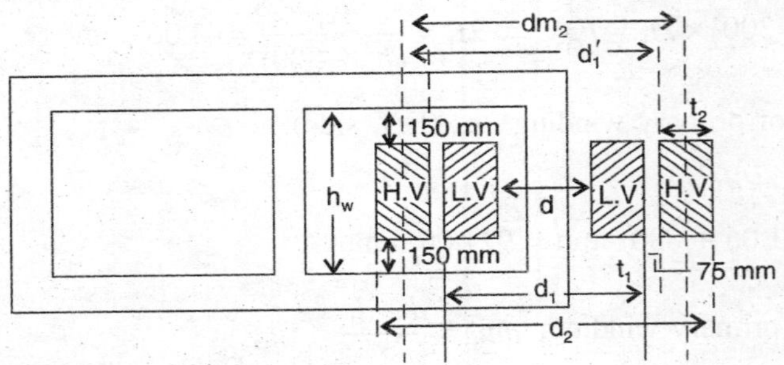


Now, from Figure,

$$(h_w - 300) \times 2t_2 = F_{c_2}$$

$$\Rightarrow\ 2t_2 = \frac{F_{c_2}}{h_w - 300} = \frac{66{,}579}{(1840 - 300)} = 43 \text{ mm}$$

Outer diameter of H.V. winding or secondary winding,

$$d_2 = d_1 + (2 \times 75) + 2t_2$$

$$= 921 + (2 \times 75) + 43$$

$$= 1114 \text{ mm}$$

Inner diameter of H.V. winding, $d_1' = d_1 + 150 = (921 + 150) = 1071$ mm

$$\therefore\ \text{Mean diameter,}\quad dM_2 = \frac{d_1' + d_2}{2}$$

$$\therefore\ L_{mt_2} = \pi\, dm_2 = 3432.1 \text{ mm} = 3.43 \text{ m}$$

**Cu-loss Estimation**

Resistance of primary winding (L.V.)

$$R_1 = 0.021 \times \frac{L_{mt_1} \cdot N_1}{F_{c_1}} = 0.12\ \Omega$$

where, $L_{mt_1}$ is in metre and $F_{c_2}$ is in mm$^2$.

Resistance of secondary winding (1 + V)

$$R_2 = 0.021 \times \frac{L_{mt_2} \cdot N_2}{F_{c_2}} = 0.61\ \Omega$$

Resistance of secondary referred to primary,

$$R_2' = R_2\left(\frac{N_1}{N_2}\right)^2 = 0.61\left(\frac{254}{533}\right) = 0.138\ \Omega$$

Total resistance referred to primary side,

$$R_T = R_1 + R_2' = 0.12 + 0.138 = 0.258\ \Omega$$

Total Cu loss = $3I_1^2 \cdot RT = 148{,}113$ watt = 0.148 MW.

Total loss = Cu loss + Iron loss

= 214,827 watt = 0.2148 MW

Efficiency at, 0.8 p.f.

$$\eta = \frac{50 \times 0.8}{(50 \times 0.8) + 0.2148} \times 100 = 99.4\%$$

**Leakage Reactance Referred to Primary**

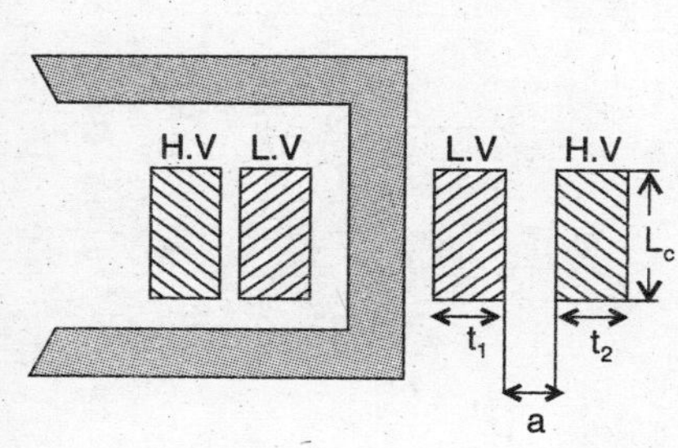


$$X_1 = 2\pi f \mu_0 N_1^2 \frac{L_{mt_1}}{L_c}\left(a + \frac{t_1 + t_2}{3}\right)$$

where, $L_{mt_1}$ = length of mean turn in primary winding

$L_c$ = axial light of winding

$t_1$ and $t_2$ = radial width of primary and secondary winding respectively

$a$ = width between H.V. and L.V. winding

$$\therefore X_1 = 2\pi \times 50 \times 4\pi \times 10^{-7} \times (254)\frac{(2.83)}{1.546}\left(0.075 + \frac{0.0205 + 0.0265}{3}\right)$$

$$= 4.149\ \Omega$$

### Voltage Regulation

P.U. resistance of transformer,

$$\varepsilon_r = \frac{I_1 . R_T}{V_{1Ph}} = 0.0029 \text{ P.U.}$$

P.U. leakage reactance,

$$\varepsilon_x = \frac{I_1}{V_{1Ph}} . X_1 = 0.047 \text{ P.U.}$$

P.U. regulation $\varepsilon = \varepsilon_r \cos\phi + \varepsilon_x \sin\phi$

Let p.f. = 0.8 lagging

$\therefore \quad \cos\phi = 0.8$

$\sin\phi = 0.6$

$$\therefore \quad \varepsilon = (0.0029 \times 0.8) + (0.047 \times 0.6)$$

$$= 0.03054$$

$\therefore$ Regulation = 3.05%

## DESIGN SHEET

1. Rating : 50 MVA
2. Voltage ratio → 66/132 kV
3. Connection → Y/Y
4. Tapping → 5%

### Main Dimension

1. Winding diameter, $d$ → 0.88 m
2. Sectional area of core, $A_i$ → 0.4359 m$^2$
3. Sectional area of yoke, $A_y$ → 0.505 m$^2$
4. Sectional area of window, $A_w$ → 1.136 m$^2$
5. Height of window, $h_w$ → 1.846 m
6. Distance between two adjacent cores, D → 1.407 m
7. Overall width, $W$ → 3.599 m

### Other Datas

1. Flux density in core, $B_m$ → 1.55 T
2. Flux density in yoke, $B_y$ → 1.336 T
3. Magnetising component of no load current, $I_m$ → 3 ampere
4. Core loss component of no load current, $I_c$ → 0.583 ampere

5. No load P.f. → 0.19
6. Total iron loss → 66,714 watt
7. Total Cu loss → 148,113 watt
8. P.f. → 0.8
9. Efficiency → 99.4%
10. Resistance of L.V. winding → 0.12 Ω
11. Resistance of H.V. winding → 0.61 Ω
12. Leakage reactance → 4.1 Ω
13. Voltage regulation → 3.05%

## QUESTIONS WITH SHORT ANSWERS

**Q. 1 Show the arrangement of sandwich coils in a shell type transformer and explain whether the leakage reactance in this case is low or high compared to that obtained in the arrangement of coils in a core type transformer.**

**Ans.** Figure below shows the cross-sectional view of a single phase shell type transformer. The low voltage (L.V.) and the high voltage (H.V.) coils are sandwiched between each other.

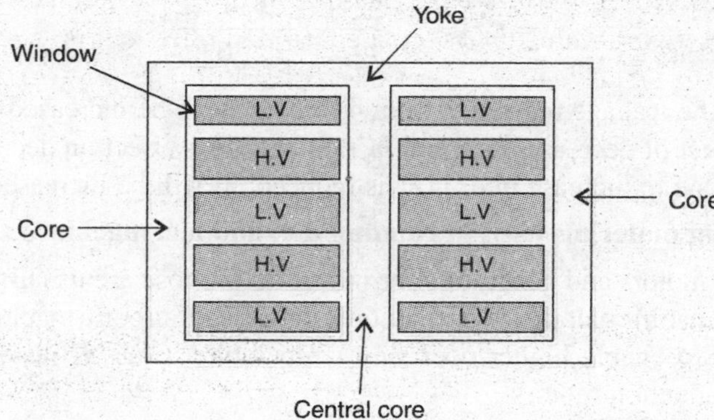


As shown, both H.V. and L.V. windings are divided into a number of coils. The H.V. and L.V. coils are shaped like pancakes and are arranged longitudinally along the core alternately. This gives rise to a sandwich winding with H.V. coils sandwiched between L.V. coils.

Due to large space required between the high and low voltage windings, it is not possible to subdivide the windings to a great extent in the case of core type transformers, while, in the shell type of windings can be easily subdivided by using sandwich coils. Thus, leakage reactance of shell type transformer is low as compared to core type transformer.

**Q. 2 Why is oil used in transformer? What are the properties required of transformer oil?**

**Ans.** Transformer oil is a mineral oil obtained by refining crude petroleum. It serves the double purpose of cooling and insulating. Vegetable and animal oils are not used in transformers as they form fatty acids which attack the fibrous insulating materials like cotton etc. Mineral oil has excellent insulating properties, low loss due to evaporation and its tendency to form sludge is much less. Sludging means the slow formation of solid hydrocarbons due to heating and oxidation. The sludge deposits itself on winding and in cooling ducts producing overheating. This makes the transformer still hotter producing more sludge. This process may continue till the transformer becomes unusable due to overheating. So, contact of oil with air should be avoided as the air contains oxygen and can enhance the oxidation.

The dielectric strength of oil is greatly deteriorated by moisture and so care should be taken to keep the oil as free from moisture as possible.

**Q. 3 Explain why tappings are provided on the H.V. side and not on the L.V. side of a transformer?**

**Ans.** The tappings are provided on the high voltage winding side because a small voltage variation can be obtained owing to large number of turns of H.V. winding. It is difficult to obtain voltage variation within close percentage limits in low voltage winding as there are few turns and voltage per turn is a large percentage of the total voltage. That is why the tappings are provided by H.V. winding side.

**Q. 4 A transformer is to be designed for a 5% leakage reactance. After working out the design details, the leakage reactance has been found to be 4%. What changes in the design be done to obtain 5% leakage reactance?**

**Ans.** Leakage reactance of a transformer of core type referred to L.V. side is given by

$$X_l = \pi f \mu_0 N_l^2 \frac{L_{mt}}{L_c}\left(a + \frac{b_1 + b_2}{6}\right)$$

Thus to increase the leakage reactance from 4 to 5%, $a$ can be increased which will decrease $b_1$ and $b_2$ but the effect of decrease of $b_1$ and $b_2$ will have less effect on decrease of $X_l$ as compared to increase of $X_l$ due to increase in $a$. If $L_c$ is reduced, $X_l$ will be increased.

**Q. 5 List the insulating materials used in rotating d.c. and a.c. machines and transformers.**

**Ans.** (*i*) D.C. and A.C. motors and generators for industrial purpose are usually insulated with class *A* or *E* materials, but turbo-alternators, traction motors and aircraft machines are insulated with class *B* materials to enable higher operating temperature. Class *E* insulation material is commonly used for induction motors.

Fibrous (class *A*) materials are usually employed for both air cooled and oil cooled transformers. Materials commonly used are cotton, synthetic-resin bonded paper, treated pressboard, high grade manilla paper, cotton tape, etc.

**Q. 6 Why cold rolled grain-oriented steel (CRGO) is preferred, for use in transformers over non-oriented sheet steel?**

**Ans.** The magnetic properties of iron can be improved by adding a certain percentage of silicon. When silicon is added, the electrical resistivity of iron increases, and hence the eddy current loss of the material is decreased. The magnetic permeability of the material increases, when the percentage of silicon above 1.8% is added to it.

However, the addition of silicon, greatly deteriorate the mechanical properties of iron. For small machines, material containing 0.5% silicon is used while for large turbo-alternators and transformers high content silicon steel is used. The maximum percentage of silicon is limited to 4.5%, beyond which steel becomes brittle. In transformers 0.35 mm and 0.5 mm thick laminations are used. For high frequency electrical equipment 0.1 to 0.2 mm thick laminations are preferred. These laminations are reduced to these thickness either by cold process or hot process.

In cold reduction process, after reducing to a thickness of 0.35 mm, the steel is annealed at about 1100°C, in order to obtain grain oriented material called CRGO. CRGO silicon steel has much better magnetic properties as compared to the hot rolled silicon steel. CRGO can be worked at much higher flux densities due to the improved magnetic properties. For flux density of 1 tesla the specific loss of CRGO is 0.5 W/kg as compared to 0.9 W/kg for hot rolled steel of best grade. Therefore, there is large reduction in losses when CRGO steel is used. So, machines can be operated at large value of specific magnetic loading, which will result in reduction of weight and size of the machine.

**Q. 7 Write short note on comparison of core and shell type transformers.**

**Ans.** Following are the points of comparison between core type and shell type transformer:

(*i*) *Construction*: Core type transformers have much simpler design and have easier assembly and insulation of windings. The core type transformers are easier to dismantle for repair work.

(*ii*) Windings in a shell type transformer have greater capability of with-standing forces produced under short-circuit conditions since these windings are surrounded and thus braced by a core over large portion of length. On the other hand, the windings in core type construction have a poor mechanical strength because windings are not braced or supported by the core.

(*iii*) *Leakage reactance*: Leakage reactance of shell type transformer winding can be reduced to low value as the windings can easily subdivided by using sandwich coils.

(*iv*) *Cooling*: In the case of core type transformer, the winding surrounds the core. The windings are better exposed and thereby cooling is better in winding of the core type transformers. In the case of shell type transformers, the core is better exposed as core surrounds the winding, therefore cooling of core is better than windings. So, the cooling of winding in case of shell type transformer is less as compared with cooling of winding of core type transformer.

**Q. 8 Why L.V. winding is provided near to the core of the transformer?**

**Ans.** L.V. winding is provided near the core of the transformer due to the reason that the insulation thickness required depends upon the voltage difference. When L.V. winding is provided near the core, the insulation to be provided on core will be for L.V. winding voltage only hence will have lesser thickness as compared to that of higher thickness of insulation if H.V. is put near the core.

**Q. 9 Why circular coils are preferred over rectangular coils for winding of a transformer?**

**Ans.** Circular coils are preferred for winding of the transformers because they can be easily wound on machines, conductors can easily be bent and windings can be properly placed on core so that they do not bulge out due to radial forces produced during working.

**Q. 10 Why power transformers are designed to have their maximum efficiency at or near full load whereas distribution transformers are designed to have their maximum efficiency at loads quite lower than the full load?**

**Ans.** Power transformers are mostly used near full load conditions and hence designed for maximum efficiency at or near full load. Distribution transformers are put in service for 24 hours and work under varying load conditions. For most of the time they are working below the full load and hence are designed to have maximum efficiency at a load much below the full load. Power transformers are put in service during load periods and are disconnected during light load periods.

## UNSOLVED PROBLEMS

1. Explain relative advantages and disadvantages of using silicon steel for transformer.
2. Why distribution transformers are designed for higher all day efficiency? How a higher value of all day efficiency is achieved during the design?
3. Prove that for a transformer:

   Voltage per turn $= K_t\sqrt{kVA \text{ rating}}$

   where, $K_t$ is constant.

   State the assumptions made.
4. Derive the output equation for a 3-phase shell type transformer. Explain why the stepped core is used in transformers?
5. Explain the forces which affect the transformer winding.
6. Explain cold rolled grain oriented (C.R.G.O.) sheat steel and its usefulness for core construction.
7. Why the yoke section in a transformer is made approximately 15% larger than the core section? Also discuss the usefulness of cold rolled grain oriented (C.R.G.O.) sheet steel for core consideration.
8. Discuss the effect of stepping of core of transformer on its performance.
9. Explain leakage reactance in three-phase core type transformer.
10. Discuss arrangement of low voltage and high voltage winding on core of 3-phase core type and 3-phase shell type transformers.
11. Explain the forces which affect the core of the transformer. Discuss the condition when core of the transformer can be crushed.
12. Derive an output equation in terms of design constants for a 3-phase core type transformer. What values of flux density and current density are assumed in designing a transformer and why?
13. Determine the main dimensions of the core, the number of turns and the cross-section of conductors for a 50 kVA, 11000/400 volts, 50 Hz, single phase core type distribution transformer.

    The net copper area in the window is 0.6 times the net cross section of iron in the core. Assume a square cross-section for the core, a flux density 1.0 Wb/m$^2$, a current density 1.4 amperes per mm$^2$ and a window space factor 0.2. The height of window is 3 times its width.

    Leakage reactance of sandwich winding is less than cylindrical core winding. Why?

    **(DDU, Univ. 2000)**
14. Estimate the main core dimensions, the number of turns in the two windings and the conductor sections in a 25 kVA, 3-phase 6600/400 volts, delta/star, 50 Hz, core type transformer with the following data:

    Stepped core for which area factor is 0.56.
    Space factor for window = 0.25
    Voltage per turn = 2.1 volts
    Current density = 2.36 A/mm$^2$.
    Maximum flux density = 1.1 T.

    **(DDU, Univ. 1997)**
15. A 3-phase, 50 Hz, oil cooled, core type transformer has the following dimensions:

    Distance between core centres = 0.2 metre
    Height of window = 0.24 metre
    Diameter of circumscribing circle for a two stepped core = 0.14 metre
    The flux density in the core is 1.25 Wb/m$^2$ and the current density in the conductor is $2.5 \times 10^6$ ampere/m$^2$.
    Estimate the kVA rating of the transformer.
    Window space factor = 0.2 and core area factor = 0.56
    Assume other suitable data if needed.

    **(DDU, Univ. 1995)**

**16.** A 300 kVA, 3-phase, 50 Hz, 6600/400 volts, delta/star, core type transformer intended for lighting load is to be designed with approximately 8.5 volts per turn and a flux density of 1.35 tesla. Take a three-stepped core and yoke area 15 per cent more than core area.

Calculate:

(*i*) Core section and yoke section.

(*ii*) Primary and secondary turns per phase. **(DDU, Univ. 1996)**

**17.** Calculate the main dimensions and winding details of 150 kVA, 2000/400 volts, 50 Hz, single-phase, shell type, oil immersed, self cooled transformer. Voltage per turn, 10 volts, flux density in core 1.1 Wb/m$^2$, current density 2.2 A/mm$^2$, window space factor, 0.32, assume:

$$\frac{\text{Window height}}{\text{Window width}} = 3 \qquad \frac{\text{Core depth}}{\text{Width of central limb}} = 2.5$$

**18.** Derive an expression between voltage per turn and kVA rating of a transformer. What values of flux density and current density are assumed in designing a transformer and why? How do these values change with the change in magnetic material for the core?

# Section III: Computer Aided Design of Electrical Machines

# 9

# Principles of Computer Aided Design

## 9.1 INTRODUCTION

It has been seen in previous chapters of designs of transformers and other rotating machines that the manual calculations are done to find out the different steps and hence the different dimensions and performance calculations of the machines. It is discussed in the design chapters that the output equations in terms of fundamental values (like voltage and current) are being converted into the output equations in terms of the machine dimensions (like diameter and length in case of rotating machines and window details in case of transformers). Design of machine is proceeded with output equation based upon the specifications and other parameters of the machine. The different variables and parameters are inter-related and the variables and performance of the machine have non-linear relationships. The design also depends upon several other factors like mechanical stresses, magnetic forces, temperature rise and cooling medium adopted for the machine.

Taking into consideration the different parameters and operating conditions, the designer has to design a machine which should be best suitable. It should fulfill all the specifications. The machine designed should be rugged, simple, efficient, economic and safe. Designers should have sufficient technical knowledge and able to visualize the simultaneous inter-relationships of all the parameters and their effect. The visualization can be based on wide practical experiences. Sometimes machine may have to operate in isolation but sometimes it is used in a system with so many machines. The machine which has to operate in such a system, cannot be designed in isolation but the design of machine depends upon the optimization of the system performance.

For a particular application, several set of design of machine may need to be done to find the optimum designed machines. In finding the optimum design of the machine many iterations may be required to incorporate the changes in parameters till the satisfactory performance design is obtained. These calculations with indefinite iterations are manually not possible. These calculations can be easily done by means of a digital computers. By using computer the data can be easily varied many times to get the optimum design. So, the computer has become a powerful and important tool in designing the electrical machines.

## 9.2 ADVANTAGES AND LIMITATIONS OF COMPUTER AIDED DESIGN

### 9.2.1 Advantages of Computer Aided Design

The use of digital computers in designing of electrical machines has changed the scene completely. The use of computer reduces the time and money used for the tedious and lengthy hand calculations. This time can be utilized by the designer in developing some kind of ideology. Several characteristics of computer aided designs are given below.

The above characteristics can be explained as:

(*i*) The computer has large memory so it can store maximum data, tables and other information required in the design.

(*ii*) Computer can do the calculations in short time.

(*iii*) It takes very less time to take the logical decisions.

(*iv*) Change in large number of parameters can be modelled, simultaneously using loops in computer aided design.

(*v*) In manual calculations, there may be errors but there are less chance of error in computer calculations.

(*vi*) Optimization of design can be obtained by computers hence it reduces the cost on fabrication.

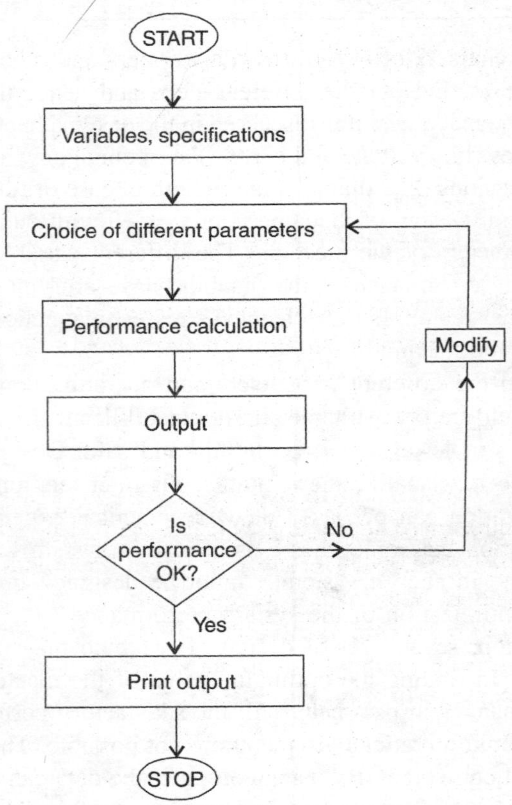


**Fig. 9.1** Flow chart.

(*vii*) Due to less error involved in calculations, high speed, fast decisions and better optimization the design becomes accurate, reliable and cost effective. The different steps in an iteration can be shown by a flow chart given in Fig. 9.1. There can be a number of iterations as it can also be seen by Fig. 9.1.

### 9.2.2 Limitations of CAD

The use of digital computers involve high initial investment and also a considerable annual expense. The computers may be beneficial for an agency producing large number of machines. It is not suitable for smaller companies due to huge investment. Small companies can share time on large computers available in other places for their design work. Some of the limitations of computer aided design can be given below.

(*i*) Memory and storage limitations: It makes computer aided design costlier.

(*ii*) Inter-operability between two computers aided design system is not available.

(*iii*) Interface problem between drawing tools and design and analysis tools.

(*iv*) Verification of feasibility of results obtained by computer aided design difficult.

Inspite of huge investment, computers should be used for research, development and economic studies.

## 9.3 DIFFERENT APPROACHES FOR COMPUTER AIDED DESIGN

There are different approaches for computer aided design. They can be categorised as below:

### 9.3.1 Analysis Method of Design

In this method, the computer is used only for the purpose of analysis only and all the decisions are taken manually by the designer.

In analysis method, the choice of basic parameters and types of construction are done by the designer. There are given as input to the computer for estimation of different machine dimensions and performance calculations. The performance calculated by computer is then critically examined by the designer according to the specifications, to be achieved by the machine. If the design is satisfactory, the output given is the final design detail. If the design is not satisfactory, the designer can make other suitable choice of the parameters to recalculate the performance. This process can be repeated till the performance calculated above are satisfactory. Considerable time is saved by calculating the performance with the help of computer inspite of by human calculations. The analysis method can be initially applicable in design of electrical machines. The flow chart for analysis method is given in Fig. 9.2.

The program for analysis method are simple, easy to understand and use. It saves much calculation times and can be referred or used in larger and sophisticated design programs. The analysis method gives widely acceptable results. The advantage in this method is that the logical decisions are taken by the designers and hence the simple programs can be designed.

### 9.3.2 Synthesis Method of Design

In analysis method, we have seen that the logical decisions are taken by the designers *i.e.*, when the output is not matching with desired specifications then designers select the suitable parameters and again the performance analysis is done.

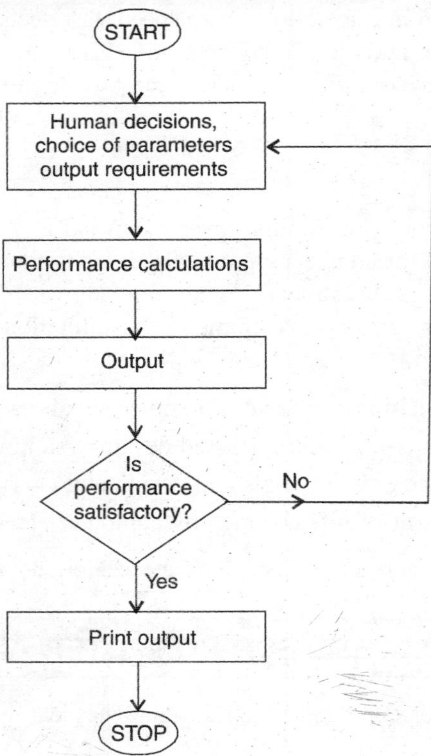


**Fig. 9.2** Analysis method.

In synthesis method, the logical decisions are taken by the computers. The logical decisions include the suitable change in parameters to achieve desired design and performance. The logical decisions are considered as a set of instructions in the program. The flow chart for synthesis method is given in Fig. 9.3. In synthesis method, the analysis and decision both are done by the computers so the time required in this method is very less. In design of electrical machines, there are several aspects such as

(*i*) Conducting, magnetic, insulating materials, manufacturing techniques, standard specifications are always continued and hence there may be changes which must be incorporated in the programs.

(*ii*) When a machine is to be designed, there are many parameters to be considered in design procedure and hence to get an optimized design there may be so many combinations of parameters which can cause difficult logic decisions and bulky programs.

In view of the above aspects, the synthesis method may have following several disadvantages:

(*a*) Large number of logical decisions to be taken by computers.

(*b*) Large number of instructions make the program complicated and lengthy.

(*c*) Complicated and lengthy programs need large computers which require huge investment.

(*d*) The programs need changes to incorporate the latest changes in materials and techniques.

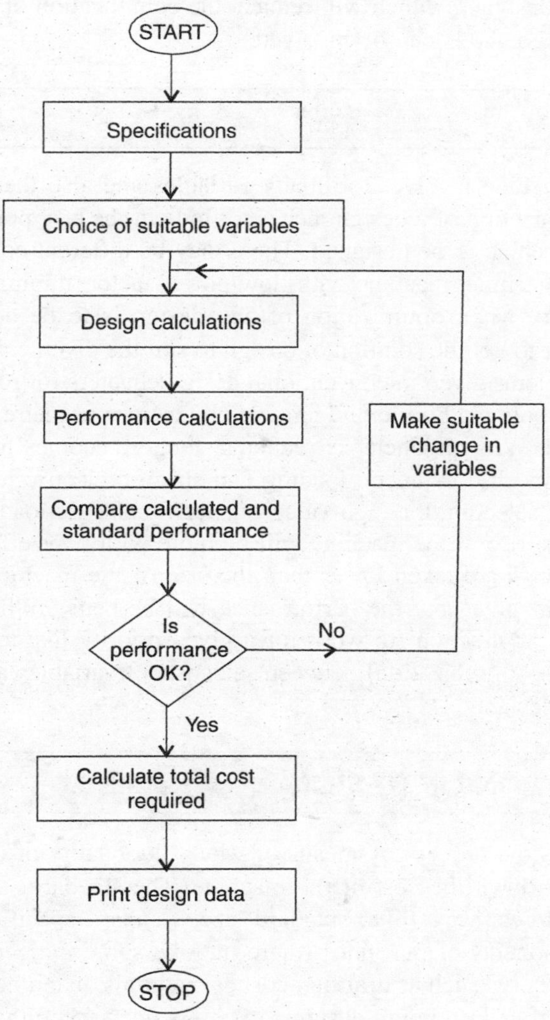


**Fig. 9.3** Synthesis method.

### 9.3.3 Hybrid Method of Design

When both analysis and synthesis methods are used simultaneously to design a machine then this is called hybrid method. It has been discussed in analysis and synthesis methods that the logical decisions require large number of instructions. The instructions should be considered in programs. The program which is made for design should have proper set of instructions to take the logical decisions. It is important that the program used for synthesis method should be prepared on the basis of good technical knowledge, experience and collective decisions of design engineers. These may involve huge investment and sometimes complicated. So, to make the design procedure easier, the logical decisions can be partly taken by the designers and partly by the computers. By using both the methods together means that the computational work can largely be done by computer programmes and some design decisions may be taken by designers which will reduce the complication of design programmes. Thus both analysis and synthesis methods can be involved.

## 9.4 OPTIMIZATION

The design of electrical machine involves too many variables and also there may be many solutions possible for a design. Optimization of a design means to find out the best possible solution (design) for a given condition of the machine to be designed. There may be different condition for the machine to be designed like, cheaper machine, machine with, low power factor, minimum losses, minimum size, overall good design etc. So, when optimization of any design is to be done, the above conditions become the important factor to get the solution of design to suit the given condition. It is clear from the solutions of the design problems given in conventional design chapters for rotating machines that when the variables like length to pole pitch ratio and ampere-conductor per metre are changed, the diameter and the length are changed. When diameter is reducing, length becomes more. It has also been discussed in the selection of specific magnetic loading and specific electric loading that the size of the machine can be reduced by selecting the high value of specific electric loadings but the losses will be more and hence temperature rise in the machine will be more with poor efficiency. When the specific electric and magnetic loadings are taken lower than the size of the machine will increase and hence more material and cost is involved but the performance will be better. When any machine has to be designed for any particular application (to work for a given priority like technical, economic, operational etc.) then compromise (optimization) between selection of variables and given situation is done to achieve a best possible design.

## 9.5 STANDARDISATION OF DESIGN

Standardisation of design is the process by means of which the quality of the machine design can be improved and also keeping down the cost of the machine. The standardisation of electrical machine design can be done in different ways like, standard size, standard specifications (ratings), standard materials of different components of machines (there may be some indigenous item to be used) etc. The cost of preliminary expenses such as drafting, cost of materials, manufacturing cost, computational cost etc should be optimized and optimum design can be obtained when the total cost of the machine is to be minimized. The standardisation can be applied in the industries where large number of machines are being produced to get the benefits of standardisation to produce the machines with better quality and reduced cost.

British Standard Specifications and Indian Standard Institute has laid down certain standards which make the design economical to a great extent. Different standards are available and the designers can use any one of them. The sizes of certain parts such as nuts, bolts armature stampings, brushes etc. have been standardised for universal applications. The design of insulation and mechanical parts are based on the empirical formulae derived from the test results.

# 10

# Computer Aided Design of Electrical Machines

## 10.1 INTRODUCTION

The importance of computer aided design and different approaches which can be followed to design any machine are discussed in chapter 9. Machine dimensions (size of the different portions) can be calculated for given specifications of the machine by selecting suitable parameters. After estimating the machine dimensions and other details, the performance such as losses, efficiency, regulation, temperature rise etc. can be calculated. The performance of the machine also depends upon several other factors like ventilation and cooling hence these factors should also be taken into consideration. If the performance of the machine is satisfactory, the design is proper. But the different parameters should be suitably changed to get the satisfactory performance of the machine. In this chapter, computer aided design of transformers, synchronous machines, three-phase induction machine and d.c. machines are discussed. Discrete problems on main dimensions, no load current, efficiency, estimation of total m.m.f. required, damper winding design, tank design etc. are solved by computers. The computer programmes in *C* are developed for each problem.

The results are discussed wherever it is needed. The corresponding flow charts are also being provided for respective problems. The flow chart for overall design and performance estimation of different machines are also provided.

## 10.2 COMPUTER AIDED DESIGN OF 3-PHASE ALTERNATOR

**P.1** Calculate inner diameter and axial length of 4.5 MVA, 50 Hz, 3-phase, 16-pole water wheel generator. Specific magnetic loading is 0.86 $Wb/m^2$ and specific electric loading is 32,000 A/m. Pole arc to pole pitch ratio is 0.71.

**Solution:**

$$Q = CD^2LN \qquad ...(1)$$

Output coefficient, $C = 11.1 \times 10^{-5} \times 0.86 \times 32{,}000$

$= 3.0547$

$$N = \frac{120 \times 50}{16} = 375$$

given, $Q = 4.5 \times 10^3$ kilovolt ampere

$$D^2 L = \frac{Q}{CN} = \frac{4.5 \times 10^3}{3.0547 \times 375}$$

$$= 3.928 \qquad \text{...(2)}$$

Considering round pole construction,

$$\frac{b_P}{\tau_P} = \frac{L}{\tau_P} = 0.71$$

So, $$\frac{L}{\tau_P} = 0.71 \text{ or } \frac{L}{\frac{\pi D}{P}} = 0.71$$

or $$L = \frac{0.71 \times \pi D}{P} = \frac{\text{Pole arc/Pole pitch ratio} \times 3.14 \times D}{P} = \frac{0.71 \times 3.14 \times D}{16}$$

or $$L = 0.139D \qquad \text{...(3)}$$

From equations (2) and (3), we get

$$D^3 = \frac{Q \cdot P}{\text{Pole arc/Pole pitch ratio} \times 3.14 \times C \times N}$$

$$0.139\, D^3 = 3.928$$

$$D^3 = 28.258$$

$$D = 3.04 \text{ m}$$

But, from equation (3), $L = 0.42$ m

We can check the feasibility of $D$ and $L$ calculated as above by calculating peripheral speed.

$$V_p = \frac{\pi DN}{60} = 59.66$$

Peripheral speed is within specified limit so we can design alternator. A computer programme can be developed in language and can be implemented for the solution of this problem.

**P.2** Following specifications are given for a 3-phase, 50 Hz, star connected synchronous generator

kVA = 7800

Voltage = 6.6 kV

Number of poles = 12

Specific magnetic loading = 0.88 T

Specific electric loading = 24,000 A/m

Pole arc to pole pitch ratio = 0.69

Calculate air gap diameter and length of alternator. Calculate peripheral speed of alternator. Peripheral speed should be below 80 m/sec. Modify specific electric loading and calculate air gap diameter and length corresponding to peripheral length below 80 m/sec. Specific electric loading can be taken up to 34,000 A/m.

**Solution:**

$$\text{Output in kVA, } Q = CD^2LN \qquad \text{...(1)}$$

$$\text{Output coefficient, } C = 11.1\times10^{-5}\times0.88\times24{,}000$$

$$= 2.344$$

$$\text{Speed, } N = \frac{120\times50}{12} = 500 \text{ rpm}$$

$$\text{From equation (1), } D^2L = \frac{Q}{CN} = \frac{7800}{2.344\times500} = 6.655 \qquad \text{...(2)}$$

Assume round pole construction,

$$\text{So, } \frac{b_P}{\tau_P} = \frac{L}{\tau_P} = 0.69$$

$$\text{So, } L = \frac{0.69\cdot\pi D}{P} = 0.18D \qquad \text{...(3)}$$

From equations (2) and (3), $D^3 = 36.972$

$D = 3.33$ m

Also, from equation (3), length, $L = 0.599$ m.

$$\text{Peripheral speed, } V_P = \frac{\pi DN}{60} = \frac{3.14\times3.33\times500}{60} = 87.13$$

With this specification the peripheral speed is not within limit. So, we will have to change specific electric loading. It is given that it can be taken up to 34,000. We can increase in the step of 500 *i.e.*, we can select 24,500, calculate diameter and then corresponding peripheral speed. We can keep on increasing this in the step of 500 and calculate diameter and corresponding peripheral speed till peripheral speed becomes below 80 m/sec. Let us assume specific electric loading of 31,000. Corresponding output coefficient, $C = 3.08$.

$$D^2L = 5.15 \qquad \text{...(4)}$$

From equations (3) and (4), $D^3 = 28.62$

$D = 3.058$

Chapter 10

So, $V_P = 8.04$

Hence specific electric loading of 31,500 can be selected and then air gap diameter and length can be calculated. With specific electric loading of 31,500.

$$C = 3.076$$

$$D^2L = 5.071 \quad ...(5)$$

From equations (3) and (5), $D^3 = 28.175$

$\Rightarrow$ $D = 3.042$

So, $V_p = 79.62$

So, $D = 3.042$ m

and $L = 0.547$ m

**P.3** Estimate the main dimensions of 1800 kVA, 50 Hz, 3-phase, 187.5 rpm, water wheel generator. The specific magnetic loading is 0.8 Wb/m$^2$ and the specific electric loading is 26,000 ampere-conductor/m. Pole arc to pole pitch ratio is 0.66. Calculate peripheral speed. Round pole construction can be considered.

**Solution:**

Computer programme is developed in *C*. The main dimensions *i.e.*, *D* and *L* can be estimated. The peripheral speed can also be calculated to see whether the peripheral speed is within the specified limit. The peripheral speed may be within the specified general limits for the alternators or some specific value in that range can be fixed. If peripheral speed is not within limit it can be marked that *D* and *L* calculated are not justified. In this particular program there is no iteration to modify the diameter to bring the peripheral speed within the specified limit. The corresponding flow chart is also prepared and given in Fig. 10.1. The value of diameter, length and corresponding value of peripheral speed can be calculated. These values can be calculated for any alternator with their specifications just like given in problem 10.2.3.

```
/*Calculation of Length and Diameter of Alternator */
#include <stdio.h>
#include <math.h>
main ()
{
    int q, f, ac;
    float n, b1, ppratio;
    float l, d, dia, C, p, Vp;
    q = 1800;
    f = 50;
    n = 187.5;
    b1 = 0.8;
    ac = 26,000;
    ppratio = 0.66;
```

```
C = (11.1e – 5)*b1*ac;
p = (120*f)/n;
dia = (q*p)/(3.14*c*ppratio*n);
d = pow(dia, 1/3);
l = q/(c*n*d*d);
Vp = (3.14*d*n)/60;
printf("Length = %f Diameter = %f peripheral speed = %f, l, d, Vp);
```

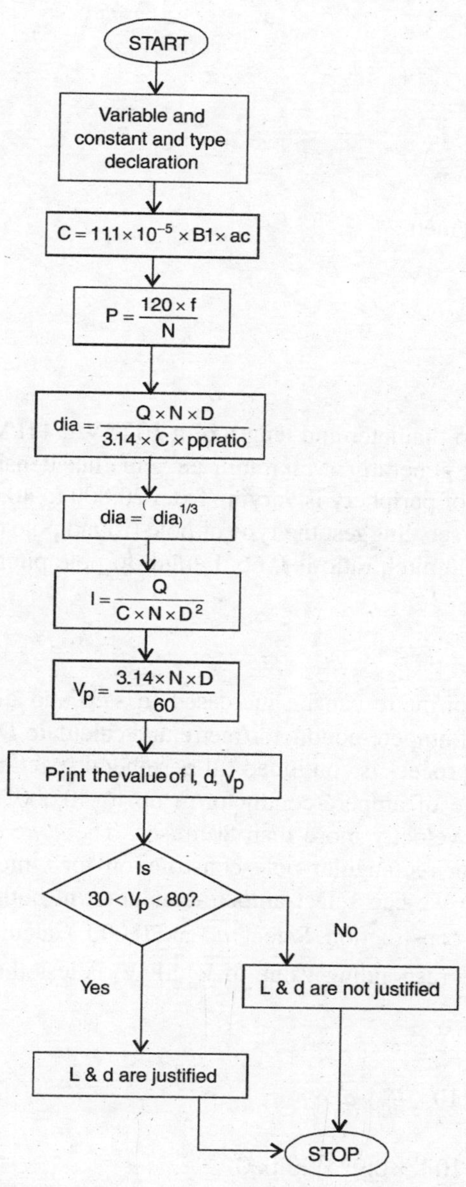


**Fig. 10.1** Flow chart without iterations.

```
    if (Vp > 30 & Vp < 80)
    {
    print f ("Length and diameter are justified");
    }
    else
    {
    printf ("Length and diameter are not justified");
    }
}
```

Q = 1800 kVA

f = 50 Hz

N = 187.5 rpm

$B_1$ = 0.8 Wb/m$^2$

ac = 26,000 ampere-conductor/metre

ppratio = 0.66

Length = l

Diameter = d

Peripheral speed = $V_p$

**P.4** Estimate suitable air gap diameter and length of a 10 MVA, 11kV, 8-pole, 3-phase, 50 Hz, star-connected synchronous generator. Maximum air gap flux density is 0.92 tesla. The ampere-conductor/metre length of periphery is varying from 20,000 to 40,000 A/m. The peripheral speed should not exceed 80 m/sec. Suggest the type of pole (round pole or long pole) construction to be selected. Pole arc to pole pitch ratio is 0.66. Length to pole pitch ratio can be taken between 1 and 5.

## Solution:

The value of ampere-conductor/metre can be increased in step and diameter can be calculated. We keep on increasing the value of ampere-conductor/metre and calculate $D$ for that value and calculate $V_p$ corresponding to this $D$. This process is continued till peripheral speed regarding $D$ is below 80 m/sec. We can select maximum value of ampere-conductor/m up to 40,000. If diameter D is calculated at 40,000 results in a peripheral velocity more than 80 m/sec. Then, we cannot design with round pole construction. We can select for rectangular pole construction for which length to pole pitch ratio is taken between 1 and 5. Again we can select ampere-conductor/m both 20,000 to 40,000. Repeat the same process as in round pole construction. Select $\overline{ac}$ at 20,000. Calculate $D$ and then $V_p$. Increase $\overline{ac}$ in step and calculate $D$ and corresponding value of $V_p$ till $V_p$ is less than 80 m/s.

Output in kVA, $\quad Q = CD^2LN \qquad$ ...(1)

Output coefficient, $C = 11.1\times10^{-5} B_{\delta_1} \cdot \overline{ac}$

$$= 11.1\times10^{-5}\times0.92\times20{,}000$$

$$= 2.042$$

$$N = \frac{120 \times 50}{8} = 750$$

$$Q = 10 \times 10^3 \text{ kVA}$$

$$D^2 L = \frac{Q}{CN} = \frac{10 \times 10^3}{2.042 \times 750}$$

$$D^2 L = 6.529 \qquad \text{...(2)}$$

First considering round pole construction,

So, $$\frac{L}{\tau_P} = \frac{b_P}{\tau_P} = 0.66$$

or $$\frac{L}{\tau_P} = 0.66$$

$\Rightarrow$ $$L = \frac{0.66 \times \pi D}{P}$$

$$= 0.259D \qquad \text{...(3)}$$

From equations (2) and (3), we get

$$D^3 = 25.20$$

or $$D = 2.932 \text{ m}$$

Peripheral velocity, $$V_P = \frac{\pi D N}{60}$$

$$= 115.08 \text{ m/sec}$$

For round pole construction,

Select $\overline{ac} = 40{,}000$.

$$C = 11.1 \times 10^{-5} \times 0.92 \times 40{,}000$$

$$= 4.084$$

$$N = 750 \text{ r.p.m}$$

$$D^2 L = \frac{Q}{CN}$$

$$= \frac{10 \times 10^3}{4.084 \times 750} = 3.626 \qquad \text{...(4)}$$

From eqns. (3) and (4), we get,

$$D^3 = 12.586$$

$$D = 2.326$$

$$V_p = \frac{\pi D N}{60} = \frac{3.14 \times 2.326 \times 750}{60} = 91.29$$

We see that even if we select $\overline{ac}$ as 40,000, peripheral speed is above 80 m/sec. We will select rectangular pole and select $\frac{L}{\tau_P}$ ratio between 1 and 5.

$$\frac{L}{\tau_P} = 1$$

$$L = \frac{\pi D}{P} = \frac{3.14 \times D}{8} = 0.392 \quad \text{...(5)}$$

From eqns. (2) and (5), we get,

$$D^3 = 16.655$$

$$D = 2.553$$

$$V_p = 100.205$$

$\frac{L}{\tau_P}$ and $\overline{ac}$ both can be increased in steps. Let us suppose that we select some value, say $\frac{L}{\tau_P} = 1.5$ and

$\overline{ac} = 30,000$

for $\overline{ac} = 30,000$

$$C = 11.1 \times 10^{-5} \times 0.92 \times 30,000$$

$$= 3.063$$

$$D^2 L = \frac{10 \times 10^3}{3.063 \times 750} = 4.35 \quad \text{...(6)}$$

Now $\frac{L}{\tau_P} = 1.5$

So, $$L = \frac{1.5 \times \pi D}{P} = 0.588\, D \quad \text{...(7)}$$

From eqns. (6) and (7), we get, $D^3 = 7.397$

$\Rightarrow$ $D = 1.948$

Corresponding, $V_p = 76.459$ m/sec

So, $D$ = 1.948 m and corresponding length $L$ is 1.145 m.

### 10.2.1 Programme and Flow Chart

The diameter of alternator is calculated by taking the round pole construction. With the round pole construction, the specific electric loading is varied to its maximum limit. But the peripheral speed for the diameter at 40,000 A/m with round pole is more than 80 m/sec. Hence, the rectangular pole construction is selected and the main dimensions are again calculated. The programme is developed in *C*. There will be iterations to calculate the suitable value of diameter and length so that the peripheral velocity is within the given limit. The construction may be round pole or long pole. The specific electric loading may also vary in the given range. All these values can be taken and then suitable value of diameter and length can be calculated to get the peripheral speed within the given limit.The flow chart for this problem is given in Fig. 10.2.

```
/* Calculation of Length and Diameter of Alternator */
#include <stdio.h>
#include <math.h>
main ()
{
    int f, p, ac;
    float q, b1, ppratio, n, dia, d, l, Vp, nppr;
    q = 10 e 3;
    f = 50;
    p = 8;
    b1 = 0.92;
    ppratio = 0.66;
    n = (120*f)/p;
    ac = 20,000;
One: dia = (q*p)/(11.1e – 5*3.14*b1*ac*n*ppratio);
    d = pow(dia, 1/3);
    Vp = (3.14*d*n)/60;
    if (Vp < 80)
    {
      l = (3.14*d*ppratio)/p;
      printf("Length = %f Diameter = %f peripheral speed = %u, l, d, Vp);
      printf ("This is round pole design");
      goto end;
    }
     else
     {
     if (ac ≤ 40,000)
```

```
{
ac = ac + 500;
goto one;
}
else
```

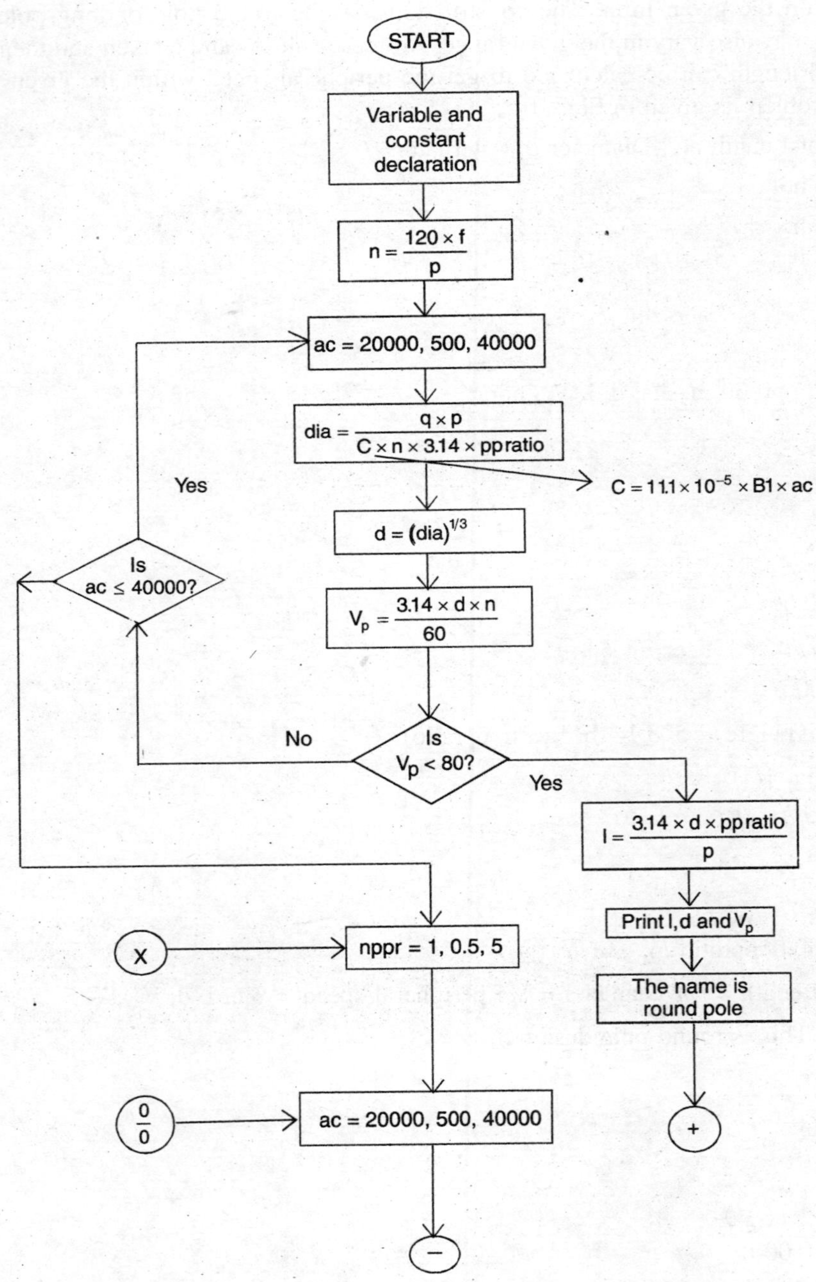

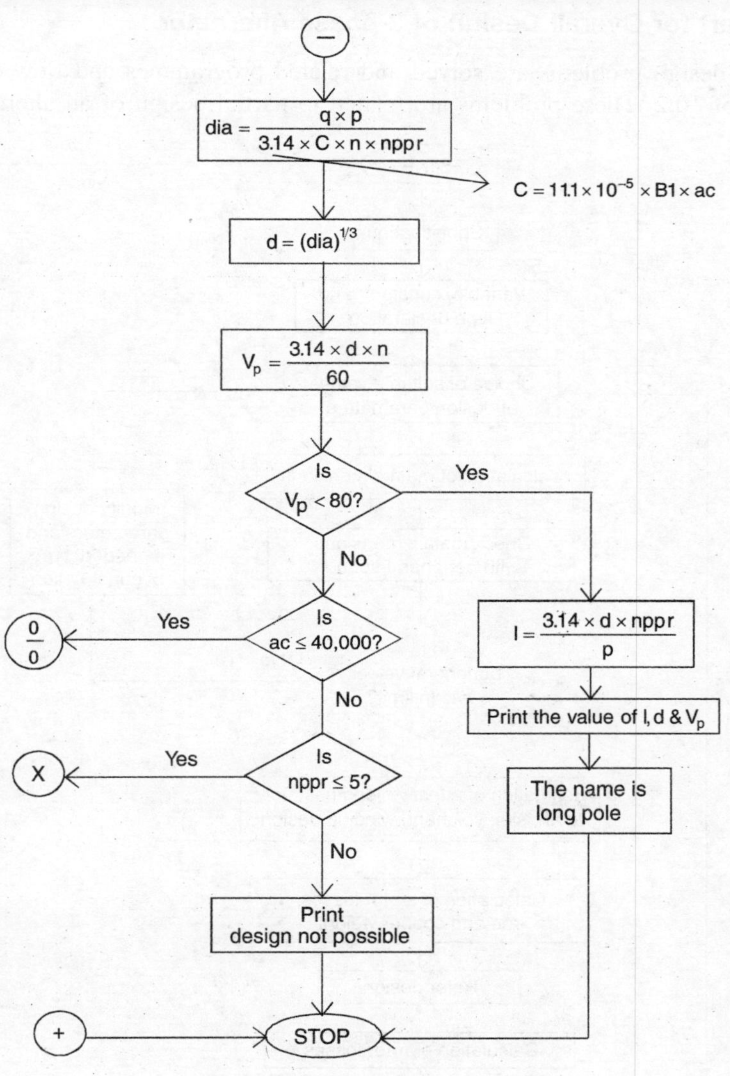


**Fig. 10.2** Flow chart with iterations.

```
        {
            goto two;
        }
    }
two: nppr = 1.0
    ac = 20,000;
three: dia = (q*p)/(11.1e – 5*3.14*b1*ac*n*nppr);
    d = pow (dia, 1/3);
    Vp = (3.14*d*n)/60
```

### 10.2.2 Flow Chart for Overall Design of 3-phase Alternator

Different types of design problems are solved and related programmes and flow charts are given in previous sections of 10.2. These problems are related to partial design of an alternator *i.e.*, only few

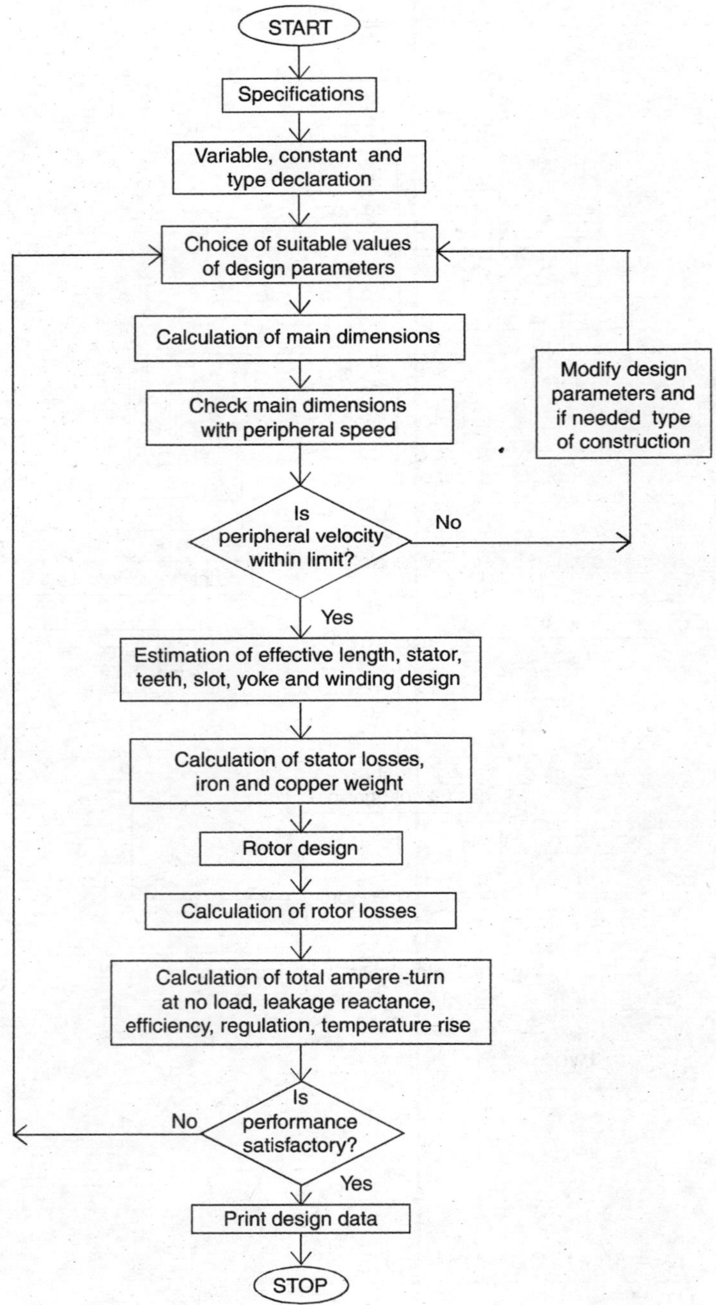


**Fig. 10.3** Flow chart for overall design of 3-phase alternator.

parameters are being calculated. A 3-phase alternator can be designed by estimating dimensions/sizes of different portions. The performance can also be calculated by calculating losses, efficiency, regulation, temperature rise etc. The overall design of a 3-phase alternator can be done with the help of related flow chart given in Fig. 10.3.

```
        if (Vp < 80)
        {
        l = (3.14*d*nppr)/p
        print ("Length = %f Diameter = %f peripheral speed = %f", l, d, Vp);
                                print ("This is long pole design");
                                go to end;

        }
        else
{
        if (ac ≤ 40,000)
        {
         ac = ac + 500;
         go to three;
        }
        else
        {
         if (nppr ≤ 5)
        {
         nppr = nppr + 0.5;
         go to three;
        }
else
        { printf ("Design is not possible");
         go to end;
         }
        }
}
end : printf ("Design work complete);
}
```

## 10.3 COMPUTER AIDED DESIGN OF 3-PHASE INDUCTION MOTOR

**P.1** Calculate the main dimensions, turns per phase, total number of slots, sectional area of conductor and area of slot to accommodate the conductors and insulation in the slot for a 200 kW,

3-phase, 50 Hz, 440 V, 1440 rpm, star-connected, slip ring induction motor. Average flux density is 0.52 Wb/m$^2$, $\overline{ac}$ is 32,000 A/m, efficiency is 0.88, winding space factor is 0.46, winding factor is 0.96, power factor 0.86 and current density in winding conductor is 4 A/mm$^2$. The design should be cheaper and hence $L/\tau_p$ ratio is taken 1.5 slot pitch is 20 mm.

**Solution:**

```
/*Calculation of turns per phase, total no. of slots and cross-sectional area of conductor in an Induction Motor */
#include <stdio.h>
#include <math.h>
main ()
{
    /* Constant and variable declaration */
    int q, f, V, N, ac, J, slot pitch, Ns, P;
    float S, B1, eff, kws, kw, pf, C, d, l, flux,
    nph, ncs, nss, nps, iph, fc, acs, Ast, dia;
    q = 200;
    f = 50;
    V = 440;
    S = 0.04;
    N = 1440;
    B1 = 0.52;
    ac = 32,000;
    eff = 0.88;
    kws = 0.46;
    kw = 0.96;
    pf = 0.86;
    J = 4;
    slot pitch = 20 e – 3;
    ns = n/(1–s);
    p = (120*f)/ns;
    C = 18.3 e–5*kw*B1*ac*eff*pf;
    dia = (P*q)/(3.14*1.5*c*ns);
    d = pow (dia, 1/3);
    l = (1.5*3.14*d)/p;
    flux = (3.14*B1*D*l)/p;
    nph = V/(1.732*4.44*f*kw*flux);
    ncs = 3*2*nph;
```

START

Constant and variable declaration

Calculation of synchronous speed

No. of poles $P = \frac{120 \times f}{N_s}$

$C = 18.3 \times 10^{-5} \times K_w \times B1 \times ac \times eff \times pf$

$D = \left( \frac{P_1 Q}{3.14 \times 1.5 \times C \times N_s} \right)^{1/3}$

$L = \frac{1.5 \times 3.14 \times D}{P}$

$\phi = \frac{B1 \times 3.14 \times D \times L}{P}$ ← Flux

$N_{ph} = \frac{(V/1.732)}{4.44 \times f \times K_w \times Flux}$

$N_{cs} = 3 \times 2 \times N_{ph}$ ← Total No. of Stator Conductors

$N_{ss} = \frac{3.14 \times D}{\text{Slot pitch}}$ ← Total No. of Stator Slots

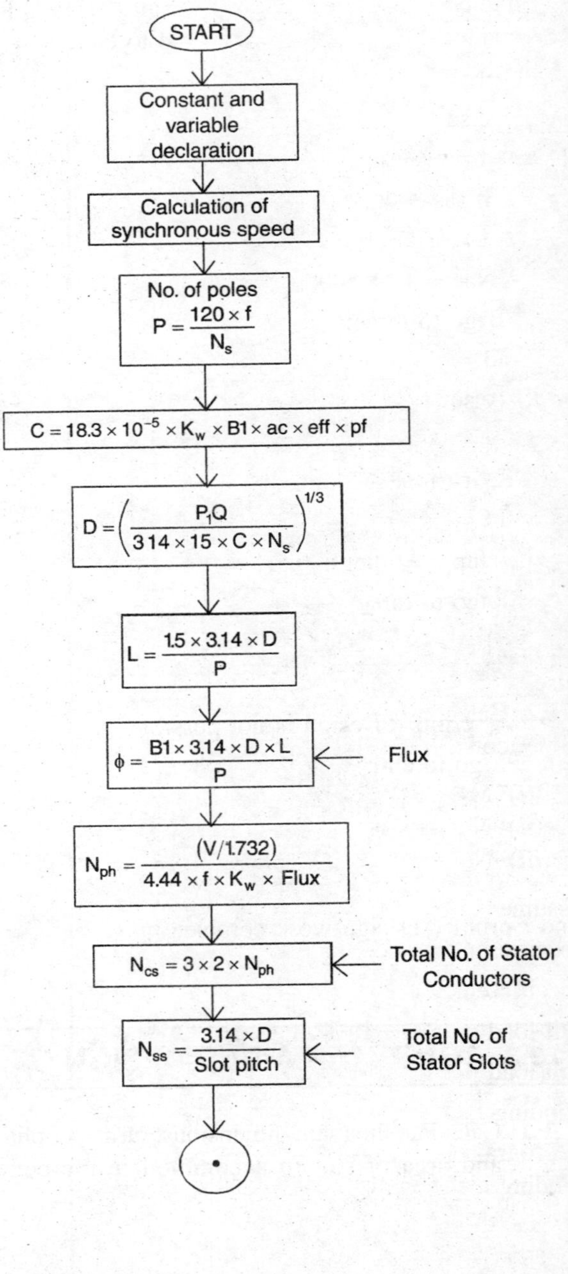

```
nss = (3.14*d)/slot pitch;
nps = abs (ncs/nss);
iph = q/(1.732*V*eff*pf);
fc = iph/s;
acs = 2*fc;
ast = acs/kws;
printf ("Turns per phase = % f total
number of slots = %f X-sectional area of
conductor = %f", nph, nss, fc);
}
```

$Q = 200$ kW

$f = 50$ Hz

$V = 440$ V

$s = 0.04$

$N = 1440$ rpm

$B_1 = 0.52$ Wb/m$^2$

ac = 32000 A/m

Eff = 0.88

$K_{ws} = 0.46$

$K_w = 0.96$

Pf = 0.86

$J = 4$ A/m$^2$

Slot pitch = $20 \times 10^{-3}$ m

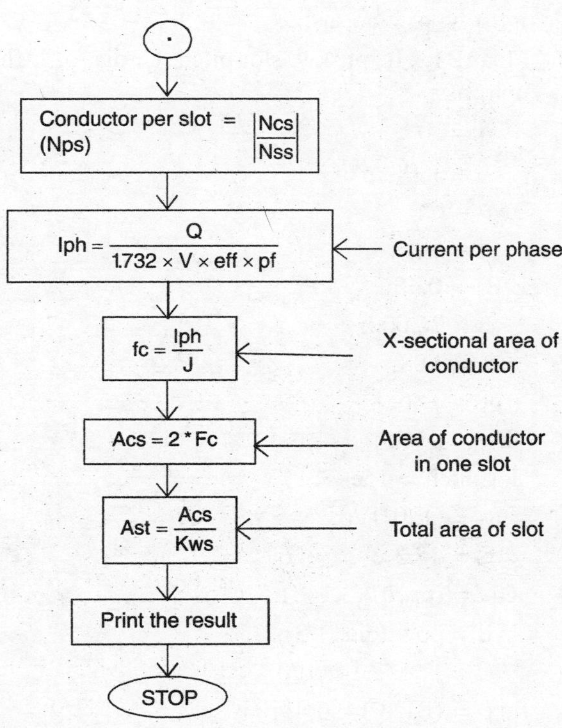


**Fig. 10.4** Flow chart for calculation of turns/phase, number slots and area of slot.

**P.2** Calculate the following design information for a 30 kW, 440 V, 3-phase, 6-pole, 50 Hz, delta connected squirrel cage induction motor

(*i*) Main dimensions of stator frame.

(*ii*) Number of turns per phase in stator winding.

(*iii*) Number of stator slots.

Assume

Specific magnetic loading = 0.48 T

Specific electric loading = 26,000 ampere-conductor/m

Full load efficiency, $\eta = 0.88$

Full load p.f. = 0.86

Winding factor = 0.955

/* Program for calculation of dimensions of stator frame, number of turns per phase in stator winding and number of stator slots*/

```
# include <stdio.h>
# include <math.h>
main ( )
{
    Int q, v, p, f, ac, ns;
    Float B1, eff, pf, kw, slot pitch, C, dia, d, l, flux,
    nph, nss;
        q = 30
        v = 440;
        p = 6;
        f = 50;
      B1 = 0.48;
      ac = 26,000;
     eff = 0.88;
      pf = 0.86;
     kw = 0.955;
    slot pitch = 15e–3;
      ns = (120*f)/p;
       C = 18.3 e–5*w*B1*ac*eff*pf;
    Dia = (p . q)/(3.14*1.5*c*ns);
       d = pow (dia, 1/3);
       l = (1.5*3.14*d)/p;
    flux  = (BI*3.14*d*l)/p;
     nph = V/(4.44*f*kw*flux);
     nss = (3.14*d)/slot pitch;
    Printf ("diameter = %f length = %f number of
    turns per phase = %f number of stator slots =
    %f", d, l, nph, nss);

}
```

q = 30 kW

V = 440 V

p = 6

f = 50 Hz

$B_1$ = 0.48 T

ac = 26,000 A-c/m

Eff = 0.88

p.f. = 0.86

kW = 0.955

Slot pitch = 15 × $10^{-3}$ m

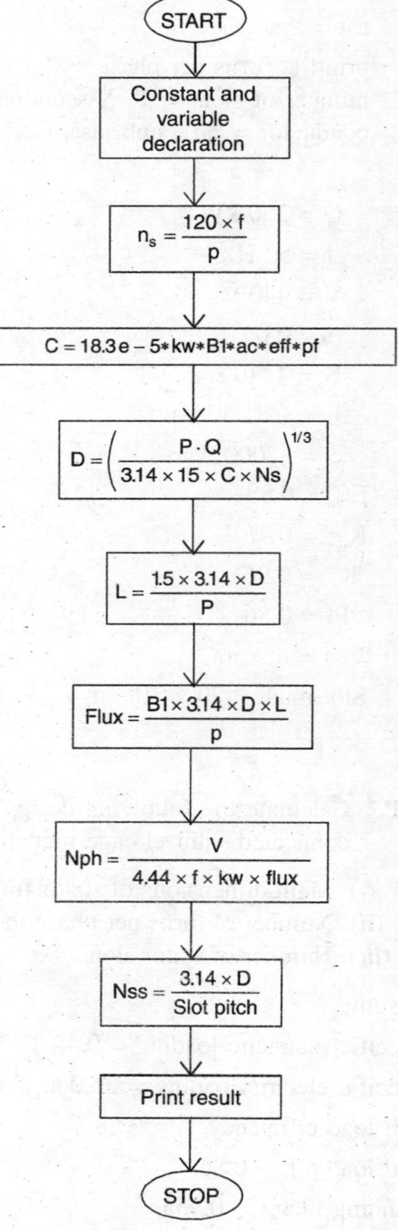


**Fig. 10.5** Flow chart.

**P.3** Explain the main dimensions, number of stator slots and number of stator conductors per slot for a 140 h.p., 3300 volt, 50 c/sec., 12 poles, star-connected slip ring induction motor. Assume

Average gap density = 0.4 $Wb/m^2$

Ampere-conductor per metre = 25,000 A/m

Efficiency = 90%

Power factor = 0.9

Winding factor = 0.96

```
#Include <stdio.h>
#Include <math.h>
main ( )
{
      int, q, v, p, f, ac, ns;
      float B1, eff, pf, kw, slot pitch, c, dia, d, l, flux, nps, nss, ncs, nph
      q = 140*0.746;
      V = 3300;
      f = 50;
      p = 12;
      B1 = 0.4;
      ac = 25000;
      eff = 0.9;
      pf = 0.9;
      kw = 0.96;
      slot pitch = 15e – 3;
      ns = (120*f)/p;
      c = 18.3e – 5*kw*B1*ac*eff*pf;
      dia = (p*q)/(3.14*1.1*c*ns);
      d = pow (dia, 1/3);
      l = (1.1*3.14*D)/p;
      flux = (B1*3.14*d*l)/p;
      nph = V/(1.732*4.44*f*kw*flux);
      ncs = 3*2*nph;
      nss = (3.14*d)/slot pitch;
      nps = abs (ncs/nss);
      printf (“Total number of stator conductor = %f Total number of stator slots = %f”, ncs ,nss);
}
```

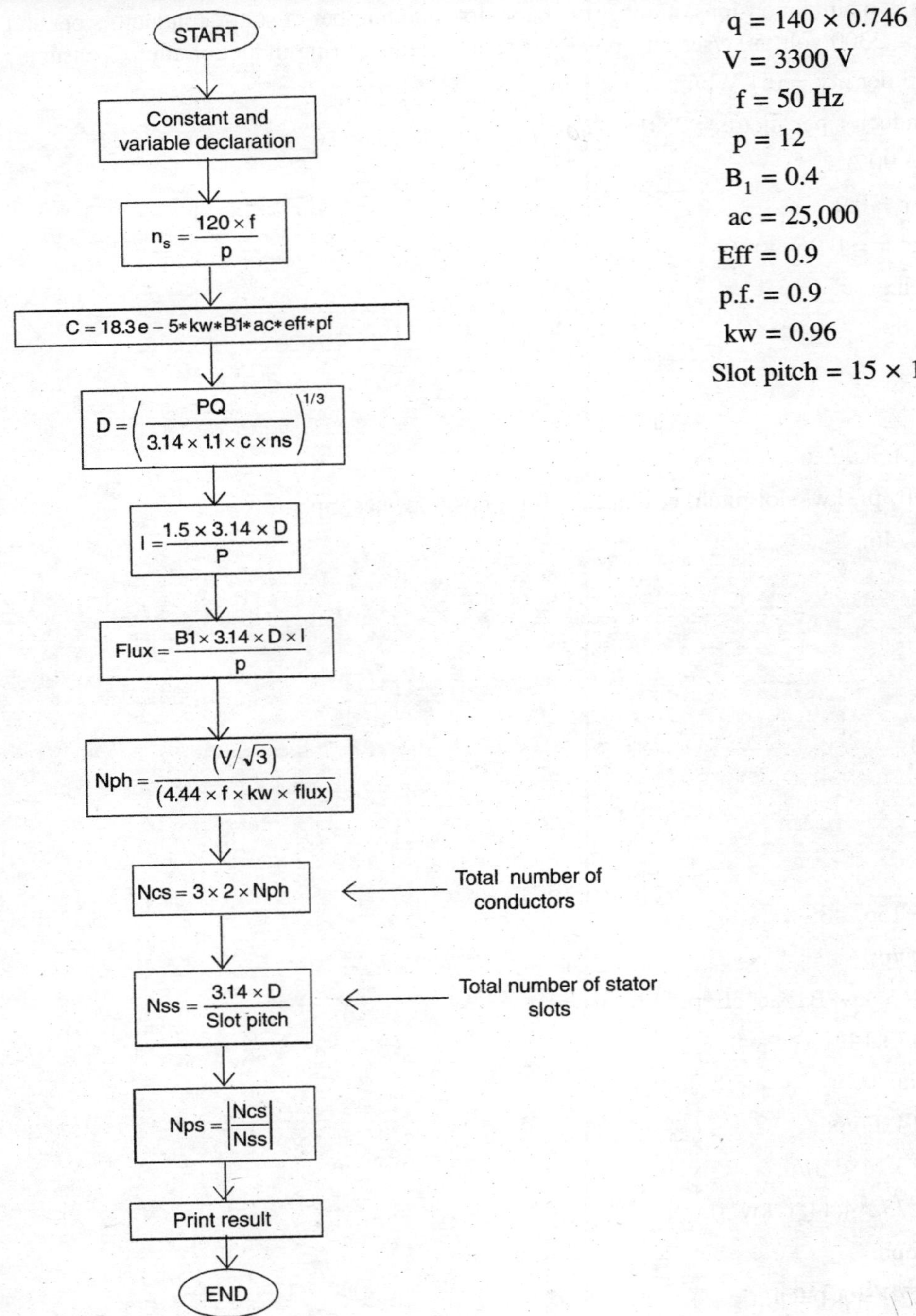


Fig. 10.6 Flow chart.

$q = 140 \times 0.746$ kW

$V = 3300$ V

$f = 50$ Hz

$p = 12$

$B_1 = 0.4$

$ac = 25{,}000$

$Eff = 0.9$

$p.f. = 0.9$

$kw = 0.96$

Slot pitch $= 15 \times 10^{-3}$ m

### 10.3.1 Flow Chart for Overall Design of 3-phase Induction Motor

A 3-phase induction motor can be designed and its performance can be calculated. Flow chart for overall design of 3-phase induction motor is given in Fig. 10.7.

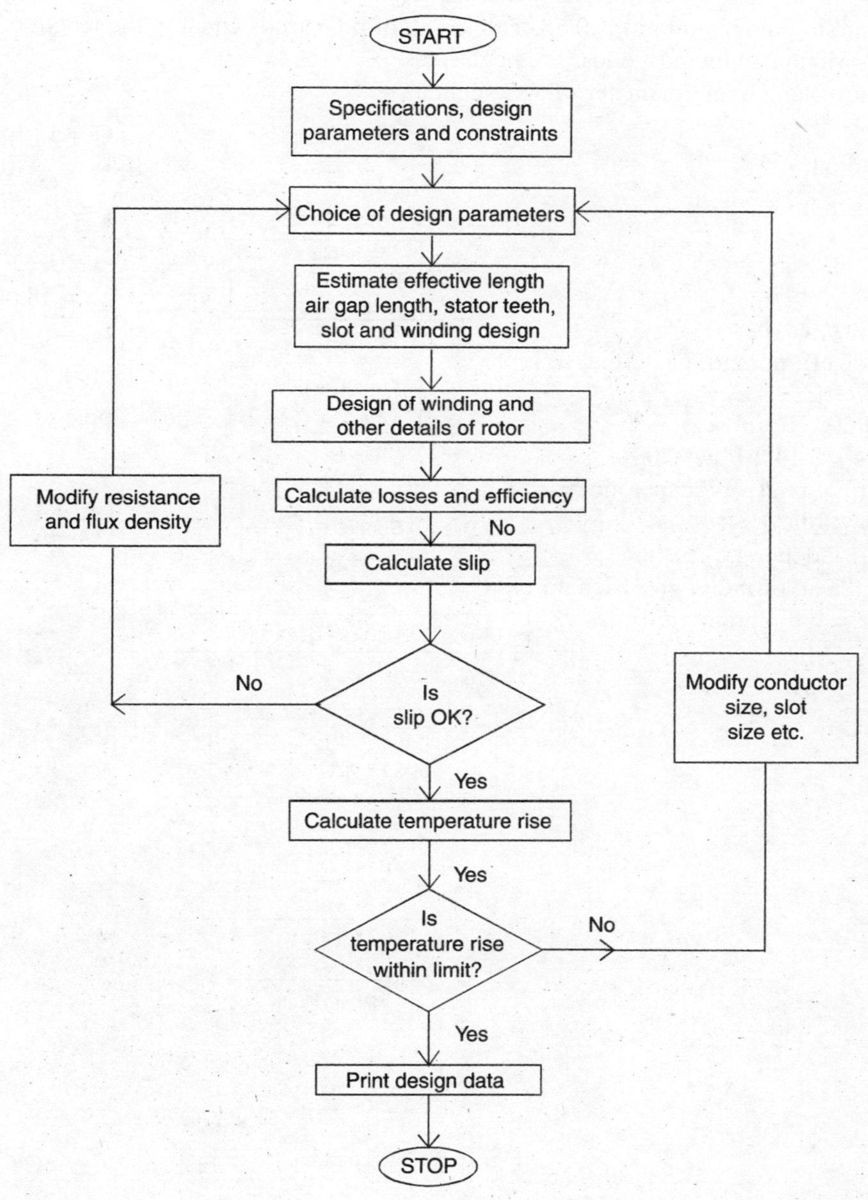


Fig. 10.7 Flow chart for overall design of 3-phase induction motor.

## 10.4 COMPUTER AIDED DESIGN OF D.C. MACHINES

**P.1** Calculate the main dimensions of a 12 kW, 220 V, 4-pole, 1200 rpm shunt generator. The specific electric loading is 18,000, average flux density is 0.46 Wb/m$^2$, full load efficiency is 0.86 and pole arc to pole pitch ratio is 0.68. The length of pole arc is equal to the length of the armature core. Friction and windage loss are neglected.

```
/*Calculation of length and diameter of DC shunt
generator*/
#include <stdio.h>
#include <math.h>
main ()
{
    int q, v, n, ac, ns;
    float b1, eff, ppratio, pa, c, dia, d, l, p;
    pa = p/eff;
    ns = n/60;
    c = 3.14*3.14*b1*ac*0.001;
    dia = (p x pa)/(3.14*c*ppratio*ns);
    d = pow (dia, 1/3);
    l = (3.14*d*ppratio)/p;
    Printf ("Dia. of motor = %f length of
     motor = %f" ppratio, d, l);
}
```

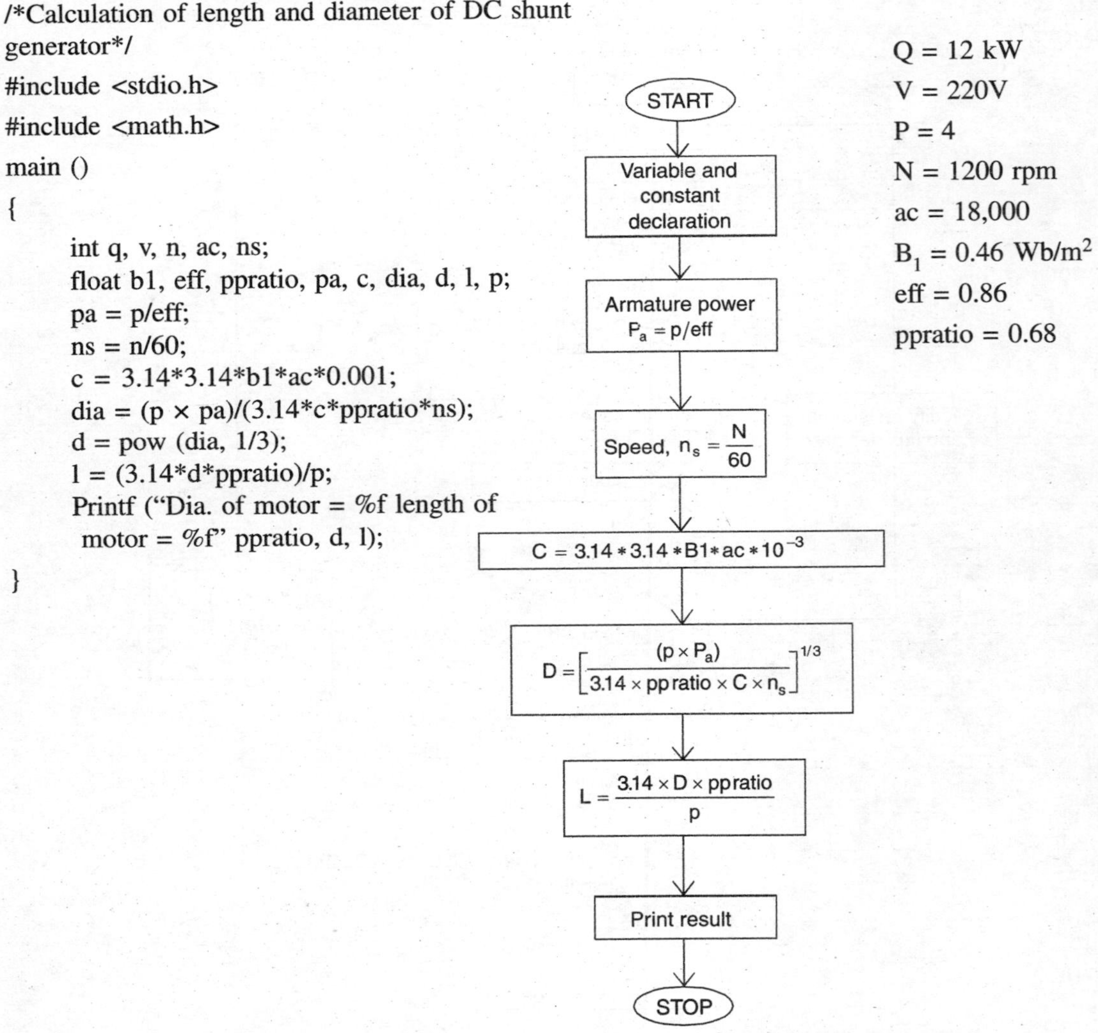


Q = 12 kW

V = 220V

P = 4

N = 1200 rpm

ac = 18,000

$B_1$ = 0.46 Wb/m$^2$

eff = 0.86

ppratio = 0.68

**Fig. 10.8** Flow chart.

**P.2** Calculate the total m.m.f. required per pole to produce a flux of 0.12 Wb in a machine with following details:

The length of machine is 0.31 m and pole arc to pole pitch ratio is 0.64. There are 12 slots per pole. The width and height of tooth are 12 mm and 36 mm respectively. Length of flux path between two adjacent poles in core is 0.20 m sectional area of core is 0.38 $m^2$, area of pole face is 0.10 $m^2$. Length of air gap 4.80 mm, sectional area of pole body is 0.076 $m^2$, height of pole 0.22 m, sectional area of yoke 0.046 $m^2$, mean length of flux path in yoke is 0.43 m between two adjacent poles. The relative permeability of teeth is 88 and for rest of magnetic circuit is 1150. Stacking factor is considered as 0.92 and leakage flux is neglected.

```
/*Calculation of total m.m.f. required per pole*/
#include <std.in.h>
#include <math.h>
Main ( )
{
    int s, t_permittivity, permittivity;
    float m_flux, p_flux, l,
    ppratio, wt, nt, lc, ac, ad, la,
    ap, hp, ay, ly, ki, atd, atp, c_flux,
    atc, aty, tpa, _flux, att,
    attotal;
    m_flux = 0.12;
    p_flux = 0.12;
    l = 0.31;
    ppratio = 0.64;
    s = 12;
    wt = 12e –3;
    ht = 36e –3;
    lc = 0.20;
    ac = 0.038;
    ad = 0.10;
    la = 4.8e –3;
    ap = 0.076;
    hp = 0.22;
    ay = 0.046;
    ly = 0.43;
    t_permittivity = 88;
    permittivity = 1150;
    ki = 0.92;
    atd = (m_flux*la)/(4*3.14e–7*ad);
    atp = (p_flux*hp)/(4*3.14–7*permittivity*ap);
    c-flux = p_flux/2;
    atc = (c_flux*lc)/(4*3.14e–7*permittivity*ac);
    aty = (p_flux*ly)/(2*4*3.14e–7*permittivity*ay);
    tpa = ppratio*s;
    att = (t_flux*ht)/(4*3.14e–7*t_permittivity*wt*ki*l)
    attotal = atd + atp + atc + aty + att;
    Printf ("Total mmf required for pole = %f", attotal);
}
```

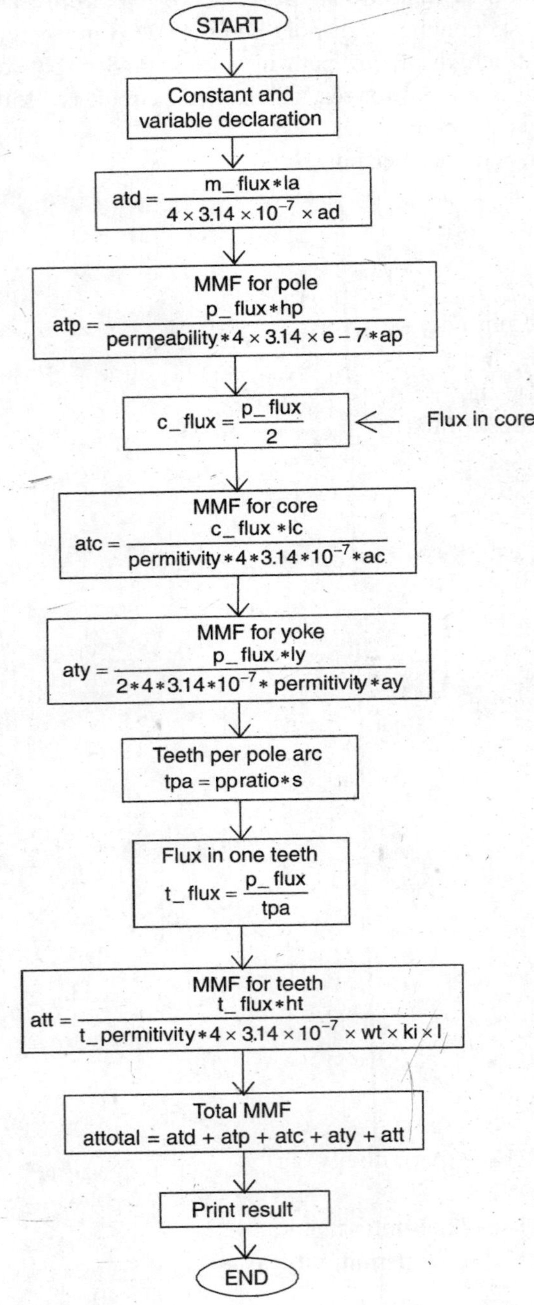


**Fig. 10.9** Flow chart.

**P.3** If the m.m.f. per pole estimated is 10 is assumed to be referred to shunt field coil of a 350 V, 4-pole generator with other details as below. Calculate diameter of conductor height of coil, number of turns and field current.

m.m.f. per pole = 6861 A, depth of winding = 42 mm, length of inner turn is 0.98 m, length of outer turn = 1.26 m. Loss dissipated from outer surface excluding ends is 1250 watt/$m^2$, space factor = 0.60, resistivity is 0.021Ω/m and $mm^2$. Consider a voltage drop of 60 V across the field regulator.

```
/*Calculation of dia, height of coil, number of turns and field current*/
#include <stdio.h>
#include <math.h>
main ( )
{
    int V, p, wl, vf, vsn, ef;
    float atfl, df, lti, lto, ks, resistivity, height, lmt, ff, dia, d, hf, tf, rf, if;
    atfl = 6861.0;
    V = 350;
    p = 4;
    df = 42e–3;
    lti = 0.98;
    lto = 1.26;
    wl = 1250;
    ks = 0.6;
    resistivity = 0.021;
    Vf = 60;
    vsh = V–Vf;
    ef = Vsh/p;
    lmt = (lti + lto)/2;
    ff = (atfl*resistivity*lmt)/ef;
    dia = (4*ff)/3.14;
    d = pow (dia, 1/2);
    height = (atfl*ef*ff*1e–6)/(wl*lto*ks*df);
    hf = pow (height, 1/2);
    tf = (ks*ht*df)/(ff*1e–6);
    rf = (tf*resistivity)/ff;
    if = ef/rf;
    printf = ("Dia = %f height of coil = %f field
                    current = %f", d, hf, if);
}
```

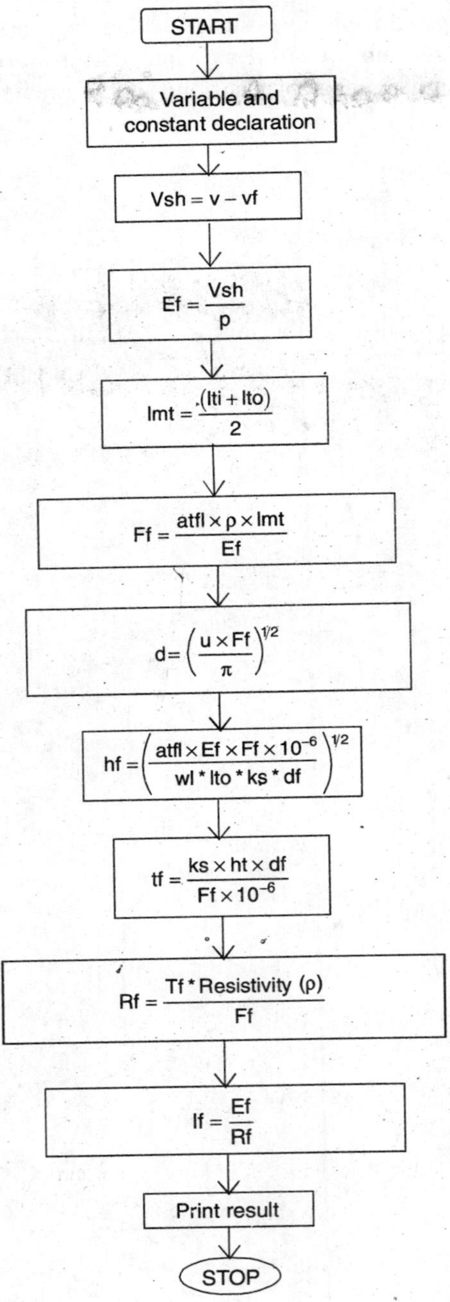


**Fig. 10.10** Flow chart.

### 10.4.1 Flow chart for overall Design of d.c. Machines

Flow chart for overall design of d.c. machine is given in Fig. 10.11. It will be helpful in the process to design a d.c. machine and to find out its performance. Several iteration may be required to design a machine with suitable size so as to give optimum performance for specific required purposes.

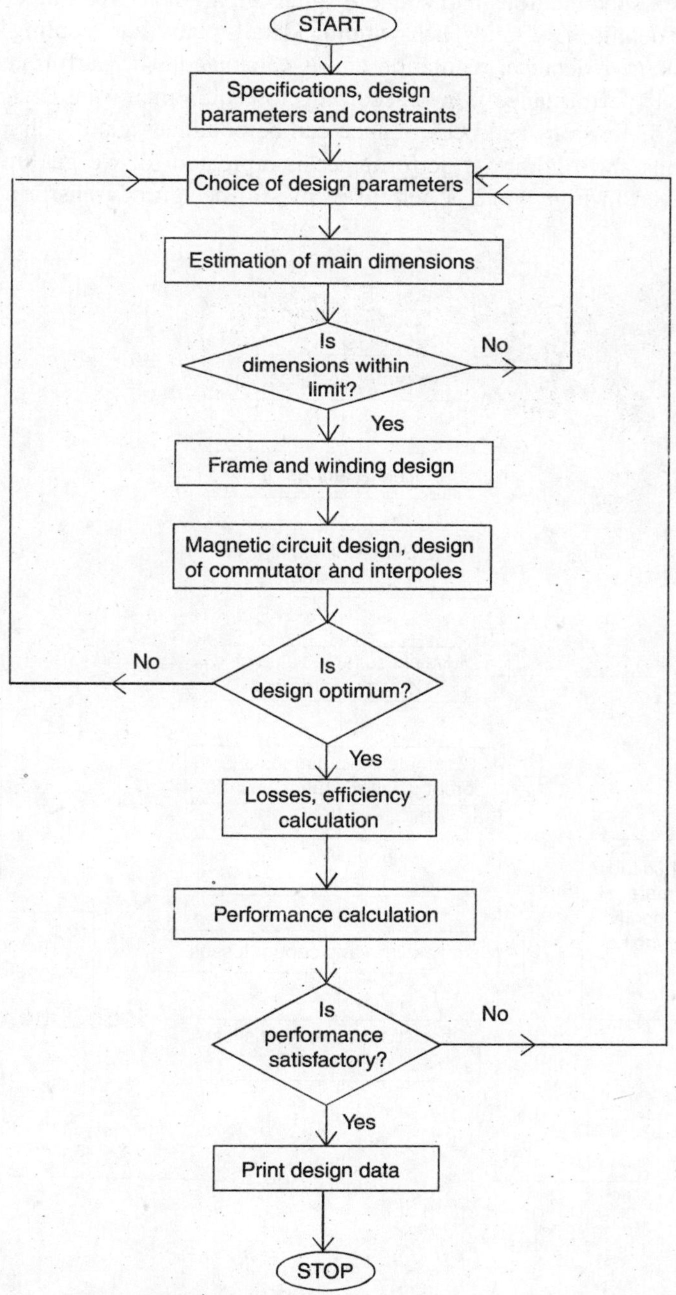


**Fig. 10.11** Flow chart for overall design of d.c. machines.

## 10.5 COMPUTER AIDED DESIGN OF TRANSFORMER

### 10.5.1 Flow Chart for Overall Design of Transformers

A transformer for given specifications and suitable value of parameters can be designed. Main dimensions, core and yoke details, L.V. and H.V. winding details, tank and cooling system is estimated. Losser, efficiency, leakage reactance, regulation etc. is calculated. The performance of transformer so designed is obtained. If performance is not according to requirement, the design parameters can be changed and different dimensions and performance can be obtained again. This process of calculating the different diemnsions and finding its performance is repreated till we get suitably designed transformer. Flow chart given in Fig. 10.12 is helpful in overall design of transformer and to find out its performance.

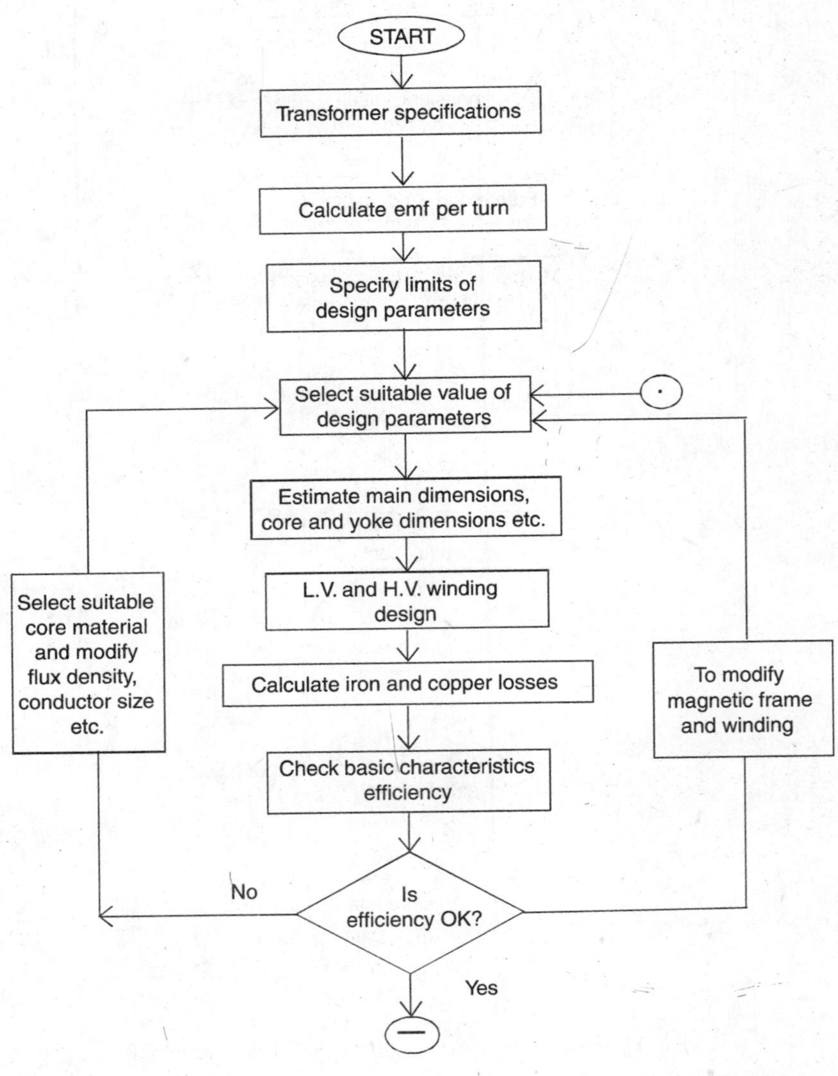

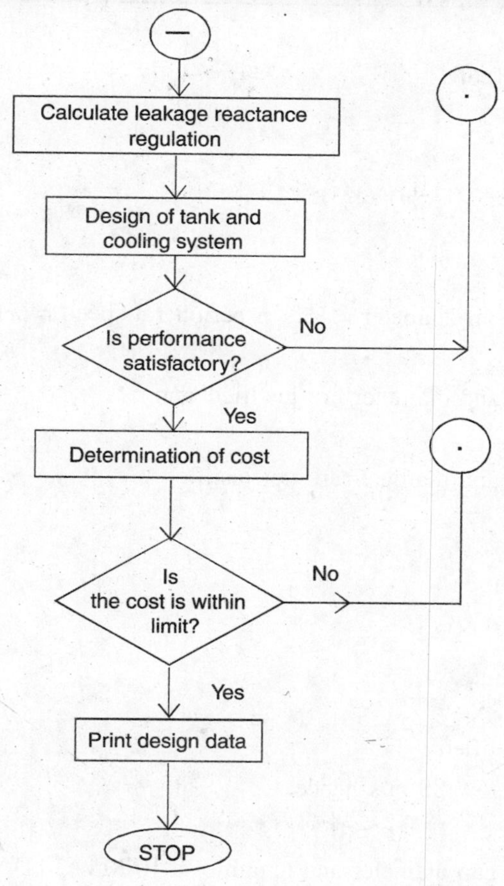


**Fig. 10.12** Flow chart for overall design of transformer.



## MATLAB PROGRAM OF DESIGN PROBLEM, ANSWER AND DISCUSSION

MATLAB programs are given for some problems for alternators and induction motors. The programs are run and the result obtained is also given at the end of the problem. The results are also discussed wherever necessary.

**Q. 1.** Estimate the main dimnesions of 1800 kVA, 50 Hz, 3-phase, 187.5 rpm, water wheel generator. The specific magnetic loading is 0.8 Wb/m$^2$ and the specific electric loading is 26,000 ampere-conductor/m. Pole arc to pole pitch ratio is 0.66.

### Solution program of question 1:

```
clear all;

clc;

% calculation of length and diameter of alternator %
```

```
q=1800;f=50;n=187.5;
b1=0.8;ac=26,000;pp_ratio=0.66;
p=(120*f)/n;
c=(11.1e–5)*b1*ac;
dia=(q/(3.14*11.1e–5*b1*ac*0.02.*n)^(1/3);
1=q/(c*n*(dia)^2);
vp=(3.14*dia*n)/60;
fprintf('\n\n\n\n length = %f, \n diameter = %f, \n peripheral speed = %f',1,dia,vp);
if(vp > 30 & vp < 80)
fprintf('\n\n\n\n\t\t\t\t length and diameter are justified \n\n');
else
fprintf('\n\n\n\n\t\t\t\t length and diameter are not justified \n\n`);
end
```

**Answer of question 1:**

Length = 0.254059,

Diameter = 4.045527,

Peripheral speed = 39.696733

Length and diameter are justified

Since peripheral speed is below 80 m/s hence.

*********************************************************************************

**Q. 2.** Estimate suitable air gap diameter and length of a 10 MVA, 11kV, 8-pole, 3-phase, 50 Hz, star connected synchronous generator. Maximum air gap flux density is 0.92 tesla. The ampere-conductor/m length of periphery is varying from 20,000 to 40,000 A/m. Pole arc to pole pitch ratio is 0.66. Peripheral speed should be less than 80 m/s. Mention the type of construction to be adopted.

**Solution program of question 2:**

```
clear all;
clc;
f=50;p=8;b1=0.92;q=10e+3;
n=(120*f)/p;
ppratio=0.66;
for ac=20000:500:40000;
    d=((q*p)/((11.1e-5)*(3.14*b1*ac*n*ppratio)))^(1/3);
    vp=(3.14*d*n)/60;
    if (vp<80)
        l=(3.14*d*ppratio)/p;
        fprintf('\n\n\n\n length = %f \n diameter = %f \n peripheral speed = %f \n\n', l,d,vp);
        fprintf('\n\n\n\n this is round pole construction \n\n');
```

```
else
    l=(3.14*d*ppratio)/p;
    fprint('\n\n\n\n length = %f \n diameter = %f \n peripheral speed = %f', l,d,vp);
    if(ac >= 40000)
    fprintf('\n\\n\n\n\t\t since vp is greater than 80 m/sec, when ac is 40000, so for this design data round pole construction is impossible. \n\t\tselect Length to pole pitch ratio separately between 1 and 5 and calculate for different values of ac');
        end
    end
end
fprintf('\n\n\n\n')
```

**Answer of question 2:**

Length = 0.759489
Diameter = 2.931825
Peripheral speed = 115.074150

Length = 0.753264
Diameter = 2.907793
Peripheral speed = 114.130876

Length = 0.747237
Diameter = 2.884530
Peripheral speed = 113.217790

Length = 0.741399
Diameter = 2.861993
Peripheral speed = 112.333240

Length = 735740
Diameter = 2.840145
Peripheral speed = 111.475701

Length = 0.730249
Diameter = 2.818949
Peripheral speed = 110.643762

Length = 0.724918
Diameter = 2.798372
Peripheral speed = 109.836114

Length = 0.719740
Diameter = 2.778383
Peripheral speed = 109.051544

Length = 0.714707
Diameter = 2.758953
Peripheral speed = 108.288921

Length = 0.709811
Diameter = 2.740056
Peripheral speed = 107.547192

Length = 0.705047
Diameter = 2.721666
Peripheral speed = 106.825377

Length = 0.700409
Diameter = 2.703759
Peripheral speed = 106.122559

Length = 0.695890
Diameter = 2.686315
Peripheral speed = 105.437878

Length = 0.691486
Diameter = 2.669313
Peripheral speed = 104.770532

Length = 0.687190
Diameter = 2.652733
Peripheral speed = 104.119766

Length = 0.683000
Diameter = 2.636557
Peripheral speed = 103.484874

Length = 0.678910
Diameter = 2.620769
Peripheral speed = 102.865189

Length = 0.674917
Diameter = 2.605353
Peripheral speed = 102.260086

Length = 0.671015
Diameter = 2.590292
Peripheral speed = 101.668974

Length = 0.667203

Diameter = 2.575574
Peripheral speed = 101.091297

Length = 0.663475
Diameter = 2.561185
Peripheral speed = 100.526529

Length = 0.659830
Diameter = 2.547113
Peripheral speed = 99.974174

Length = 0.656263
Diameter = 2.533344
Peripheral speed = 99.433763

Length = 0.652772
Diameter = 2.519869
Peripheral speed = 98.904850

Length = 0.649354
Diameter = 2.506676
Peripheral speed = 98.387014

Length = 0.646007
Diameter = 2.493754
Peripheral speed = 97.879855

Length = 0.642728
Diameter = 2.481095
Peripheral speed = 97.382995

Length = 0.639514
Diameter = 2.468690
Peripheral speed = 96.896072

Length = 0.636364
Diameter = 2.456528
Peripheral speed = 96.418743

Length = 0.633275
Diameter = 2.444603
Peripheral speed = 95.950684

Length = 0.630244
Diameter = 2.432907
Peripheral speed = 95.491583

Length = 0.627272
Diameter = 2.421430
Peripheral speed = 95.041144

Length = 0.624354
Diameter = 2.410168
Peripheral speed = 94.599086

Length = 0.621490
Diameter = 2.399112
Peripheral speed = 94.165139
Length = 0.618678
Diameter = 2.388256
Peripheral speed = 93.739046

Length = 0.615916
Diameter = 2.377594
Peripheral speed = 93.320563

Length = 0.613202
Diameter = 2.367120
Peripheral speed = 92.909454

Length = 0.610536
Diameter = 2.356828
Peripheral speed = 92.505495

Length = 0.607916
Diameter = 2.346713
Peripheral speed = 92.108470

Length = 0.605340
Diameter = 2.336769
Peripheral speed = 91.718176

Length = 0.602807
Diameter = 2.326991
Peripheral speed = 91.334413

Since $V_p$ is greater than 80 m/sec, when $\overline{ac}$ is selected to its maximum value of 40,000, so, for this design data round pole construction is not possible.

Select length to pole pitch ratio separately between 1 and 5 and calculate for different value of $\overline{ac}$.

**Discussion:** The program is written to design the machine initially in round pole. Ampere-conductor is increased from 20,000 to 40,000 in step of 500. At each step length, diameter and peripheral speed is calculated. It is seen that even at 40,000 ampere-conductor loading the peripheral speed is above 80 m/s. This means that we can not design the alternator with round pole construction.

We can further modify the program and length to pole pitch ratio can be separately selected between 1 and 5. Again $\overline{ac}$ can be selected between 20,000 and 40,000 and can be increased in step of 500. This means that length to pole pitch ratio and ampere-conductor both can vary. An optimum value of diameter and length can be calculated corresponding to which peripheral speed should be less than 80 m/s.

**************************************************************************************

**Q. 3.** Calculate the main dimensions, turns/phase, total number of slots, sectional area of conductor and area of slot to accommodate the conductors and insulation in the slot for a 200 kW, 3-phase, 50 Hz, 440V, 1440rpm, star-connected, slip ring induction motor. Average flux density is 0.52 Wb/m$^2$, $\overline{ac}$ is 32,000 A/m, efficiency is 0.88, winding space factor is 0.46, winding factor is 0.96, power factor 0.86 and current density in winding conductor is 4A/mm$^2$. The design should be cheaper and hence ratio is taken 1.5 slot pitch is 20 mm.

**Solution program of question 3:**

```
clear all;
clc;
%calculation of turns per phase, total number of slot and cross-sectional area
%of conductor in an I.M.
q=200;f=50;v=440;s=0.04;
n=1440;b1=0.52;ac=32000;eff=0.88;
kws=0.46;kw=0.96;pf=0.86;j=4;
slot_pitch=20e–3;
ns=n/(1–s);
p=(120*f)/ns;
c=(18.3e–5*kw*b1*ac*eff*pf);
dia=(p*q)/(3.14*1.5*c*ns);
d=(dia)^(1/3)
l=(1.5*3.14*d)/p
flux=(3.14*b1*d*1)/p;
nph=v/(1.732*4.44*f*kw*flux);
ncs=3*2*nph;
nss=(3.14*d)/slot_pitch;
nps=abs(ncs/nss);
iph=q/(1.732*v*eff*pf);
fc=iph/s;
acs=2*fc;
ast=acs/kws;
fprintf('\n\n\n\n turn per phase = %f, \n total number of slots = %f of X-sectional area of conductor = %f \n\n',nph,nss,fc);
```

Chapter 10

### Answer of question 3:

d = 0.3713
l = 0.4372
turn per phase = 17.990061,
total number of slots = 58.291693 of X-sectional area of conductor = 8.669386
**************************************************************************************

**Q. 4.** Calculate the following design information for a 30kW, 440V, 3-phase, 6-pole, 50Hz, delta connected squirrel cage induction motor:

(*i*) Main dimensions of stator frame

(*ii*) Number of turns per phase in stator winding

(*iii*) Number of stator slots

Assume

Specific magnetic loading = 0.48 T

Specific electric loading is 26,000 ampere-conductor/m

Full load efficiency = 0.88

Full load p.f. = 0.86

Winding factor = 0.955

### Solution program of question 4:

```
clear all;
clc;
q=30;v=440;p=6;f=50;
b1=0.48;ac=26,000;eff=0.88;pf=0.86;kw=0.955;
slot_pitch=15e–3;
ns=(120*f)/p;
c=18.3e–5*kw*b1*ac*eff*pf;
dia=(p*q)/(3.14*1.5*c*ns);
d=dia^(1/3);
l=(1.5*3.14*d)/p;
flux=(b1*3.14*d*l)/p;
nph=v/(4.44*f*kw*flux);
nss=(3.14*d)/slot_pitch;
fprintf('\n\n\n\n diameter = %f, \n length=%f,\n number of turns per phase =%f,\n no. of stator
slot=%f \n\n',d,l,nph,nss);
```

### Answer of question 4:

diameter = 0.285015,

length=0.223736,

number of turns per phase=129.560541,

no. of stator slot=59.663061

**Q. 5.** Estimate the main dimensions, number of stator slots and number of stator conductors per slot for a 140 h.p., 3300volt, 50c/s, 12 poles, star-connected slip ring induction motor. Assume,

Average gap density = 0.4Wb/m$^2$

Ampere-conductor/m length of periphery=25,000A/m

efficiency = 90%, power factor = 0.9, winding factor = 0.96

**Solution program of question 5:**

```
clear all;
clc;
q=140*0.746;
v=3300;f=50;p=12;b1=0.4;
ac=25,000;eff=0.9;pf=0.9;kw=0.96;
slot_pitch=15e–3;
ns(120*f)/p;
c=18.3e–5*kw*b1*ac*eff*pf;
dia=(p*q)/(3.14*1.1*c*ns);
d=dia^(1/3);
l=(1.1*3.14*d)/p;
flux=(b1*3.14*d*l)/p;
nph=v/(1.732*f*kw*flux);
nss=(3.14*d)/slot_pitch;
ncs=3*2*nph;
nps=abs(ncs/nss);
fprintf('\n\n\n\n diameter=%f, \n length=%f, \n total number of stator conductor=%f, \n total no. of stator slots=%f \n\n',d,l,ncs,nss);
```

**Answer of question 5:**

Diameter=0.798944,

Length=0.229963,

Total number of stator conductor=12384.951152,

Total number of stator slots=167.245535

# Appendices

## APPENDIX A

### TABLES FOR STANDARD SIZES AND INSULATION COVERING

**Table A.1** Different sizes of rectangular copper conductor (approximate cross-sectional area in $mm^2$)

| *Thickness (in mm)* | *0.8* | *0.9* | *1.0* | *1.1* | *1.2* | *1.3* | *1.4* | *1.5* | *1.6* | *1.7* | *1.8* |
|---|---|---|---|---|---|---|---|---|---|---|---|
| *Width (in mm)* | | | | | | | | | | | |
| 1.0 | – | – | 0.86 | – | – | – | – | – | – | – | – |
| 1.2 | – | – | – | – | 1.3 | – | – | – | – | – | – |
| 1.4 | 0.98 | 1.12 | 1.26 | 1.4 | – | – | 1.75 | – | – | – | – |
| 1.6 | 1.14 | 1.30 | 1.46 | 1.62 | 1.78 | 1.87 | – | – | 2.35 | – | – |
| 1.8 | 1.3 | 1.48 | 1.66 | 1.84 | 2.02 | 2.13 | 2.31 | 2.49 | – | – | 2.93 |
| 2.0 | 1.46 | 1.66 | 1.86 | 2.06 | 2.19 | 2.39 | 2.59 | 2.79 | 2.99 | 3.19 | – |
| 2.2 | 1.62 | 1.84 | 2.06 | 2.21 | 2.43 | 2.65 | 2.87 | 3.09 | 3.31 | 3.53 | 3.65 |
| 2.5 | 1.86 | 2.11 | 2.36 | 2.54 | 2.79 | 3.04 | 3.29 | 3.54 | 3.79 | 4.04 | 4.19 |
| 2.8 | 2.10 | 2.38 | 2.66 | 2.87 | 3.15 | 3.43 | 3.71 | 3.99 | 4.17 | 4.45 | 4.73 |
| 3.0 | 2.26 | 2.56 | 2.86 | 3.09 | 3.39 | 3.69 | 3.99 | 4.19 | 4.49 | 4.79 | 5.09 |
| 3.2 | 2.42 | 2.74 | 3.06 | 3.31 | 3.63 | 3.95 | 4.27 | 4.49 | 4.81 | 5.13 | 5.45 |
| 3.5 | 2.66 | 3.01 | 3.36 | 3.64 | 3.99 | 4.34 | 4.69 | 4.94 | 5.29 | 5.64 | 5.99 |
| 3.8 | 2.90 | 3.28 | 3.66 | 3.97 | 4.35 | 4.73 | 5.11 | 5.39 | 5.77 | 6.15 | 6.53 |
| 4.0 | 3.06 | 3.46 | 3.86 | 4.19 | 4.59 | 4.99 | 5.39 | 5.69 | 6.09 | 6.49 | 6.89 |
| 4.5 | 3.46 | 3.91 | 4.36 | 4.74 | 5.19 | 5.64 | 6.09 | 6.44 | 6.89 | 7.34 | 7.79 |

*(Contd.)*

| | | | | | | | | | | | |
|---|---|---|---|---|---|---|---|---|---|---|---|
| 5.0 | 3.86 | 4.36 | 4.86 | 5.29 | 5.79 | 6.29 | 6.79 | 7.19 | 7.69 | 8.19 | 8.69 |
| 5.5 | 4.26 | 4.81 | 5.36 | 5.84 | 6.39 | 6.94 | 7.49 | 7.94 | 8.49 | 9.04 | 9.59 |
| 6.0 | 4.66 | 5.26 | 5.86 | 6.39 | 6.99 | 7.59 | 8.19 | 8.69 | 9.29 | 9.89 | 10.5 |
| 6.5 | 5.06 | 5.71 | 6.36 | 6.94 | 7.59 | 8.24 | 8.89 | 9.44 | 10.1 | 10.7 | 11.4 |
| 7.0 | 5.46 | 6.16 | 6.86 | 7.49 | 8.19 | 8.89 | 9.59 | 10.2 | 10.9 | 11.6 | 12.3 |
| 7.5 | 5.86 | 6.61 | 7.36 | 8.04 | 8.79 | 9.54 | 10.3 | 10.9 | 11.7 | 12.4 | 13.2 |
| 8.0 | 6.26 | 7.06 | 7.86 | 8.59 | 9.59 | 10.2 | 11.0 | 11.7 | 12.5 | 13.3 | 14.1 |
| 9.0 | 7.06 | 7.96 | 8.86 | 9.69 | 10.6 | 11.5 | 12.4 | 13.2 | 14.1 | 15.0 | 15.9 |
| 10.0 | 7.86 | 8.86 | 9.86 | 10.8 | 11.8 | 12.8 | 13.8 | 14.7 | 15.7 | 16.7 | 17.7 |
| 11.0 | 8.66 | 9.76 | 10.9 | 11.9 | 13.0 | 14.1 | 15.2 | 16.2 | 17.3 | 18.1 | 19.5 |
| 12.0 | 9.46 | 10.7 | 11.9 | 13.0 | 14.2 | 15.4 | 16.6 | 17.7 | 18.9 | 20.1 | 21.3 |
| 13.0 | 10.3 | 11.6 | 12.9 | 14.1 | 15.4 | 16.7 | 18.0 | 19.2 | 20.5 | 21.8 | 23.1 |
| 14.0 | 11.1 | 12.5 | 13.9 | 15.2 | 16.6 | 18.0 | 19.4 | 20.7 | 22.1 | 23.5 | 24.9 |
| 15.0 | 11.9 | 13.4 | 14.9 | 16.3 | 17.8 | 19.3 | 20.8 | 22.2 | 23.7 | 25.3 | 26.7 |
| 16.0 | 12.7 | 14.3 | 15.9 | 17.4 | 19.0 | 20.6 | 22.2 | 23.7 | 25.3 | 26.9 | 28.5 |
| 18.0 | – | 16.1 | 17.9 | 19.6 | 21.4 | 23.2 | 25.0 | 26.7 | 28.5 | 30.3 | 32.1 |
| 20.0 | – | – | 19.0 | 21.8 | 23.8 | 25.8 | 27.8 | 29.7 | 31.7 | 33.7 | 35.7 |
| 22.0 | – | – | – | 24.0 | 26.2 | 28.4 | 30.6 | 32.7 | 34.9 | 37.1 | 39.3 |
| 25.0 | – | – | – | – | – | 32.3 | 34.8 | 37.2 | 39.7 | 42.2 | 44.7 |

Table A.1 Contd...

| *Thickness (in mm)* / *Width (in mm)* | *1.9* | *2.0* | *2.2* | *2.5* | *2.8* | *3.0* | *3.2* | *3.5* | *3.8* | *4.0* |
|---|---|---|---|---|---|---|---|---|---|---|
| 0.9 | – | – | – | – | – | – | – | – | – | – |
| 1.0 | – | – | – | – | – | – | – | – | – | – |
| 1.2 | – | – | – | – | – | – | – | – | – | – |
| 1.4 | – | – | – | – | – | – | – | – | – | – |
| 1.6 | – | – | – | – | – | – | – | – | – | – |
| 1.8 | – | – | – | – | – | – | – | – | – | – |
| 2.0 | – | 3.69 | – | – | – | – | – | – | – | – |
| 2.2 | – | – | 4.58 | – | – | – | – | – | – | – |
| 2.5 | 4.44 | 4.69 | – | 7.94 | – | – | – | – | – | – |
| 2.8 | 5.01 | 5.29 | 5.85 | – | 7.28 | – | – | – | – | – |
| 3.0 | 5.39 | 5.69 | 6.29 | 7.19 | – | 8.44 | – | – | – | – |
| 3.2 | 5.77 | 6.09 | 6.73 | 7.69 | 8.41 | – | 9.68 | – | – | – |
| 3.5 | 6.34 | 6.69 | 7.39 | 8.20 | 9.25 | 9.95 | – | 11.69 | – | – |
| 3.8 | 6.91 | 7.29 | 8.05 | 8.95 | 10.1 | 10.9 | 11.6 | – | 13.88 | – |
| 4.0 | 7.29 | 7.69 | 8.25 | 9.45 | 10.7 | 11.5 | 12.3 | 13.5 | – | 15.44 |
| 4.5 | 8.24 | 8.45 | 9.35 | 10.7 | 12.1 | 13.0 | 13.9 | 15.2 | – | 17.1 |
| 5.0 | 9.1 | 9.45 | 10.5 | 12.0 | 13.5 | 14.5 | 15.5 | 16.6 | – | 19.1 |
| 5.5 | 10.1 | 10.5 | 11.6 | 13.2 | 14.9 | 16.0 | 16.7 | 18.4 | – | 21.1 |
| 6.0 | 11.1 | 11.5 | 12.7 | 14.5 | 15.3 | 17.5 | 18.3 | 20.1 | – | 23.1 |
| 6.5 | 12.0 | 12.5 | 13.8 | 15.7 | 17.7 | 19.0 | 19.9 | 21.9 | – | 25.1 |
| 7.0 | 13.0 | 13.5 | 14.9 | 17.0 | 19.1 | 20.5 | 21.5 | 23.6 | – | 27.1 |

| | | | | | | | | | | |
|---|---|---|---|---|---|---|---|---|---|---|
| 7.5 | 13.9 | 14.5 | 16.0 | 18.2 | 20.5 | 22.0 | 23.1 | 25.4 | – | 29.1 |
| 8.0 | 14.9 | 15.5 | 7.1 | 19.5 | 21.9 | 23.5 | 24.7 | 27.1 | – | 31.1 |
| 9.0 | 16.8 | 17.5 | 19.3 | 22.9 | 24.7 | 26.5 | 27.9 | 30.6 | – | 35.1 |
| 10.0 | 18.7 | 19.5 | 21.5 | 24.5 | 27.3 | 29.5 | 31.1 | 34.1 | – | 39.1 |
| 11.0 | 20.6 | 21.5 | 23.7 | 27.0 | 30.3 | 32.5 | 34.3 | 37.6 | – | 43.1 |
| 12.0 | 22.5 | 23.5 | 25.9 | 29.5 | 33.1 | 35.5 | 37.5 | 41.1 | – | 47.1 |
| 13.0 | 24.4 | 25.5 | 28.1 | 32.0 | 35.9 | 38.5 | 40.7 | 44.6 | – | 51.1 |
| 14.0 | 26.3 | 27.5 | 30.3 | 34.5 | 38.7 | 41.5 | 43.9 | 48.1 | – | 55.1 |
| 15.0 | 28.2 | 29.5 | 32.5 | 37.0 | 41.5 | 44.5 | 47.1 | 51.6 | – | 59.1 |
| 16.0 | 30.1 | 31.5 | 34.7 | 39.5 | 44.3 | 47.5 | 50.3 | 55.1 | – | 63.1 |
| 18.0 | 33.9 | 35.5 | 39.1 | 44.5 | 49.9 | 53.5 | 56.7 | 62.1 | – | 71.1 |
| 20.0 | 37.7 | 39.5 | 43.5 | 49.5 | 55.5 | 59.5 | 63.1 | 69.1 | – | 79.1 |
| 22.0 | 41.5 | 43.5 | 47.9 | 54.5 | 61.1 | 65.5 | 69.5 | 76.1 | – | 87.1 |
| 25.0 | 47.2 | 49.5 | 54.5 | 62.0 | 69.5 | 74.5 | 79.1 | 86.6 | – | 99.1 |

*Note:*

1. Transformers for special purposes may need the size other than those given in above table.
2. The reduction in area due to rounding off the edges have been taken into consideration for the areas given above.

**Table A.2** Standard size of round conductors according to SWG.

| *SWG* | *Diameter (mm)* | *Area ($mm^2$)* |
|---|---|---|
| 1/0 | 8.2296 | 53.1921 |
| 1/0 1/2 | 7.9248 | 49.3249 |
| 1 | 7.6200 | 45.6037 |
| 1 1/4 | 7.4676 | 43.7978 |
| 1 1/2 | 7.3152 | 42.0284 |
| 1 3/4 | 7.1628 | 40.2954 |
| 2 | 7.0104 | 38.5990 |
| 2 1/4 | 6.8580 | 36.9390 |
| 2 1/2 | 6.7056 | 35.3155 |
| 2 3/4 | 6.5532 | 33.7284 |
| 3 | 6.4008 | 32.1779 |
| 3 1/4 | 6.2738 | 30.9137 |
| 3 1/2 | 6.1468 | 29.6748 |
| 3 3/4 | 6.0198 | 28.4612 |
| 4 | 5.8928 | 27.2730 |
| 4 1/4 | 5.7658 | 26.1101 |
| 4 1/2 | 5.6388 | 24.9726 |
| 4 3/4 | 5.5118 | 23.8603 |
| 5 | 5.3848 | 22.7734 |
| 5 1/4 | 5.2578 | 21.7119 |

| | | |
|---|---|---|
| 5 1/2 | 5.1308 | 20.6757 |
| 5 3/4 | 5.0038 | 19.6648 |
| 6 | 4.8768 | 18.6793 |
| 6 1/4 | 4.7752 | 17.9091 |
| 6 1/2 | 4.6736 | 17.1551 |
| 6 3/4 | 4.5730 | 16.4173 |
| 6/0 | 11.7856 | 109.092 |
| 6/0 1/2 | 11.3792 | 101.699 |
| 5/0 | 10.9728 | 94.5637 |
| 5/0 1/2 | 10.5664 | 87.6889 |
| 4/0 | 10.1600 | 81.0734 |
| 4/0 1/2 | 9.8044 | 75.4973 |
| 3/0 | 9.4488 | 70.1205 |
| 3/0 1/2 | 9.1440 | 65.6695 |
| 2/0 | 8.8392 | 61.3643 |
| 2/0 1/2 | 8.5344 | 53.2052 |
| 7/0 | 12.7000 | 126.677 |
| 7/0 1/2 | 12.2428 | 117.720 |
| 7 | 4.4704 | 15.6958 |
| 7 1/4 | 4.3638 | 14.9904 |
| 7 1/2 | 4.2672 | 14.3013 |
| 7 3/4 | 4.1656 | 13.6284 |
| 8 | 4.0640 | 12.9717 |
| 8 1/4 | 3.9624 | 12.3312 |
| 8 1/2 | 3.8608 | 11.7069 |
| 8 3/4 | 3.7592 | 11.0989 |
| 9 | 3.6576 | 10.5071 |
| 9 1/4 | 3.5560 | 9.93146 |
| 9 1/2 | 3.4544 | 9.37205 |
| 9 3/4 | 3.3528 | 8.82889 |
| 10 | 3.2512 | 8.30192 |
| 10 1/4 | 3.1750 | 7.91727 |
| 10 1/2 | 3.0988 | 7.54186 |
| 10 3/4 | 3.0226 | 7.17547 |
| 11 | 2.9464 | 6.81824 |
| 11 1/4 | 2.8702 | 6.47012 |
| 11 1/2 | 2.7940 | 6.13116 |
| 11 3/4 | 2.7178 | 5.80129 |
| 12 | 2.6416 | 5.48054 |
| 12 1/4 | 2.5654 | 5.16892 |

*(Contd.)*

| | | |
|---|---|---|
| 12 1/2 | 2.4892 | 4.86642 |
| 12 3/4 | 2.4130 | 4.57304 |
| 13 | 2.3368 | 4.28877 |
| 13 1/4 | 2.2606 | 4.01363 |
| 13 1/2 | 2.1844 | 3.74761 |
| 14 | 2.0320 | 3.24293 |
| 14 1/4 | 1.9812 | 3.08281 |
| 14 1/2 | 1.9304 | 2.92674 |
| 14 3/4 | 1.8796 | 2.77473 |
| 15 | 1.8288 | 2.62677 |
| 15 1/4 | 1.7780 | 2.48287 |
| 15 1/2 | 1.7272 | 2.34301 |
| 15 3/4 | 1.6764 | 2.20721 |
| 16 | 1.6256 | 2.07547 |
| 16 1/4 | 1.5743 | 1.94778 |
| 16 1/2 | 1.5240 | 1.82414 |
| 16 3/4 | 1.4732 | 1.70456 |
| 17 | 1.4224 | 1.58904 |
| 17 1/4 | 1.3716 | 1.47756 |
| 17 1/2 | 1.3208 | 1.37014 |
| 17 3/4 | 1.2700 | 1.26677 |
| 18 | 1.2192 | 1.16746 |
| 18 1/4 | 1.1684 | 1.07219 |
| 18 1/2 | 1.1176 | 0.98099 |
| 18 3/4 | 1.0668 | 0.89383 |
| 19 | 1.0160 | 0.81073 |
| 19 1/4 | 1.0160 | 0.77070 |
| 19 1/4 | 0.9906 | 0.77070 |
| 19 1/2 | 0.9652 | 0.73168 |
| 19 3/4 | 0.9398 | 0.69368 |
| 20 | 0.9144 | 0.65670 |
| 20 1/4 | 0.8890 | 0.62072 |
| 20 1/2 | 0.8638 | 0.58575 |
| 20 3/4 | 0.8382 | 0.55180 |
| 21 | 0.8128 | 0.51887 |
| 21 1/4 | 0.7874 | 0.48695 |
| 21 1/2 | 0.7620 | 0.45604 |
| 21 3/4 | 0.7366 | 0.42614 |
| 22 | 0.7112 | 0.39726 |
| 22 1/4 | 0.6858 | 0.36939 |
| 22 1/2 | 0.6604 | 0.34253 |
| 22 3/4 | 0.6350 | 0.31669 |
| 23 | 0.6096 | 0.29186 |
| 23 1/2 | 0.5842 | 0.26805 |
| 24 | 0.5588 | 0.24525 |
| 24 1/2 | 0.5334 | 0.22346 |

| | | |
|---|---|---|
| 25 | 0.5080 | 0.20268 |
| 25 1/2 | 0.4826 | 0.18292 |
| 26 | 0.4572 | 0.16417 |
| 27 | 0.4166 | 0.13628 |
| 28 | 0.3759 | 0.11099 |
| 29 | 0.3454 | 0.09372 |
| 30 | 0.3150 | 0.07791 |
| 31 | 0.2946 | 0.06818 |
| 32 | 0.2743 | 0.05910 |
| 33 | 0.2540 | 0.05067 |
| 34 | 0.2337 | 0.04289 |
| 35 | 0.2134 | 0.03575 |
| 36 | 0.1930 | 0.02927 |
| 37 | 0.1727 | 0.02343 |
| 38 | 0.1524 | 0.01824 |
| 39 | 0.1321 | 0.01370 |
| 40 | 0.1219 | 0.101167 |
| 41 | 0.1118 | 0.00981 |
| 42 | 0.1016 | 0.00811 |
| 43 | 0.0914 | 0.00657 |
| 44 | 0.0813 | 0.00519 |
| 45 | 0.0711 | 0.00397 |
| 46 | 0.0610 | 0.00292 |
| 47 | 0.0508 | 0.00203 |
| 48 | 0.0406 | 0.00130 |
| 49 | 0.0305 | 0.00073 |
| 50 | 0.0254 | 0.00051 |

**Table A.3** Round conductor covering as per IS specifications with enamel and paper covering.

| *Diameter of bare conductor* | *Diameter with enamel covering* | | *Diameter with paper covering* | |
|---|---|---|---|---|
| | *With medium covering* | *With fine covering* | *With ordinary covering* | *With fine covering* |
| *(mm)* | *(mm)* | *(mm)* | *(mm)* | *(mm)* |
| 1.00 | 1.095 | 1.070 | 1.275 | 1.200 |
| 1.06 | 1.155 | 1.130 | 1.335 | 1.260 |
| 1.12 | 1.215 | 1.190 | 1.395 | 1.320 |
| 1.18 | 1.278 | 1.253 | 1.455 | 1.380 |
| 1.25 | 1.350 | 1.325 | 1.525 | 1.450 |
| 1.32 | 1.420 | 1.395 | 1.595 | 1.520 |
| 1.40 | 1.505 | 1.480 | 1.700 | 1.575 |
| 1.50 | 1.606 | 1.580 | 1.800 | 1.675 |
| 1.60 | 1.710 | 1.680 | 1.900 | 1.775 |
| 1.70 | 1.810 | 1.785 | 2.000 | 1.875 |
| 1.80 | 1.915 | 1.885 | 2.100 | 1.975 |

*(Contd.)*

| | | | | |
|---|---|---|---|---|
| 1.90 | 2.015 | 1.990 | 2.200 | 2.075 |
| 2.00 | 2.180 | 2.150 | 2.350 | 2.250 |
| 2.12 | 2.241 | 2.211 | 2.470 | 2.370 |
| 2.24 | 2.365 | 2.335 | 2.590 | 2.490 |
| 2.36 | 2.488 | 2.485 | 2.710 | 2.610 |
| 2.50 | 2.630 | 2.600 | 2.850 | 2.725 |
| 2.65 | 2.758 | 2.752 | 3.000 | 2.875 |
| 2.80 | 2.935 | 2.905 | 3.150 | 3.025 |
| 2.90 | 3.040 | 3.010 | 3.250 | 3.125 |
| 3.00 | 3.140 | 3.110 | 3.350 | 3.225 |
| 3.15 | 3.295 | 3.262 | 3.500 | 3.375 |
| 3.25 | 3.395 | 3.365 | 3.600 | 3.475 |
| 3.35 | 3.497 | 3.465 | 3.700 | 3.575 |
| 3.45 | 3.600 | 3.567 | 3.800 | 3.675 |
| 3.55 | 3.700 | 3.670 | 3.900 | 3.775 |
| 3.65 | 3.800 | 3.770 | 4.000 | 3.875 |
| 3.75 | 3.902 | 3.872 | 4.100 | 3.975 |
| 3.85 | 4.003 | 3.973 | 4.200 | 4.075 |
| 4.00 | 4.155 | 4.125 | 4.350 | 4.300 |

**Table A.4** Round conductor covering with different cotton covering.

| *Diameter of bare conductor (mm)* | *Diameter with single cotton covering (S.C.C.)* | | *Diameter with double cotton covering (D.C.C.)* | |
|---|---|---|---|---|
| | *Ordinary covering (mm)* | *Fine covering (mm)* | *Ordinary covering (mm)* | *Fine covering (mm)* |
| 1.00 | 1.162 | 1.135 | 1.290 | 1.215 |
| 1.06 | 1.225 | 1.195 | 1.350 | 1.275 |
| 1.12 | 1.285 | 1.255 | 1.410 | 1.335 |
| 1.18 | 1.345 | 1.315 | 1.470 | 1.395 |
| 1.25 | 1.415 | 1.385 | 1.540 | 1.465 |
| 1.32 | 1.485 | 1.460 | 1.615 | 1.535 |
| 1.40 | 1.590 | 1.565 | 1.720 | 1.645 |
| 1.50 | 1.695 | 1.665 | 1.820 | 1.745 |
| 1.60 | 1.795 | 1.765 | 1.920 | 1.845 |
| 1.70 | 1.895 | 1.865 | 2.020 | 1.945 |
| 1.80 | 1.992 | 1.970 | 2.125 | 2.045 |
| 1.90 | 2.095 | 2.070 | 2.225 | 2.150 |
| 2.00 | 2.225 | 2.195 | 2.375 | 2.275 |
| 2.12 | 2.345 | 2.315 | 2.495 | 2.395 |
| 2.24 | 2.465 | 2.440 | 2.620 | 2.575 |
| 2.36 | 2.585 | 2.560 | 2.740 | 2.635 |
| 2.50 | 2.730 | 2.700 | 2.880 | 2.780 |
| 2.65 | 2.880 | 2.850 | 3.035 | 2.930 |

| | | | | |
|---|---|---|---|---|
| 2.80 | 3.030 | 3.005 | 3.185 | 3.080 |
| 2.90 | 3.105 | 3.105 | 3.285 | 3.185 |
| 3.00 | 3.235 | 3.205 | 3.385 | 3.285 |
| 3.15 | 3.385 | 3.355 | 3.540 | 3.435 |
| 3.25 | 3.485 | 3.455 | 3.640 | 3.535 |
| 3.25 | 3.385 | 3.355 | 3.540 | 3.435 |
| 3.35 | 3.585 | 3.560 | 3.740 | 3.635 |
| 3.45 | 3.680 | 3.630 | 3.821 | 3.739 |
| 3.55 | 3.790 | 3.760 | 3.940 | 3.840 |
| 3.65 | 3.890 | 3.865 | 3.840 | 3.740 |
| 3.75 | 3.990 | 3.965 | 4.145 | 4.040 |
| 3.85 | 4.140 | 4.110 | 4.245 | 4.140 |
| 4.00 | 4.245 | 4.215 | 4.395 | 4.295 |

# APPENDIX B

# STANDARD SPECIFICATIONS

## B.1 Standard Specifications for Transformers

(1) IS: 3639-1966. Fittings and accessories for power transformer.

(2) IS: 1180-1972. Specification for outdoor type 3-phase distribution transformers up to and including 100 kVA, 11 kV.

(3) IS: 660-1972. Guide for loading of oil immersed transformers.

(4) IS: 1885 (Part XXVIIII). 1973. Electrotechnical vocabulary — Transformer.

(5) IS: 2026 (Part I) 1977. Specifications for power transformers: Part I: General.

(6) IS: 2026 (Part II) 1977. Specifications for power transformers: Part II: Temperature rise.

(7) IS: 2026 (Part III): Specifications for power transformers Part III. Insulation levels and dielectric test.

(8) IS: 2026 (Part IV) 1977. Specifications for power transformers Part IV. Terminal markings, tappings and connections.

## B.2 Standard Specifications for Rotating Electrical Machines

(1) IS: 7132-1973. Guide for testing synchronous machines.

(2) IS: 7306-1974. Methods for determining synchronous machine quantities from tests.

(3) IS: 7816-1975. Guide for testing insulation resistance of rotating machines.

(4) IS: 4722 D.C. Motors.

(5) IS: 900-1965. Code of practice for installation and maintenance of induction motors.

(6) IS: 2254-1965. Dimensions of vertical shaft motors for pumps.

(7) IS: 4029-1967. Guide for testing 3-phase induction motors.

(8) IS: 4691-1968. Degrees of protection provided by enclosures for rotating electrical machinery.

(9) IS: 4728-1968. Terminal marking for rotating electrical machinery.

(10) IS: 4729-1968. Measurement and evaluation of vibration of rotating electrical machines.

(11) IS: 4889-1968. Methods of determination of efficiency of rotating electrical machines.

(12) IS: 6362-1971. Designation of methods of cooling for rotating electrical machines.

(13) IS: 1885 (Part XXXV) – 1973. Electrotechnical vocabulary – Rotating machinery.

(14) IS: 7538-1975. Specification for 3-phase squirrel cage induction motors for centrifugal pumps for agricultural application.

(15) IS: 325-1978. Specification for 3-phase induction motors.

(16) IS: 8798-1978. Values of performance characteristics for 3-phase induction motors.

# Bibliography

1. **G.C. Jain,** "*Design, Operation and Testing of Synchronous Machines*", Asia Publishing House, Bombay, 1961.
2. **M.K. Sibbal,** "*Electrical Machine Design and Machine Drawing*", Khanna Publishers, Delhi, 1970.
3. **A.E. Clayton and N.N. Hancock,** "*Performance and Design of D.C. Machines*", The English Language Book Society and Sir Isaac Pitman and Sons Ltd., London, 1969.
4. **M.V. Deshpande,** "*Design and Testing of Electrical Machines*", A.H. Wheeler & Company Ltd. Allahabad, 1983.
5. **R.K. Agrawal,** "*Principles of Electrical Machine Design*", S.K. Kataria & Sons, Delhi, 2000.
6. **A.K. Sawhney,** "*A Course in Electrical Machine Design*", Dhanpat Rai & Company (P) Ltd., Delhi, 2001.
7. **M. Ramamoorty,** "*Computer Aided Design of Electrical Equipment*", Affiliated East-West Press Pvt. Ltd., New Delhi, 1987.
8. **V.N. Mittle and A. Mittle,** "*Design of Electrical Machines*", Standard Publishers Distributors, Delhi, 1996.
9. **S.N. Singh,** "*Electric Power Generation, Transmission and Distribution*", Prentice Hall of India Private Limited, New Delhi, 2003.
10. **S.L. Bhatia,** "*Hand book of Electrical Engineering*", Khanna Publishers, Delhi, 1997.
11. **S.P. Seth and P.V. Gupta,** "*Physics Properties and Applications of Electrical Engineering Materials*", Dhanpat Rai & Sons, 1985.
12. **Malika Jain and Priyanka Jain,** "*ABC of Electrical Engineering*" Dhanpat Rai & Sons, Delhi, 1996.
13. **P.S. Bhimbra,** "*Electrical Machinery*", Khanna Publishers, Delhi, 1989.
14. **M.G. Say,** "*The Performance and Design of Alternating Current Machines*", Pitman, London, 1970.
15. **A.V. Narlikar and S.N. Ekbote,** "*Superconductivity and Superconducting Materials*", South Asian Publishers, New Delhi, Madras, 1983.
16. **Mervyn Lovell, Alan Avery and Michael Vernon,** "*Physical Properties of Materials*", ELBS and Van Nostrand U.K., 1983.
17. **Yu Koritsky,** "*Electrical Engineering Materials*", Mir Publishers, Moscow.
18. **Shanmuga Sundaram A., Gangadharan G. and Palani R.,** "*Electrical Machine Design Data Book*", Wiley Eastern, 1979.
19. **Nagrath, I.J. and Kothari D.P.,** "*Electrical Machines*", Tata McGraw-Hill, 1985.